The Biology of Cells

THE BIOLOGY OF CELLS
Herbert Stern and David L. Nanney

THE BIOLOGY OF ORGANISMS
William H. Telfer and Donald Kennedy

THE BIOLOGY OF POPULATIONS
Robert H. MacArthur and Joseph H. Connell

THE BIOLOGY OF CELLS

HERBERT STERN University of Illinois

DAVID L. NANNEY University of Illinois

John Wiley & Sons, Inc., New York · London · Sydney

SECOND PRINTING, JULY, 1967

Library of Congress Catalog Card Number: 65-19473
Printed in the United States of America

FOREWORD

The idea for this series of three books was conceived in June 1960 at a meeting sponsored by John Wiley & Sons and involving Professors David Nanney (University of Illinois), Robert MacArthur (University of Pennsylvania), Joseph Gall (Yale University), Peter Ray (University of Michigan), William Van der Kloot (New York University Medical School), and Clifford Grobstein (Stanford University). A year later the three books were outlined, discussed, and interrelated during a second meeting held at the Morris Arboretum of the University of Pennsylvania. Most of the authors participated in this second meeting, which profited from the courtesy and counsel of Professor David Goddard (University of Pennsylvania). A sense of common objective emerged from the two meetings, and it is hoped that the books reflect it.

What was this common objective? In broadest terms it was to demonstrate the conviction that the teaching of biology could be reoriented to convey more effectively and forcefully the intellectual revolution which biology was undergoing. We were delighted to find that we shared not only the conviction but also a vision of how to realize it. It is now clear—but was not then—that this vision was common to many biologists. Revision of biological curricula at the university level is rapid and widespread because teachers (many of whom also are investigators) have almost simultaneously seen how far pedagogy lags behind the swift advance of research.

The vision was embodied in several basic decisions on organization of the three volumes. The first decision was adoption of the "levels" approach. This recognizes that the living world is composed of a hierarchy of organizational patterns that can best be encompassed by explicit discussions of cells, organisms, and populations. In parallel with the various levels of organization treated in the physical sciences, each level has its own set of characteristics. These compel the biologist to be concerned with the relation of properties and behavior at one level to each of the other levels. We believe (unlike the vitalists of an earlier era) that much of the behavior of living things is referable to the properties of their components—although we also recognize that many of the properties of the components cannot be ascertained in isolated systems alone. We recognize, also, that biological entities at lower levels of organization are markedly dependent upon the properties of those at higher levels of which they are part. The properties of biological systems have developed from both molecular mechanisms and evolutionary forces; these opposing poles account for much of the pattern and unity of life.

A second decision was to emphasize the uniformities of nature as opposed to its variety. The great generalizations of biology are those that apply to life in all (or many) of its manifestations. The endless diversity of life is a fact, but its endless documentation is inappropriate as an introduction where the major focus should be on life *in toto*.

A third decision concerned assumptions about the previous background of the student. How much information and experience relevant to biology are students entering college likely to have? Improvements in secondary school preparation have been dramatic, not only in biology but also in mathematics and the physical sciences. The prospect is that we shall see increasing numbers of well-prepared students for whom most past college introductions to science are inappropriate and obsolete. Accepting the near-impossibility of accommodating all levels of prior training, we need deliberate attempts to build upon the enriched experience of these students who have already been effectively exposed to some of the basic concepts of physics, chemistry, mathematics, and biology.

Finally, we decided that—although science is frequently regarded as a "body of knowledge," and presented as an up-to-date summary of that body—any such treatment is doomed to early obsolescence. Science is more than a body of knowledge; it also is a process. In the long run, comprehension of the process is more significant than knowledge of any of its particular products. Therefore, it is important to present the intellectual roots of as many topics as possible and the elements essential to their growth.

These were the decisions and the intentions. Their validity, and the effectiveness of our execution, must now be judged in practice. Each of these books may be used alone as a presentation of its particular level. Each, however, has been written under the explicit assumption that no one level of biological organization is fully comprehended without consideration of the other two. Accordingly, the three books will be most meaningful if used together and in their intended sequence—Cell, Organism, Population. Only in this way is the student likely to benefit from the great continuity and cohesiveness that are the products of the last two decades of biological investigation.

The Authors and Clifford Grobstein, *Series Consultant*

PREFACE

This is an introductory textbook in cell biology. Its emphasis is on underlying principles which we have tried to communicate without oversimplification. Our aim is to explain rather than to describe these principles and, in some parts, we have leaned heavily on the history of biology to achieve this end. We have woven some discussions of physics and chemistry into the subject matter of the text. We have done so not because these limited excursions into the physical sciences are sufficient for a comprehension of cell biology, but because without such excursions our presentation would have been false to the history of the subject.

In Part I we consider the foundations of certain great generalizations which have become an integral part of the structure of biology—the cell doctrine, the gene concept, the chromosome theory, and the physicochemical basis of life. In Part II we examine the superstructure erected on these foundations. Our approach has been to move from the general to the particular. We have organized our presentation in terms of the challenges that living systems must confront in order to persist, prosper, and evolve. For some challenges, our knowledge of cell behavior is reasonably profound and the relevant problems may be clearly formulated. For other challenges, however, our understanding of cells remains superficial and we have reserved our limited discussion of these to the final section of the text.

We hope that readers will be moved to reflect on the impressiveness of the intellectual venture in cell biology, a venture that has run the course of many generations and the end of which cannot be foreseen. Its most prominent feature is that of continuity, continuity in the diverse mental and technical skills within each generation and continuity in successive generations. This book will have served its purpose to the extent that it has made the reader aware of the achievements of the past and conscious of the possibilities of the present.

Herbert Stern
David L. Nanney

Urbana, Ill.
April 1965

CONTENTS

Section B. The Maintenance of Specificity

Section C. The Regulation of Cell Behavior

PART ONE

THE BACKGROUND OF IDEAS

SECTION A

The Cell Doctrine and Gene Concept

This book is directed toward an understanding of the cell, of its components and its architecture, of the functional relations of its parts, and of the properties which have given it a central position in biological theory. Biology existed long before the existence of cells was suspected, but only through their discovery did it become possible to consolidate biological information and to recognize the essential unity of life. Our ideas concerning the cell are sometimes briefly summarized by "the cell doctrine." In a stripped form this doctrine holds that the cell, consisting of a nucleus and cytoplasm, is the irreducible unit of biological activities; the cell is the unit of biological structure, of function, and of reproduction.

We cannot consider the cell doctrine in isolation, however, for its full elaboration depended on correlated advances in several areas. The cell doctrine, the gene concept, and the idea of evolution through natural selection were all crystallized early in the second half of the nineteenth century. Each of these ideas contributed enormously to the unification of biology. The gene concept attributed biological specificity, heredity, and variation to specialized but universal cellular components. The idea of evolution through natural selection provided a common origin for the multiplicity and diversity of living organisms through differential survival of hereditarily distinct forms in different environments. Although today these ideas are firmly welded, they initially appeared to be unrelated, and were achieved, at least superficially, in very different ways. They offer significant contrasts in the ways by which scientific progress is achieved.

The doctrine of natural selection is associated primarily with a single man—Charles Darwin, a single book—*The Origin of Species,* and a single date—1859. While this sharp focus undoubtedly exaggerates Darwin's achievement and minimizes the contributions of his predecessors, the fact remains that a single creative mind integrated a vast amount of scattered information and set forth at a single time a compelling synthesis and rationalization.

The gene concept is also usually credited to a single man—Gregor Mendel, a single scientific paper, and a single date—1865. But Mendel and Darwin were very different kinds of men. They differed in their interests, in their approaches, and in their receptions. Darwin considered the biological universe, while Mendel confined his studies to a pea patch in a monastery garden. In contrast to Darwin's fame and notoriety, Mendel died without achieving even understanding, much less acclaim.

1

The Development of the Cell Doctrine

The cell doctrine, in an approximately modern form, did not come into being precipitously. It cannot be assigned properly to a single person or to a particular year. A major reason for its indefinite parentage and date of origin lies in the fact that the cell doctrine is not so much a single concept as a logical synthesis of many observations which were established at different times by different investigators. It was gradually developed through the steady accumulation of factual information on more and more kinds of organisms until certain common features of living systems became recognized. To set any date for the establishment of the cell doctrine would be arbitrary, for in a sense the doctrine is still being developed.

A brief review of the evolution of our modern concepts will serve as an introduction to the cell and will set the stage for more recent studies. We will have more to say about Mendel and the concept of the gene later on, but a detailed consideration of Darwin and natural selection must be sought elsewhere.

TECHNOLOGICAL PREREQUISITES

Before 1650 the very existence of cells could not be suspected. Nearly all cells are too small to be seen with the unaided eye, and effective devices for magnifying very small objects were not available. The magnifying power of glass spheres filled with water was well known before the second century, when Ptolemy discussed the phenomenon in his *Optics*. Not until 1300, however, were glass lenses brought into use through the invention of spectacles for the improvement of vision. The inventor remains unknown and, also, the circumstances which led to his discovery. But as the use of spectacles spread, curiosity was gradually aroused concerning the theory of glass lenses. In the sixteenth century treatises began to appear on the geometrical properties of spectacles in relation to the improvement of vision. By the seventeenth century lenses were being tried for the purpose of magnifying distant objects, a technique of military value. About that time someone, presumably a spectacle maker, discovered the usefulness of combining a convex and a concave lens in a tube. With the concave lens held close to the eye distant objects were made recognizable. Thus was the principle of the telescope discovered. Galileo recognized its significance and built the first scientifically used telescope, but he also discovered that by looking through his telescope in reverse, very small objects could be magnified. With this primitive microscope Galileo first discovered the compound eye of an insect. From that time on the natural curiosity to see smaller things better was a sufficient impulse for the development of the microscope.

The earliest microscopes were extremely simple. Anton van Leeuwenhoek's microscope, for example, consisted of little more than a highly polished glass bead mounted in a metal plate. It was used by holding it extremely close to the eye, for the focal point of the lens—the point where distant light rays were caused to converge—was very near the surface of the sphere. By current standards these microscopes were crude and inefficient, but some gave magnifications of 100 diameters or more, and they opened a new realm of nature which greatly excited the curiosity and imagination of investigators.

There is probably no better example of the dependence of cell biology on physics than the history of the microscope. The principles of optics had to be worked out before the potentialities of lens systems could be fully exploited. Distinguished figures like Sir Isaac Newton, Christian Huygens, and René Descartes centered considerable attention on optical problems. To discuss these principles in detail would lead us too deeply into a very broad area of physics, but some appreciation of the problems of microscope construction is necessary if we are to understand the pace of advance in microscopic science.

The behavior of light may be described in one of two ways. The first of these, which is a relatively coarse description of optical phenomena, is called "geometrical optics" and is adequate for outlining the basic principles of light microscopy. The second of these, which treats light as a form of electromagnetic radiation, will be discussed in connection with the use of light energy for photosynthesis (p. 22).

Reflection and Refraction. A light ray may be either reflected or refracted (Fig. 1-1*a*) at the boundary of two media. For reflected rays the angle of

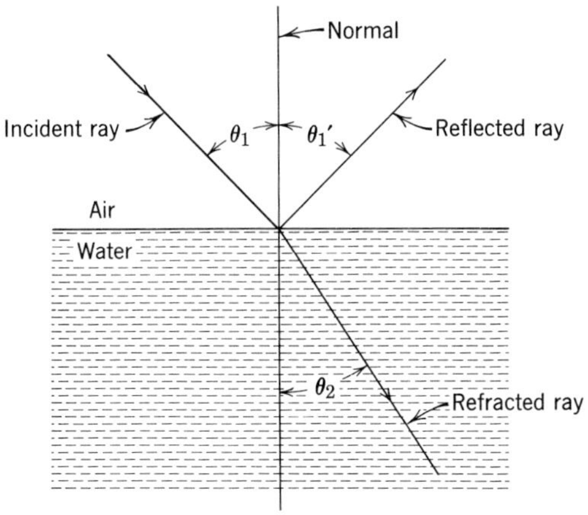

(a)

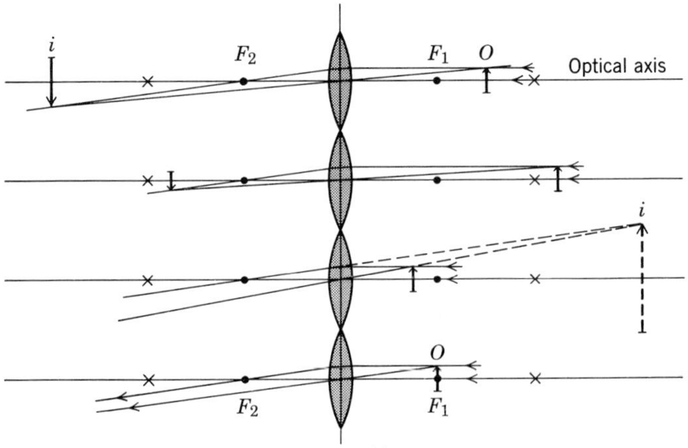

(b)

Fig. 1-1. Illustrations of optical principles.

incidence (θ_1) equals the angle of reflection. Angle θ_1 is defined as the angle formed by an incident ray and a line perpendicular or "normal" to the plane of the interface. The fraction of incident light which is reflected depends on the nature of the surface and the angle of incidence. Highly polished surfaces, such as mirrors, reflect all incident light; highly transparent materials, on the other hand, reflect very little. In all cases, however, as the angle of incidence increases, a limit is reached beyond which all incident rays are reflected. This angle is known as the "critical angle." The behavior of refracted light rays is expressed by the equation $\sin\theta_1/\sin\theta_2 = n$, where n is the refractive index of medium 2 relative to medium 1. For any two media there is a constant which can be used to calculate the degree of refraction of an incident light ray. The constant is determined experimentally, and is expressed for any particular substance as the "refractive index" of that substance relative to a vacuum. Air has a value of 1.0003, water 1.33. Four types of glass used in microscope lenses—fluorite, fused quartz, crown glass, and flint glass—have the respective values of 1.43, 1.46, 1.52, and 1.66. These indices apply to a wavelength of 5890 Å; for other wavelengths there are correspondingly different refractive indices. The relationship between wavelength and refractive index is highly important in lens design.

Optical Properties of Lenses. The capacity of lenses to magnify arises from two properties: the high refractive index of glass relative to air and the curvature of the glass surface. The high refractive index causes bending of all rays which do not strike the lens at an angle normal to the surface. The curvature of the lens affects the angle of incidence and hence the angle of refraction. For a particular lens two factors determine the degree of magnification and the nature of the image. The first is the "focal point," which is a function of the refractive index and the radius of curvature. It may be determined experimentally by shining parallel rays of light on the lens and finding the point at which all the rays converge. The second factor is the position of the object. The manner in which position affects magnification is illustrated in Fig. 1-1*b*. An object placed in the light path of such a lens will produce a more or less sharply defined image at a particular point along the optical axis. The position and size of the image may be determined by a few simple geometrical rules. (1) A light ray coming from any point in the object which runs parallel to the optical axis is so bent as to pass through the focal point on the distant side of the lens. (2) Conversely, a light ray passing through the focal point of the lens emerges parallel to the optical axis on the distant side of the lens. (3) A light ray passing through the optical center of the lens does not change direction. By drawing any two such rays from a single point in the object, the position of the image is located at the point of intersection. Such "paper optics" reveals certain characteristics of lenses which can be verified experimentally. (1) Objects placed at a distance greater than twice the focal length of the lens (x in Fig. 1-1*b* marks twice the focal length) form images smaller than the objects. (2) Objects placed between the focal point and the lens do not produce a "real" image because the rays diverge. If a ground-glass screen is used to locate the position of the image in an experimental setup, no such position will be found. If, however, an observer looks through the lens, he will perceive an image behind the object as illustrated in the diagram. Such an image is called "virtual." A screen placed in the position of the virtual image will not reveal it; the light rays do not bounce back from the lens. The image is revealed to the human eye because the eye acts as a second lens which focuses the divergent rays. Paper optics describes the apparent size and position of the image as perceived by the human eye. Optical systems which produce virtual images are particularly useful for microscopy. If an observer were to record real images, he would have to position his eye at the exact position of the image which would vary with the lens used. Since magnification is equal to i/o, where i and o are distances from the lens center to image and object respectively, magnifications of the order of 1000 × would require the observer to view the image at relatively long distances from the object. Images thus viewed would be relatively dim because the intensity of light decreases as the inverse square of the distance. (3) To produce a real and magnified image by a convex lens, the object must lie between the focal point and twice the distance from the lens center (first lens diagram in Fig. 1-1*b*). The greater the curvature of the lens, the greater the magnification which is achieved, but also the shorter the focal length. A very short focal length is advantageous because losses in light intensity are minimized, and this is a factor of primary importance. The areas viewed under a microscope are extremely small, and the amount of light which enters a lens is a very small fraction of the light shone on a specimen.

Design of the Compound Microscope. All modern microscopes have two sets of lenses. (See Fig. 1-1*c*.) Primary magnification is achieved by the equivalent of a convex lens of short focal length (the "objective") which produces a real image i_a. A second lens system (the "ocular") is so positioned

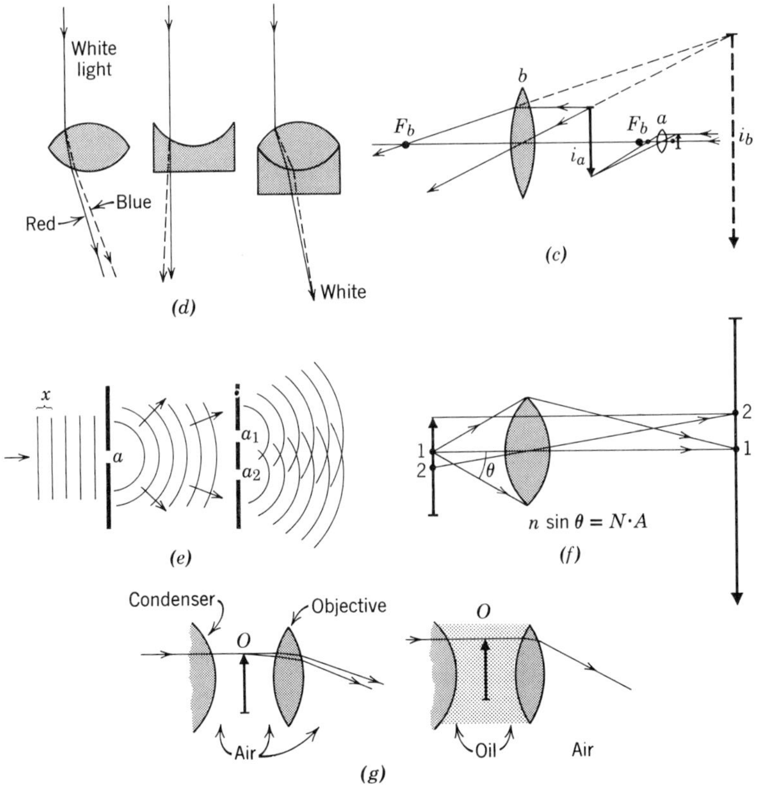

Fig. 1-1 (continued).

that i_a lies between its center and focal point, thus producing a virtual image i_b. This image appears to be about 25 cm from the lens, the distance at which the human eye is adapted to see objects most clearly. All objectives are so designed that irrespective of their magnification, the images are produced in the same location. The ocular therefore magnifies all images to the same extent. To achieve this, and also to correct for the aberrations discussed below, lenses of several kinds are cemented together in each lens system. Although not shown in the diagram, microscopes also have a lens system (condenser) to focus light on the specimen.

Aberrations. Since the smaller the focal length, the greater the curvature of a lens, problems hitherto not considered arise. These problems perplexed the designers of microscopes for some 200 years. The first problem arises from the fact that the angle of refraction is a function of the wavelength. As illustrated in Fig. 1-1*d*, the shorter the wavelength the greater the angle of refraction. Differences between the angles of refraction for different wavelengths become increasingly evident as the curvature of the lens becomes greater; for lenses of high curvature, such as those required for microscope objectives, these differences result in prominent distortions of the image usually referred to as "chromatic aberrations." Newton pondered the problem and concluded that a lens could not focus the component colors of a white ray on a single point; in his opinion, therefore, magnification by a lens of an object illuminated with white light had to give rise to a fuzzy multicolored image. Seventy years later an amateur discovered that by putting together a concave and a convex lens of different refractive indices the problem could be overcome (Fig. 1-1*d*). John Dollond, a celebrated manufacturer of optical instruments, put this discovery of the "achromatic lens" to commercial use about 1760. Even so, achromatic objectives were not incorporated into microscopes until another seventy years had gone by. Throughout this period biologists used microscopes which produced images too distorted for accurate study. Magnifications were therefore limited and ill-suited to a detailed study of cells. Only a genius as Leeuwenhoek was able to grind his own lenses to such perfection that with his simple microscope he was able to observe spermatozoa and some bacteria. Shorter wavelengths of light were refracted more than longer ones. In convex lenses the blue component of white light is deflected inward to a greater extent than the red component, and each color therefore has a different focal point. In concave lenses the deflection is outward so that blue light diverges more than red light from the optical axis. By combining a convex and concave lens, chromatic aberration may be eliminated. However, lens shape alone is not the sole factor in such a system. The purpose of the convex lens is to magnify, but if two lenses refract equally, though in opposite directions, no magnification is achieved. To make a double-lens system effective, different types of glass are used for each of the lenses. In the arrangement illustrated above, the glass in the concave lens has a much lower refractive index than that in the convex one. The net result is a correction of chromatic aberration without a cancellation of magnifying power.

The technology of glass manufacture, lens grinding, and mounting ultimately developed so that other problems confronting the perfection of the light microscope were largely overcome. Of these, the most outstanding relates to the properties of light which are not fully covered by geometrical optics. The graphical method used in Fig. 1-1 to locate images does not reveal anything about the clarity or "definition" of the image. Apart from chromatic aberration, poor definition may also arise from the failure of those light rays which strike the peripheral region of a lens to be focused in the same plane as those which strike near the center. The phenomenon is known as "spherical aberration." The geometric rule for locating images is an approximation valid only if the rays of light entering a lens are parallel, or nearly so, to the optical axis. This ideal is rarely achieved in practice. Without entering into details of how problems of spherical aberration were eventually resolved, it may be summarily stated that the only useful portion of a lens is that which focuses all incident rays in the same plane. For lenses with short focal lengths (and high curvature) the extent of useful area becomes a major factor. As the area decreases, the amount of entering light decreases; and to achieve perfect definition by making the image nearly invisible is of no practical use. The development of strong light sources and an effective system of condenser lenses to concentrate light on the microscopic specimen helped greatly to increase image definition without undue loss of intensity.

However, an even more fundamental problem is associated with the useful area of a lens, and is entirely removed from geometrical optics. A microscope which provided an extremely sharp definition of the contours of a cell, or of a subcellular component, but which revealed nothing of its interior organization, would have limited application. The value of a microscope finally depends on its ability to reveal details of structure; this quality is called "resolution." The closer two structures can be and remain evident as such in the image is a measure of the resolving power of a microscope. Theoretical as well as practical limitations are imposed by the behavior of light rays. Figure 1-1*e* illustrates what happens when perfectly parallel rays of light pass through an opening which has the dimensions of a light wave.

A ray of light composed of parallel waves (wavelength x) and passing through an opening a of diameter equal to x loses its parallel pattern and assumes a spherical one. The thin lines in the figure represent the crests of the light waves as would be seen by an observer if he could look down on a thin cross section of the light ray. If the wave crests in all cross-sectional planes are superimposed and of equal intensity, the ray will appear optically uniform. If the wave pattern passing through a were unchanged, the observer would see a uniformly illuminated circle. Such behavior has never been observed. For light passing through openings with magnitudes of the order of wavelengths we see a series of concentric rings alternately bright and dark. The ring pattern arises from the interaction of diffracted waves; where the crests of light waves coincide, intensity is increased, and where crests coincide with troughs, intensity is decreased. From the standpoint of microscopy the principal point of interest in diffraction is its effect on visualizing adjacent points in an object, since the size of some subcellular structures is of the order of magnitude of wavelengths of light. A simplified illustration of the effect of diffraction is given in the second drawing where a light ray passes through two adjacent apertures. The shape of the wave front indicates that although the observer might perceive two regions in the light beam of somewhat higher intensity, the definition of the apertures would be lost. The effect would be exactly the same if light were passing through a portion of a cell which had pores or grooves separated from one another by distances of the order of the wave length of light.

Figure 1-1*e* thus diagrams the theoretical basis for the limitations of even perfect lens systems in revealing the adjacent parts of an object. The limits may be calculated from the $h = \lambda/2NA$ formula, where h represents the distance between the closest points in an object evident as two points in the image, λ is the wavelength of light, and NA (numerical aperture) $= n \sin\theta$ (n, refractive index of medium). The diagram in Fig. 1-1*f* indicates the nature of θ. The greater the angle formed between a point in the specimen and the periphery of the useful area of a lens (that area which focuses all incident light in the same plane), the greater the resolution. Other factors being equal, the shorter the focal length, and the higher the refractive index of the medium through which light passes between specimen and lens surface, the greater the value of NA. Although this simple formula for calculating the resolving power of a lens does not indicate how diffraction might be involved, the relationship between wavelength and NA fundamentally derives from the phenomenon of light diffraction.

Numerical aperture and magnifying power are stamped on the metal covering of most objectives. Oil immersion lenses are used for greatest magnifications because the high refractive index of oil increases resolution. An additional advantage of the oil immersion system is that oil and glass have nearly the same refractive index. Since tissue sections are usually embedded in a medium of refractive index similar to glass, light issuing from a specimen is not refracted by any of the interfaces between it and the lens. This is illustrated in Fig. 1-1g.

The technological developments in microscope design and even the elaboration of optical principles might appear to be an entirely peripheral subject in the area of biological studies. Yet, quite apart from the fact that our present information about cellular organization would have been impossible without the microscope, a consideration of microscopic technology illustrates an important general principle in scientific activity. The information made possible through applying a particular technique receives, for obvious reasons, a great deal of attention. The history of events leading to the cell doctrine is a good illustration of this. But for the users of a technique an understanding of its limitation is equally important. Unless we know precisely the point at which a technique no longer yields reliable information, fact becomes intermingled with fancy, and interpretation becomes wooly speculation. Those biologists who appreciated optical theory and who realized the inherent limitations of light microscopy were disposed to contemplate, perhaps dreamily at first, the possibility of quite different techniques which would enable them to resolve structures of molecular dimensions rather than those of bacterial size. Eventually the electron microscope took its place as a standard tool of the cell biologist, yet the impetus for its development came not only from reflections on the technical limitations of the light microscope but also from the rich store of information which this instrument provided.

The light microscope is a tool which provides both physical and chemical information about cells. The human eye is sensitive to differences in light intensity and to differences in color. The extent to which a particular area in a cell will transmit or refract light determines the relative intensity of illumination of that area in the final image. Color, however, has a specific relationship to the chemical constitution of cells. Where color is localized, as in the chloroplast for example, we not only perceive the localized regions more easily but also obtain information about their chemical constitution. Given the fact that chlorophyll is green, the localization of chlorophyll in chloroplasts is demonstrable with the light microscope. In the development of microscopy both properties of the instrument were exploited. Not only were modifications of the instrument introduced (phase and interference microscopy) to provide more physical information about cells, but specific chemical reactions were developed which, by differentially coloring the various parts of a cell, provided a more precise knowledge of chemical composition.

The early microscopists, particularly Nehemiah Grew, Marcello Malpighi, Robert Hooke, and Anton Leeuwenhoek, examined many kinds of living materials in the latter half of the seventeenth century. They were highly successful in observing a variety of simple unicellular organisms (Fig. 1-2*a*). From our perspective they were less successful with the more complex forms. They did observe that plant tissues often appeared to contain "little boxes." But they were unaware of the contents of these boxes and believed their walls to be continuous structures forming fibrous networks within which the empty spaces were established. The only cells from higher forms with which they were generally familiar were the blood cells. But these and the unicellular organisms were not related to the empty boxes in plant tissues. Similarly, animal tissues were examined and some of their grosser features pointed out, but no elementary units were identified. Plants and animals appeared to be very different in their microscopic organization.

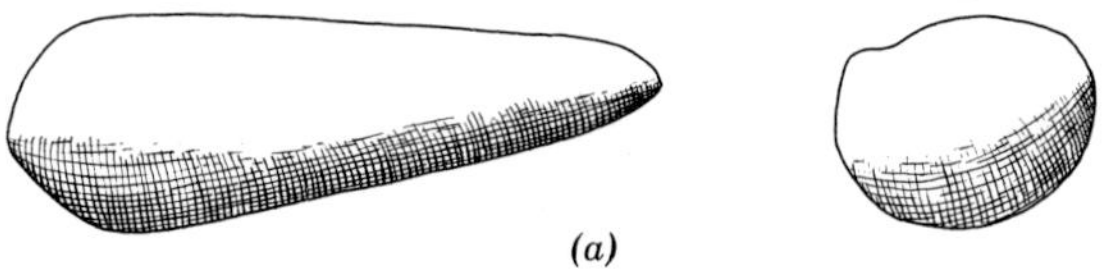

(a)

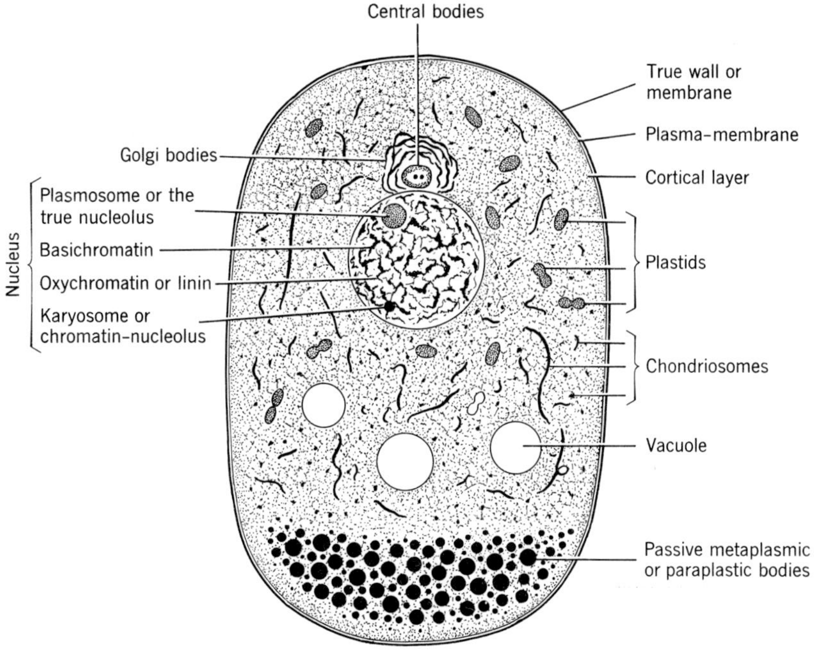

(c)

Fig. 1-2. Attempts made at various times during the development of microscopy to depict the structure of a cell. (*a*) Leeuwenhoek's pictures of the intestinal protozoa of frogs, drawn in 1683. (A) *Opalina (Cepedea) dimidiata.* (B) *Nyctotherus cordiformis.* (From C. Dobell, *Antony van Leeuwenhoek and His Little Animals,* Dover, New York, 1960.) (*b*) Drawing by M. Schleiden of a plant cell illustrating the vacuolar system and protoplasmic streaming. (From M. Schleiden, *Grandzuge der Wissenschaftlichen Botanik,* Wilhelm Engelmann, Leipzig, 1845.) (*c*) E. B. Wilson's general diagram of a cell. (From E. B. Wilson, *The Cell in Development and Heredity,* Macmillan, New York, 1925.) (*d*) A generalized diagram of the structure of a cell as revealed by studies with the electron microscope.

The failure to detect cellular organization in animals and to explore the contents of the "little boxes" in plant tissues is only partially due to the inadequacies of the microscopes. The fact is that much of the cellular organization—and an even larger part of the subcellular organization—simply was not present in the materials examined. The

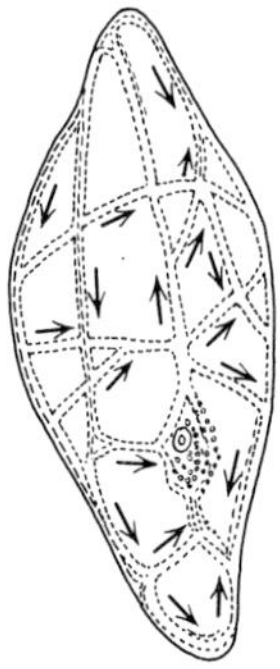

(b)

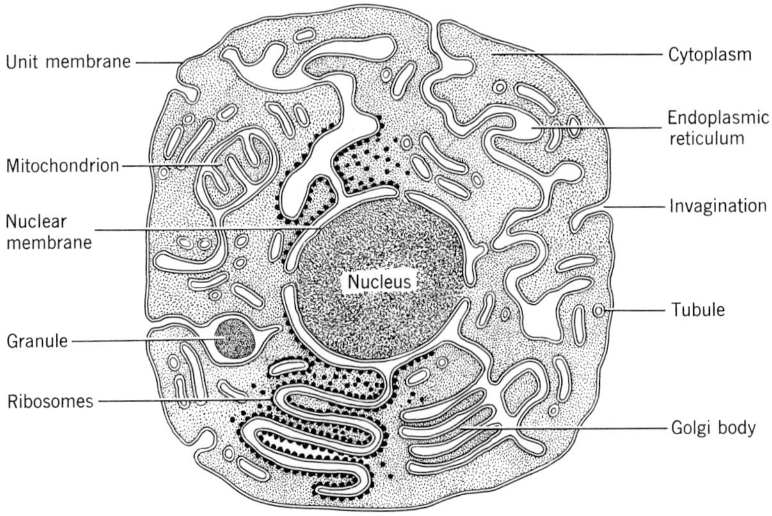

(d)

Two prominent features of cell structure which are beyond the resolving power of the light microscope were made apparent in these studies. (1) The presence of membranes throughout the cytoplasm, and their absence from the nucleus. (2) The presence of small granules ("ribosomes") demonstrated by other methods to consist of ribonucleic acid and protein. (From J. D. Robertson, in *Cellular Membranes in Development,* Symposium of the Society for the Study of Growth and Development, Academic Press, New York, 1963.)

image produced by a perfect lens system can be no better than the object magnified, and the problems involved in suitably preparing cells for microscopic study proved to be a major obstacle for investigators. One problem arises from the fact that as magnification increases, the depth of focus decreases. At very high magnifications the thickness of a specimen which can be brought into focus is of the order of the limits of resolution (about 0.2-0.3μ). A specimen with a thickness of 50μ or more would provide so much interference with light passage that high magnifications would prove useless. Unicellular organisms presented no serious problem in this respect, but most multicellular organisms did. To examine microscopically the stem of a plant or the organ of an animal required a sectioning of the tissue. Only when the art of preparing thin sections of tissue was adequately developed did microscopic study become freed from its limitations. There was nothing intellectually noble about this development; it was pure craft. Yet this craft, expressing itself in cookbook recipes and cute hints, was indispensable to progress. Indeed, the other problems were solved in much the same way, and even today the student who wishes to learn the techniques for microscopic studies will have to rest content with a bewildering array of recipes and procedures that claim no more than empirical success.

The need to kill cells for most microscopic studies presented more difficult problems than the need for thin sectioning. That living cells are not suitable for many microscopic studies is evident from simple optical considerations. The human eye distinguishes adjacent parts of an image either by virtue of a difference in intensity or as a result of differences in color. The living parts of a cell are about 80% water and generally very similar in their degree of refraction or absorption of light. Thus, except for colored inclusions such as chloroplasts, many details of cellular structure are absent in images of living specimens. Only by a highly sophisticated application of optical principles could a microscope be designed which would reveal the details of internal structure in living cells. Such a microscope ("phase contrast") has come into use during the past decade; it was unknown during the most fruitful period of cytology. Yet it does not dispense with the need for examining killed cells. Living cells present moving images and thus are not suitable for detailed studies at high magnifications. Moreover, cells of a tissue do not live happily on a microscope slide.

A quick though not always reliable way of determining whether a cell is dead or alive is by determining the extent to which it refracts light. Since a killed cell does not retain all its water, differences in concentrations of components develop at death and cause corresponding differences in light refraction. A dead cell is optically more heterogeneous than a living one. At once, however, the question arises as to the fidelity of the image produced by a dead cell. How many of the structures seen are artifacts? No theoretical answer to this question is possible, and throughout the nineteenth century the art of "fixation" (a pleasanter term than "killing") was pursued with the aim of killing cells so as to provide as little distortion of the original organization as possible. Since the "original organization" was unknown, esthetics —not logic—served as arbiter. The better a cell looked, the better the fixative was taken to be. In retrospect it can only be said that esthetic judgment went hand in hand with natural design. The feuds of the period now seem quaint, and very few merit recapitulation. The catalog of fixatives is large, and each has its peculiar virtue without benefit of a respectable theoretical basis.

Not only were the pioneers in cytology working without benefit of fixatives and the art of sectioning, they were also without a third important art—cellular staining. The introduction of a wide variety of dyes or "stains" which could be used to color the interior of cells differentially made possible an exceedingly clear visualization of cellular structures.

CYTOLOGICAL OBSERVATIONS

In spite of the defects in the microscopes and the primitive state of the techniques employed, some advances continued to be made throughout the eighteenth century. In particular, descriptions were published of granular and globular elements of various sizes in many tissues of both plants and animals. Some observers believed that animal tissues consisted largely of a frothy mass of bubbles or "bladders." The nature and properties of these elements were not clearly understood, and we are not even sure which of the structures described were in fact cells. The spherical aberrations in the microscopes often resulted in the appearance of "halos" of diffraction around many small objects which obscured both their size and shape and produced an appearance of clustered globules from materials of various sorts. Although their observations were not adequate to establish their ideas, the investigators were already, in their search for a common denominator among the bewildering array of biological materials, anticipating the cell doctrine.

Early in the nineteenth century, after the means of correcting for spherical aberrations was discovered, much more reliable observations were possible. One of the first consequences of improved observation was a de-emphasis on the differences between plants and animals. Animal tissues, like plant tissues, contained elementary units of a sort, even though these were not so clearly set apart by heavy cell walls. Moreover, and most significantly for these investigators, the cell wall was recognized as of secondary importance even in plants; embryonic and certain mature tissues of plants were found to lack conspicuous walls and to resemble closely animal tissues. In teased plant tissues the walls did not come out as continuous fibers running throughout a tissue, but appeared as discontinuous units associated with individual cells. The cell walls were not the primary structures but were only the products of more fundamental units.

Improved techniques of preservation and microscopy now permitted a closer inquiry into the nature and contents of these units. Nuclei—central bodies with optical properties different from the rest of the cell—had been occasionally seen and figured during the eighteenth century; they were now recognized as common components in both plant and animal cells. The similarity in plant and animal cells was further emphasized by the fact that the nuclei of both usually included smaller spherical structures, the nucleoli, also with somewhat different optical properties. These facts were most decisive in leading Matthias Schleiden and Theodor Schwann to conclude that the fundamental unit of both plants and animals was the cell. Their papers in 1839 are sometimes said to mark the origin of the cell doctrine, but they more properly mark the general recognition of one aspect of this doctrine: units, containing nuclei, are the structural elements of all kinds of organism.

The nature and significance of the cytoplasm, the jelly-like material lying between the nucleus and the perimeter of the cell, were gradually recognized through studies in the 1830's and 1840's. As might be expected from the difficulties in examining living cells from higher forms, the cytoplasm was first appreciated by students of protozoa. Moreover, the cells of plants—even the simpler plants—are often nearly filled with vacuoles, spaces containing extraneous materials and separated by membranes from the cytoplasm proper. The cytoplasm is often reduced to a thin layer lying just under and obscured by the cell wall and to fine strands connecting the nucleus to the peripheral cytoplasm (Fig. 1-2*b*). Eventually, however, plant cytoplasm was recognized and its similarity to animal cytoplasm was noted. And here for the first time the pertinence of chemical and physical properties in assessing similarities was noted. The consistency of plant and animal cytoplasm was similar; they behaved in a similar way in response to substances like acetic acid and alcohol; they both stained brown with iodine.

One more step remained before a reasonably modern understanding of the cell was achieved. By 1850 the cell, as a structural element containing a nucleus and cytoplasm, was recognized as a common feature in both plants and animals. But the larger significance of cells was bound up with their origin, and the origins of cells were still obscure. We are surprised to find such an important figure as T. H. Huxley writing in 1853 that cells "are not the instruments but indications, that they are no more the producers of vital phenomena than the shells scattered in orderly lines along the sea-beach are the instruments by which the gravitative force of the moon acts on the oceans. Like these, the cells mark only where the vital tides have been and how they have acted." This attitude is understandable only in terms of the ideas then prevalent concerning cellular origins.

Even the earliest observers were aware that the number of cells increased as an organism grew, and many hypotheses concerning the sources of these new cells were current. On the basis of fragmentary observations some biologists were led to conclude that cells precipitated in intercellular spaces. Some believed that new cells arose by fragmentation of old ones. Some, like Schleiden and Schwann, thought that new cells were budded off the nuclei of older cells. The facts were difficult to disentangle, particularly when attention was focused on mature tissues of higher organisms. In these tissues new cells were not arising frequently, and the details were often lost in the complexities of the tissues. However, when studies were directed to simple plants and to embryonic tissues, these matters were settled decisively; this resolution we owe largely to the patient researches of Robert Remak and the valuable discussions of Rudolph Virchow. Cells were found to arise by division of preexisting cells and only in this way. The egg was recognized as a specialized cell which by division gives rise to all the cells which make up the adult. Cells are not to be considered as by-products of the act of living, but as the very substance and means of life in all its manifold forms.

The details of the division process were not immediately established, and were not to be completely resolved until further microscope improvements and modern cytological methods—including specific biological stains—were exploited in the last quarter of the century. We will defer our discussion of cell division and related problems, and will leap ahead for a quick look at the architecture of the cell as it is revealed by the most advanced techniques (Fig. 1-2*d*). This picture of the cell is very similar to that developed by cytologists at the turn of the century, (Fig. 1-2*c*) but it includes a vastly larger amount of detail.

This detail comes from the application of a magnifying device which escapes the theoretical limitations of the light microscope—the electron microscope. The basic advantage of the electron microscope is that it uses electron waves which have lengths of the order of 0.05 Å. The geometry of magnification is the same as that in the light microscope; the essential difference between the two is that whereas visible waves are bent by an object, electrons either pass through an object or are scattered by it. The electron microscope thus distinguishes areas in a specimen according to their scattering power. The latter is a function of the mass of the atom. The bigger the nucleus of an atom and the more closely such nuclei are packed, the greater the scattering power. Since the atoms composing organic substances are light, they offer little resistance to the passage of electrons. It is therefore customary to "stain" tissues with heavy atoms such as osmium, uranium, tungsten, and manganese. Electron waves, of course, cannot be seen, but their positions may be visualized either by means of a fluorescent screen or a photographic plate. If electron waves could be focused as precisely as light waves so that the magnetic lens system of the electron microscope would have a numerical aperture of one, the resolving power would be 0.025 Å. Unfortunately, the lens systems in use today have numerical apertures of the order of .005 Å, and no one has yet been able to obtain resolutions much better than 6 Å.

We could draw a great deal of satisfaction from the fact that the electron microscope permits us to see almost a thousand times more than the light microscope; indeed, we do. Yet, even though we now can visualize a great deal more detail in cell structure, an additional and very important item has to be considered. Our goal is to be able to describe exactly how molecules are fitted together. For this reason we cannot help asking how closely electron microscope images define individual molecules. At this point we turn to the physicist and chemist to tell us about actual molecular sizes which they have been able to determine by a number of techniques. A hydrogen atom has a diameter of 1 Å; protein molecules have dimensions ranging from 20 Å to 250 Å or more. Thus, with the present limit of 6 Å resolution, we cannot distinguish individual atoms in an electron micrograph, but we can distinguish large molecules. Put differently, we can now measure the images we obtain of cells, and calculate from such measurements the probable number and arrangement of constituent molecules. We must nevertheless bear in mind that the actual resolution of the microscope may involve some deception. Such exceedingly small distances of visualization require even finer preservation of actual cellular structure during fixation. In this regard much has yet to be worked out.

With our greatly extended range of magnification, what do we see in cells? In nearly all cells we see certain regions which had long been recognized, particularly the nucleus and the cytoplasm (Figs. 1-3, 1-4). In a cell which is not dividing, the nucleus shows little internal structure, only a region of greater density (the nucleolus) and a fine granulation. We now know that the fine granulation represents the chromosomes in a greatly extended state. The granules are not actually granules at all but cross sections of the long chromosomal fibers. Before cell division the nucleolus disappears, and the chromosomal fibers coil and condense into thicker and shorter structures; they eventually reach dimensions which enable them to be seen with light optics. It is small wonder that early cytologists were puzzled by the appearance and disappearance of the chromosomes, and that the idea of the continuity of chromosomes from one cell division to the next was long in becoming established. We shall see shortly that the nucleus is the seat of the hereditary functions of the cell, and that the nucleolus seems to serve as a communication center, active in transmitting chemical messages from the nucleus to the cytoplasm.

Another thing which we see, and which can never be seen with light microscopes, is the membrane separating the nucleus from the cytoplasm (Figs. 1-3, 1-4) and that which bounds the external limits of the cell (Fig. 1-4). We can even see that these membranes are double. The individual membranes are only about 25 Å wide, far below the theoretical limits of optical microscopy. The existence of these membranes was known from indirect evidence and many of their properties had been deduced, but some of their features—such as the openings in the nuclear membranes which permit direct communication between the nucleus and the cytoplasm—had not been suspected.

Electron microscopy also permits an extension of observations to cells whose sizes are so small that even the existence of nuclei in them could not be ascertained with optical methods (Fig. 1-5). Bacteria, for example, were long considered to be exceptions to the cell theory on the grounds that they contained no nuclei. Electron micrographs (as well as optical micrographs made with improved techniques) demonstrate, however, that even bacteria contain a central region comparable in many respects to the nuclei of higher forms. Differences do exist, however; the bacterial "nuclei" are not surrounded by nuclear membranes but lie in direct contact with the cyto-

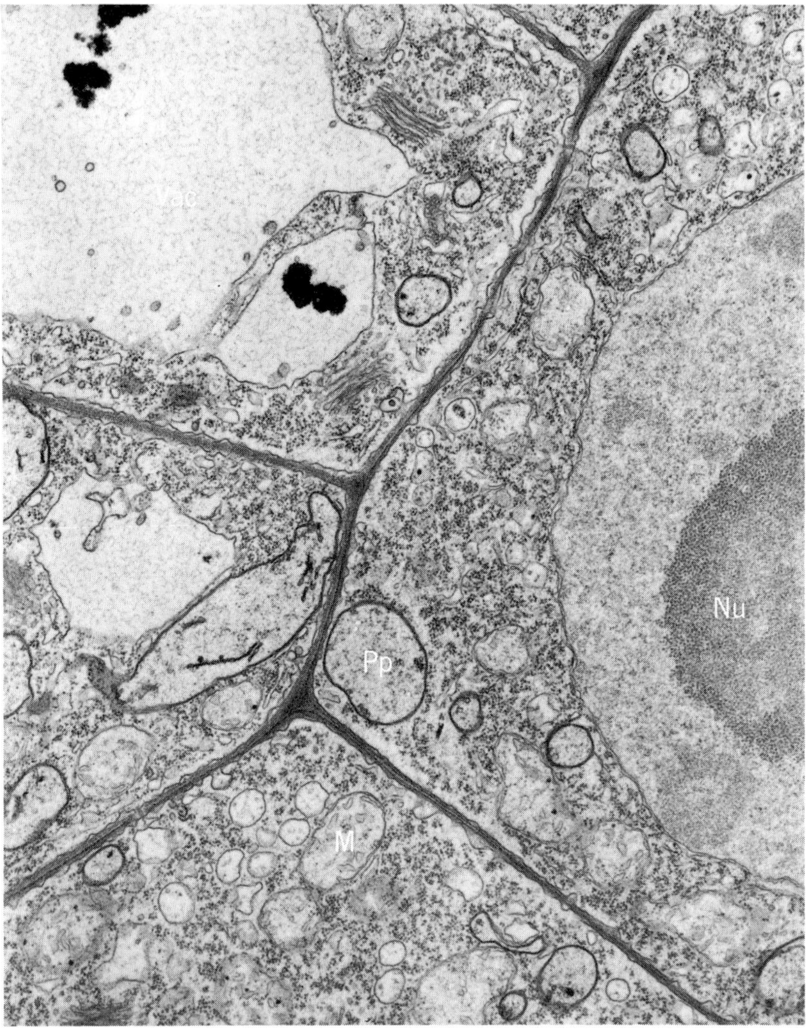

Fig. 1-3. Electron micrograph of plant cells (*Saintpaulia ionatha*). The magnification is 16,000×. The walls between the young cells are easily identified. Other features are indicated by symbols: Vac, vacuole; Nu, nucleolus. The granules within the nucleolus are similar to, though probably not identical with, the ribosomes in the cytoplasm. The remainder of the nucleus shows no defined structural pattern. Much of the interior is occupied by filaments of deoxyribonucleic acid and protein. The darker areas in the nucleus are due to regions of the chromosome which are more compact, the filaments presumably being packed more tightly; M, mitochondrion; Pp, proplastid. (Micrograph by Keith Porter.)

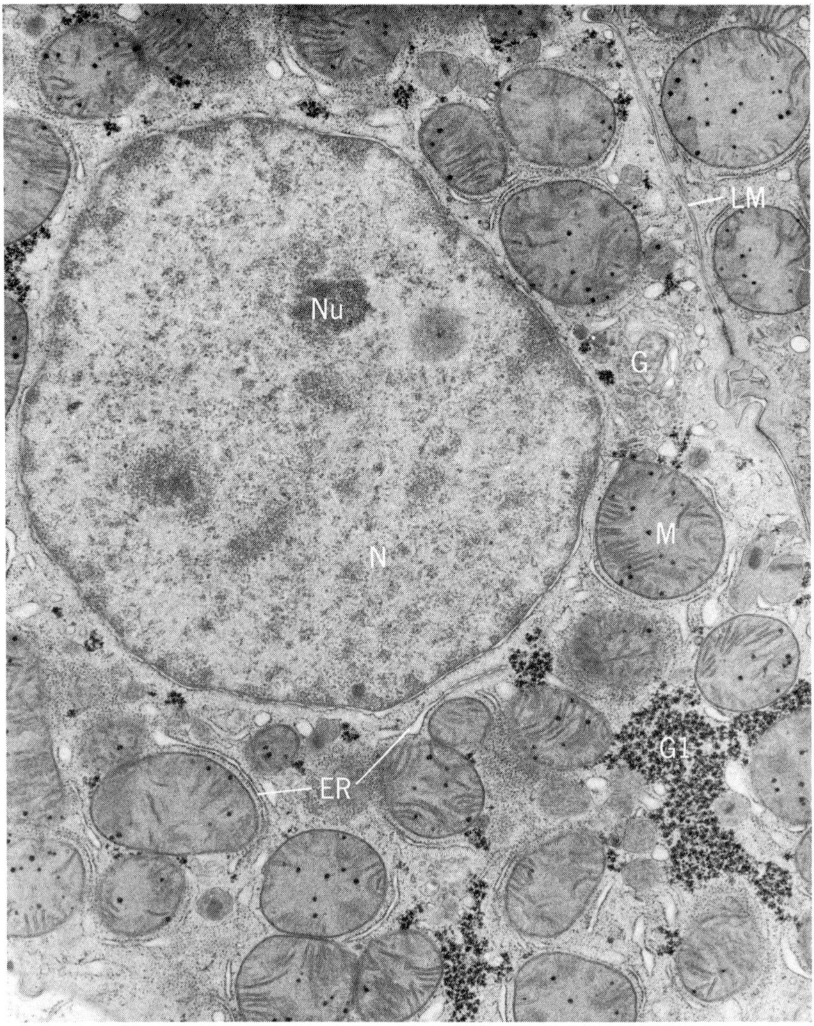

Fig. 1-4. Electron micrograph of a cell from the liver of a bat. Only a portion of the cell is seen in this photograph. The general characteristics are the same as those seen in the plant cell except that the limiting membrane (LM) in plant cells is obscured by the cell walls. In liver cells, however, glycogen granules (Gl) are prominent. The endoplasmic reticulum (ER) is more prominent, although not nearly so prominent as in a secretory cell of the pancreas. A well-developed golgi region is also evident (G). (Micrograph prepared by Keith Porter.)

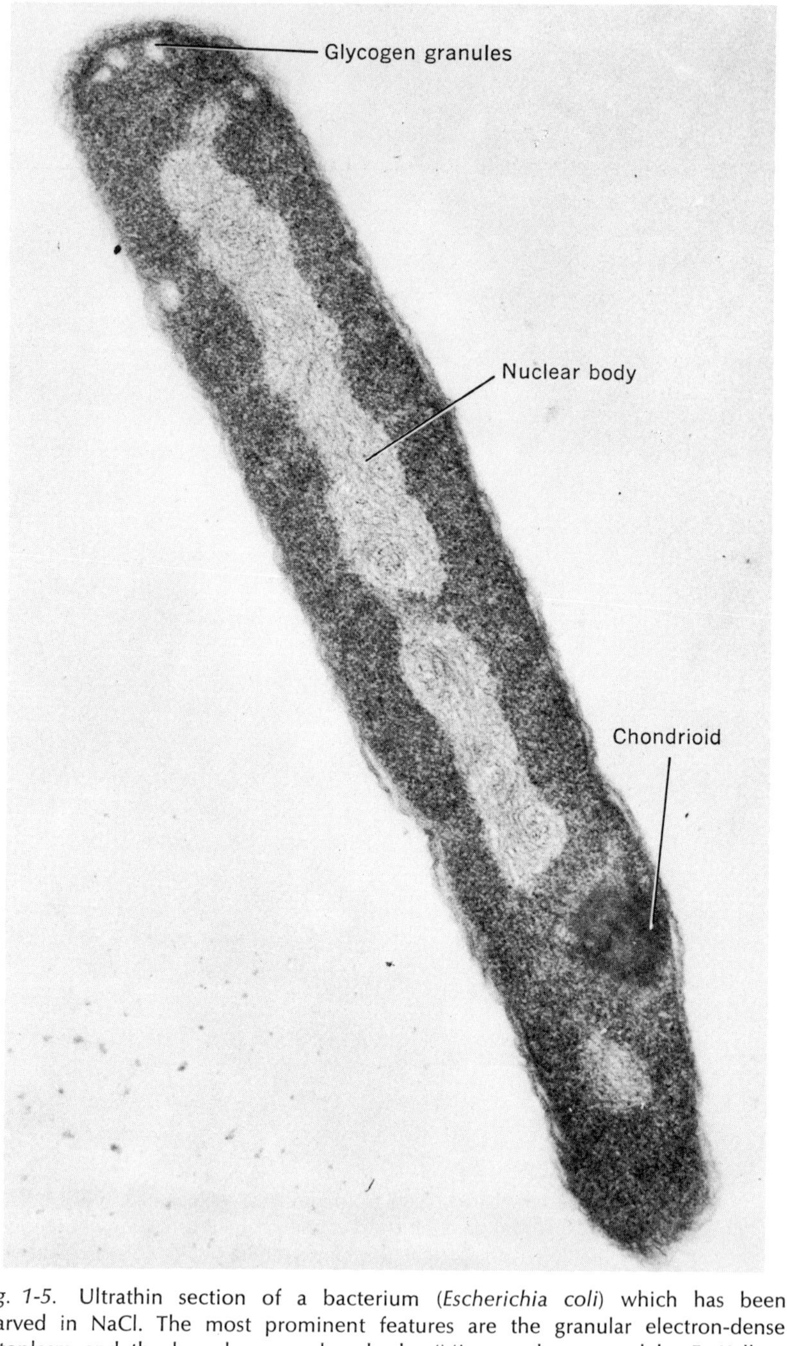

Fig. 1-5. Ultrathin section of a bacterium (*Escherichia coli*) which has been starved in NaCl. The most prominent features are the granular electron-dense cytoplasm and the less dense nuclear body. (Micrograph prepared by E. Kellenberger and A. Ryter. Published in *J. Biophys. Biochem. Cytol.*, **4,** 671–678 (1958).)

plasm, and the structure of the chromatic material of the nucleus is somewhat different from that in chromosomes. In both these respects the bacteria and the blue-green algae depart from the usual pattern and raise questions which will have to be taken up later.

In contrast to the relative uniformity of nuclear structures, cytoplasmic organization is highly variable. We have mentioned certain common differences between plant and animal cells: the usual presence of cell walls and vacuoles in plants. Perhaps, in a strict sense, these structures should not be considered as part of the cytoplasm proper but rather as products of cellular activity. The distinction between "vital" and "nonvital" parts of cells has been a subject of controversy ever since the coarser aspects of cell organization were revealed through the light microscope. During the nineteenth century the controversy was intense; but, like so many semantic issues, the topic gradually lost interest because far more challenging problems engaged the biologists' attention. A real difference, nevertheless, exists between the two categories of cellular structures, irrespective of the term used to describe them. Almost all unicellular organisms are encircled by a protective coat (usually a polysaccharide) and although such coats may function for purposes other than mechanical protection, there is little doubt that the primary processes essential to growth and reproduction occur within the confines of the cytoplasmic membrane. The delicate chemical balances essential to the integrity of the cell interior are not required for the preservation of the outer coats.

A similar consideration applies to vacuoles. Most, if not all, living cells acquire "unwanted" compounds either by absorption from their surroundings or as by-products of chemical activities. Cells would die if such compounds accumulated in the cytoplasm or the nucleus, and hence mechanisms have evolved to effect their removal. A common feature of these mechanisms is a membrane which acts as a sort of trapdoor. In multicellular animals the trapdoor mechanism is not apparent in most cells because of the complex circulatory and excretory systems which service the various tissues. In higher plants no excretory system exists, rather the cells develop a large vacuole which is little more than a membrane-bounded sac filled with water and a variety of solutes. Protozoa on the other hand have relatively small vacuoles, but the cells are capable of causing the vacuoles to contract and to eject their contents into the medium. The contents of plant vacuoles cannot be excreted and they frequently include toxic concentrations of a variety of products. Crystals of potentially toxic oxalate salts are commonly seen in many plant cells.

In addition to cell walls and vacuoles, many types of small inclusions are found in the cytoplasm. Most of these are food storage depots containing substances like starch, glycogen, fat, or protein. In all cases they may be regarded as passive components; the machinery for their degradation and synthesis is found within the cytoplasm. Nevertheless, structures such as these contribute largely to apparent cell diversity.

Although many species of organisms have distinctive cell types, characteristic differences also occur among the cells of a single multicellular individual—between leaf cells and stem cells, between muscle cells and nerve cells. Perhaps the most striking differences within a species involve the male and female sex cells. The human egg, for example, is nearly a million times as large as the human sperm, and most of this size differential is due to differences in the cytoplasm. Size is an extremely important factor in biological processes, since it is directly related to the surface area exposed to the environment; variations in size require many secondary modifications in cellular organization. When we consider the wide size ranges of cells, we may expect many cellular differences to be related more or less directly to this parameter. Table 1-1 provides a short list of cell sizes expressed in cubic microns (μ); this list effectively spans the entire range. Even if we exclude the viruses (Fig. 1-6) as not being "true" cells, a matter to be considered later, cell volumes differ by a factor of approximately 10^{17} (10 followed by 16 zeros). As may be readily imagined, a major aim in cellular biology has been to identify the common and constant components in otherwise highly variable patterns of cytoplasmic organization. These universal, or nearly universal, structures are our first concern.

Table 1-1 Mean Volumes of Representative Cells

Cell Type	*Volume (cubic μ)*
Ostrich egg	1.1×10^{15}
Hen egg	5.0×10^{13}
Human egg	1.4×10^{6}
Human sperm	1.7×10^{1}
Largest bacterium	7.0×10^{0}
Smallest bacterium	2.0×10^{-2}
Smallest virus	7.0×10^{-7}

In the photographs (Figs. 1-3, 1-4, 1-7) are seen some larger bodies, the mitochondria, with a distinctive architecture. They are surrounded

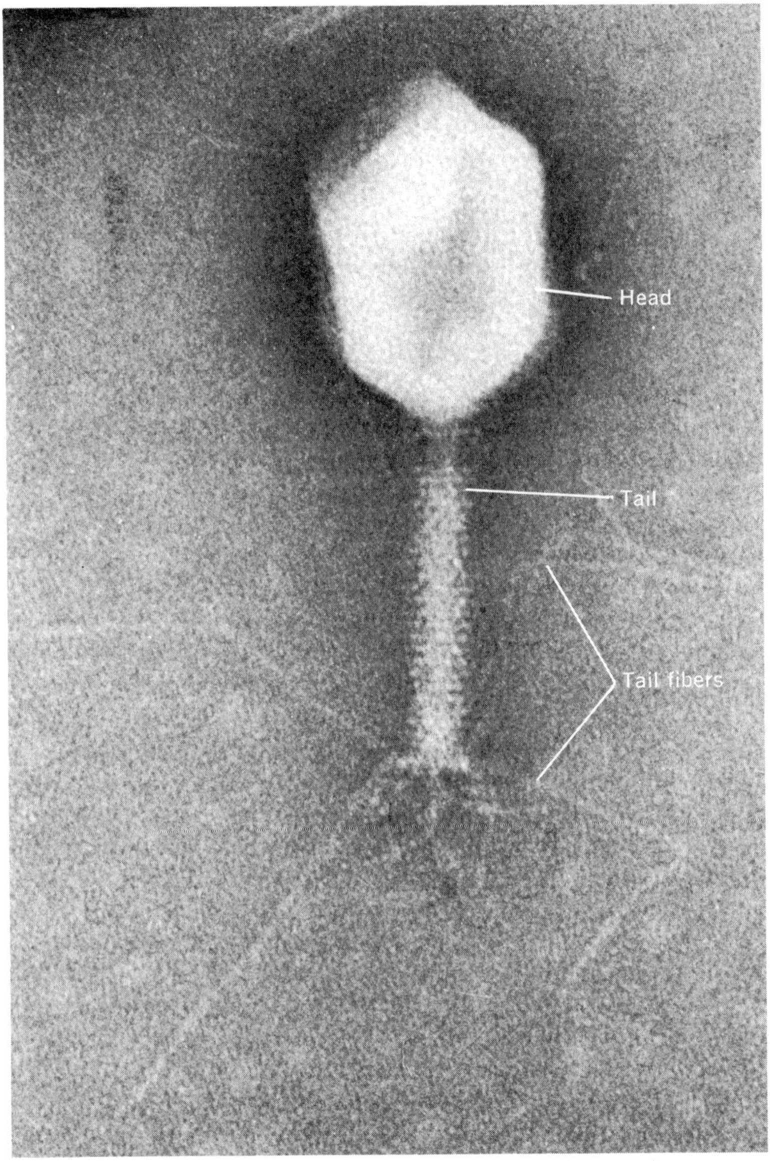

Fig. 1-6. Electron micrograph of bacteriophage T_4, showing the characteristic head and tail regions and the tail fibers. (Micrograph by E. Boy de la Tour.)

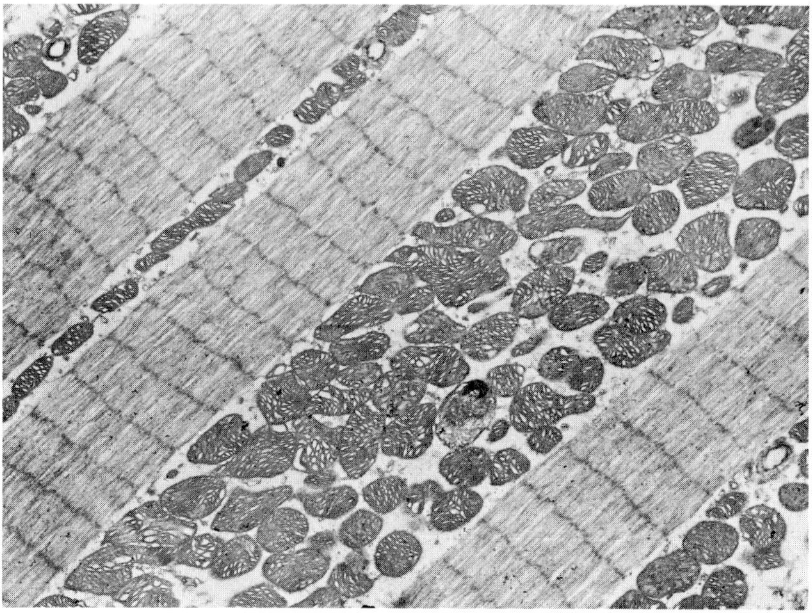

Fig. 1-7. A thin section of insect muscle. Note the numerous mitochondria lying between the muscle fibrils. This photograph may be compared with Fig. 16-5 in which details of the contractile elements are shown. The distribution of the mitochondria, the principal sources of chemical energy, among the fibrils, the principal consumers of chemical energy, illustrates one relationship between spatial organization and functional requirement. (Micrograph by D. Stegwee, Service Institute for Technical Physics in Agriculture, Wageningen, the Netherlands.)

by a double membrane, the inner member of which projects finger-like extensions toward the interior. The pattern is so distinctive that even a novice can identify mitochondria in electron micrographs of most cell types. The importance of mitochondria was long suspected by cytologists because of their nearly universal distribution. Although their fine structure remained unknown until the introduction of electron microscopy, they were large enough to be seen as particles through the microscope and could be distinguished from similar looking particles in living cells by virtue of their affinity for a dye known as Janus Green. The unique role of mitochondria in respiration has been clarified only during the past twenty years. As a result we can now better understand their distribution and architecture. Very active cells, those in the flight muscles of insects, for example, have large numbers of mitochondria. The folds of the inner membrane, which are often poorly developed in dormant or very young cells, furnish a striking example of an arrangement which provides a high surface/volume ratio. The significance of this ratio with respect to mitochondrial function will be discussed in Part II. Not all cells have mitochondria. Bacteria, which are similar in size to mitochondria, possess a much less elaborate structure for respiration. Yet, just as the "primitive" nuclei of bacteria are homologous with the nuclei of higher organisms, so are the "primitive" respiratory particles of bacteria homologous with the complex mitochondria of higher organisms.

The plastids found in many plant cells are similar in structure to the mitochondria insofar as they are both surrounded by double membranes and project membranous plates into their centers. The arrangement of the membranes is, however, as distinctive as their function. Chloroplasts, for example, are primarily concerned with the absorption of light, and the parallel layering of membranes is especially suited for maximal light absorption. When viewed through the light microscope, chloroplasts cannot be seen in blue-green algae, yet under the electron microscope parallel layers of membranes can be seen to fill the whole cell. In general, cells, whether plant or animal, which have specialized light-absorbing functions also have a system of layered membranes.

Another common component of cytoplasm is a set of membranes designated as the endoplasmic reticulum (Figs. 1-3, 1-4). These often appear in sections as lines lying parallel to the nuclear membrane or as irregularly parallel rows in the cytoplasm. In some cases they appear to be continuous with the nuclear membrane or with the external limiting membranes of the cell and may, in fact, be extensions of these other membranes (Fig. 1-2*d*). Of all the structures revealed by electron microscopy, this is perhaps the most outstanding, because its existence was hardly even suspected from studies with the light microscope. Surface and vacuolar membranes were presumed to exist because of certain physico-chemical observations, but no comparable evidence suggested the existence of internal membranes. As more and more cells were examined through the electron microscope, it became increasingly apparent that these membranes must play a very important role in cellular functions (Fig. 1-8). Cells vary markedly with respect to the concentration of the membranes; in some the endoplasmic reticulum fills a major portion of the cytoplasmic space, but in others it is scattered sparsely. The differences are not haphazard but, as will be seen, are correlated with the kinds of activities the cells are pursuing. Yet, however limited the membranes are in a cell, no cell has been found to be entirely without them. Indeed, if we take into account the fact that mitochondria and chloro-

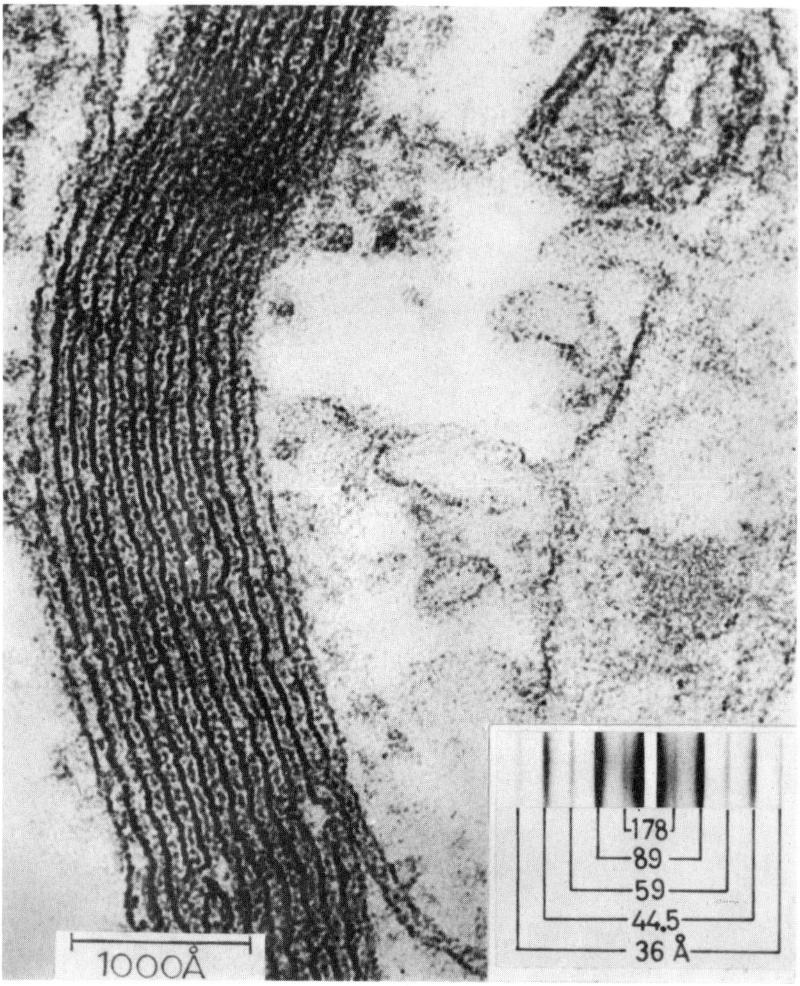

Fig. 1-8. High-resolution electron micrograph of myelin-sheath segment from transverse section of frog sciatic nerve demonstrating concentric array of dense and intermediate layers. Magnification: 217,000×. *Insert,* Low-angle x-ray diffraction pattern of fresh rat sciatic nerve recorded with Finean camera. This pattern features a fundamental period of 178 Å, with characteristic alterations of the intensities of the even and odd orders. (From H. Fernandez-Moran, "New Approaches in Correlative Studies of Biological Ultrastructures by High-resolution Electron Microscopy," the *J. Roy. Microscop. Soc.,* **83,** Pts. 1 & 2, June (1964), Plate 84.)

plasts are essentially membranous structures, the generalization can be made that membranes are one of the basic and universal patterns into which molecules are organized.

There are in fact a very limited number of structural assemblies found in cells. Leaving aside the "ground substance" of cells (that part which appears to have no defined form but acts as a sort of diffuse matrix), only three structures have been recognized—membranes, particles, and fibers. Particles (as defined by electron microscopy) are small solid bodies consisting of no more than a few large molecules; the best example of these are the "ribosomes" (so-called because of their high ribonucleic acid content) which are frequently found associated with the endoplasmic reticulum. Ribosomes, like membranes, are universal in distribution; their concentration is greatest in cells which are making a great deal of protein. It is doubtful that a cell could make protein without such particles, and, indeed, no cell is known to be without them.

Fibers, which are essentially thread-like arrangements of molecules, are not always apparent in electron micrographs. They are highly developed in a variety of specialized cells—hair, muscle, ciliated cells —but even in cells which show no visible evidence of fibers they are invariably present. The molecules of chromosomes are arranged in the form of fibers, and so too are the elements in the cytoplasm which enable cells to contract or chromosomes to be drawn to the poles during division.

This brief survey of cellular structures neglects many common components and essentially all the specialized ones, but it should serve for an introductory orientation. We have already begun to assign roles to cellular organelles, but we have not indicated how we are able to do so. Much of the remainder of this book is in fact devoted to a clarification of the relationship between the grosser organization, the chemical composition, and the biological functions of cellular components. This information has not come entirely through observation; a detailed account of the appearance of an organelle does not automatically inform us of its role in cellular processes, even though it may be suggestive. Most organelles had a variety of different functions proposed for them before decisive answers became available. The appearance of cells and of cell parts provides only the initial frame of reference from which to launch more penetrating inquiries.

Different cellular organelles have been analyzed in different ways, and various kinds of evidence have been assembled. The presence of an organelle might be correlated with a particular cellular activity; the numbers or sizes of organelles might be associated with quantitative variations in functions; activities might be localized in living cells by tracing the course of tagged chemical compounds through them; parts of cells might be isolated under conditions which permit the functions to persist. A large number of analytical devices, accumulated through a long technological history, have been employed to associate cell structures with cell functions.

As an illustration of the procedures employed to associate organelles and functions, we have chosen to consider in some detail that historical progression linking chromosomes with heredity. The "chromosome theory" provides an especially instructive example for several reasons. The central elements were established long enough ago to permit us some perspective, but recently enough for us to share in the thought processes of the investigators. Moreover, the experimental procedures were sufficiently simple that the major ideas are not obscured by technical details. Finally, the establishment of the chromosome theory was the first clear triumph of modern experimental and theoretical biology and one whose ramifications are still being explored.

2

The Gene Concept and the Chromosome Theory of Inheritance

The work of Gregor Mendel makes up one of the most fascinating chapters in the history of biology. Mendel's perception of fundamental problems, his fertile imagination, and his grasp of the nature of evidence provide a model of scientific analysis which is seldom matched. Nevertheless, we may plausibly argue that the history of biology would be little different had he never lived. Certainly he had no influence on his contemporaries and his work had to be done anew in a later generation. Partly because of the significance of this work, and partly in the hope of gaining insight into the nature and conditions of scientific activity, we should try to understand his accomplishment.

We have little personal information about Mendel. He was an instructor at a parochial high school in a small town in Austria who studied inheritance in the peas of a monastery garden. His major work was published in 1865. Most of our knowledge of Mendel comes from his few scientific papers and letters. Since the formal patterns and stereotyped language of scientific communications seldom reflect actual course of events or significant thought processes, we cannot hope to probe Mendel's mentality in detail. This is perhaps just as well, for it permits us to devise a progression much easier to follow than the probable sequence of detours and blind alleys.

To understand Mendel we must first understand something of the state of biology in the 1860's, for both his success and his failure reflect his times. We must remember, for example, that the idea that cells are the fundamental units of biological organization had been generally accepted less than thirty years earlier. Although nuclei were known to be nearly universally present in cells, their significance was guessed by very few. And chromosomes were not to be available for study until biological stains were developed in the 1880's. Thus Mendel's knowledge of the material basis of heredity was limited to an appreciation of the role of sex cells.

We must also remember that Darwin's epoch-making work on the origin of species was published only in 1859 and was the subject of intense interest and violent controversy, not only among biologists but also in many other literate circles. Darwin's contribution focused attention on cosmic problems, on the distribution of plants and animals across the face of the earth since the beginning of biological time. Perhaps this Darwinian focus provides a partial explanation of the lack of interest in events occurring in Mendel's pea patch, but it is not a complete explanation, for Darwin himself realized how important it was to understand the manner in which hereditary variations arise and how they are transmitted. Darwin, in fact, tried to develop an explanation, but in comparison with Mendel's carefully reasoned and experimental approach, Darwin's effort in this area is simply an historical curiosity.

In trying to understand Mendel, we should also not neglect events in other scientific fields. The atomic revolution in chemistry, with its emphasis on ultimate particles, had occurred half a century earlier and its success was widely appreciated. It might also be of significance to remember that the physics of Mendel's time presented an orderly mechanical universe operated by simple cause-and-effect relationships. The direct influence of developments in physics and chemistry on Mendel was probably slight, but the indirect effects may have been considerable. And certainly Mendel was aware of them, through the discussions in his local scientific society if by no other means.

Finally, we must appreciate that biology before Mendel had been largely a qualitative and descriptive science. Unlike physics and chemistry, mathematics had little place in biological studies, and quantitative experimentation was yet to come. Actually, biometry and statistical analysis in biology were not far in the future, for Sir Francis Galton, the founder of biological statistics, was born in the same year and was working at the same time as Mendel. But Mendel not only anticipated the biometricians; in the power of his analysis if not in the sophistication of his techniques, he was their master.

ELEMENTARY PROBABILITY

The mathematics employed by Mendel was quite elementary. The principles were understood by mathematicians, employed by physical scientists, and exploited by gamers and gamblers. It is neither necessary nor appropriate at this point to set forth the elements of mathematical probability, but some understanding of the principles of probability is required for an appreciation of "Mendelism." We can get by with a definition and two general rules.

The definition of a "probability" is essentially identical with that of a frequency—the number of times an event occurs out of the total number of possible occurrences. The event may be of any sort, a tossed coin landing with "heads" up, a particular card drawn from a deck of cards, a die landing with a particular surface exposed. Numerically, probability is a fraction. It may be expressed as a common fraction, a decimal fraction, or a percentage. Its limits are Zero, indicating an impossible event, and One, indicating a certain event.

The two rules may be designated as "the rule of the sum" and "the rule of the product." These rules pertain to combinations of probabilities. The rule of the sum may be stated as follows: *The probability that either of two (or more) mutually exclusive events will occur is equal to the sum of their separate probabilities.* Thus the probability that a tossed coin will land either heads or tails is equal to the probability that it will land heads ($\frac{1}{2}$) plus the probability that it will land tails ($\frac{1}{2}$), or 1. The probability that one will draw either a club or a diamond from a deck of bridge cards is equal to the proba-ability of drawing a diamond ($\frac{1}{4}$) plus the probability of drawing a club ($\frac{1}{4}$), or $\frac{1}{2}$. The probability of rolling a 1, 2, or 3 with a particular die is equal to $\frac{1}{6} + \frac{1}{6} + \frac{1}{6} = \frac{1}{2}$.

The rule of the product is concerned with events which are not exclusive of each other, but which can occur together. It is stated as follows: *The probability that both of two (or more) independent events will occur is equal to the product of their separate probabilities.* Thus the probability that two coins tossed simultaneously will land heads is equal to the probability that the first will be heads ($\frac{1}{2}$) times the probability that the second will be heads ($\frac{1}{2}$), or $\frac{1}{4}$. The probability that one will draw spades from two different bridge decks becomes $\frac{1}{4} \times \frac{1}{4} = \frac{1}{16}$. And the probability that two dice rolled simultaneously will both come to rest with the upper side showing one spot is $\frac{1}{6} \times \frac{1}{6} = \frac{1}{36}$.

These two rules are often employed in conjunction. For example, what is the probability of drawing cards of the same suit from two decks—that is, two clubs or two diamonds or two hearts or two spades? Since the probability of drawing two clubs is $\frac{1}{4} \times \frac{1}{4} = \frac{1}{16}$, and the probabilities of drawing two diamonds, hearts, or spades are calculated in the same way, the probability of drawing identical suits is $\frac{1}{16} + \frac{1}{16} + \frac{1}{16} + \frac{1}{16} = \frac{1}{4}$.

Or, what is the probability that a family with three children will include one boy and two girls? This is equivalent to asking the probability that any one of the following sequences will occur.

	First Child	Second Child	Third Child
(*a*)	*boy*	girl	girl
(*b*)	girl	*boy*	girl
(*c*)	girl	girl	*boy*

We may assume for sake of simplicity (though this is not quite true) that births of boys and girls are equally frequent. Each sequence then has a probability of $\frac{1}{2} \times \frac{1}{2} \times \frac{1}{2} = \frac{1}{8}$. The probability that either (a) or (b) or (c) will occur is therefore $\frac{1}{8} + \frac{1}{8} + \frac{1}{8} = \frac{3}{8}$.

This procedure of listing all the ways in which a particular result might be achieved and summing the various probabilities is only practically feasible when the number of combinations of events is fairly small. Consider, for example, the task of listing all the sequences of sons and daughters in families of 10 children which would yield 4 sons and 6 daughters. A useful device for avoiding such tedium is the binomial expansion $(p + q)^n$. Here p and q may represent the probabilities of two mutually exclusive events which cover all possibilities. The expansion $(p + q)^2 = p^2 + 2pq + q^2$ may be used to calculate the probabilities of various combinations of boys and girls in families with two children. The symbol p can indicate the probability for a male at any birth and q the probability for a female. Then p^2 represents the probability for two boys, $2pq$ the probability for one child of each sex, and q^2 the probability for two girls.

The higher powers of the binomial expansion may be used for larger sample sizes. Some of these are listed below for the reader's convenience.

The simplest way to obtain the terms of the expansion $(p + q)^n$ is as follows. (1) The first term is p^n. (2) The second term is $np^{n-1}q$. (3) In each successive term the power of p is reduced by 1 and the power of q is increased by 1. (4) The coefficient of a term is obtained by multiplying the coefficient of the preceding term by the power of p in that same term and dividing the product by the power of q in that term plus 1.

$$(p + q)^3 = p^3 + 3p^2q + 3pq^2 + q^3$$
$$(p + q)^4 = p^4 + 4p^3q + 6p^2q^2 + 4pq^3 + q^4$$
$$(p + q)^5 + p^5 + 5p^4q + 10p^3q^2 + 10p^2q^3 + 5pq^4 + q^5$$
$$(p + q)^6 = p^6 + 6p^5q + 15p^4q^2 + 20p^3q^3 + 15p^2q^4 + 6pq^5 + q^6$$

Note that the exponents of p and q in the expansions represent the frequencies of the two events to be combined and that the coefficients of the terms indicate the number of possible ways that these events could be arranged. Thus if we wanted to know the probability that a family of five children would consist of 1 boy and 4 girls, we would select the expansion of $(p + q)^5$ and the term $5pq^4$.

EXERCISES IN PROBABILITY

1. What is the probability of drawing an Ace from each of two decks of cards? Of drawing two Aces in successive draws from a single deck?

2. What is the probability of drawing either an Ace or a Club from a deck of cards?

3. What is the probability when tossing a coin three times of obtaining the sequence T T H? If one tosses a coin twice and obtains tails both times, what is the probability that the third toss will yield heads?

4. When rolling two dice, what is the most likely figure to be obtained from summing the exposed dots?

5. Is one more likely to get a 1:1 sex ratio in a family of 4 children or 6 children? Is one more likely to have progeny of a single sex with 4 or 6 children?

6. If the frequency of blue-eyed individuals is $\frac{1}{4}$, regardless of sex, what is the probability of finding among five randomly assembled individuals, three who are blue-eyed females?

7. If in a particular population 30% of the individuals are blonde and 70% are brunette, and if the frequencies are the same with males and females, and if marriages are contracted with no reference to hair color, what fraction of the marriages in this population should involve two blondes? What fraction should consist of a brunette and a blonde?

8. If one finds in a particular population that 20% of the individuals have blue eyes and 80% have brown eyes, with equal frequencies in the two sexes, and that 25% of married couples consist of a blue-eyed and a brown-eyed individual, what might you conclude about marriage preferences in this population?

9. A sample of families with six children is collected to determine if the sexes of the children are determined at random, or if some families are predominantly "male producing" and others "female producing." The distribution of sexes in the families is as shown.

Number of male children	0	1	2	3	4	5	6
Number of families	68	373	906	1157	890	372	74

What would be the frequencies expected if no familial tendencies exist? How would you evaluate these data?

10. If a cell contains a certain number of particles which are distributed at random at cell division, what is the probability that a daughter cell will lack such a particle entirely if
   (a) Only one particle is found in each cell?
   (b) Two particles?
   (c) Four particles?
   (d) Ten particles?
   (e) 100 particles?

MENDELISM

Returning from this brief excursion into probability, we may now consider Mendel's contributions. He was fortunate in choosing garden peas for his experimental material. These plants can be crossed readily by transferring pollen from the male organs of one plant to the female organs of another. Or, the plants may be left undisturbed to undergo self-fertilization (or simply "selfing"). Mendel observed among the several varieties of peas which he cultivated a number of characteristic differences. Since these differences persisted under the usual selfing regimen, he concluded that they were hereditary and that the strains in question were "pure breeding"; several such strains were selected for detailed study. He found, for example, that some varieties produced red flowers and others white flowers; some produced round seeds and others wrinkled seeds; some of the plants were tall and others short.

His first experiments consisted of making crosses between strains which differed in a particular trait and observing the characteristics of the plants in the next generation or—in the case of seed characters—observing the seeds themselves. The crosses could be made with either kind of plant as the male parent. He did not, however, observe any differences in such "reciprocal crosses"; the contributions of the male and female parents were equivalent. He did observe that for any particular trait the offspring strongly resembled one of the parental strains. In a cross of a plant bearing red flowers with one bearing white flowers, all the F_1 (first filial generation) had red flowers, regardless of whether the red-flowered parent was used as a male or as a female. Similarly, in a cross of a plant bearing round seeds with one producing wrinkled seeds, all the seeds produced were round. In a cross of a tall plant with a short plant, all the progeny were tall.

Since such results occurred for every pair of alternative traits he studied, Mendel concluded that some general mechanism of exclusion in expression was characteristic of hereditary phenomena. He generalized his observations in his so-called principle of dominance. For our purposes it may be stated as follows: *When pure-bred strains differing in a particular trait are crossed, the F_1 individuals are uniform and express one of the traits to the exclusion of the other.* The trait of a pair which is expressed he termed the *dominant trait;* that which is not expressed he termed the *recessive trait.* We may say that red flowers are dominant to white; round seeds are dominant to wrinkled; tallness is dominant to shortness. Note that this "principle" is what might be called a "first-order" generalization—simply a summary of direct observations on several different but comparable situations.

The next step in the analysis was to see what happened in subsequent generations. Did the recessive traits disappear permanently or could they reappear later? This question could be answered by crossing the F_1 plants to yield a second, or F_2, generation. Since these plants are capable of self-fertilization the simplest means of producing the F_2 was by selfing the F_1. When this was done, Mendel observed in each case that some of the F_2 plants manifested the recessive characteristics. In the F_2 of a cross of red $\times$ white, some of the plants produced red flowers and some produced white flowers. Similarly, the seeds produced by F_1 plants from a cross of round $\times$ wrinkled consisted of some round and some wrinkled seeds. The F_2 progeny of a cross of tall $\times$ short plants included both tall and short plants.

Again, since comparable results were obtained in all his F_2 progenies, Mendel concluded that a general hereditary phenomenon was involved and he set forth the generalization referred to as the principle of segregation. In its simplest form the principle may be stated as follows: *Recessive traits, which disappear in the F_1 between pure-bred strains, reappear in a portion of the F_2.*

Neither of these sets of observations was entirely new to plant breeders, and, as we shall see later, neither of these "principles" represents an inviolable "law of nature." What was important about their formulation was that they summarized neatly a phenomenon which needed to be explained and led Mendel into the next stage of his inquiry. The observations outlined a pattern of transmission of hereditary traits which was apparently a common pattern. In seeking to explain this pattern, Mendel made some provisional hypotheses and sought to explore their consequences. Perhaps the most important assumption he made was that the traits were determined by "hereditary factors," fundamental biological elements equivalent in some ways to the chemists' atoms. He proposed that a hereditary factor—

or as we now say, a *gene*—was responsible for red flower color; this gene may be designated by the symbol *R*. The contrasting white color might then be controlled by an alternative (or *allelic*) gene represented by the symbol *r*.

Mendel then proceeded to deduce the properties of these hypothetical elements by an analysis of the patterns of inheritance. One of the first questions that required an answer was the number of genes of a kind which might occur in a cell. Could a particular cell contain only one gene of a kind or could it contain many? A related question concerned the number of kinds of genes within a single cell. Could a cell contain only *R* genes or *r* genes, or could it contain both? Partial answers to such questions could be provided by information already available. Some of the F_2 individuals were white and thus would be assumed to have *r* genes. They must have received these *r* genes from their parents, the F_1 individuals. Hence the F_1 individuals must have had *r* genes as well as the *R* genes which determined their own flower color and which they transmitted to some of their progeny. We may conclude then that the F_1 individuals contain two kinds of genes for flower color and that they must have at least two genes for flower color. The simplest way we can represent the genetic constitution (or *genotype*) of the F_1 is *Rr*.

Since the F_1 plants contain (at least) two flower color genes, we should expect similar numbers also in the original parental strains, in the F_2 progeny, and in all other plants which might be produced. This expectation is based primarily on faith in regularity or uniformity, but without such faith any kind of analysis becomes impossible. Since we know that the parental strains come from true-breeding lines, and since we are assuming at least two flower color genes in each individual, the simplest genotypic designation for the parental strains makes them *RR* and *rr*. This then gives us three genotypic designations—*RR* and *Rr* for plants producing red flowers, and *rr* for plants producing white flowers. The *RR* and *Rr* plants are not, however, distinguishable by inspection; they are said to have the same *phenotype*. Since different genotypes may underly the same phenotypes, we cannot directly assign genotypes to all the F_2 individuals. We can assign the genotype *rr* to the white-flowered plants, but the genotypes of the red-flowered plants are indeterminate. We will return to this problem in a moment.

First, however, we must consider a more general problem. If two parents contribute genes to the offspring, and if the number of genes per organism remains constant generation after generation, the parents cannot give all of their genes to any one offspring. Several theoretically possible schemes might accomplish this result, but most of these can be eliminated by the information already presented. Mendel's proposal, however, is entirely consistent with the available data. He suggested that each parent has two genes of a kind but contributes only one of these to each offspring. He proposed that the somatic (body) cells contain two genes of a kind, but the sex cells (gametes) contain only one. If this is correct, the number of genes per somatic cell remains constant in each generation; the number is reduced by one half in the formation of the sex cells and is restored to the doubled number at fertilization when the male and female sex cells are united.

Following this line of reasoning we are now in a position to describe more fully the mechanism of transmission in the classical pattern of inheritance and also to deal more fully with the genotypes in the F_2. We have postulated that the original parental strains carry two identical flower color genes; in modern genetic terminology we would say that they are *homozygous*. The *RR* parent would produce only *R* sex cells and the *rr* parent would produce only *r* sex cells. At fertilization the new organism established would be initiated with one gene of each kind and would have the genotype *Rr;* it would be *heterozygous*. If such an F_1 individual is selfed, it will produce some male sex cells containing the *R* gene and some with the *r* gene. Similarly, it should produce some female sex cells of the *R* type and some with the *r* gene. The consequences of this self-fertilization can be predicted if certain regularities are encountered. In particular, the two classes of sex cells (*R* bearing and *r* bearing) should be produced in equal frequency, and the male and female sex cells should unite in fertilization at random; that is, *R*-bearing sperm do not preferentially fertilize either *R*-bearing or *r*-bearing eggs.

Under these conditions not only the kinds of progeny produced but also their relative frequencies can be predicted.

Male Sex Cells	Female Sex Cells	Genotype of Progeny
R $(\frac{1}{2})$	R $(\frac{1}{2})$	RR $(\frac{1}{4})$
R $(\frac{1}{2})$	r $(\frac{1}{2})$	Rr $(\frac{1}{4})$
r $(\frac{1}{2})$	R $(\frac{1}{2})$	Rr $(\frac{1}{4})$
r $(\frac{1}{2})$	r $(\frac{1}{2})$	rr $(\frac{1}{4})$

The probability that any particular individual will receive an *R* from the maternal side is $\frac{1}{2}$; the probability that an individual will receive an *R* from the paternal side is also $\frac{1}{2}$. The probability that it will receive an *R* from each parent is, therefore, $\frac{1}{2} \times \frac{1}{2} = \frac{1}{4}$. A heterozygous individual may be constituted in either of two ways, each with a probability of $\frac{1}{4}$, to yield a combined probability of $\frac{1}{2}$. A homozygous recessive individual is established only when an individual receives an *r* sex cell from each parent, an event with a probability of $\frac{1}{4}$. Such considerations lead to the expectation of three classes of F_2 genotypes—two of which are phenotypically indistinguishable. The ratio of phenotypically dominant to phenotypically recessive individuals should be 3:1.

To verify this quantitative prediction, Mendel counted the red- and white- flowered plants in the F_2 and found approximately the proportions expected. He carried out similar studies on other pairs of alternative traits and repeatedly verified the 3:1 ratio. Thus his interpretation provided not only a qualitative explanation of the principle of segregation, but also a quantitative and verifiable prediction concerning the ratios to be expected.

Several additional tests of the hypothesis were feasible, and these were carried out in convincing detail. For example, Mendel predicted that the F_2 reds would be of two kinds—homozygotes and heterozygotes in a ratio of 1:2. This prediction could not be tested by looking at the F_2 plants, but it could be tested by further breeding studies, specifically by selfing each of the F_2 reds. The homozygotes should breed true, but the heterozygotes should repeat the behavior of the F_1, producing both red and white flowered plants in a ratio of 3:1. In such tests on several pairs of traits, Mendel confirmed the existence of two types of phenotypically dominant individuals in the F_2, and also demonstrated that twice as many heterozygotes as homozygotes were produced.

Having found an interpretation of hereditary patterns which explained their chief features and which was capable of quantitative verification, Mendel proceeded to another question: the relationship in transmission of genes controlling different traits. Instead of following a single pair of alternatives in a cross, he could follow two or more pairs. Thus he could cross a tall plant with red flowers to a short plant with white flowers. The F_1 generation would be expected to be tall and red under any circumstances, and the critical results would be obtained from the F_2. From previous experiments he knew that some of the F_2 would be red and some white, and that some would be tall and some short. But, would the red plants all be tall and the white plants all short, or would the original parental combinations be broken up to yield some tall whites and some short reds? If the parental combinations can be broken, and particularly if the two sets of traits are transmitted independently of each other, the relative frequencies of the various possible combinations can be calculated. Thus, three-fourths of the F_2 plants should be tall and three-fourths should be red. If these determinations are independent, the frequency of tall red plants should be $\frac{3}{4} \times \frac{3}{4}$ or $\frac{9}{16}$. Similarly, the frequency of tall white plants should be $\frac{3}{4} \times \frac{1}{4}$ or $\frac{3}{16}$, and the frequency of short red plants should be $\frac{1}{4} \times \frac{3}{4}$, or $\frac{3}{16}$. The frequency of the double recessive plants, those which are short and white, should be $\frac{1}{4} \times \frac{1}{4} = \frac{1}{16}$. In a similar manner we may calculate the probabilities for the various genotypes, or we may extend such calculations to include any number of segregating genes in the F_2. As more pairs of genes are followed, the results become progressively more compound, but the elementary principles remain the same. We simply calculate separately the results expected for each pair of genes and combine them by the rules of probability already discussed. For a single pair of genes only six different crosses are possible, those listed in Table 2-1.

Table 2-1 Summary of Genetic Ratios for Single-Factor Crosses

	Genotypes of Progeny		
Crosses	TT	Tt	tt
TT × *TT*	1	0	0
TT × *Tt*	½	½	0
TT × *tt*	0	1	0
Tt × *Tt*	¼	½	¼
Tt × *tt*	0	½	½
tt × *tt*	0	0	1

For several combinations of traits, Mendel studied the phenotypic ratios in the F_2. In each case he found the expected 9:3:3:1 ratio. From such studies he concluded that the genes for different traits of an organism were transmitted independently of each other, and this generalization was formulated into his third principle—the principle of independent assortment.

These observations, along with their interpretations, constitute Mendelism. It can be applied in a wide variety of situations in many organisms, with any number of hereditary factors and with any parental constitutions. Mendelism consists of two sets of generalizations: the first-order generalizations represented by the principles which summarize the observations collected in a variety of crosses, and the second-order generalizations which are concerned with the behavior of hypothetical elements which cannot be observed and whose properties can only be inferred. We may summarize the second-order generalizations as follows:

1. Hereditary traits are controlled by hereditary factors, or genes, which are transmitted from a parent to its progeny.

2. These genes exist in pairs in the cells of adult organisms, but each sex cell receives only one of the two.

3. At fertilization the sex cells unite at random; that is, the sex cells carrying a particular kind of gene do not preferentially unite with other sex cells containing particular genes.

4. Genes controlling different traits are transmitted independently; in the formation of sex cells, particular genes do not remain preferentially associated.

5. When an organism is constituted with different genes for a particular trait, one is usually expressed and the other, though maintained and later transmitted, does not manifest its effects.

EXERCISES IN MENDELISM

1. Calculate the probabilities for the following kinds of offspring from the cross of *AabbCc* × *aaBBCc*.

(a) *AaBbCc* (b) *aaBbcc* (c) *AabbCc*

2. In man albinism (lack of pigment) is determined by a recessive gene *a*. In a particular population assume that these genes have a frequency of 10% and are distributed at random.

(a) What proportion of the population will be phenotypically albino?

(b) What proportion will be heterozygous?

(c) What fraction of the marriages in the population will be between two heterozygotes?

(d) Calculate the distributions of albino children in marriages of heterozygotes which yield four offspring.

Number of albinos	Fraction of marriages
0	
1	
2	
3	
4	

3. A common exception to Mendel's principle of dominance is illustrated by the blood group factors in man. Instead of one gene completely masking its allele, both may be expressed. In addition, this system of inheritance illustrates another complication which Mendel did not encounter. Instead of only two alternative forms of a gene, the blood types are determined by at least three. However, the transmission of these

factors follows Mendelian patterns and a single individual has no more than two kinds of gene. The gene and genotype designations commonly employed are set forth below:

Genes	Genotypes	Phenotypes
I^A	I^AI^A	*A*
	I^Ai	
I^B	I^BI^B	*B*
	I^Bi	
i	I^AI^B	*AB*
	ii	*O*

(a) In a particular population 20% of the genes are I^A, 10% are I^B, and 70% are *i*. What is the frequency of type *A* individuals? What is the frequency of type *O* individuals?

(b) What fraction of the marriages between type *A* individuals are potentially capable of producing type *O* children?

(c) A type *A* man marries a type *B* woman and their first child is type *O*. What is the probability that their second child will be type *AB*?

(d) A type *B* man marries a type *A* woman and their first child is type *B*. What is the genotype of the child?

4. Several different strains of the lamarckian malarky are maintained in laboratories. One has green eyes and another has purple eyes. When crosses are made between the strains, all the offspring have brown eyes. In the F_2, however, some have green, some purple, and some brown eyes, and in addition a few have blue eyes. Devise an explanation for these results, predict the phenotypic ratio in the F_2, and describe how you might further test your interpretation.

5. We have discussed the reasons for concluding that at least two genes of a kind are found in the somatic cells of higher organisms. We have further shown that the breeding results are explicable on the basis of just two genes of a kind. We have not rigorously demonstrated that the existence of more than two genes is excluded by the data. Assume, for present purposes, that four genes of a kind are characteristic of somatic cells, and that reproductive cells receive two of these in a random assortment. Then, predict the numerical consequences of such a scheme in the F_2 of a cross between a red- and a white-flowered pea plant.

CHROMOSOME BEHAVIOR

In sharp contrast to its initial reception, Mendelism provoked an immediate and widespread response when it was rediscovered in 1900. There are several reasons for this. We may note, for example, that Mendelism was rediscovered not once, but three times—by Karl Correns in Germany, by Gustav Tschermak in Austria, and by Hugo De Vries in Holland. For 35 years Mendel's work was ignored and then within a single year three individuals, largely independently, recognized the pertinence of his ideas and called them to the attention of the scientific community.

This incident illustrates in a dramatic way the problem of evaluating the relative significance of the individual and the society in which he operates. The scientific community has a direction and a momentum within which the individual must operate if he is to be effective. Mendel ranks as one of the giants of biology in his intellectual accomplishment; but because of his cultural isolation, he may be ranked as a near failure, of lesser importance than many of his far less accomplished contemporaries. The simultaneity of discoveries serves to emphasize the social aspects of science and the dispensability of an individual in the long-range processes.

In a sense the individual is of lesser importance in science than in other aspects of human culture. Biology today would not be much different if Mendel or Darwin had never lived; Mendel's rediscoverers discovered not only Mendel's papers, but also his phenomena, and Darwin was being closely followed by Alfred Russell Wallace. In contrast, had Beethoven or Michelangelo never lived, the music and the sculpture and paintings of today might not be very different, but our musical repertory and our museums would be incomparably poorer. The accomplishments of science are not artifacts of permanent significance, but are contributions to a moving stream which quickly lose their individual identity.

On a less philosophical level we may note that since the time of Mendel, biologists had come a long way in experimental approaches and quantitative thinking. Mendel's way of reasoning was no longer bizarre and incomprehensible, but appeared as an approach of choice to biological problems. Perhaps even more important, information on cell structure and behavior acquired during the intervening years made it possible to identify a material counterpart to Mendel's theoretical entities. Mendelism was not simply accepted; it was appropriated as the starting point for a new development in biology, in a way which would have been impossible in 1865. To understand the assimilation of Mendelism we must consider in some detail the advances made in cell biology during the last quarter of the nineteenth century.

At the time of Mendel's studies, the nucleus was recognized as a nearly universal component of cells, but little was understood of its structure and behavior during cell multiplication. Such understanding could only be achieved once satisfactory techniques were devised for identifying the component structures of nuclei. When suitable microscopic stains did become available, resolution of structural changes during cell division followed almost as a matter of course. Prior to cell division the nucleus which, except for its nucleolus, was hitherto optically homogeneous took on the appearance of a tangled mass of threads. Because these threads stained particularly well with certain dyes, they were designated as *chromosomes* (colored bodies).

Several important generalizations concerning chromosomes were established through careful microscopic study. First, in all cases where chromosomes could be reliably counted, their number was found to be constant within a single kind of organism. This was true not only for all the cells of an individual, but also for the cells of different individuals of the same species. Secondly, the chromosomes in a single cell differed from each other and could often be distinguished on the basis of size or shape or staining properties. The chromosome set of a cell was not a collection of identical elements, but a complex of individualized organelles. And all the cells of the organism had not only the same number of chromosomes but the same assortment of chromosomes. Finally, cytologists discovered that, with a few significant exceptions, each somatic cell of an organism contained not one but two of each kind of chromosome. Such pairs of identical chromosomes are commonly described as "homologous."

The uniformity in the numbers and kinds of chromosomes in related cells indicated the operation of a precise mechanism of replication and distribution of chromosomes at cell division. This mechanism was described in considerable detail and the process was termed *mitosis.* Its essential features may be summarized as follows. Chromosomes are linear structures and during most of the cell cycle they are extended into extremely fine fibers whose narrow width makes them invisible to optical methods, and whose lengths are far greater than the diameter of the cell containing them. Prior to cell division, however, the fibers coil into more compact structures which can be seen and counted. One segment of each chromosome is usually distinguished by its staining properties and subsequent behavior as the *centromere.* The centromere may be near the end of the chromosome or it may be near the middle, but its position is constant for a particular chromosome type. In conjunction with other features its position may be useful in identifying particular chromosomes.

To illustrate the subsequent behavior of the condensed chromosomes we will consider an organism with only two pairs of chromosomes, one with a median centromere and the other a longer kind with a subterminal centromere (Fig. 2-1). At the time of condensation prior to cellular division (*prophase* stage), the chromosomes may be seen to be already double. It thus appears that the actual manufacture of new chromosomes occurs while the chromosomes are in the extended state between divisions (*interphase* stage). Subsequent chemical studies have reinforced this conclusion. The two strands of a single chromosome are, however, still connected firmly in the centromere

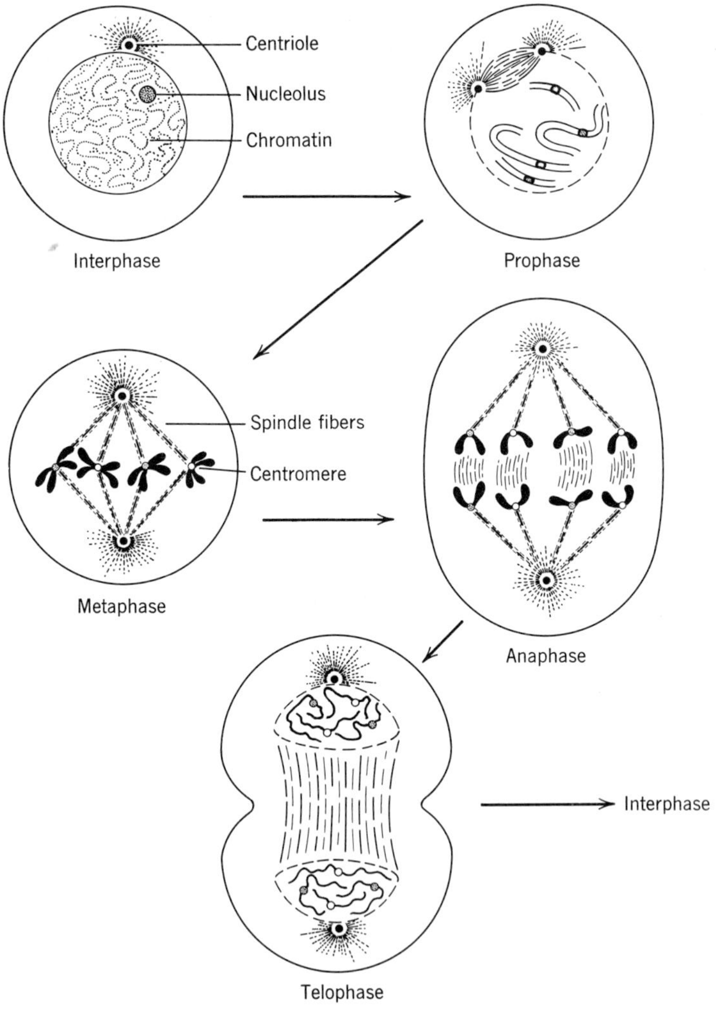

Fig. 2-1. Mitosis.

regions. These regions are important in the orientation and separation of the duplicate strands.

While the chromosomes are undergoing their condensation, other changes are occurring in the cell. In particular, from a center outside the nucleus (the centriole) a spindle-shaped apparatus is organized from oriented protein fibers. The nuclear membrane disappears, and the chromosomes come to lie on a plane across the equator of the spindle (*metaphase stage*). The centromeres then divide and begin to move in opposite directions to the two poles of the spindle (*anaphase* stage) with the chromosome arms following behind. As they approach the poles the spindle elongates, the chromosomes begin to uncoil, the nuclear membrane is reestablished, and the cell as a whole is separated into two daughter cells (*telophase*). The consequence of this sequence of events is the production of two cells which have the same numbers and kinds of chromosomes as the single cell from which they were derived.

The understanding of mitosis permitted an important generalization. Earlier investigators had established that life is always derived from preexisting life, and that cells are derived from preexisting cells. Now the continuity of biological structures was extended downward from the cellular level to include the nucleus and chromosomes.

Investigators of this period were also concerned with the processes involved in the formation of the sex cells—sperm and egg—distinctively associated with fertilization and the formation of new individuals. Again they focused attention on the chromosomes and observed some highly significant deviations from the usual behavior. The cells destined to become sex cells undergo a series of divisions during which the chromosomes are distributed in a manner different from that in ordinary mitosis. This special sequence of events is designated as *meiosis* (Fig. 2-2).

In essence, meiosis is a process whereby cells undergo two divisions while their chromosomes undergo only one. As a result, the presumptive sex cells come to have only one rather than two chromosomes of each kind. This is brought about as follows. As in ordinary mitosis the chromosomes emerge from the interphase condition by a process of coiling and condensing. And as in mitosis the new chromosomal materials are synthesized either in the interphase condition or shortly after condensation begins. But, instead of homologous chromosomes remaining separate as in mitosis, they line up side by side in pairs (synapsis). Since the chromosomes which come together are already duplicated, a total of four chromosome strands (*chromatids*) are held together in close association with various entanglements apparent.

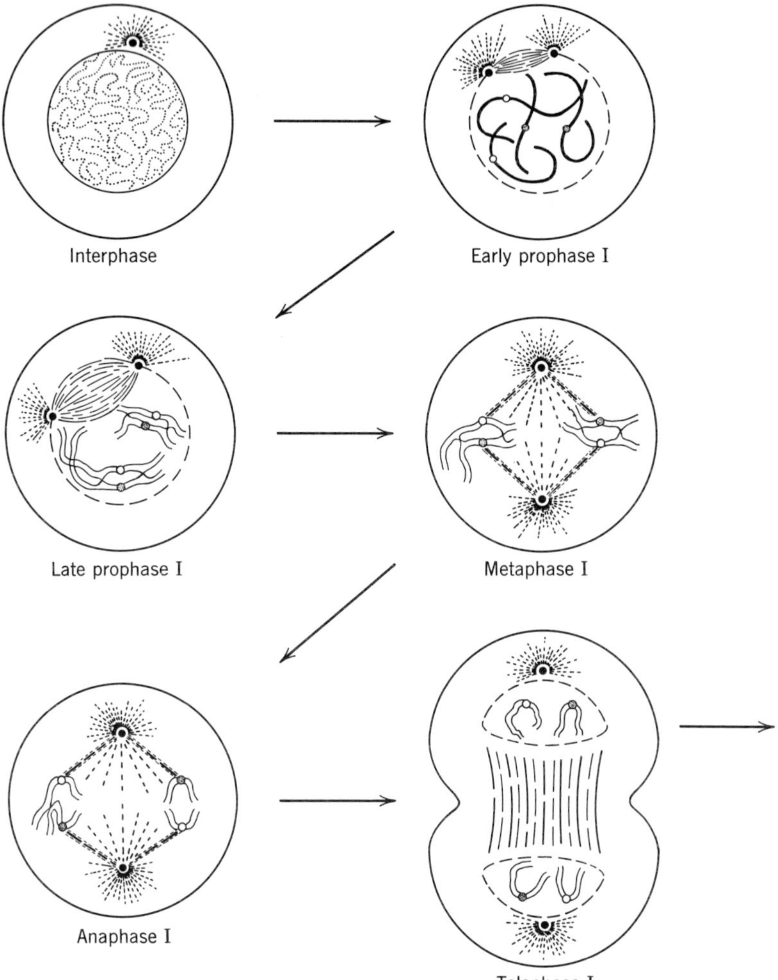

Fig. 2-2. Meiosis.

During the process of replicating and pairing only the centromeres appear to remain single. Thus at the metaphase of meiosis the undivided centromere carries two chromatids both of which move along with it at anaphase. At telophase of the first meiotic division, therefore, each daughter nucleus receives the usual number of chromosome strands, but only half as many centromeres.

During the second meiotic division no new chromosomal material

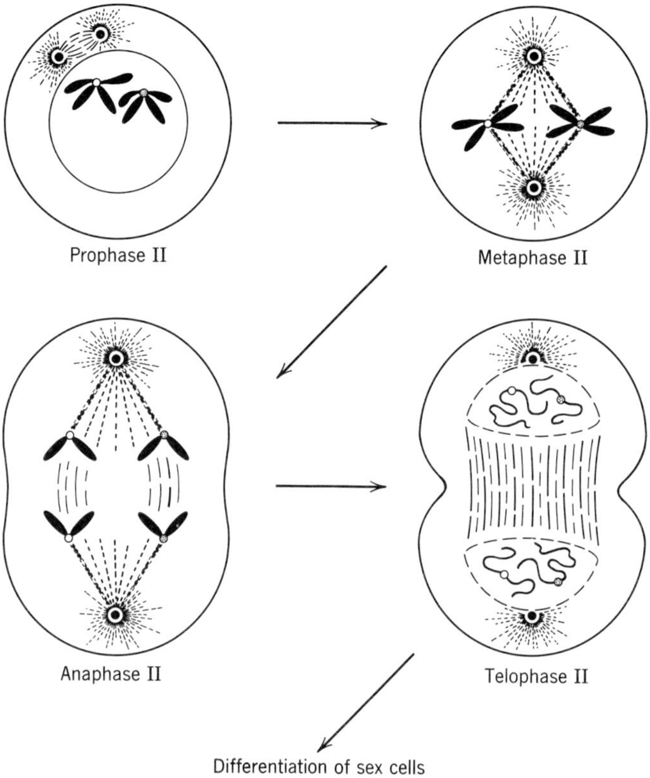

Fig. 2-2 (continued).

is synthesized, but the dual strands separate and each centromere produces two daughter centromeres. The chromosomes again condense and move to the metaphase plate. The centromeres divide and move apart, now carrying with them only a single chromatid. Finally, at telophase of the second meiotic division, and only then, are nuclei reformed which have half the usual chromosome number and one strand of each kind of chromosome.

The cells produced by the meiotic divisions are thus very different from the ordinary body (or somatic) cells, but they are not yet definitive sex cells. Those destined to become sperm cells, for example, usually lose a large amount of their cytoplasm and develop a propulsive organ to render them motile. In contrast, those which become egg cells usually accumulate large supplies of nutrient in their greatly expanded cytoplasm. But the nuclear processes in the development of eggs, sperms, and pollen are almost indistinguishable. At fertilization, when sperm and egg unite, the sperm nucleus enters the egg, often leaving the tail behind, and fuses with the egg nucleus, thus reestablishing the characteristic double (or *diploid*) chromosome number.

The facts of mitosis and meiosis strongly impressed biologists with the singularity of chromosome behavior, especially insofar as the chromosomes appeared to be the agents of biological continuity. Their numbers and kinds were precisely regulated both in the development of the organism and in the critical events between generations. Even before the turn of the century a number of workers, particularly August Weismann and E. B. Wilson, had cogently argued that the chromosomes must be the material basis for heredity. We may readily understand the reception which Mendelism received in 1900 and the almost immediate attempts to relate Mendel's genes to the chromosomes seen with the microscope.

CYTOGENETIC CORRELATIONS

Certain "numerological" correspondences were immediately apparent, and these had much more force than they might have had a few decades earlier. Mendel had deduced that two genes of a kind occur in somatic cells, but only one of a kind in germ cells. The cytologists had correspondingly shown that two chromosomes of a kind occur in somatic cells, but only one of a kind in germ cells. The "reduction" in chromosome numbers during meiosis paralleled precisely Mendel's postulated reduction in the number of genes. Indeed, a detailed explanation of Mendelism entirely in terms of chromosome behavior was set forth in 1902, again not once but twice, by W. S. Sutton in America and Theodor Boveri in Germany.

This synthesis of cytological observations and breeding analysis was not, however, immediately and universally accepted. Some of the strongest supporters of Mendelism refused over a period of many years to attach any significance to the correlation between genes and chromosomes. There were several reasons for this attitude. Many biologists had learned from hard experience to suspect simple interpretations; the numerical correspondence between genes and chromosomes could have been entirely coincidental. And immediately after Mendelism was rediscovered a number of hereditary phenomena were discovered which Mendel had not considered and which did not fit obviously into a chromosomal interpretation.

Much of the subsequent history of genetics has consisted of attempts to relate in a more direct way these two different kinds of biological elements—the one studied through breeding analysis and the other through cytological observation. We cannot give a complete account of these studies, but a discussion of a few of the more important ones will illustrate how a theory, through the accumulation of more and more evidence, becomes gradually established as an interpretation valid beyond reasonable doubt.

One of the first direct evidences of an hereditary function of the chromosomes came from a study of chromosomes which differed systematically in animals of different sexes. Specifically, males of certain species were observed to have among their pairs of otherwise morphologically similar chromosomes an "odd pair," two chromosomes which differed in size and structure. One of these, the larger, was designated as the X chromosome and the smaller as the Y chromosome. Female animals of these same species had two X chromosomes and no Y chromosomes. These observations suggested that "sex" was an hereditary characteristic determined by the individual's chromosomal constitution. The male could be considered to be a chromosomally heterozygous individual and the female to be a homozygote. This system of XY sex determination is by no means universal, but it is common in mammals and in many insects. Several other systems of chromosomal sex determination were also worked out which differ in detail from the XY system but are similar in principle. In each case there is a direct correlation between the chromosomes an individual receives from its parents and the sex it develops. At least for these special traits, therefore, the chromosomes are shown to have a role in inheritance.

Perhaps more convincing in the long run, however, was an analysis of an important exception to Mendel's principle of random assortment of genes discovered by William Bateson, and thoroughly exploited by the Drosophila workers—T. H. Morgan, A. H. Sturtevant, C. B. Bridges and H. J. Muller. Studies of inheritance in a number of organisms made it evident that nonrandom assortment occurred regularly though infrequently. The aberrations, nevertheless, conformed to a pattern of their own. The pattern may be shown by the following hypothetical example: Two pure-bred strains differing in two traits are crossed; this cross may be designated as $AABB \times aabb$. The F_1 would then be expected to have the constitution $AaBb$. If such an organism is crossed to a homozygous recessive (*aabb*), the following results would be expected under independent assortment.

Sex Cells of Heterozygote	Sex Cells of Homozygote	Genotypes	Exp.	Found
AB ($\frac{1}{4}$)	*ab*	*AaBb*	$\frac{1}{4}$	40%
Ab ($\frac{1}{4}$)	*ab*	*Aabb*	$\frac{1}{4}$	10%
aB ($\frac{1}{4}$)	*ab*	*aaBb*	$\frac{1}{4}$	10%
ab ($\frac{1}{4}$)	*ab*	*aabb*	$\frac{1}{4}$	40%

Such a cross (a test cross) is especially useful since it restricts variation to one source; one parent contributes several kinds of sex cells and the other produces a "blank" against which any differences among the sex cells of the other parent may be detected. Thus, when departures from expectation occur among the genotypes of the progeny, they are known to reflect departures from expectations among the sex cells of the heterozygous parent. The failure to find a 1:1:1:1 ratio among the progeny indicates that the different kinds of sex cells are not produced in the standard ratio. The problem is to determine the basis for the excessive production of certain classes of sex cells.

A further exploration of this phenomenon shows that the particular departures from expectation found in this case do not occur with all *AaBb* heterozygotes. If we cross a strain of genotype *AAbb* with one which is *aaBB*, the heterozygous F_1 appears to be identical to that in the last cross. But, if this *AaBb* individual is test-crossed to an *aabb* individual, the genotypic ratio among the progeny is

AaBb	10%
Aabb	40%
aaBb	40%
aabb	10%

The numerical values in the two crosses are identical, but the numbers apply to different classes.

This nonrandom distribution of genes indicates that some kind of preferential association ("coupling" or "linkage") must be occurring. The difference between the two crosses just presented lies in the manner of origin of the F_1. In the first case the sex cells which formed the F_1 had the composition *AB* and *ab;* in the second case they had the composition *Ab* and *aB*. The quantitative differences in the sex cells produced *by* the F_1 are directly related to the differences in the kinds of sex cells going *into* the F_1. If *A* and *B* enter the F_1 together, they tend to remain together in the sex cells of the next generation. But if *A* and *b* enter the F_1 in the same sex cell, they tend to remain associated in the sex cells produced by the F_1.

The particular ratios observed when "linked" genes are studied vary from one set of genes to another. If, for example, the *A* genes are also linked to the *C* genes, the ratios obtained in test crosses might be very different from the 40:10:10:40 ratio observed with the *A-B* set. We might obtain the following results.

AACC × *aacc*
↓
AaCc × *aacc*
↓

AaCc	47%
Aacc	3%
aaCc	3%
aacc	47%

AAcc × *aaCC*
↓
AaCc × *aacc*
↓

AaCc	3%
Aacc	47%
aaCc	47%
aacc	3%

Again, the gene combinations entering the F_1 occur in excess among the sex cells of the F_1, and the probability of retaining the "old combination" of genes is characteristic of the set of genes being studied. We could say that different sets of genes show different "strengths of linkage"; some combinations break up frequently and some only rarely. As a measure of the strength of linkage, the frequency of sex cells with "new combinations" of the genes was chosen as an arbitrary standard. Thus the genes of the *A* and *B* sets show a frequency of new combinations, or "recombinations" of 20%, while the genes of the *A* and *C* sets show a recombination frequency of 6%. In both these cases the low percentage of recombinations constitutes an exception to Mendel's principle of independent assortment. Note that when sets of genes assort independently the frequency of new combinations is 50%, the same as the frequency of old combinations; inasmuch as linkage inhibits random assortment, the frequencies of new combinations for different sets of genes might vary between 0 and 50%.

Another observation of significance concerned genes which are linked to a common gene. In the present example, not only are the *B* set and the *C* set linked to the *A* genes, but the *B* and *C* genes are also found to be linked to each other. In extensively studied organisms it is often possible to identify several groups of genes, with linkage among all members of each group. Members of one linkage group, however, assort independently of the genes in all other linkage groups.

The next question concerns the significance of linkage values within a linkage group. Can these values be used to investigate the physical relationship of the genes? As soon as data became available for several linkage groups, certain regularities were encountered. In the examples, genes *A*, *B*, and *C* are linked; if the linkage value of *A-B* is expressed as 20 to correspond with the recombination value, and that for *A-C* as 6, the linkage value of *B-C* is not free to take just any value between 0 and 50. Indeed, experimentally the value of *B-C* would always be approximately the sum or the difference of the values for *A-B* and *A-C*. *B-C* is, therefore, either 14 or 26. For the hypothetical case under consideration, let us assume that the linkage value for *B-C* is 14.

An understanding of the meaning of such results comes from a geometrical analogy. Regardless of whether genes are spatially distributed in one dimension, two dimensions, or three dimensions, we may construct a "map" of linked genes by making a simple assumption. We may assume that genes in the same linkage group are distributed at different distances from each other, and that the linkage values are a measure of the distance between genes. Genes which are close to each other might be expected to get separated (form new combinations) less frequently than genes which are physically far apart. Thus *A* would be judged to be closer to *C* than to *B*.

What kind of geometry is then implied by the linkage values? We may construct a genetic map using a ruler and compass (Fig. 2-3). Gene *A* is placed at some point on a piece of paper. Gene *B* is then placed at any point 20 units away from *A*. Circles are drawn representing the possible loci for *C*, with a radius of 6 units from *A* and 14 units from *B*. Notice that the two circles are tangential. Only one point is available for *C* which satisfies the two sets of data, and this point lies on a line connecting *A* and *B*. Taking a value of *B-C* = 26,

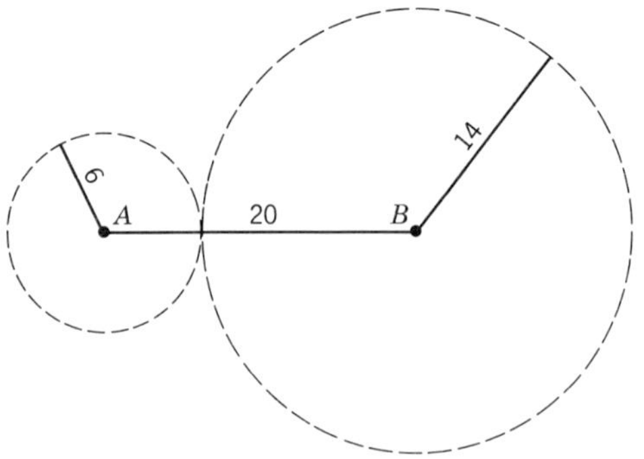

Fig. 2-3. An illustration of genetic mapping with recombination frequencies.

a similar linear array is generated, providing verification of the first construction.

With only three points it is not possible to discriminate between a two- and a three-dimensional geometry, since any three points may be made to lie on a single plane, but even with four and more points the genetic geometry is always approximately linear. The distances are measured in units whose correspondence to known physical units is not established, and which may vary from organism to organism or even within a single organism. But the topology of the genetic material is reasonably clear; the genes appear to be arranged in a linear order, much like beads on a string.

The studies on linkage thus establish another significant correspondence between genes and chromosomes. Breeding studies indicate that genes are associated in linear arrays; and the chromosomes are the only apparent linear structures within the nucleus. This correspondence suggests that a linkage group represents a set of genes associated with a particular chromosome. This interpretation has certain verifiable consequences. The number of linkage groups in a particular organism should, for example, correspond to the number of pairs of chromosomes in that organism. But we would not expect to detect all the linkage groups in an organism, particularly one with many pairs of chromosomes, unless a very large number of gene differences were available for a study. A verification of this corollary requires extensive study on a particular organism, and especially on an organism with a small number of chromosomes. Yet verification has been obtained in several investigations.

The studies on linkage in higher organisms provided a detailed explanation of breeding results in terms of chromosome behavior during meiosis, but they were incomplete in one important respect. They were statistical analyses of numerous individual meiotic events, and the consequences of recombination during a particular meiotic sequence were only inferred. Several ingenious techniques were developed to validate these inferences, but later studies with certain microorganisms allowed the events in single meiotic sequences to be studied in a direct and simple manner. The red bread mold *Neurospora crassa*, which we shall consider in other connections, provides a particularly diagrammatic illustration of such analyses.

The major structural element in this organism, as in other fungi, is the *hypha*, a thread-like element which may branch and anastomose with others to form a mat of fibers designated as a *mycelium*. The culture may be propagated by transferring a piece of the mycelium to new nutrient, or through the use of reproductive spores (*conidia*) which are budded off the hyphae. Neurospora does not have a strict "cellular" organization in the sense that each nucleus is surrounded by its own private cytoplasm, but many nuclei inhabit a common cytoplasm which is only partially compartmentalized by perforated septa. Even the conidia usually contain several nuclei.

Although this organism may be propagated indefinitely by vegetative means, it is also capable of sexual processes. A mature culture produces specialized reproductive structures (*protoperithecia*) which function as "female" organs. The fruiting structure consists of a ball covered by a husk of hyphae; extending from its top are several specialized hyphae (*trichogynes*) which are the receptive structures. Inside the ball are certain other specialized hyphae making up an *ascogonium* and containing the female nuclei.

Although Neurospora engages in sexual activity, it does not have "sexes" in the same sense as many higher organisms. All strains produce protoperithecia and any particular strain may function as both a male and a female. The male elements are usually the conidia, but any piece of vegetative hypha may also serve in the same manner. These may in isolation yield an entire new culture; but if they make contact with a trichogyne, one of the nuclei may migrate down this specialized hypha into the protoperithecium and accomplish a fertilization. Although most strains produce functional male and female structures, a strain does not ordinarily cross with itself; each strain is self-sterile. And not all strains of different origin can be crossed. Two classes of strains, or "mating types," may be distinguished by their mating patterns. These classes are designated as types *A* and *a*. Two strains mate only if they are of different classes. The only way to identify the mating type of a strain is to attempt matings with standard strains of known mating type.

Neurospora, like many other fungi, also differs from most higher organisms in the placement of meiosis and fertilization in the life cycle. In higher animals, for example, meiosis occurs in the formation of the sex cells, and fertilization occurs immediately thereafter. The haploid stage occurs only in the sperm and the egg, and the haploid nucleus undergoes no division. The diploid stage is restored by fertilization and is maintained until the sex cells are produced in the next generation. In higher plants the situation is not much different. The nuclei of the pollen and the egg usually undergo a few divisions prior to fertilization, but the haploid phase of the life cycle is a very small portion of the total. In Neurospora, on the other hand, fertilization is followed immediately by meiosis and the haploid condition is the characteristic state of the nuclei through nearly all the life cycle. These facts must, of course, be taken into consideration in making a genetic analysis.

The events which occur when a male nucleus enters the protoperithecium are of special interest in our current discussion. Usually only one nucleus enters the female structure and comes to lie beside a nucleus of the other mating type. These two nuclei then proceed to undergo a series of divisions in synchrony (conjugate division) until a large number of pairs of nuclei are established. These pairs of nuclei are cut off into separate cytoplasmic units where they fuse in fertilization. The many fertilization nuclei then undergo meiosis in special elongated sacs (*asci*) in a manner which does not permit the nuclei to slip past each other. The two meiotic divisions, yielding four nuclei from each fertilization nucleus, are followed by a mitotic division to make a total of eight haploid nuclei within the ascus. Walls form around these nuclei to make a series of sexual spores (*ascospores*) in a linear array. The eight-spored asci may be removed from the mature perithecium and the ascospores may be isolated in order (Fig. 2-4). Thus all the products of a single meiotic process may be studied, and the lineages of the various nuclei can be known with certainty. This situation is to be contrasted with that in most higher organisms. Among the latter the products of a particular meiotic sequence cannot be isolated in a functional state, and for females three of the four nuclei formed usually disintegrate. In Neurospora, on the other hand, all the meiotic products give rise directly to cultures which can be studied without an intervening fertilization.

We may now consider the distribution of hereditary traits among the products of a single meiotic sequence. We will use for illustrative purposes the mating types *A* and *a* and assume that these are determined by genes *A* and *a* respectively. When cultures are grown from the haploid products of a single ascus, what are the mating types of these cultures? First we may note that the eight spores in an ascus may be reduced to four sets, because the spores produced by the last division in the ascus are the result of a mitotic division and, as would be expected, the traits of the resulting cultures are identical. Different asci are found now to contain different distributions of traits among the four sets of spore pairs. All the possible arrangements are listed below:

1	2	3	4	5	6
A	*a*	*A*	*a*	*A*	*a*
A	*a*	*a*	*A*	*a*	*A*
a	*A*	*A*	*a*	*a*	*A*
a	*A*	*a*	*A*	*A*	*a*
I		II			

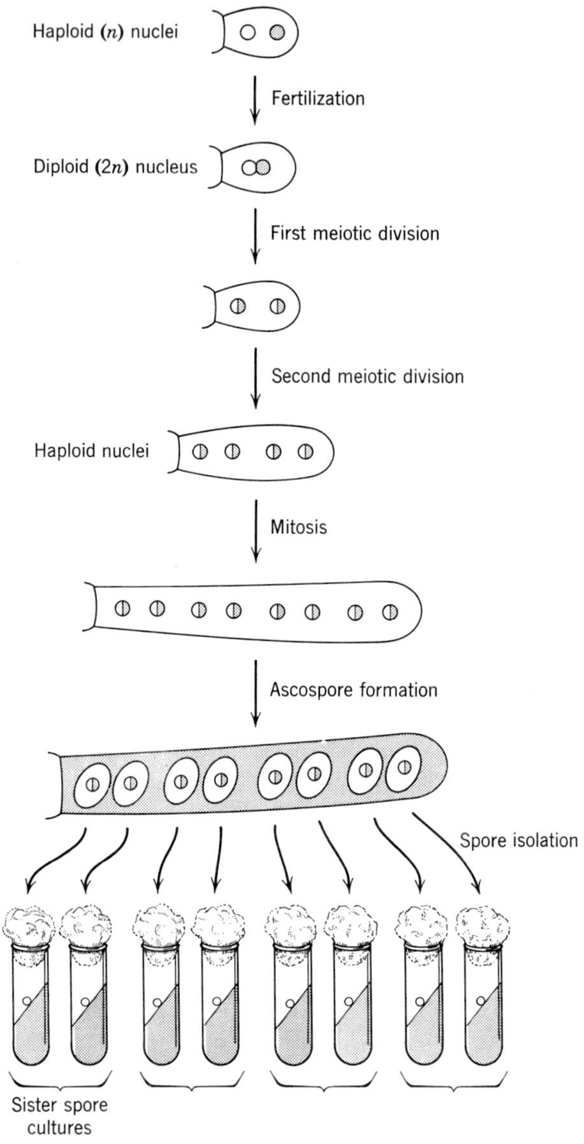

Fig. 2-4. Genetic analysis in *Neurospora crassa*.

First, note that regardless of the distribution of the traits, each ascus produces precisely two cultures of type *A* and two of type *a*. Thus the 1:1 segregation of allelic genes usually inferred on statistical grounds is directly demonstrated when all the meiotic products of a single sequence are recovered.

Second, the allelic genes may separate from each other either at the first nuclear division in meiosis (MI) or in the second (MII). This signifies that in certain cases the two attached chromatids going to the same pole in the first meiotic division carry different alleles, whereas in other cases they carry identical alleles. From the schematic representation of meiosis we would expect the attached chromatids to be identical since they are replicas of the same chromosome held together by the original centromere. However, an event occurring in some but not all fertilization nuclei alters the expected relationships and results in nonidentical chromatids, or sections of these, being attached to a single centromere. This event is designated as "crossing-over" and may be visualized as follows. At fertilization each nucleus acquires two homologous chromosome strands, one bearing the gene *A* and the other the gene *a*, connected to different centromeres. Subsequently each strand replicates. During or immediately following this replication, sections of chromatids are exchanged so that dissimilar strands may be connected to the same centromere. At anaphase such centromeres separate and carry with them one chromatid with an *A* gene and one with an *a* gene. At the second meiotic division, since the centromeres divide but the strands do not replicate, each daughter nucleus receives either an *A* or an *a* strand. The orientations of the centromeres at this division determine whether the spores of the resulting ascus are arranged as *A-a-A-a*, *A-a-a-A*, etc.

If the probability of a crossing-over event is roughly the same at different points in the chromosome, the frequency of crossing-over should provide a measure of the distance between two points. Here the frequency of crossing-over between the mating type locus and the centromere determines the frequency of MII segregations and those at greater distances should show a higher frequency. Unfortunately, mapping with respect to the centromere is possible only in organisms which permit the recovery of all meiotic products, and we should return to more conventional organisms to explore the modalities of crossing-over under more usual situations.

One question we might examine is the reason for the limitation of 50% recombinant sex cells for any two linked factors. For exploratory purposes let us assume two pairs of linked genes, *Aa* and *Bb*, arranged in the same arm of a chromosome. Since different crossing-over events occur in different cells, and since the products of each meiotic sequence cannot be separately identified, we must treat recombinant production as a population problem. Nevertheless, the characteristics of a population of sex cells are determined by individual events and may be analyzed by considering the separate events contributing to the final population structure. For example, we may consider first of all a subpopulation of cells in which no crossing-over occurs in the relevant region—the region between *A* and *B*. Each meiotic sequence will yield four sex cells, all of the "old combinations" *AB* and *ab*. The frequency of recombination among such sequences will be 0.

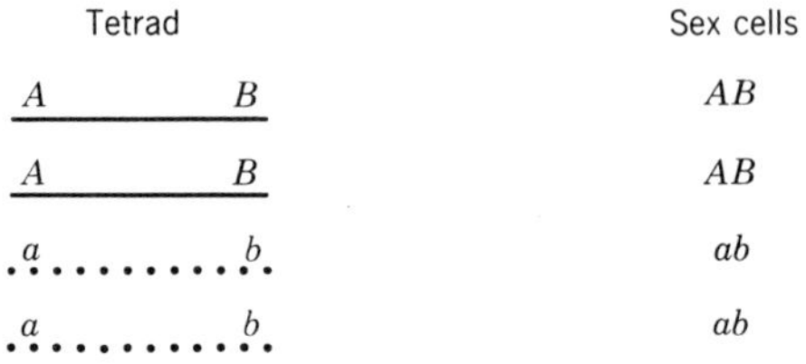

Next we may consider a subpopulation of cells in which each tetrad has one and only one crossover in the region between *A* and *B*. From each such tetrad will emerge two sex cells with the old combinations and two with the new combinations. The frequency of recombinations will be 50%.

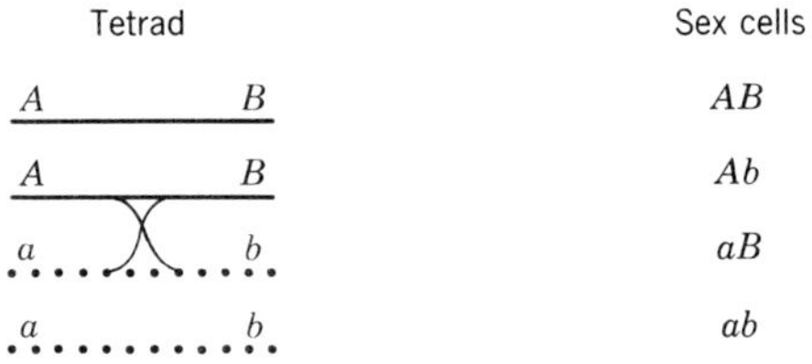

Third, we may consider what happens in a subpopulation when two crossovers occur within the relevant region. By arbitrarily fixing the pattern of the first crossover, we may then explore the consequences of additional crossover events on the genotypes of the sex cells. A second crossover may occur between the two strands which were not involved in the first, yielding a four-strand double crossover. In this

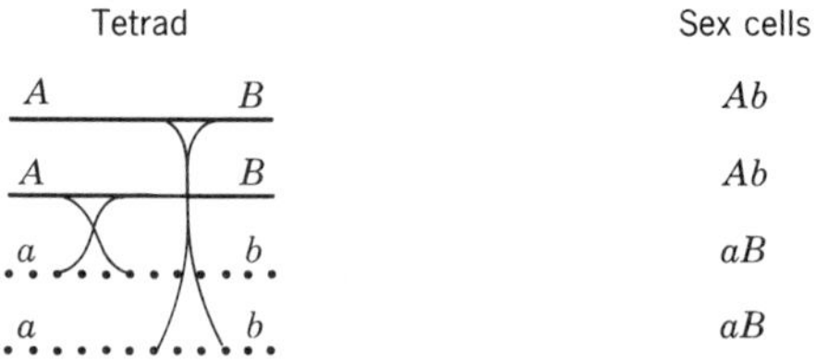

case all the sex cells produced (100%) will be recombinant. If, however, the second crossover is independent of the first crossover, it may equally likely involve the same strands undergoing the first crossover. And here, although two crossovers have occurred, no recombinants have been produced. Two-strand double crossovers simply return the genes to their original associations, and the frequency of recombination in the subpopulation is 0.

Tetrad	Sex cells
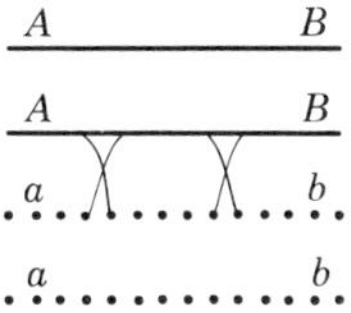	*AB* *AB* *ab* *ab*

Two other kinds of double crossovers may occur, and on the hypotheses of independence they should occur equally as frequently as the two-strand and the four-strand doubles. These are the three-strand doubles in which one of the original strands is involved in a crossover with one of the strands previously left intact. Each configuration results in two old combinations and two recombinations to yield a frequency of 50% recombinants.

Tetrad	Sex cells		
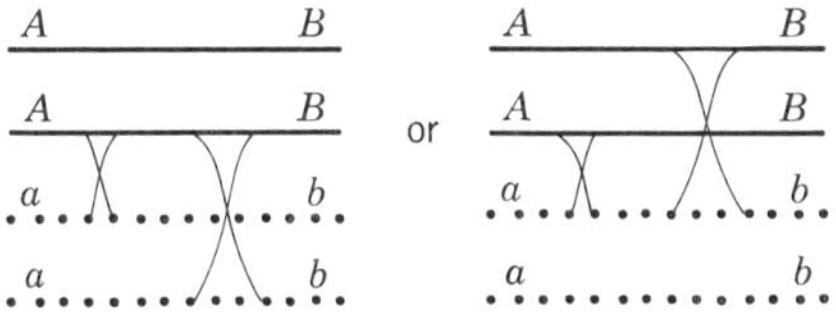	*AB* *Ab* *ab* *aB*	or	*Ab* *AB* *aB* *ab*

Hence with two crossovers per tetrad, assuming an equality of two-strand and four-strand double crossovers, the sex cells produced are 50% recombinant, precisely the same value as is obtained with one crossover per tetrad. And increasing the number of crossover events to three or four does not change the value. If two genes are on the same chromosome and manifest linkage (i.e., they produce less than 50% recombinants), less than one crossover per tetrad must occur on the average, for if every tetrad of a chromosome type had a crossover in the relevant region, the genes would be assorting independently.

More specifically, if one-half of the tetrads have no crossover (thus yielding 0% recombinants) and one-half have one crossover (yielding 50% recombinants), the total frequency of recombinants becomes 25%. Similarly, if only 10% of tetrads have a crossover, the frequency of recombinants is 5%.

Such considerations open the way for a thorough analysis of chromosome mechanics, but this subject will not be pursued beyond this point. The importance of such studies for our purpose lies in their detailed correlation of breeding data and chromosome behavior, which allows a better understanding of both.

The studies just outlined demonstrated that the behavior of genes in breeding studies, even under exceptional circumstances, could be understood by reference to the behavior of chromosomes. But such studies did not permit an identification of specific linkage groups with particular chromosomes or a localization of specific genes within particular chromosomes. This was accomplished through a different kind of analysis. The first association of a linkage group with a particular chromosome came through studies on the fruitfly, *Drosophila melanogaster*. Variant strains were established which had white eyes instead of the usual red eyes and these traits were transmitted in an unusual way.

One characteristic of this inheritance pattern was a difference in reciprocal crosses. If a red-eyed female from a pure breeding strain is crossed to a white-eyed male, all the progeny are red-eyed; this suggests that red-eyes are dominant over white eyes. But in the reciprocal cross —that of a white-eyed female to a red-eyed male—all the female progeny had red eyes and all the males had white eyes. A further analysis demonstrated that the females in the F_1 from either cross were heterozygous; in crosses of these F_1 red-eyed females to red-eyed males about half of the males were white-eyed. That is, the females must have had genes for both red eyes and white eyes. But the F_1 males did not appear to have concealed determinants; a red-eyed F_1 male crossed to a purebred red-eyed female did not yield any white-eyed progeny, and a white-eyed F_1 male crossed to a white-eyed female did not produce any red-eyed progeny. The males behaved as if they had only one kind of gene—that which they expressed. Moreover, the pattern of transmission indicated that the male parent never contributed an eye color to his sons at all, but only to his daughters.

These observations raised the question what factor can be present singly in the male but doubly in the female and be transmitted from a father to his daughters only. The X chromosome clearly fulfills this requirement and is the only cytological element known to have these properties. This then is the basis of "sex-linked inheritance", and the crosses just described may be diagramed in the following fashion:

	$X^rX^r \times X^RY$		$X^RX^R \times X^rY$	
F_1	X^RX^r	Red-eyed females	X^RX^r	Red-eyed females
	X^rY	White-eyed males	X^RY	Red-eyed males

This pattern of inheritance has been found for a number of traits in several organisms with XY sex determinants. Examples include color blindness and hemophilia in man. The theoretical significance of sex linkage lies primarily in the association of a particular chromosome with a particular hereditary trait. It is also important as an exception which "proves the rule"; not only are genes and chromosomes transmitted in similar patterns from generation to generation but, when a particular chromosome is transmited in an unusual pattern, certain traits show corresponding peculiarities in transmission.

The association of genes and linkage groups with chromosomes other than sex chromosomes was somewhat more difficult, but was ultimately accomplished when it became possible to modify chromosome structures by means of external agents and to examine the consequences of these modifications on hereditary transmission. By using x-rays, for example, H. J. Muller was able to bring about a variety of microscopically visible changes in chromosome architecture. Pieces of chromosomes might be lost entirely (deficiencies); sections lost from one chromosome might become attached to another (translocations); or a lost piece might be reinserted into the original chromosome but in a reversed position (inversion). In each case the transmission of certain traits was modified: genes which had previously been linked were found to be transmitted independently, or genes which had previously been unlinked were now transmitted together, or "map distances" between genes were altered. Not only did these studies establish beyond any reasonable doubt the validity of the chromosome theory of inheritance, but they also enabled the identification of particular chromosome regions with particular genes. This was the culmination of classical cytogenetics—the development of cytological maps which correspond point by point with the linkage maps established through breeding analysis.

EXERCISES IN CYTOGENETICS

1. (a) Gene loci for *A* and *B* are 10 map units apart. An individual of type *AAbb* is crossed to another of type *aaBB*. The F_1 individuals are then crossed to yield the F_2. Calculate the phenotypic ratio in the F_2.

(b) An individual of genotype *aabb* is crossed to one of type *AABB*. The F_1 is now crossed to the F_1 in the previous cross. Calculate the phenotypic ratio in the next generation.

(c) In some organisms the crossover frequencies for the same genes may vary in the two sexes, and in certain cases crossing-over occurs only in one sex. In Drosophila, for example, crossing-over does not occur in the male. Assume that genes *B* and *C* are linked with 15% crossing-over. A cross of *BBCC* × *bbcc* yields an F_1 generation. If no crossing-over occurs in the male, what will be the F_2 phenotypic ratio?

2. In a certain organism the recombination frequencies were experimentally determined for a series of paired traits, and the results summarized below were obtained.

(a) *A-B*	8%	(b) *F-G*	42%	(c) *K-L*	18%
A-C	22%	*FH*	30%	*K-M*	50%
A-D	5%	*F-I*	50%	*K-N*	13%
BC	14%	*G-H*	50%	*LM*	50%
BD	3%	*G-I*	20%	*L-N*	31%
CD	17%	*H-I*	50%	*M-N*	50%

Construct genetic maps which account for these data.

3. Sex-linked inheritance superficially resembles a phenomenon called sex-influenced inheritance. The first, however, is exceptional in the transmission pattern of genes (one parent has only one gene of a kind); the second is conventional in genic transmission but is unconventional in genic expression. The differences may be illustrated by the following genotype-phenotype correlations:

Color blindness	Sex linkage
Genotype	*Phenotype*
X^CX^C }	Female, normal vision
X^CX^c }	
X^cX^c	Female, color blind
X^CY	Male, normal vision
X^cY	Male, color-blind

Pattern baldness	Sex-influenced hypothetical	
	Phenotype	
Genotype	*Male*	*Female*
BB	Bald	Bald
Bb	Bald	Nonbald
bb	Nonbald	Nonbald

(a) Two nonbald parents produce a bald child. What is the sex of the child?

(b) In a certain population 19% of the males show a trait which is present in 3.6% of the females. Would sex linkage or sex influence more readily explain this distribution?

(c) A bald color blind man whose father was not bald marries a nonbald, non-color blind woman whose mother was bald and whose father was color blind. What kinds of children might they be expected to produce? With what probabilities?

4. When homologous chromosomes pair in prophase of the first meiotic division, they do so with a point for point specificity, even when one of the chromosomes has undergone a structural rearrangement. Thus in a

cell with one chromosome manifesting the normal arrangement of segments (*A B C · D E F*) and its homologue with an inverted region (*A′ B′ / D′ · C′ / E′ F′*), the *A* region will pair with the *A′*, the *B* with the *B′*, the *C* with the *C′*, etc.

(a) Draw the configuration of the paired chromosomes in such an "inversion heterozygote." Don't forget that at this stage four strands (chromatids) are present.

(b) Show what kinds of sex cells would be produced if a crossover occurred within the inverted region.

(c) Consider a similar situation for a chromosome pair in which the centromere is located outside the inverted region (between *A* and *B*, for example). What kinds of sex cells would be produced if a crossover occurred within the inverted region?

5. An organism with the genotype *AABBCC* is crossed to one with the genotype *aabbcc*. The offspring are test-crossed to a strain with the genotype *aabbcc* with the following results:

Genotypes	*Number observed*	*Genotypes*	*Number observed*
AaBbCc	240	*aaBbCc*	37
AaBbcc	140	*aaBbcc*	98
AabbCc	92	*aabbCc*	150
Aabbcc	23	*aabbcc*	220

(a) Determine which if any of the genes are linked, and construct a genetic map which accounts for all the data.

(b) What is the effect of double crossovers in estimating the distance between widely separated genes?

(c) Determine whether double crossovers occur as frequently as expected by chance.

ANSWERS TO EXERCISES IN PROBABILITY

1. $(1/13)(1/13)$; $(1/13)(3/51)$
2. $1/13 + \frac{1}{4} - (\frac{1}{4})(1/13)$
3. $(\frac{1}{2})^3$; $\frac{1}{2}$
4. 7
5. $6(\frac{1}{2})^2(\frac{1}{2})^2 \gg 20(\frac{1}{2})^3(\frac{1}{2})^3$; $(\frac{1}{2})^4 + (\frac{1}{2})^4 \gg (\frac{1}{2})^6 + (\frac{1}{2})^6$
6. $10(\frac{1}{8})^3(\frac{7}{8})^2$
7. $(.3)(.3)$; $2(.7)(.3)$
8. 25% is less than the expected 32%; individuals tend to marry others with the same eye color.
9. —
10. (a) $\frac{1}{2}$ (b) $(\frac{1}{2})^2$ (c) $(\frac{1}{2})^4$ (d) $(\frac{1}{2})^{10}$ (e) $(\frac{1}{2})^{100}$

ANSWERS TO EXERCISES ON MENDELISM

1. (a) $(\frac{1}{2})(1)(\frac{1}{2})$ (b) $(\frac{1}{2})(1)(\frac{1}{4})$ (c) $(\frac{1}{2})(0)(\frac{1}{2})$
2. (a) $(.1)(.1)$ (b) $2(.1)(.9)$ (c) $[2(.1)(.9)]^2$
   (d) $(\frac{3}{4} + \frac{1}{4})^4$
3. (a) $(.2)(.2) + 2(.2)(.7)$; $(.7)(.7)$
   (b) $4(.1)^2(.7)^2/(.1)^4 + 4(.1)^3(.7) + 4(.1)^2(.7)^2$
   (c) $\frac{1}{4}$ (d) $I^B i$
4. —
5. —

ANSWERS TO EXERCISES ON CYTOGENETICS

1. (a) 0.5025:0.2475:0.2475:0.0025
   (b) 0.5225:0.2275:0.2275:0.0225
   (c) 0.7125:0.0375:0.0375:0.2125
2. (a) *ADBC* (b) *HFGI* (c) *LKN........M*
3. (a) male (b) sex linkage (c) —
4. —
5. —

SECTION B

The Search for Physicochemical Mechanisms

Alongside the attempts to describe accurately the microscopic appearance of cells and their behavior in reproduction and fertilization, an equally intensive effort was made to explain cellular activities in physicochemical terms. The results achieved in pursuit of these two different but related objectives provide a striking contrast. The cell theory, the account of mitosis and fertilization, stand as definitive and final answers to the questions which were then being raised; a student who learns these answers today does so as a definitive part of his education in cell biology. He may dig deeper, but the broad frame of reference defined in the preceding section provides him with a foundation which requires no modification. This whole facet of nineteenth and early twentieth century history provides an excellent example of how a defined, and necessarily restricted, objective can be achieved in science if the experimental tools are adequate. This was not so for those who sought to explain the behavior of living cells in terms of physics and chemistry. They were motivated by a diffuse though lofty goal: to prove that the process of life was a necessary consequence of physicochemical laws. Theirs was a faith that led them to use every resource to discredit the "vitalistic" view that life was an expression of some nonmaterial principle. They were not necessarily different from those who used the microscope to describe cells; many, indeed, followed both courses. But insofar as they sought to explain the whole of life, they were missionaries more than craftsmen. They sowed the seeds of research eagerly and applied whatever they could from the then current advances in physics and chemistry. It is not surprising that the harvests turned out to be scattered. In retrospect their efforts present a mosaic of poor and luxuriant patches, some of the latter being replete with pertinent and lasting information. Many individual pieces of research are models of scientific analysis, but the full goal of the investigators—a physicochemical explanation of life—was not attained. If today we appear to be moving much closer to that goal it is because so very much of the ground has been cleared for us. To be sure, we have not inherited a self-contained scheme in which to fit the underlying mechanisms of cell behavior, but we have inherited a knowledge of many physicochemical properties which guides every student of the cell in his pursuit of fundamental questions. The purpose of this section is to acquaint the student with these elementary cellular properties.

The properties to be discussed are answers to questions raised by cell investigators mainly during the eighteenth and nineteenth centuries. It is not difficult to trace the origins of these questions since they were, on the whole, directly associated with contemporary studies in physics and chemistry. One important field of activity was the properties of aqueous solutions; another was the analysis and synthesis of organic compounds; still another was the phenomenon of catalysis, the mechanism by which chemical reactions were accelerated; finally, there was the physicist's preoccupation with the nature of energy. Each of these fields had an important and distinctive influence on studies of cell behavior, and although it is convenient to follow the different lines of research separately, they periodically crossed and eventually fused with one another.

3

The Properties of Aqueous Solutions

The importance of water in the maintenance of life was recognized long before biologists considered the physicochemical properties of solutions. With the exception of dormant organisms such as seeds or spores, cells are about 80% water and their environment is directly or indirectly a watery solution. Unicellular organisms live in aqueous media; in multicellular terrestrial organisms only those cells present at the body surface are partly exposed to a nonwatery environment. Indeed, the principal function of such cells, whether plant or animal in origin, is to shield the rest of the cellular population from water loss. Early observers of living plants were aware of the fact that when leaves lost undue amounts of water they became flaccid, and that when the water balance was restored the leaves regained their characteristic stiffness. Once it was realized that living tissues consisted of cells, it must have become apparent that some type of force was developed in leaf cells which enabled them to draw in enough water to render them turgid. Plant cells, because of their strong outer wall, readily absorb water without bursting, and although today this is commonplace information, it was a challenging mystery in the nineteenth century. To realize this, we need only read a few remarks of René Dutrochet written in 1826 after he had observed the movement of water into the capsule of a fungus: "Whence came this water? What was the force that made it enter the capsule? It was necessary for me to place this phenomenon in the category of those of which the cause is entirely unknown."

OSMOTIC PRESSURE

As far back as 1748 the Abbé Nollet, an astronomer among other things, performed an unusual experiment which marked the beginning of a series of studies destined to explain the unknown force pondered by Dutrochet. He capped a cylinder full of wine with an animal bladder and then immersed it in water. The result, to say the least, was intriguing: the bladder began to swell and, in some of the many tries, even burst due to the uptake of water. What this experiment revealed about wines and bladders in general is not too obvious, but the phenomenon clearly merited a name. And it was called osmotic pressure, derived from the Greek word meaning "impulse." More than one hundred years later a systematic study of the phenomenon began with the observation that when plant cells were exposed to concentrated solutions of nontoxic substances (sugar, for example), the living portion of the cells ("protoplasts") contracted away from the nearly rigid outer wall. If cells thus treated were returned to dilute solutions of pure water, the protoplasts again expanded. The only consistent explanation which could be given for these large volume changes was either the inflow or outflow of water between cell interior and external solution. For even if the protoplast contraction were due to a loss of solids, protoplast expansion had to be attributed to an inflow of water, since the expansion took place on exposure of cells to distilled water.

Plant cells were particularly well suited for making such observations because their fixed outer wall made it possible to identify protoplast contraction (or "plasmolysis") under the microscope without the use of any measuring devices. It was easy to compare different solutes with respect to the concentration necessary in each case to produce a similar degree of plasmolysis in the same plant cell. As a result of all these studies two conclusions were drawn. (1) The difference in concentrations of solutes inside and outside the cell determined the direction in which water would move. (2) Some kind of membrane existed at the surface of protoplasts which permitted the passage of water but not that of solutes. The "force" which Dutrochet sought was, therefore, to be found in the simple fact that water would move into a cell if the intracellular concentration of solutes exceeded the external one. This was the same force which the Abbé Nollet presumed to be effective in his game with the wine bottle.

With this background in mind, the botanist Wilhelm Pfeffer decided to measure the magnitude of such forces, and to do so he constructed artificial membranes according to a technique discovered by Ludwig

Traube in 1867. He filled a clay jar with potassium ferrocyanide solution and immersed it in a solution of copper sulfate. By causing a current to flow through the porous walls of the jar, copper ferrocyanide was precipitated in the pores. Thus treated, the walls of the jar were permeable to water but not to solutes such as sodium chloride or sucrose. Pfeffer then filled the jar with different concentrations of a number of solutes and by arranging the apparatus in the manner illustrated (Fig. 3-1)

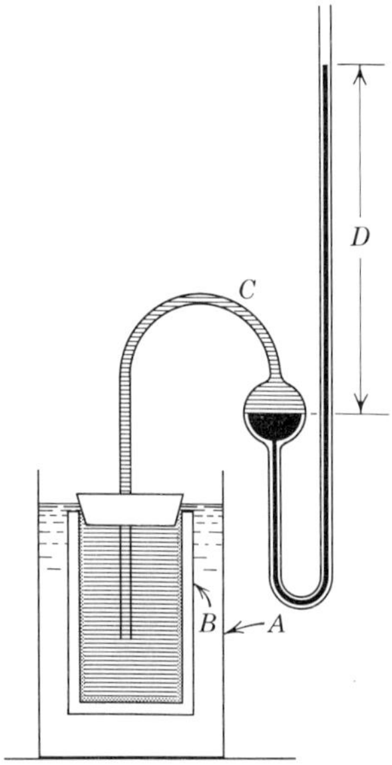

Fig. 3-1. Pfeffer's apparatus. *A*, a vessel of water; *B*, a porous cup with semipermeable membrane deposited in it; *C*, a mercury manometer. *D*, height of mercury column indicates the osmotic pressure. (From N. A. Maximov, *Plant Physiology*, McGraw-Hill, 1938.)

he measured the pressure generated inside the filled jar when it was immersed in distilled water. By this means he discovered the pressure to be directly proportional to the concentration of solute inside the jar.

The physicist J. H. van't Hoff became very much interested in the fact that solutions of many kinds would suck in water through a semipermeable membrane with a pressure proportional to the difference in concentration between the inner and outer solutions. Analyzing the data, he hit on a very simple relationship governing the pressure exerted and the concentration difference, one that was identical with the formula applied to the behavior of gases. He drew an analogy between the apparatus used by Pfeffer and that used by physicists for measuring gas pressures. He pictured the water as empty space, the solute molecules as gas molecules behaving in accordance with the gas law $PV = nRT$, where P stands for osmotic pressure, V for volume, n for the number of moles of solute contained in V, R for the gas constant, and T the absolute temperature. If the equation is rewritten in the form $P = (n/V)RT$, n/V represents the concentration in moles per liter and the osmotic pressure of a molar solution may be calculated accordingly:

$$\text{O.P.} = C \times 0.082 \times 273$$

R has a numerical value of 0.082 if P is expressed in atmospheres, C in moles/liter, and T in degrees Kelvin. For gases one mole occupies 22.4 liters at standard temperature and pressure (273°K, one atmosphere). For solutions a molar solution exerts an osmotic pressure of 22.4 atmospheres at 0°C or 24 atmospheres at 20°C. Values obtained in this way are only approximately correct for certain types of solutes; the limitations of the equation will be discussed later.

The analogy between solutes and gases is intended to illustrate but not to explain the behavior of solutions enclosed within semipermeable membranes. The immediate cause of osmotic pressure, whether observed in a living cell or in a jar with semipermeable walls, is the movement of water from a dilute to a concentrated solution. The phenomenon may be tied to the general proposition that nature abhors inequalities and that water moves in a direction which tends to equalize concentrations. However, the analogy used by van't Hoff to represent the inequality, though brilliant, can be misleading. Osmotic effects of solutes are closely related to their effects on freezing and boiling points in which the gas law analogy does not apply. A quantitative relationship exists between osmotic pressure, depression of freezing point, and raising of boiling point. (One mole of solute in 1000 g of water raises the boiling point by 0.52°C and depresses the freezing point by 1.86°C.) These properties would appear to have a common origin and they are most simply explained in terms of solvent concentration. It will be sufficient for our purpose to state that whenever a solute is added to water, the concentration of water molecules is decreased. When two solutions are separated by a semipermeable membrane, a difference in the concentration of water molecules which is inversely proportional to the difference in solute concentration will be established. Since water molecules can pass freely through the membrane, they are more likely to cross the membrane in the direction of greater solute concentration. The net result of such a difference is the accumulation of water in the more concentrated solution. Were a membrane not present between adjacent solutions of different concentration, the inequality would adjust itself by a diffusion of solute molecules; since the solute molecules cannot pass the semipermeable membrane, the only adjustment possible is for water to enter the more concentrated solution and thereby dilute it. The pressure which must be exerted on the more concentrated solution to prevent any increase in its volume is a measure of the osmotic pressure.

Table 3-1 Osmotic Pressures of Various Solutions (20°C)

Substance	Molar Conc.	Atmospheres
Glucose	0.01	0.24
	0.1	2.45
	0.5	13.22
	1.0	27.67
Sucrose	0.01	0.24
	0.1	2.51
	0.5	13.77
	1.0	32.89
Sodium chloride	0.01	0.47
	0.1	4.51
	0.5	22.25
	1.0	46.5

Departures from the values predicted by the equation as shown in Table 3-1 may be noted. For glucose and sucrose the factor mainly responsible for the deviations is the actual volume of water in each solution. As the concentration is increased the amount of water present per unit volume of solution is decreased, and the measured osmotic pressure is higher than the one calculated from the formula.

For sodium chloride an additional factor comes into play—the degree of dissociation. Dissociation effectively decreases with increasing concentration due to ionic interactions, and as a result yields osmotic pressures lower than ideal theoretical values. It may be seen, however, that at a 1.0*M* concentration the effect of a decrease in water volume is becoming apparent.

DIFFUSION

Osmosis is a special case of diffusion in which water is the only mobile component. The phenomenon occurs because of the interposition of a semipermeable barrier between two solutions of different concentration. Since water movement occurs because its own concentration is affected by the presence of other solutes, the only relevance which the chemical nature of the solutes has to osmosis is the extent to which they affect the concentration of water molecules. If, however, the semipermeable barrier between two solutions is removed, the nature of molecular movement alters considerably. Each species of molecule behaves independently due to the tendency of each particular substance to achieve a uniform distribution through the body of the liquid. If a molar solution of sucrose is juxtaposed to a molar solution of sodium chloride, salt and sugar will diffuse in opposite directions. Their rates of movement will depend on their molecular size and shape, their solubility, and the concentration gradient. The smaller and more spherical the molecule, the less frictional resistance will it encounter and the quicker will it diffuse. The more soluble the molecule, the more readily will it enter into new regions of the solvent. The steeper the concentration gradient, the greater will be the tendency toward equalization of concentrations.

The concentration gradient, the difference in concentration of a solute between one region of the solvent and another, is the force promoting diffusion. Whereas the other factors affect only the rate, the concentration gradient determines *direction.* To picture how concentration gradient affects rate, consider the situation in a beaker half-filled with a 1*M* solution of sucrose. If water is carefully layered over the sucrose solution, a boundary between the solution and water will be established. At "zero time" the lower surface of the boundary will contain a 1*M* concentration of sucrose, whereas the upper surface will contain none. Since all molecules, because of their kinetic energy, move about randomly, a large number of sucrose molecules would cross the boundary on the basis of random movement without compensatory movement in the opposite direction. If, however, the upper solution contained some sucrose, compensatory movement would occur, and the net rate of sucrose movement in the direction of the more dilute region would be decreased.

Quantitative measurements of diffusion are difficult. Even if a perfect molecular separation of two layers could be achieved at zero time so that the initial concentration difference could be accurately known, the sharp difference in concentration would disappear as soon as diffusion commences. Nevertheless, once A. E. Fick developed his law of diffusion, techniques for measuring diffusion were progressively improved. Fick based his law on the following experimental situation. A tube of cross-sectional area 1 cm^2 contains an unequilibrated solution. In any region of the tube the difference in concentration between two adjacent layers can be expressed as dc/dx, where dc represents the change in concentration and dx the distance over which this change occurs. The rate of diffusion s (the amount of solute moving across the tube area in 1 sec over the distance dx) is then equal to $-D\ dc/dx$. The diffusion constant D figures importantly. Its magnitude is specific for the molecule tested. Whereas area and concentration difference have the same effect on diffusion rate regardless of the kind of molecule involved, D is characteristic for each substance; it depends on the size, shape, and solubility of the molecule. Given standard conditions of cross-sectional area, temperature, and concentration difference, a value of D may be obtained for any substance which characterizes the rate at which it will diffuse through water.

Were cells bounded by truly semipermeable membranes, the phenomenon of solute diffusion would be of secondary interest. But even without an intimate knowledge of cell behavior, it should be apparent that substances must move across cell boundaries. And since both the interior and exterior of a cell are aqueous, the nature of solute movements in water is a question of fundamental interest. Do solutes move into cells by virtue of concentration differences? Do they move about within a cell by virtue of diffusion forces? Do small molecules move much more rapidly than larger ones? These questions were not fully considered when the osmotic properties of cells were first studied. Gradually, however, the knowledge accumulated about the diffusion of solutes in water led to more refined studies of solute movements in cells and to a better understanding of the nature of the cell membrane.

Once Fick's law had been formulated, there was a great deal of interest in determining D values and in finding a relationship between D and molecular weight. The assumption that the rate of diffusion would be inversely proportional to molecular weight seemed reasonable, but the relationship did not turn out to be a simple one. Molecular shape and its bearing on frictional forces had to be taken into account. For gases these forces were negligible, and the constant D was found to be inversely proportional to the square of the molecular weight. For dissolved molecules, especially the larger ones, frictional forces were also a function of both the viscosity of the medium and the shape of the molecule. Spheres would be expected to encounter less frictional resistance than ellipsoids. Specific molecular shapes could not be taken into account, however, until adequate techniques were available for their determination. Although such techniques have been largely developed in recent years, a physical chemist, W. Sutherland, had in 1905 determined that for most solutes the value D was nearly inversely proportional to the cube root of the molecular weight. From the standpoint of the movement of solutes across the cell membrane and also within the cell, this meant that the value D was much the same for most molecules. A 27-fold increase in molecular weight, for example, would decrease the diffusion coefficient by only one-third. If no other forces affected the movement of solutes into cells, the principal factor governing the movement of solutes would be the concentration gradient.

Given this quantitative statement, it is possible to test cells with respect to factors governing the flow of solutes. If glucose (or any other molecule) entered the cell solely by virtue of a concentration gradient, then cells should absorb more glucose as the external concentration is increased. We will defer discussion of experiments such as these to a later section for they involve mechanisms which are not covered by the phenomenon of diffusion and osmosis. At this point the student will be troubled by a contradiction between studying the cell on the one hand as though it were lined by a semipermeable membrane, and on the other as though its surface were permeable to solute molecules. The contradiction is real even though cells display both properties. It is impossible, however, to understand the mechanisms which govern the movements of solutes across cell membranes without first understanding something about the physical nature of these membranes. Experiments based on the assumption that cell membranes are more or less semipermeable are not only easy to perform, but despite the obvious limitations of that assumption the studies make evident a number of fundamental characteristics of cell membranes in general. Historically the approach could not have been otherwise. The cell physiologist was applying the laws of solutions to cells almost as quickly as they were being enunciated. Whereas studies of solute uptake would have required a knowledge of metabolism, studies of the cell membrane as a mechanical barrier required little more than direct microscopic observation.

DISSOCIATION

In contrast to the forces governing solute diffusion, those governing osmotic pressure are indifferent to the chemical nature of the solute molecules. If we ignore the deviations from the law of van't Hoff which may be traced to molecular interactions or to the volume occupied by the solute molecules (Table 3-1), then a solution of one mole of glucose in a liter has the same osmotic pressure as a half mole of glucose plus a half mole of sucrose mixed in the same volume. In general, the important factor is the number of "particles" per unit volume. Those substances which dissociate in water (sodium chloride, for example) exert an osmotic pressure proportional to the number of particles into which they dissociate. Theoretically a molar solution of sodium chloride at 20°C has an osmotic pressure of approximately 48 atmospheres, and that of sulfuric acid approximately 72. The realization that certain substances dissociated in water whereas others did not had an important influence on the direction of physicochemical studies of the cell.

Although it is difficult, and perhaps unnecessary, to recapture the enthusiasm with which many nineteenth century biologists received the new science of solutions, it is worth recognizing their reaction in order to appreciate the extensive studies which then took place. Experiments, particularly by Michael Faraday, established two kinds of water-soluble substances: nonelectrolytes and electrolytes. Electrolytes were compounds which behaved in solution as though they carried an electrical charge. They were distinctive not only because they were good conductors of electricity when dissolved in water, but also because of their tendency to be decomposed at electrodes. Hydrochloric acid, for example, yielded hydrogen at the negative pole and chlorine at the positive one. The full implications of these observations were not immediately recognized, but eventually an adequate concept of solute behavior emerged. Faraday's experiments, for example, had two possible interpretations. The first was that electric current induced molecules to break up into charged particles which would migrate to the oppositely charged electrodes, become electrically neutralized, and hence chemically transformed. The second interpretation held that the charged particles preexisted in solution; supporting this interpretation was the fact that the amount of current passing through a solution of electrolyte was the same from the very beginning. If the particles were formed by the current, then the current should increase with time. But this idea was not readily accepted until comparisons of the osmotic pressures of electrolytes and nonelectrolytes led Svante Arrhenius to conclude that even in the absence of electric current, electrolytes behaved as though they were dissociated in solution.

The realization that the three classes of electrolytes—acids, bases, and salts—dissociated into charged particles, or ions, was bound to have immediate significance for those using a physicochemical approach to the functions of cells. The toxic effects of acids and alkalis on life were long known; the possibility of explaining these effects in terms of common ions (hydrogens and hydroxyls) simplified the phenomenon. Moreover, when Arrhenius formulated his theory of dissociation (1887), the effects of salt solutions on living cells were being extensively studied. Around 1870 Jules Raulin, in studying the growth habits of the fungus *Aspergillus* discovered that traces of zinc and manganese salts (e.g., 2 parts of zinc per million of water) were essential to cell growth. Although today we have a good understanding of the need for such traces, it was a complete mystery then. It could hardly be otherwise in the many instances where the biologist grabbed whatever the chemist could offer.

Much the same random experimentation characterized the important discovery that metallic traces (copper, mercury, lead, tin, iron, or silver) would kill various plant cells. Sydney Ringer, among others, made such observations on animal cells, but one of his discoveries overshadows all others. He was preoccupied with the problem of providing a solution in which isolated frog hearts would continue to beat normally. Since sodium chloride was known to be the principal salt component of blood, he began his experiments with a sodium chloride solution and, recognizing the importance of osmotic pressure, used a concentration which exerted an osmotic pressure equal to that of the cell interior (such solutions are called "isotonic"). Under these conditions he found that the heart beat ceased within twenty minutes and could not be restored by electrical stimulation (a technique commonly used to stimulate heart beat). If, however, he added a little calcium chloride, the beating resumed, although irregularly. The addition of some potassium chloride to the mixture restored normal beating. On the basis of these experiments Ringer formulated and used a solution of salts which would maintain normal beating of an isolated frog heart for a considerable period of time. The theory of electrolyte dissociation was then unknown to Ringer. In fact, he was not too careful about the kind of anions he introduced along with the cations. Although he consistently used sodium chloride, he added calcium in a variety of forms as sulfate, carbonate, phosphate, or chloride. As a refinement he also added a little sodium bicarbonate, and although he recognized that this made the solution slightly alkaline, he was still without the concept of hydrogen ion concentration which would have made his formulation much more intelligible to him. Ringer's discovery had implications beyond those pertaining to heart beat. It became apparent that the proportions of sodium, potassium, and calcium ions which he used were generally suitable for the maintenance of viability in many animal cells. The fact that sea water contained these ions in identical proportion, though in higher concentration, led many biologists to believe that life originated in the sea. Subsequent experiments all pointed in one direction—that living matter was so organized as to have specific requirements for salts, both with respect to kind and to concentration. We will see later why ions are so essential, but without an understanding of the basic features of electrolyte dissociation, such knowledge could not have been gained.

We might assume that all electrolytes dissociate completely into their constituent ions when dissolved in water, but a variety of experiments argue against this. Equimolar concentrations of NaCl and HCl yield approximately the same osmotic pressure (about twice the pressure of an equimolar concentration of a nondissociating substance such as glucose); $C_2H_4O_2$ (acetic acid) or NH_4OH (ammonium hydroxide), on the other hand, yield appreciably lower values than the first-named electrolytes but higher values than glucose. Equally striking differences are found when the conductivities of electrolytes are compared. NaCl and HCl permit more current to pass through solution than do $C_2H_4O_2$ or NH_4OH even though the same number of molecules are present per unit volume of water. The one explanation which accounts for the various observations is that electrolytes differ in the degree to which they dissociate in solution. Substances such as HCl and NaCl appear to dissociate completely into their constituent ions and hence display about twice the osmotic pressure of nonelectrolytes at the same concentration. By contrast, electrolytes such as $C_2H_4O_2$ and NH_4OH dissociate to such a limited extent that they have fewer charged particles in solution to carry current and fewer total particles to contribute to osmotic pressure. It is customary to refer to the first group of electrolytes as "strong" and to the second group as "weak."

A qualitative distinction, however enlightening, has a limited application in scientific research. In living systems, for example, differences between different cells or between the same cells under different conditions are rarely absolute except for such contrasts as "alive" or "dead." Most commonly we observe differences in degree, but unless we have a way of measuring the degree of difference and of evaluating the measurement in terms of a general relationship, we cannot usefully interpret what we observe. It is for this reason that biology owes much to the genius of Arrhenius, who provided a quantitative formulation of electrolyte behavior. Although the Arrhenius equation has been subject to a number of refinements, its basic feature remains valid and useful today. To arrive at this expression, we may use the example of acetic acid and first write the chemical equation for its dissociation:

$$CH_3COOH \leftrightharpoons CH_3COO^- + H^+$$

The equation simply states that when acetic acid is dissolved in water, its molecules will tend to dissociate into ions and that, when the component ions of acetic acid are present in water they will tend to associate into molecules. Arrhenius obviously realized that the behavior of individual molecules or ions could not be determined, but that he could characterize the behavior of the population of ions and molecules as a whole. He did so by assuming that in any solution the number of molecules dissociating into ions in unit time would eventually equal the numbers of sets of ions associating into molecules. This state was termed the equilibrium point; once equilibrium was reached the relative numbers of ions and molecules would remain constant. Arrhenius then showed that for any particular concentration of electrolyte the equilibrium point could be determined by a single equation. If the concentrations of components within a solution are expressed in terms of the relative numbers of molecules or ions (molar concentrations, for example), the following formula sums up the relationship:

$$\frac{[CH_3COO^-]\,[H^+]}{[CH_3COOH]} = K$$

The square bracket signifies molar concentrations and *K* is called the dissociation constant; the higher the value of *K*, the greater the degree of dissociation. Acetic acid, for example, has a value of 1.75×10^{-5} at 25°C (the dissociation constants for a variety of electrolytes are listed in many chemical handbooks). The constant permits us to calculate the dissociation of a substance. It is important to recognize that the *proportion* of solute molecules which dissociate is a function of concentration. This may be clearly seen if the law is expressed algebraically. Let us dissolve x molecules of a weak electrolyte in one liter of water; at equilibrium we find that y molecules have dissociated. If the hypothetical electrolyte dissociates into two ions, each of these will be y in number, and the undissociated molecules will be $x - y$. The equation would read:

$$\frac{y^2}{x - y} = K \qquad \text{or} \qquad \frac{y^2}{K} + y = x$$

If we select an arbitrary value for *K* and construct a table or a graph to compare the changes in x as y is changed, we see that the smaller the value of x (that is, the greater the dilution), the higher the proportion of y/x. Thus as we lower the concentration of a weak electrolyte, a larger proportion of molecules dissociate; by contrast, strong electrolytes dissociate completely over a broad range of concentrations. At relatively high concentrations ionic interactions become appreciable so that weak electrolytes do not behave precisely as the formula indicates, nor do strong electrolytes remain completely dissociated. Important though these departures are in considering the internal environment of cells where numerous interactions occur, they need not concern us in our general survey.

pH AND BUFFERS

One kind of ionic interaction, which is extremely important in interpreting the behavior of cells, concerns the regulation of hydrogen ions. When Ringer devised his physiological solution consisting of a fixed proportion of Ca, Na, and K, he pointed out that adding some sodium bicarbonate to the mixture in order to make it slightly alkaline rendered the solution much more favorable to the maintenance of heart beat. It was, of course, long recognized that living cells were killed under strongly alkaline or acidic conditions. Ringer's observation, however, had much more meaning because he was measuring a specific cellular process (contraction) and was thus able to discover that small changes in acidity had deep-seated effects on cell function even though such changes did not produce the extreme disorganization associated with death. It is idle though interesting to speculate on how far Ringer would have carried his studies had the theory of dissociation been available to him. Soon after the elaboration of this theory biologists demonstrated in a variety of ways the general importance of hydrogen ion concentration to vital functions. Today we take it as a matter of course that ion concentrations must be regulated in all physiological or biochemical experiments. Essentially all cellular processes are influenced by hydrogen ions; the full significance of ionic effects, however, can only be appreciated after considering all of the many different activities, structural and functional, of cells.

A question of primary interest to biologists is the regulation of hydrogen ion concentration in aqueous solutions. Strong electrolytes, because of their complete dissociation, do not appear to be susceptible to regulation unless their production and removal are directly controlled. We might suppose that such a mechanism exists in cells, but there is no evidence for its general operation. A far more plausible mechanism may be derived directly from Arrhenius' theory of dissociation, and is based on the characteristics of the weak electrolytes so abundant in living cells. In the dissociation of acetic acid, for example, the theory requires a constant value for the product of hydrogen and acetate ion concentrations divided by the concentration of undissociated molecules. It does not specify the source of hydrogen ions. To be sure, if acetic acid is present alone it is the only source of hydrogen ions. But if another acid, say, HCl, is added to a solution of acetic acid, then hydrogen ions would enter into the equilibrium equation for acetic acid (although chloride ions would not). If no changes occurred, the product of hydrogen and acetate ions would be increased because of the additional hydrogen ions present. However, changes do occur and in the direction predicted by the Arrhenius equation: some of the added hydrogen ions combine with acetate ions to form molecules of acetic acid, thus simultaneously increasing the value of the denominator and decreasing that of the numerator. When equilibrium is reached, the product of hydrogen and acetate ions divided by acetic acid molecules is identical with the value obtained with acetic acid alone. The addition of one of the components of a system in equilibrium will push it in the direction which reduces the concentration of the component added.

The example just cited illustrates a general principle. When a strong acid is added to a weak one, the resulting hydrogen ion concentration is less than that expected if all of the strong acid ionized completely. A similar relationship holds for bases and for mixtures of acids and bases. The operation of this principle was of obvious interest to cell biologists, and once it was understood, the significant relationships were soon formulated in quantitative terms. By 1909 Jönen Sørensen introduced the "pH scale" as a convenient device for expressing hydrogen ion concentrations. To understand the nature of this scale, we must begin by considering some characteristics of pure water.

Before the turn of the century physical chemists established that water dissociated. The most direct evidence for this was the capacity of water to conduct an electrical current. In early experiments the conductivity was thought to be due to contaminating ions, but this view was abandoned when even the most extensive procedures for purification could not reduce conductivity below a certain value. To explain this property of water, its molecules were assumed to dissociate into hydrogen and hydroxyl ions. Since Faraday had already clarified the relationship between conductivity and the number of ions, the degree of dissociation could be calculated. As would be expected from the low conductivity of pure water, the degree of dissociation was found to be very small. At 25°C the concentrations of hydrogen and hydroxyl ions were each $10^{-7}N$. The concentration of undissociated water molecules is easily calculated.

One liter of water weighs approximately 1000 g. Since the molecular weight of water is 18, the number of moles of water per liter is 1000/18, or roughly 56. The actual number of water molecules would be $56 \times 6.06 \times 10^{23}$ (Avogadro's number, i.e., the number of molecules in one mole of substance) and that of the hydrogen or hydroxyl ions would be $6.06 \times 10^{23} \times 10^{-7}$, or 6.06×10^{16}. In other words, there would be 56×10^{7} water molecules for each hydrogen ion. Such a low degree of dissociation means that water behaves as a weak electrolyte, and must have a characteristic dissociation constant. The constant is calculated from the Arrhenius formula:

$$\frac{[H^+]\,[OH^-]}{[H_2O]} = \frac{10^{-7} \times 10^{-7}}{56} = \frac{10^{-14}}{56} = K_W$$

The pH scale is based essentially on the response of the weak electrolyte water to the presence of acids or bases. There are two reasons for devising a hydrogen ion scale which uses water as a reference. First, electrolytes are usually studied in aqueous solutions, and, as pointed out earlier, cells are largely aqueous systems. Second, the mathematics of hydrogen ion behavior in relation to water turns out to be very simple because of the unusually low dissociation constant of water.

The fundamental point in setting up a pH scale is that the dissociation constant of water, like that of any other substance, must remain fixed at any given temperature regardless of the numbers of hydrogen or hydroxyl ions present. Using this relationship, the concentrations of hydrogen and hydroxyl ions resulting when a strong acid or base is added to water may be predicted. To simplify the problem of prediction we need only assume that we do not change the concentration of water by adding acid or base. This is equivalent to saying that enough room exists between molecules in pure water to accommodate the added solute molecules, and in dilute solutions this is virtually the case. Suppose we add 10^{-1} moles of HCl to a liter of pure water. Since HCl dissociates completely, the concentration of added hydrogen ions is $10^{-1}N$. We know, however, that the dissociation constant of water is fixed at $10^{-14}/56$, so that adding hydrogen ions must cause some of the hydrogen and hydroxyl ions to reaggregate into water molecules. Since 560,000,000 water molecules are present for each hydroxyl or hydrogen ion, even if all the hydroxyl ions present were combined with the excess of hydrogen ions the increase in number of water molecules would be negligible. For all practical purposes then, we need only consider the numerator of the Arrhenius equation in predicting the relative numbers of hydrogen and hydroxyl ions. This numerator must equal 10^{-14} because the value of 56 in the denominator will remain unchanged. Thus if we increase ion concentration to 10^{-1}, the hydroxyl ion concentration must fall to 10^{-13}. No matter what combination we choose the product can only have a single value; if the hydrogen ion concentration is known, the hydroxyl ion concentration is also known.

This relationship was particularly appealing to Sørensen because acidity and alkalinity could be expressed in terms of the hydrogen ion concentration alone. Whenever the hydrogen ion concentration fell below 10^{-7} the solution acquired alkaline properties, and whenever it rose above 10^{-7} it acquired acidic properties. Sørensen, however, did not like minus signs or power indices. His preference for simple positive numbers led him to devise the familiar pH scale. Since concentrations are expressed in terms of decimals, any particular concentration can be written in the form 10^x. For any concentration lower than 1, x would have a negative value and because hydrogen ion concentrations commonly encountered are less than 1N, x is almost always negative. To avoid negative expressions Sørensen arbitrarily set up the rule that the hydrogen ion concentration should be written as $-\log_{10}$ (H+) and that in this form it should be called pH. Log_{10} of any power of 10 is equal to that power, and the minus sign gets rid of the negative sign in the power. Thus $10^{-7}N$ hydrogen ion is equal to a pH of 7; $10^{-2}N$ to a pH of 2 and $10^{-12}N$ to a pH of 12. Sørensen must have had an intuitive sense of human taste, because the pH scale was quickly and widely adopted. All commercial devices which are specifically designed to measure hydrogen ion concentrations use this scale; indeed, to many laboratory technicians the term pH is as familiar as its derivation is unfamiliar.

On the whole, cells cannot tolerate external hydrogen ion concentrations outside the pH range 6–8, although some fungi can grow at pHs of the order of 2.0 and others manage well at pHs as high as 11.0. The internal pH of cells is not necessarily similar to that of the exterior. And even within the interior, cells often segregate acidic solutions into membrane-bounded vacuoles. Plant cells are prominent in this respect, for usually the high acidity of various plant juices originates in the large vacuoles. The extent to which the living fabric itself can tolerate variations in pH is open to some question because of the difficulties in determining the pH of cytoplasm or nucleus. When indicators have been injected or allowed to diffuse into cells, the colors observed point to a pH of about 6.8 in the cytoplasm. These are nevertheless rough pointers, and perhaps the best indication of possible limits to pH variation within the interior of living systems comes from studies of blood. In humans the pH of the blood is maintained close to 7.4; if it decreases to 7.0 or increases to 7.8, death results. The same range of tolerance cannot be presumed for all cells, but many pieces of scattered evidence indicate that the vital machinery of cells must operate within narrow limits of pH; little motive now exists for extending or deepening the evidence.

It is understandable that at the turn of the century investigators were very much interested in mechanisms which would regulate the pH of living systems. The topic was popular, and enthusiasm outdid itself in a search to explain a number of cellular phenomena in terms of pH changes. The principal result of these many studies was to make cell biologists aware of the general importance of pH in maintaining the integrity of cellular structures. Yet this could never have been achieved without the understanding which chemical studies provided about the behavior of weak electrolytes in solution with respect to a property we call "buffering."

Well before Arrhenius proposed his theory of dissociation, biologists were familiar with the fact that blood could withstand the effects of appreciable additions of acid or alkali. The phenomenon was intriguing because even then blood was known to be a delicately balanced fluid which ceased to be effective when the balance was tilted. To those who preferred mechanistic rather than vitalistic interpretations it seemed reasonable to suppose that blood contained certain chemical substances which enabled it to withstand the effects of additions of acids or bases. And it was correctly suggested that the phosphates and carbonates in the blood were the responsible agents. Once the theory of dissociation was elaborated, the mechanism by which phosphates and carbonates could operate became apparent. In 1900 two French biologists, A. Fernbach and L. Hubert, reported experiments in which naturally occurring substances such as phosphates, carbonates, and amino acids were found to be capable of removing hydrogen or hydroxyl ions from solution when these were added in the form of a strong acid or alkali. They considered the phenomenon worthy of a christening and so proposed the descriptive word "tampons" to cover such substances. Sørensen liked the word, probably as much as he disliked negative exponents in expressing hydrogen ion concentrations, and adopted it. When his works were translated into German, the word "Puffer" was used; the second act of translation resulted in the English word "buffer." Despite the fact that the famous English physiologist, Sir William Bayliss, found the word undesirable as late as 1918, and did not use it, the term "buffer" acquired a secure place in the lexicon of science.

The nature of buffering may be illustrated by two titration experiments (Fig. 3-2). In the first we begin with a 0.1*N* solution of HCl, say, 10 ml, and gradually add an equal volume of 0.1*N* NaOH. If the pH of the solution is measured during each addition, it will be found to change only slightly until virtually all of the NaOH has been added. At that point the smallest excess of alkali will sharply raise the pH to a value of 12 or more. The student could predict this pattern of behavior by assuming that the two electrolytes dissociate completely. Bearing in mind the definition of pH, he will find that to increase the pH of the solution from one to two (that is, from the original hydrogen ion con-

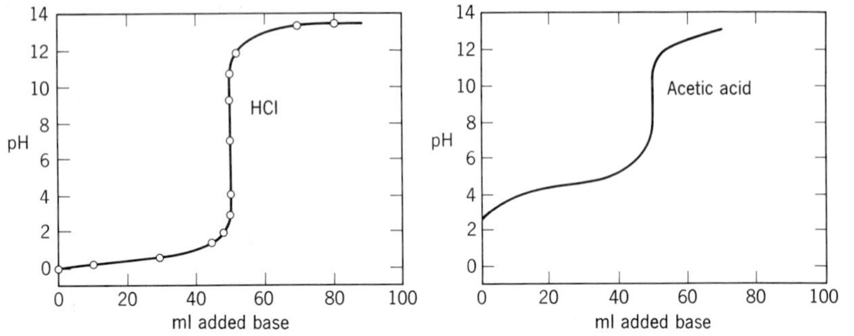

Fig. 3-2. Titration curve. (From Andrews and Kokes, *Fundamental Chemistry,* Wiley, 1962.)

centration of $0.1N$ to $0.01N$) requires the removal of about nine times as many hydrogen ions as the increase from pH two to seven.

In the second titration experiment acetic acid is substituted for HCl. Since acetic acid is a weak electrolyte the starting pH will be higher than that of the HCl solution, but the most striking difference is in the shape of the titration curve. On first adding NaOH the rise in pH is comparatively steep, but as titration is continued the rate of change in pH slackens and remains low until about 8 ml of $0.1N$ NaOH have been added. Careful inspection of the curve will show that the mid-point of the "horizontal" portion is close to a pH of 4.7, and is the point at which 5 ml of alkali have been added. Clearly, in this system, as in all systems involving weak electrolytes, hydrogen ions are not removed at a steady rate on gradual addition of a neutralizing agent. This behavior indicates that the incompletely dissociated acetic acid acts as a source of hydrogen ions which are released as the equilibrium is upset on addition of NaOH. The general relationship may be formulated quantitatively.

Designating the anion of a weak acid as A^-, the equation for dissociation reads

$$\frac{[H^+]\ [A^-]}{[HA]} = K_a \quad \text{or} \quad [H^+] = K_a \left[\frac{HA}{A^-}\right]$$

If the latter equation is converted to negative logarithms, it becomes

$$-\log\,[H^+] = -\log K_a - \log\,[HA/A^-]$$

Now $-\log\,[H^+]$ equals pH, and by granting ourselves the same privilege of convenience with respect to dissociation constants as we have to hydrogen ion concentrations, $-\log K_a$ becomes pK. The equation now reads pH $=$ pK $-\log\,[HA/A^-]$. To eliminate the remaining negative we simply invert the fraction so that pH $=$ pK $+ \log\,[A^-/HA]$.

One approximation brings this equation into a form that is exceedingly useful in biological studies. When a weak acid, such as acetic, is being neutralized, the resultant salt (e.g., sodium acetate) remains dissociated; the evidence that led to the characterization of strong electrolytes points to a more or less complete dissociation of salts even if they are derived from weak acids or bases. In our particular titration the anions (A^-) present are largely derived from sodium acetate since the acetic acid molecules dissociate only to a very slight extent. This is apparent when we consider that at the half-way point of titration, the hydrogen ion concentration is roughly $10^{-5}N$ whereas the concentration of acetic acid is $0.05N$. It is possible to apply the equation just given to a mixture of a weak acid and its salt by letting the term A^- equal the concentration of salt in solution, the equation then becomes $pH = pK_a + \log \text{salt/acid}$. Weak acids and bases show the greatest resistance to pH change at the point where the concentration of acid or base equals that of salt. Since $\log_{10} 1 = 0$, a weak electrolyte will show greatest buffering characteristics at a pH equal to the pK. This in itself is a point of some convenience because if we wish to make up a solution with a fairly constant pH, tables of dissociation constants may be consulted and the appropriate weak electrolyte chosen.

The relationship between dissociation constant and buffering capacity also enables us to make certain predictions about cellular regulation of pH. Many of the soluble components of cells are weak electrolytes and therefore are capable of exercising buffering effects. In a yeast cell, for example, 40% of the anions are bicarbonate and 43% are phosphate. Both of these are anions of weak acids. As far back as 1908 the physiologist L. J. Henderson noted that of a large number of acids tested, these had among the highest capacities to resist pH change on addition of alkali. Furthermore, the pK values of biological interest are 6.34 for carbonate and 7.2 for phosphate. (The reason for selecting pK values is that bicarbonate has two dissociable hydrogen ions and phosphoric acid has three; it is characteristic of such molecules that each stage of dissociation has a different dissociation constant. Thus $H_2CO_3 \rightleftharpoons H^+ + HCO_3^-$ and $HCO_3^- \rightleftharpoons H^+ + CO_3^=$ have pK values of 6.34 and 10.25 respectively, whereas $H_3PO_4 \rightleftharpoons H^+ + H_2PO_4^-$, $H_2PO_4^- \rightleftharpoons H^+ + HPO_4^=$, and $HPO_4^= \rightleftharpoons H^+ + PO_4$ have the respective values 2.12, 7.20 and 12.66.) Quite apart from other ions that might influence cellular pH, the two commonest ones are components of choice with respect to both buffering capacity and the range in which cells require their pH to be regulated.

So extensive were studies on this subject at the beginning of the century that in 1923 a two-volume monograph on the hydrogen ion content of cells was published. Few cell biologists would care to study this monograph today—and for good reason. Although the history of this subject is punctuated by some salient pieces of research, such as the resolution of the mechanism by which mammalian blood maintains a constant pH, the totality of the studies represented a grand scouting operation. The ground was unfamiliar and had to be probed; the probes were those supplied by physics and chemistry. Our principal inheritance from that excursion into cell biology is the realization, now taken for granted, that most living processes are in one way or another dependent on pH conditions. To be sure, we do not fully know how cells regulate pH any more that we know exactly how warm-blooded mammals maintain a body temperature of 37°C. And undoubtedly if we pursued all the circumstances surrounding pH regulation deeply enough, we would probably uncover some still unknown properties of living matter. The art of successful science, however, is not merely the posing of unanswered questions, but the selection of those questions which under prevailing circumstances can yield significant answers. On that count cell biologists of the last two decades have declared pH a loser.

PERMEABILITY OF CELL MEMBRANES

To round off the account of what nineteenth century interest in the properties of solutions contributed to our knowledge of the cell, we may return to one of the starting points in this chapter—the semipermeable nature of the cell membrane. With our present perspective we can see that this early period of cell physiology led to an account of the cell surface which, so far as it went, was complete, and within its self-imposed limits, lasting. By about 1925 the cell surface was known to oppose the passage of certain kinds of solutes and to be indifferent to the passage of others. A distinction could be made with respect to the penetration of electrolytes and nonelectrolytes. Although "hypertonic concentrations" (greater than that of the cell interior) of nonelectrolytes like sucrose would plasmolyze cells and keep them so for long periods of time, other nonelectrolytes, like urea, would produce only a temporary plasmolysis. Extensive studies demonstrated that the smaller the molecule, the more readily it penetrated into the cell.

A physical explanation of such observations is simple. Suppose a cell is exposed to a hypertonic concentration of urea. The immediate effect is a shrinkage of the whole cell if no rigid wall is present or, if one is, a shrinkage of the protoplast away from the wall. The shrinkage is understandable in terms of osmotic pressure. The next observed effect is an expansion of the shrunken protoplast. Since the expansion must be due to the intake of water, something must have happened to increase the osmotic pressure within the cell. The only sources of such increase is urea, which must have gradually diffused through the cell membrane. Any doubts about this may be dispelled by removing the cells from the medium and analyzing their contents for urea. The sequence of effects can be best explained by postulating that water molecules diffuse much more rapidly through the membrane than do those of urea. Initially, therefore, water moves out faster than urea moves in and plasmolysis results. Gradually, however, urea diffuses in, driven by the force of diffusion arising from the urea concentration gradient. A diffusion force does not act to decrease the concentration of other intracellular molecules since the surface membrane blocks their outward passage.

Molecular weight is one of the major factors bearing on the permeability of substances. Sucrose (M.W. 342) is too large to enter cells without the intervention of special transferring mechanisms; and urea (M.W. 60) serves as an example from the lower end of the molecular scale at which passage is relatively free. As expected, gases penetrate readily. The extensive studies of the relationship between molecular size and permeability led to a picture of the cell surface as a membrane containing different sized pores of diameters commensurate with the sizes of the penetrating molecules. This picture is only a partial one, but insofar as it defines a particular condition for the penetration of molecules its usefulness persists to this day.

At the same time that nonelectrolytes were thus being studied, investigators were also following the behavior of dissociable compounds. Compounds which dissociated completely did not penetrate readily into living cells regardless of the sizes of their constitutent ions. Plant cells, for example, would stay plasmolyzed indefinitely in a hypertonic solution of calcium chloride. Moreover, weak and strong electrolytes have different capacities to penetrate cells. To demonstrate this, two experimental techniques were usually employed. In the one, plant cells containing natural acid-base indicators were used, and in the other, both plant and animal cells were first exposed to indicator dyes which accumulate in living cells. With neither experimental system would strong acids or bases alter the internal dye color, but weak acids or bases would. Most investigators concluded that, quite apart from the membranes having pores, they must also contain charges which would prevent passage even of ions of small diameter by electrical repulsion. One kind of charge on the membrane would be theoretically sufficient to exclude both anions and cations; the attraction between oppositely charged ions is so great that if a membrane were to repel cations, anions could not move very far before the very strong forces of attraction of the lagging cations would hold them back. The truth of this proposition became evident when other techniques made it possible to show that different ions of the same charge could exchange with each other across some cell membranes. Red blood cells, for example, when exposed to a solution of sodium sulfate will lose their own chloride ions in exchange for sulfate ions; cations, however, move in very slowly.

We need not review the very many studies of such exchanges in different kinds of cells simply because this phenomenon was being studied in a framework wherein the question of ion movement had the earmarks of a fiction. For ions do move into cells; if they did not, the cells would perish. The early investigators were treating cells as if their membranes had the same static properties of charge and composition as do artificially constructed membranes. The treatment has obviously a limited validity, and within these limits the important information which emerged was the charged nature of cell membranes. The poor penetration of ions into living cells is a useful rule of thumb for physiologists when considering the appropriateness of different substances for testing various intracellular reactions. Conditions which depress the ionization of organic compounds generally favor their penetration; weak acids do better at acidic pH's, weak bases at alkaline pH's. But beyond this coarse, though important, indication, the deeper factors associated with ion penetration must be deferred to a later discussion.

We have still to consider another contribution which together with the two preceding ones has provided us with a rough and essentially correct sketch of the cellular surface. Studies of polar solutes were paralleled by studies of apolar solutes (those which readily dissolved in organic solvents such as benzene and were poorly soluble in water). Apolar substances penetrated far more readily than polar ones. The relationship between the solubility of a substance and its permeability was carried to a refined level of analysis by characterizing each of the substances examined in terms of its "partition coefficient." This is simply the ratio of concentrations found in an organic solvent (ether, benzene, oil, etc.) and water when a particular compound is shaken up in equal volumes of both and the liquids are allowed to separate after equilibrium has been achieved. This type of characterization was common among organic chemists of the nineteenth century. It was a quick way of determining the net effect of the different carbon atom groupings within a molecule on its solubility. For example, a grouping of carbon atoms with hydrogens (e.g., —CH_3) interacted very little with water, whereas a grouping with oxygen atoms (e.g., —COH, —COOH) interacted strongly. Molecules with only carbon and hydrogen atoms were thus poorly soluble, if at all, in water, whereas those containing carbon, hydrogen, and oxygen were water soluble in proportion to the relative numbers of the two groupings within the molecule. The relationship is illustrated by a comparison of a group of alcohols in Table 3-2.

Table 3-2 Partition Coefficients for Some Common Alcohols

Alcohol	Formula	Solubility in Water	Partition Coefficient (olive oil/water)
Methyl	CH_3OH	Infinite	0.0097
Ethyl	CH_3CH_2OH	Infinite	0.0357
Propyl	$CH_3(CH_2)_2OH$	Infinite	0.156
Butyl	$CH_3(CH_2)_3OH$	7.9 g/100 ml	0.588
Amyl	$CH_3(CH_2)_4OH$	Slight	2.13

Cell physiologists were certain that polar and apolar properties of molecules bore important relationships to cell behavior. Virtually all molecules in the cellular framework consisted of continuous or interrupted chains of carbon atoms. And, although these molecules were surrounded by water, some of them were apolar in nature. The finding that the ease with which a particular compound penetrated a cell was in direct proportion to its apolar characteristics suggested that the membrane must, in part at least, be constructed out of fats. To be sure, the relationship was not entirely consistent when tested in various kinds of cells. Molecular size had a bearing, and in some cells at least molecular size was the deciding factor for any group of compounds having similar partition coefficients. But, as with ion movement, these studies could not be pushed too far. One important property of the partition coefficient not covered by permeability characteristics was its relationship to toxicity. If a column indicating the concentrations of different alcohols necessary to kill cells were added to Table 3-2, we would see that the more apolar a compound is the higher its toxicity. In fact, amyl alcohol is more than a hundred times as effective as methyl alcohol in poisoning a cell. So effective are apolar compounds in killing cells, that a few drops of toluene are often added to stored solutions to prevent the growth of organisms.

These studies of permeability provided a picture of a patchwork membrane containing uncharged areas of fat and areas of polar substances which were charged and punctuated with different sized pores. As a detailed picture of the cell membrane, this is wanting. But as a first approximation it is proving to be excellent. The hopes of some that investigations of permeability would provide an answer to many facets of cell behavior remained unfulfilled, but their contribution is unique in one respect. For the first time physicochemical studies alone led to a molecular model for a specific, and at that time indiscernible, cellular structure.

At this point we may draw together the essentials of the various studies discussed in this section. We could readily justify discussing these particular studies first, but we could also rationalize other sequences. The many cell biologists who sought to interpret cell behavior in terms of physics and chemistry had a long-range motive but not a long-range plan. We have shown in this section how interests and developments in the physical chemistry of solutions led to the discovery of a number of physiological characteristics of cells. This point is emphasized because we are only 50–60 years away from the peak of that era. Since that time we have been slowly moving away from a romantic fascination with the possibilities inherent in a physicochemical approach to cells, and turning toward a crisper intellectual approach in which we seek out physicochemical methods that serve to answer specific biological questions. In this and associated sections we are therefore familiarizing ourselves with the ground. We have applied somewhat sophisticated concepts of physics and chemistry to a rather unsophisticated concept of cell behavior. Perhaps the chief justification for beginning with the properties of solutions is the one given at the beginning of this section—namely, that water is so much a part of cell organization as to make any study of a cell impossible without some understanding of how different kinds of molecules behave in the presence of water.

4

The Impact of Organic Chemistry

No more obvious example could be chosen of the indispensability of both theory and measurement to the development of a science than the history surrounding the origins of modern chemistry. The idea that natural diversity represented a variable blending of a limited number of universal elements goes back at least as far as the Greeks, but for about 2000 years the idea remained little more than a fuzzy philosophical prejudice. Less than 300 years ago when a few chemical arts, such as soap manufacture and alcohol distillation, were being widely practiced, chemistry had many of the airs of witchcraft and all of its limitations. Only for amusement or out of serious interest in the history of thought would we examine this period for its contributions to the understanding of living matter. In 1680 a distinguished Dutch physician, Cornelius Bontekoe, was still prescribing 100–200 cups of tea per day for good health. Milk was considered suitable only for the very young or very old, and vegetables were distrusted because they engendered "melancholy." Despite occasional and significant thrusts into a more realistic analysis of foods and food requirements, Galen's theory of nutrition, governed by the four "humours"—hot, moist, dry, and cold—was dominant nearly throughout the seventeenth century. Foods were analyzed according to their watery, oily, and saline contents. In brief, virtually nothing was known about the chemical nature of biological compounds.

One brief jump in history to a period beginning about 1750 brings us into an era so revolutionary that a simple line of developments cannot be followed. Black, Priestley, Lavoisier, Berzelius, Dalton, Avogadro, Volta, and Faraday are only a partial list of those who contributed to a distinguished 100 year period. Broadly speaking, they threw mysticism out of chemistry by showing that chemical elements were real and identifiable, that compounds were characteristic combinations of elements in fixed proportions, and that the proportions were equal to the relative numbers of atoms, each atomic element having its own characteristic weight. To say that these achievements opened the dyke which had so long held back useful chemical analysis is probably an understatement. The analogy will suffice, however, to indicate how a host of talents was suddenly unleashed in pursuit of a single challenge—to determine the chemical composition of compounds. In such an atmosphere it would have been impossible for organic compounds, the stuff of living matter, to escape being a target. Just as the later interests of scientists in the properties of solutions gave rise to much of our background knowledge of the physicochemical behavior of cells, so did the prevailing interests in chemical composition give rise to our basic knowledge of the chemical nature of living material. We shall see, however, that the effect was a snow-balling one; once compounds could be accurately identified, the path was open to studying relationships between them.

By the middle of the nineteenth century it was common knowledge that all the substances constituting living organisms contained carbon, hydrogen, and oxygen; and that a high proportion of living material also contained nitrogen. The realization that the whole of living diversity could be reduced to combinations of four elements must have been startling. A few other elements, those of inorganic salts among them, were known to be essential to life in minor proportions, but this fact could hardly reduce the challenge of explaining living organization out of so parsimonious a collection of elements. Small wonder that organic chemistry evoked strong interests, and that all kinds of rationalizations were sought to escape the restrictive implications of the new chemistry. If organic compounds were mere chemical compounds, then where could a boundary be established between living and nonliving? No indisputable answers to such a question could be given. Those who intuitively felt that the organization of life must involve something beyond the reach of mere physics and chemistry injected the principle of "vitalism" into their biological conceptions. They were as ready as their ideological opponents to use the new science to describe life, but they eagerly sought and readily found reasons why such descriptions could not explain it. The mechanists, on the other hand, did not immediately find ultimate explanations, but they could consistently show one advantage—a step-wise success in resolving specific biological problems.

From that period we have inherited a fundamental conception which today appears to be reaching its fullest development—the conception of the molecule as the unit of cellular structure and function. Essentially, this conception maintained that once a number of elements were combined into a compound, the molecules thus formed had two distinctive characteristics. (1) The forces holding the atoms together were relatively strong so that under normal conditions molecules were stable units. (2) The properties of a molecule were not an arithmetic sum of the properties of individual atoms but a blending which made impossible a prediction of the behavior of a molecule purely from a consideration of the behavior of its constituent atoms. This is a sweeping statement requiring a number of qualifications, but even as it stands, it is accurate enough to give the student the gist of the philosophy underlying the molecular approach to cell biology.

There always were, and still are, those who maintained that life could not be understood by means of chemical techniques because cells had first to be destroyed in order to be studied chemically. A cell could easily be broken up into molecules, so the argument ran, but molecules could not be put together into a cell. The idea that the parts of a cell could be studied in much the same way as parts of a watch was rejected on the grounds that a cell did not consist of discrete mechanical parts. Such a dispute cannot be settled at the theoretical level. The principal point at issue was whether molecules could be regarded as discrete units despite the interactions that must occur within the living cell. Those who assumed that cellular molecules could be so regarded proceeded to study the properties of the isolated molecules, and from these inferred the kinds of interactions molecules might undergo within the cellular framework. Today we accept the correctness of this view because it has been fruitful. We have learned that the molecules composing the cells of bacteria are much the same as those composing the cells of man, and that the major differences among living things arise from slight differences among similar molecules. We are probably as amazed today that the whole of natural diversity should be vested in a few kinds of molecules as early organic chemists were that the world of organic molecules is an assortment of four or five elements.

ATOMIC STRUCTURE AND THE FORMATION OF MOLECULES

We could begin chronologically with a description of what biologists found in trying to satisfy their curiosity about the chemical composition of living substances. We stand to lose, however, by not first considering what other chemists and physicists had come to learn about atoms. For, vigorous as was the intellectual drive to discover the composition of compounds, an even more vigorous drive was ordering the patterns into which elements combine. Why did oxygen almost always combine with two atoms of hydrogen, whereas chlorine combined with only one? Why did some elements have a valence of one, others of two or three, and still others more than one valence? And why did certain combinations of elements ionize in water whereas others did not? In brief, could any set of principles be found that would account for the specific properties of the elements? The existence of such principles was suspected if only because elements could be classified into groups showing certain common chemical properties. But even after Dmitri Mendeleev provided a sound and useful classification of the elements, the principles did not become apparent until some basic features of atomic structure were disclosed. This disclosure did not take long. Mendeleev published his work around 1870; by 1910 models of the atom were highly sophisticated architectural pieces.

Inasmuch as our objective is to understand the basic factors involved in the formation of chemical compounds, we need concern ourselves only with the outer portion of atoms. For a fundamental premise in atomic theory is that the chemical behavior of atoms is a function of their peripheral electrons. Mendeleev's periodic chart of the elements is understandable if we ignore the insides of the atom and consider only the disposition of electrons in the outer regions. To be sure, there is a very big conceptual jump between the idea of atoms as the elementary units of matter and the image of an atom as a planetary system. To trace the evolution of this image would involve us too deeply in physics, for although we may readily point to the experiments of Faraday as establishing the operation of electrical forces in matter, it is difficult to select model experiments which, without profound interpretation, would at once make evident the reality of electrons spinning about atomic nuclei.

Theory of atomic structure is a product of the highest levels of abstract thinking. Ordinarily, we think of structures in terms of three-dimensional mechanical models in which the component pieces have defined positions. Furthermore, we expect that if a particular structure exists, clear proof of the spatial distribution of its parts should be obtainable. Nothing, however, could meet these expectations less than the modern concept of the atom. The only spatial arrangement we can consider in familiar terms is that of a dense positively charged center bounded by a relatively diffuse negatively charged cloud of electrons. Beyond that, any model is no more than a symbolic representation of a number of abstract forces. There is no doubt about the reality of electrons as almost weightless negatively charged particles, nor of protons as dense positively charged ones. The discovery of radioactivity and the experimental smashing of atoms have adequately demonstrated the existence of these units. For the physicists of this century, however, one of the greatest intellectual challenges has been to arrange the electrons in such a way as to account for the manifold physical and chemical properties which the elements display.

The mere fact that the number of electrons per atom increases with atomic weight is a major source of complexity. Electrons have the same charge and should, therefore, repel each other; how, then, do they associate without pushing one another apart? Alternatively, if the nucleus of an atom is positively charged why aren't the electrons sucked into its center? To answer these questions we first need a model which is stable, since the fact that an element can exist as such is proof of such stability. The earliest models pictured the atom as a solar system in which the centrifugal force of the orbiting electron balanced the force of electrical attraction between it and the atomic center. But the problem of arranging the orbits of all the electrons in a particular atom still remained. Hydrogen, with only 1 electron, posed no such problem, but uranium with 92 posed a staggering one. If electrons did not have different individual orbits, they would be constantly colliding. Thus any acceptable theory of the atom had to account not only for the stable orbiting of an individual electron, but also for the stability of orbits between electrons.

Physicists were by no means free to design orbital patterns solely on the basis of electrical and centrifugal forces. Mendeleev's system of classifying the elements had already revealed that certain groups of elements had common chemical characteristics even though each group showed a broad range of atomic weights. Fluorine, chlorine, bromine, and iodine, for example, had properties in common even though their atomic weights ranged from 19 to 127, and the numbers of their electrons from 9 to 53. That chemical properties could not possibly involve the atomic nucleus was apparent to Mendeleev; although he lacked any detailed knowledge of atomic structure, he indicated that the one thing which does not change when an element enters into chemical combination is its atomic weight. To rationalize the fact that the chemical properties of an element with 9 electrons was similar to one with 53, the assumption had to be made that the outer electrons, which were the ones likeliest to interact, were similarly arranged. If the same line of reasoning is applied to all the other groups of elements, the number of possible arrangements diminishes. The chemical properties of the elements are, however, only one of several sources of information which physicists had at their disposal. Far more understanding of the detailed relationships between electrons was gained from a variety of physical studies. And although our purpose will be served by symbolizing these relationships in simple diagrams, they cannot be considered as real mechanical models of the atom.

The scheme which best explains chemical behavior is one in which electrons are pictured as being confined to particular shells surrounding the atomic nucleus. These shells are designated by the letters *K*, *L*, *M*, *N*, *O*, and *P* from the inner to the outermost shell. Hydrogen, the lightest element, has only one shell; uranium, the heaviest element, has all six. By assigning two electrons to the *K* shell, eight to *L*, 18 to *M*, 32 to *N*, 50 to *O*, 72 to *P*, and by following the rule that in going up the scale of atomic weights existing shells are filled before new ones are begun, a relationship can be observed between electron patterns and chemical properties. Lithium, sodium, and potassium have one electron in their outer shells, whereas fluorine, chlorine, bromine, and iodine are one electron short of the full number. Thus by taking two demonstrable physical facts—the atomic weights of the elements and the existence of electrons—and by assuming that successive elements on the atomic weight scale increase their number of electrons by one, a purely imaginary pattern of electron arrangements could be set up which would conform to chemical behavior. The pattern as here described proves to be inadequate in a number of respects but these inadequacies have been met by a simple refinement. All shells except *K* were divided into subshells; *L* was given two, *M* three, and so on. The purpose of this modification was to rewrite the requirement that each shell be filled before another is begun. The new rule stated that subshells had to be filled, but that in elements having many electrons certain subshells could be only partly filled or skipped and a new one started.

A number of elements had specific chemical properties which made this change of rule necessary. Among these were the "inert gases" which have an important general bearing in the translation of electron patterns into chemical behavior. Inert gases, as their name implies, do not commonly undergo chemical reactions. Inasmuch as chemical reactions are considered to be interactions between the peripheral electrons of the elements, the relative chemical inertness of these gases must reflect highly stable electron patterns. To rationalize the stability of electron patterns we would have to understand the forces operating on the different electrons within an atom. Such understanding requires the kind of experimental evidence which modern physics has been acquiring and for which a purely mechanical model of the atom is no longer suitable. But for understanding most of the basic features of chemical behavior, we may accept the correlation between the inertness of the inert gases and their proposed electron pattern. One of the reasons for introducing the subshell rule will be evident on examining Table 4-1.

Table 4-1 Electron Patterns of Major Elements in Living Matter and of the Inert Gases

Shell	*K*	*L*		*M*			*N*			
Subshell	1	1	2	1	2	3	1	2	3	4
Maximum No. of Electrons	2	2	6	2	6	10	2	6	10	14
Element	*Number of Electrons*									
Hydrogen	1									
Helium	2									
Carbon	2	2	2							
Nitrogen	2	2	3							
Oxygen	2	2	4							
Neon	2	2	6							
Sodium	2	2	6	1						
Phosphorus	2	2	6	2	3					
Sulfur	2	2	6	2	4					
Chlorine	2	2	6	2	5					
Argon	2	2	6	2	6					
Iron	2	2	6	2	6	6	2			
Krypton	2	2	6	2	6	10	2	6		

The inert gases are helium, neon, argon, and krypton (xenon and radon not shown). Note that subshell M_3 is unfilled in the argon atom as are subshells N_3 and N_4 in the krypton atom. In recent years the long-held view of absolute chemical inertness has been upset by demonstrations that some of these elements could form compounds under special conditions. Nevertheless, the fact remains that these elements are stable and virtually inert under ordinary conditions. A metallic element such as iron, provides a more extreme contrast with the inert elements than do the nonmetallic ones. Note that in iron M_3 remains unfilled but that N_1 is occupied. At a later point in the book the distinctive role of the iron atom in biological systems will be considered.

Equipped with a theory of atomic structure, we may now rationalize the chemical behavior of the elements. The guiding principle is that under favorable conditions elements will undergo those interactions that will result in an electron pattern resembling one of the inert gases. An element may achieve this by either losing or gaining electrons. Since electrons do not normally exist outside of the atoms, the gain of an electron by one atom requires the loss of an electron by another. Whenever such a complete transfer occurs, the electrical neutrality of the atoms is destroyed and the atoms become positively or negatively charged. Oppositely charged atoms are always closely associated because the forces of electrical attraction are overwhelming; no one has yet physically separated positive and negative ions into two groups and maintained them as such. Sodium and chlorine have been included in Table 4-1 to illustrate this type of interaction. Sodium is similar to the inert gas neon in its electron pattern except for the presence of one additional electron in the *M* shell. To acquire a pattern identical with neon, the outer electron must be removed; a similar argument would apply to other elements such as potassium, rubidium, and cesium, all of which commonly occur as monovalent cations. The opposite argument would apply to an element such as chlorine which requires one electron to fill its outer orbit to give it the configuration of argon. Atomic theory accordingly explains the molecular structure of a large variety of salts. The bonds holding the atoms together are called "electrovalent bonds." Such bonds are characteristically formed between electropositive atoms (those with a marked tendency to lose electrons) and electronegative atoms (those with a similar tendency to acquire electrons). A dry crystal of salt is thus a balanced assembly of positive and negative ions; when dissolved in water, the ions, as Faraday discovered, are free to migrate. Now the question arises why the strong electrical forces holding the ions together in the dry crystal do not continue to do so when the salt is dissolved.

Dissociation of electrolytes is generally explained on the basis of water molecules getting between the associated ions. For reasons to be considered later, water molecules are assumed to behave as though they were slightly positive in charge at the hydrogen end and slightly negative at the oxygen end. The positive end would be attracted to anions, the negative end to cations. Individual ions are therefore pictured as being surrounded by a shell of water molecules which acts as a barrier to close contact and thus reduces the force of electrical attraction. This picture is, in fact, essential to explain certain properties of ions which were not discussed in our general consideration of electrolytes. Sodium ions, for example, diffuse more slowly in water than do potassium ions, even though the latter have a higher atomic weight and would be expected to be bigger and hence encounter greater frictional resistance to movement. But since the sodium ion has only 3 subshells with a total of 10 electrons, and the potassium ion has 5 subshells with a total of 18 electrons, it follows that the single positive charge each carries must be more concentrated in the smaller volume of the sodium ion. A more concentrated charge, because it is stronger, would be expected to attract a larger number of water molecules. It is actually possible to measure the radii of ions both in the solid and dissolved states. In solids the ions of lithium, sodium, and potassium have radii of 0.78 and 0.98, and 1.33Å respectively, and these parallel their increasing atomic weight. In solution, however, the radii of the hydrated ions are 3.66, 2.81, and 1.88Å, an order which parallels their decreasing charge density. A comparison of the radii of solid and dissolved lithium ions illustrates forcibly the extent to which water molecules may cluster about a single charged atom. Although the thickness of hydration shells varies widely with intensity of atomic charge, it is generally true that water markedly decreases the effectiveness of electrovalent bonds. Thus in living systems, where water is a major component, the electrovalent bonds cannot be expected to maintain the intactness of molecules.

The transfer of electrons is only one means by which atoms can achieve stable electron configurations. An alternative is for two interacting atoms to share their electrons. Oxygen, for example, lacks two electrons in its *L* shell, whereas carbon may be regarded as having an excess or a lack of four. One carbon and two oxygen atoms can so arrange themselves that at one instant the carbon atom lends two electrons to each of the oxygen atoms, and at another the two oxygen atoms lend a total of four to the carbon atom. Such an alternation of electron positions is presumed to represent the basic feature of electron sharing. With refinements in atomic theory all sharing phenomena are pictured as involving pairs of electrons. No matter how many electrons are involved in a chemical bond, each electron of one element must pair with one of the second element. In a perfect bond of this type we picture the electron pair as orbiting with equal frequency about each of the atoms involved; the net result is that neither atom acquires a charge but that the stability acquired through this alternation in electron pair position gives the bond between the atoms its known strength. Bonds thus formed are called "covalent."

The concept of the covalent bond underlies our belief that compounds isolated from living cells consist of the molecules originally present in the intact system. To be sure, all covalent bonds are not equally strong. Some, because of their weakness, are easily degraded, but even their disruption may be avoided by isolating the compounds concerned under suitable conditions. Indeed, successful biochemistry is frequently the art of discovering such suitable conditions.

We have thus far described two ways by which atoms could alter their electron configurations so as to resemble those of the inert gases. One way is by the addition or removal of electrons; elements in this category are described as electronegative or electropositive according to whether they are stabilized by losing or gaining electrons. The second way is a sharing of electrons in which the atoms involved remained neutral. Is there any way of determining which behavior a particular element will follow? The answer to this question is not entirely simple. The electron patterns of sodium and chlorine may be converted to stable types by the addition or subtraction of a single electron. An atom like carbon, on the other hand, could just as easily achieve the configuration of helium by losing four electrons as it could achieve that of neon by gaining four. The forces acting either way are equal, and the net result is that carbon oscillates equally between the two conditions by electron sharing. Such an atom is classified as a neutral atom, and carbon is indeed the most neutral of all the elements composing living matter. In general, the smaller the number of electrons which an element needs to lose or gain in order to achieve the inert gas configuration, the greater will its tendency be to form electrovalent bonds. Where the number to be gained is smaller than the number to be lost, the element tends to be electronegative and vice versa for the electropositive tendency. On the other hand, where the number to be gained or lost is equal, the element tends to be neutral. Thus two other major elements of organic compounds, oxygen and nitrogen, are electronegative because they can more easily achieve stability by gaining electrons. Since oxygen requires two electrons whereas nitrogen requires three, and since the atoms are similar in size, oxygen is the more electronegative element. Oxygen is, in fact, the most electronegative element found in organic matter.

If we had no knowledge of the actual composition of molecules, we might suppose that the only possible combinations are either between electropositive and electronegative elements or between complementary neutral atoms. But we know this is not the case. How do we characterize bonding between an electronegative atom and a neutral one? We do so by a compromise. Whenever atoms having both covalent and electrovalent tendencies are joined, we envisage the bond as being of a mixed character, partly one and partly the other. The proportions of each are presumed to depend on the relative magnitudes of the forces tending toward a particular bond type. A strongly neutral atom such as carbon will largely, though not entirely, counteract the electrovalent tendencies of the strongly electronegative oxygen. This description, however, does not clarify the exact form of such mixed bonding. We could very simply picture the mixed bond as representing a mixture of bonds, some covalent, others electrovalent. A more useful picture is of a bond in which the shared electrons favor one atom over another. If the mixed bond is only slightly electrovalent, we imagine that the shared electrons spend only slightly more time in the orbit of one of the atoms than in that of the other. Thus whenever oxygen is in chemical combination, it will tend to have a greater or lesser excess of electrons surrounding its nucleus. This is the basis for the earlier statement that water molecules have an asymmetric charge distribution, the oxygen end of the molecule being negative. The correlation between the polar character of organic compounds (i.e., their solubility in water) and their relative content of oxygen atoms is explained by the tendency of oxygen regions to be negative and hence to attract the positive end of water molecules.

The hydrogen atom is unique among the elements in that either the gain or loss of a single electron renders it internally stable; in gaining an electron it achieves the configuration of helium; in losing an electron it becomes the smallest of atomic structures, a "proton" without any surrounding electrons. In association with carbon, hydrogen thus forms covalent linkages; in association with oxygen, on the other hand, the bonds are of mixed character. Taken together, the four principal elements of living matter which we have discussed encompass a wide range of possibilities. At the center is the neutral atom carbon, on the electronegative side are oxygen and, to a lesser degree, nitrogen; on the electropositive side is hydrogen. Although hydrogen can also behave as an electronegative element because of its unique structure, it is always electropositive in combination with oxygen or nitrogen.

THE CONCEPT OF LARGE MOLECULES

Virtually nothing of these theories was known to people like Liebig who were busy determining the molecular constitution of living matter during the middle of the nineteenth century. They recognized that molecules represented stable arrangements of atoms and therefore sought to analyze the kinds of molecules which made up the stuff of life. A highly sophisticated theory of the atom would have been of little help at the time. So essential did it then seem to determine accurately the composition and purity of substances that Baron Justus von Liebig inscribed over the door of his laboratory "God has ordered all His creation by weight and measure." Not until 1838 did the word "protein" appear in biological literature when a Dutch chemist, G. J. Mulder, applied it to a substance he had isolated and to which he gave the empirical formula $C_{40}H_{62}N_{10}O_{12}$. The compound was, in fact, a mixture of things, but so overwhelming was the evidence accumulated over the years that living matter contained distinctive nitrogen compounds, that the term protein never dropped out of use. By 1862 the mess of data was sufficiently tidied, especially through the efforts of Liebig, that the pathologist Rudolf Wagner included in his *Handbook of General Pathology* a statement that living matter consisted of an enormously complex mixture of three types of substances: carbohydrates, lipoids (fats), and proteins. It would be hard to find a more primitive statement about the chemical composition of living systems; yet in a little more than the span of a single lifetime, biologists were able not only to separate these "enormously complex mixtures" but also to provide a great deal of information as to how they contribute to the vital processes of cells.

Anyone who examines a list of natural organic compounds would be staggered by its diversity, though the list would very likely be incomplete. If we knew no more than Liebig and his contemporaries about the function of naturally occurring molecules, we would probably continue to classify organic compounds into carbohydrates, lipoids, and proteins. That we do not do so is due to those who sought to characterize not only the chemistry of organic compounds but also their function. Although the living world is an almost inexhaustible source of novel compounds, the principal kinds of molecules which are basic to the functions of cells are few in number. The newcomer to biology may be cheered to know that, even though the details of living processes require familiarity with very many kinds of compounds, the principles of living processes require familiarity with only a few.

It might now seem appropriate to list at once the "key" cellular components. This is not so. Before biologists could confidently speak of key compounds two streams of study had to merge, each of which was concerned with a distinctive aspect of the molecular constitution of cells. One of these we have referred to briefly—the painstaking and slow effort to determine the atomic composition of organic molecules. The other, which was of a far more general nature, addressed itself to the question of the size of cellular molecules. Like so many of the questions concerning the chemistry and physics of the cell, this was not preconceived. No one purposefully set out to determine the molecular weights of cell components as though such measurements had spectacular information in store. Molecular size happened to become a topic of interest because some chemists were applying the arts of their profession to the utterly strange materials of the living world.

In 1858 Faraday reported an interesting experiment. He dissolved some phosphorus in carbon disulfide and added it to a solution of gold chloride. When the mixture was shaken, the gold solution turned red. Few experimenters would have noted anything more. But Faraday was the keenest of observers and additionally noted that this red solution of gold was not a true solution by accepted criteria of physical chemistry. When a beam of light was passed through the red solution, the solution appeared turbid. Colored solutions do not ordinarily appear turbid when placed in a beam of light because molecules are generally too small to deflect light rays. Faraday therefore explained the turbidity of the "solution" by postulating that it consisted of particles large enough to scatter the incident light. The idea that molecules could aggregate into relatively large particles was new to physical chemistry.

Faraday did not devote his talents to the phenomenon, although his student John Tyndall contributed greatly to its elucidation. The principal figure in subsequent studies of molecular aggregation was the chemist Thomas Graham. Like many of his contemporaries, Graham was interested in the properties of solutions. But he did not devote his time to inorganic substances. Instead, he began to experiment with rather poorly defined organic compounds such as gelatin (a protein obtained by boiling the hooves of horses) albumin (egg white protein), starch, and some highly complex inorganic substances like silicic acid. He paid little attention to the chemical class of compound studied, for what he observed was so unusual that chemical composition appeared to be largely irrelevant. His suspicions were aroused when he performed what would have otherwise been an orthodox experiment—a determination of the rates of diffusion of the substances listed above. The rates proved to be extremely low, and the only plausible explanation for such behavior was that these substances either had exceedingly large molecular weights or were molecular aggregates such as Faraday had observed in his red solutions of gold chloride. Graham enthusiastically chose the latter explanation. He pointed out that since none of the organic substances which he tested could be crystallized, they could not have existed as free molecules in solution. The more he studied these substances, the more he become convinced that they were molecules uniquely aggregated into supermolecular particles. To crown this discovery he put forward the idea that matter existed in three states of aggregation—atomic, molecular, and "colloidal" (glue-like). He also introduced a second term "crystalloidal" to cover the many substances ordinarily studied which could be crystallized, and then proceeded to argue that colloids were the basic units of living substance. His views gained very many followers because the behavior of substances like albumin or gelatin had many parallels in the behavior of living cells. Cytoplasm, especially in its outer zone, is jelly-like; and colloids gelled under appropriate conditions. Acids and salts precipitated colloids and also coagulated the cytoplasm. Colloidal gels could be stained in much the same way as cells. As time went on more and more parallels were found; for one group of biologists the colloidal state was a fact, and as recently as the 1920's textbooks dealing with cell physiology gave colloidal theory major attention.

On one point no argument was possible. Substances such as albumin and gelatin isolated from living material did behave in solution as if they were particles of a size much larger than organic compounds like sucrose. But the assumption that such large particles were aggregates of molecules was unwarranted because the molecular weights of the constituents of colloids had not been determined. The decisive answer to the question of molecular weights did not, however, originate in studies of organic chemistry. That same discipline, physical chemistry, which had served Graham so well was put to different use by Theodor Svedberg at the very time that colloidal theory was at the height of popularity among cell physiologists. Svedberg's effectiveness in solving the problem lay in his development of an entirely new instrument—the "ultracentrifuge." This instrument could expose solutions to very high centrifugal forces, and under these conditions solute molecules would sediment at rates governed by factors previously discussed in connection with diffusion—molecular size and shape being the principal ones. The ultracentrifuge operated on the principle of gravitational force, rather than concentration difference, to generate diffusion. Thus ultracentrifuge experiments made possible a much more accurate determination of molecular size as inferred from the rate at which a solute molecule moved from the centripetal to the centrifugal end of a tube. By analyzing the sedimentation of different proteins during centrifugation, Svedberg was able to establish that such proteins had fixed and characteristic molecular weights. The discovery did not at once clear the air, but this technique of analysis, which permitted the chemist to determine whether a protein preparation contained one or several sizes of molecules, laid the groundwork for a variety of chemical studies.

Today we know that 90–95% of cellular material is made up of large molecules. And 70% of cellular substance is protein, and although a few proteins have unusually low molecular weights, of the order of

3000–5000, the average molecular weight is about 60,000 and some reach over 1,000,000. Comparing these molecular weights with those of the organic compounds studied by chemists during the nineteenth century (of the order of 100–300), it is not surprising that Graham turned to colloids rather than large molecules as an explanation of the physical behavior of the carbohydrates and proteins. Even today we are not entirely accustomed to large sizes for molecules. In the past few years many biologists were stunned to learn that the chromosomal molecules of viruses had molecular weights of about 120,000,000. But if molecules reach that size in a tiny virus, what are their dimensions in higher organisms? A molecular weight of one billion now sounds fantastic, but perhaps in the near future it will seem as commonplace as the one million value for some proteins.

What are the implications of large molecular size? The safest answer is that if cells are built out of large molecules, such molecules must have properties which are denied to small ones. This, to be sure, is not a satisfactory answer, but it does serve to emphasize the point that in studying how cells behave we are also compelled to study how large molecules behave. Just as the properties of large molecules must be related to their constituent atoms, so must the properties of cells be related to their large molecules. To anyone aware of the complexities of cell organization this represents a formidable challenge. On the other hand, the challenge to the cell biologist is clear, "No matter what phenomenon you observe in studying cells, you must eventually explain it in terms of the properties of one or more kinds of large molecules; if more than one kind of molecule is responsible for a particular phenomenon, then you must be able to show how the molecules interact to produce the phenomenon." At the turn of the century this point of view was already implicit, and occasionally explicit, in the studies of gifted biologists. But not until the chemical nature of large molecules was clarified could this approach exercise a full sweep in cellular studies.

CARBOHYDRATES

The chemists who tackled the problem of defining the chemical constitution of large molecules did not do so in any order of biological priority. They had no information about vital activities which would equip them to assign such an order. More important, however, was their need for pure compounds, and they naturally preferred to analyze those substances which were available in pure form. The isolation and purification of large molecules was an art slowly acquired; indeed, only in the past decade have biochemists devised satisfactory tools to accomplish the task.

Carbohydrates were an attractive target for chemists of the nineteenth century. Crystalline sucrose was already being prepared from the sugar cane plant in 300 A.D. In 1800 sugar refineries were common throughout Europe. Organic chemists thus had a pure substance which could serve as a chemical model of carbohydrates. More than that, two other carbohydrates of larger molecular weight were available in fairly pure form: cellulose from cotton, and starch from the many plants which stored it in the form of large granules. Elemental analysis of these and other substances showed them all to have a uniform composition. For each atom of carbon there were two atoms of hydrogen and one of oxygen. The basic formula was thus CH_2O, and since at the time this was discovered nothing was known about the way in which these atoms fitted together, the substances were called carbohydrates to indicate compounds of carbon and water.

A major advance in the understanding of carbohydrates was made by Gustav Kirchhoff in 1811 when he boiled starch in acid and isolated a crystalline sugar as a product of the reaction. This technique should be considered briefly because it illustrates a chemical principle widely used in determining the structures of large molecules. In the presence of acid certain covalent bonds in the starch molecule are weaker than others; otherwise Kirchhoff would not have isolated a crystalline product. If all the bonds were of equal strength, they would either not have disrupted at all, or they would have broken randomly until nothing but the free elements remained. The fact that he isolated a crystalline product meant that the starch molecule was composed of much smaller molecular units, all of the same kind, and all tied to one another by a bond which was significantly weaker in acid than any of the bonds linking the atoms within the small molecular units. In this respect the behavior of starch is paralleled by large molecules other than carbohydrates which under proper conditions break up into molecular subunits. Thus instead of a mixture of elements, the organic chemist has available the intact units which comprise the large molecule.

Many chemists adopted the technique of heating carbohydrates in acid to determine their composition, and by 1840 it was clearly established that starch, cellulose, and grape juice all contained the same sugar, glucose. Once techniques were elaborated to determine how the hydrogen and oxygen atoms were arranged with respect to carbon, carbohydrate chemistry was on sure footing. Before the end of the nineteenth century, one of the most outstanding organic chemists,

Emil Fischer, had succeeded in synthesizing the two principal natural sugars, glucose and fructose. By then the distinctive chemical feature of carbohydrates was recognized as a chain of carbon atoms flanked by hydroxyl (OH) groups and hydrogen atoms. These groups were relatively inert chemically, and only the first or second carbon atom of the known sugars had different groupings which rendered them more susceptible to chemical reactions. The sugars were represented as straight chains, but ultimately it was realized that they generally were arranged in a ring structure (Fig. 4-1).

Exact methods introduced into the analysis of high molecular weight carbohydrates demonstrated that for every sugar molecule set free, one molecule of water was consumed. The breakdown of carbohydrates in this way became known as "hydrolysis," signifying the addition of water; as will be seen, the same process occurs with the other large molecules of the cell. The addition of water on breakdown implies that whenever these units are put together, one molecule of water must be

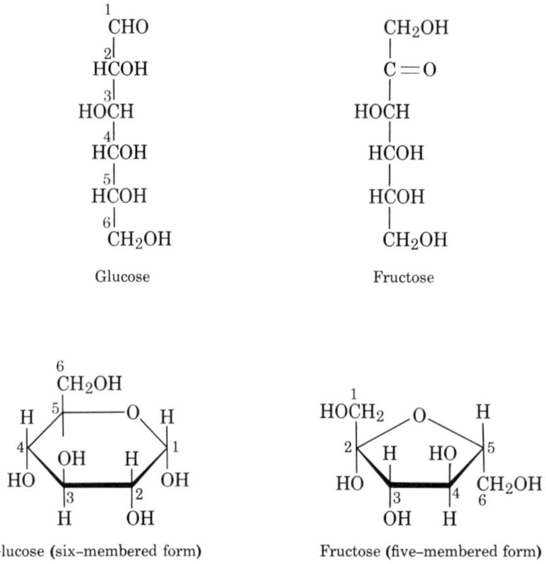

Fig. 4-1. Structural formulas of two common monosaccharides. In aqueous solutions 99.9% of the sugars are in the ring form. Actually, the six-membered ring permits two isomeric forms which are not possible in the straight chain compound. For glucose, the position of the hydroxyl in C_1 may lie either above or below the plane of the ring; for fructose, the hydroxyl on C_2 may lie in either plane. The monosaccharides can also form five-membered rings, and such a ring form is illustrated for fructose. Generally, free hexoses do not form five-membered rings because they are unstable. However, when in combination with one another, this may occur. In sucrose, for example, the fructose is in the five-membered form.

removed from each pair of adjoining molecules. We could suppose, of course, that in the cell starch is not formed by linking glucose molecules. But this is not the case; large molecules are invariably formed from the same molecular subunits that appear as a consequence of hydrolysis (Fig. 4-2).

Not all sugars consist of six carbon atoms, nor do all polysaccharides (large molecules of many sugars) consist of glucose units. A wide variety of carbohydrates is known, and some have a few nitrogen atoms included. Generally, however, carbohydrates are accessories of cell organization rather than vital components. Within cells polysaccharides are generally present as food storage forms (starch, glycogen), but their most prominent location is along the exterior of cells where in various

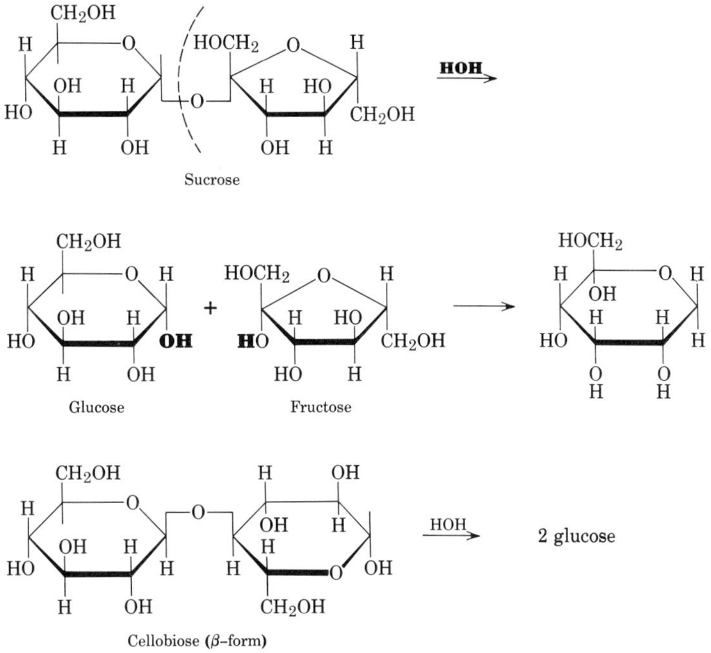

Fig. 4-2. Two types of disaccharides and their products of hydrolysis. A common characteristic of the breakdown of larger organic molecules into their constituent subunits is the uptake of one molecule of water for each inter-unit bond split. The reverse process, linking of subunits, must therefore involve the elimination of one molecule of water. This fact is of cardinal importance in cellular synthesis of organic molecules, and will be considered at length in Part II. Note that the fructose moiety of sucrose exists as a five-membered ring in the disaccharide, but changes to the six-membered form after splitting from the glucose molecule. The disaccharide cellobiose is itself a subunit of cellulose from which it may be obtained after partial hydrolysis.

combinations they act chiefly, though not entirely, as a protective layer for the more delicate cytoplasm. The cellulose content of the walls of plant cells is common knowledge. Carbohydrates have other special functions, but the basic structural and functional properties of cells can be described with only secondary references to their large carbohydrate molecules.

PROTEINS

When Liebig introduced proteins as a major category of organic substances, he was well aware that all living forms possess nitrogenous compounds in abundance. Indeed, all biologists of his time who believed in a physicochemical interpretation of life were agreed that proteins must play a major role in vital behavior. But organic chemistry was just at its beginning, and experimental procedures for extracting, purifying, and separating large protein molecules were well beyond the reach of the times. Fortunately, nature provided certain "sample" proteins; egg white (albumin) and gelatin. These could be analyzed in the same way as the carbohydrates, and organic chemists of the nineteenth century lost no time in doing this. Their work had none of the spectacular qualities which colloidal studies had at the time. In retrospect, however, we can see that the painstaking analyses, followed by equally painstaking searches for purification procedures, had the more durable value. For today we no longer question the idea that the basic unit of protein is a molecule, and we take it as a matter of course that each kind of protein can be purified, crystallized, and examined for its specific properties.

Proteins, like polysaccharides, yield molecular subunits on hydrolysis. Unlike most polysaccharides, the subunits consist of many kinds of molecules. By the close of the century Emil Fischer had clarified exactly how these subunits were linked together in the intact protein molecule. The principal fact which Fischer had at his disposal was that all the subunits of protein had one feature in common, though they differed in most other respects. This feature is best seen in the structural formulas of the subunits; the subunits are called amino acids because each contains an "amino group" ($-NH_2$) and an acidic "carboxyl group" ($-COOH$) (Fig. 4-3). Fischer concentrated his attention on the last two carbon atoms of the amino acids, all of which had the structure $NH_2-\underset{|}{C}H-COOH$. He reasoned that since all the amino acids were released in hydrolysis they must be tied together in a similar way. The most obvious way in which this could occur was by implicat-

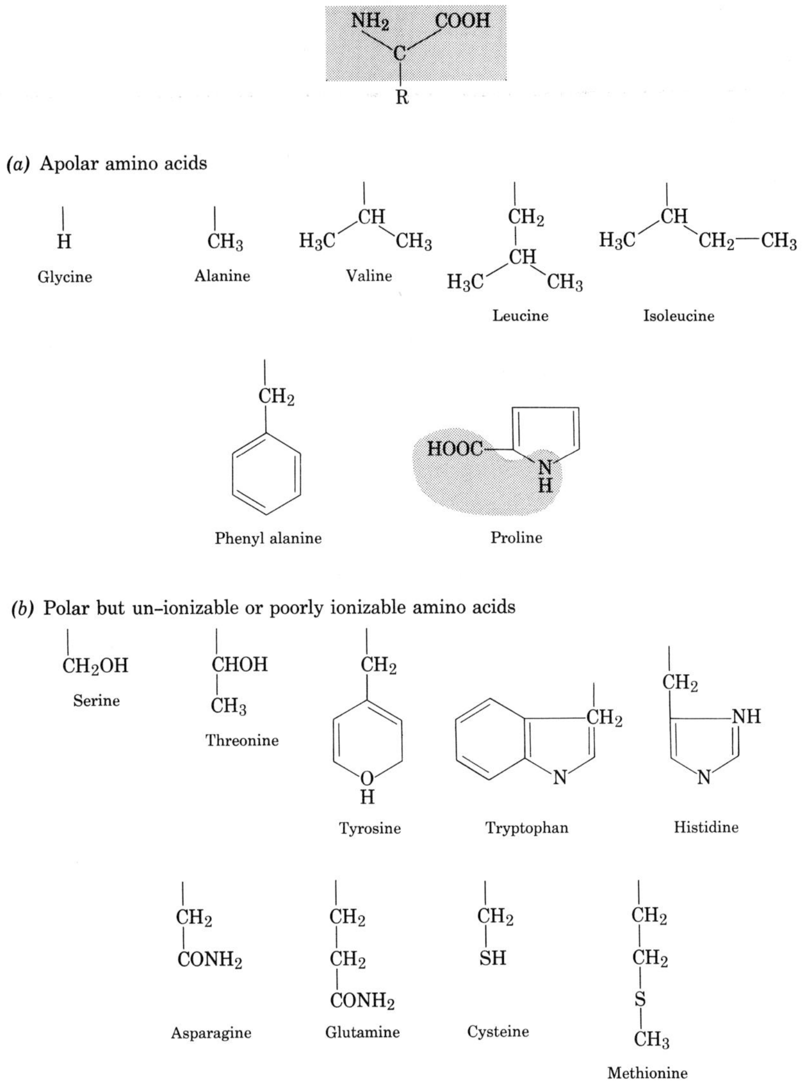

Fig. 4-3. Amino acids. The nature of their R groups in relation to water.

(*c*) Basic amino acids

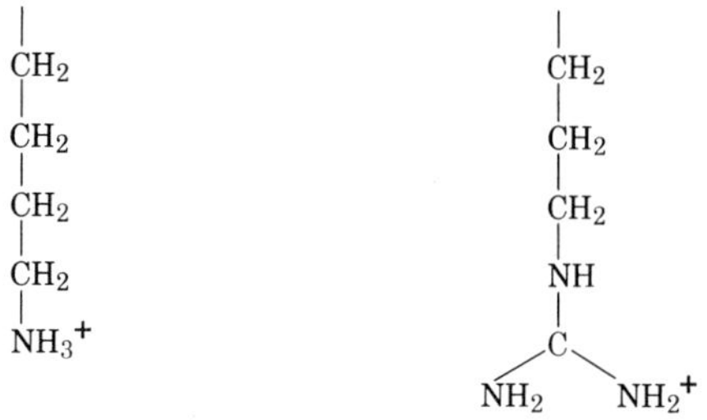

Lysine Arginine

(*d*) Acidic amino acids

Aspartic Glutamic

ing only that end of the molecule which was common to all the amino acids. He therefore pictured amino acids linking in the following way (Fig. 4-4). Fischer called such bonds "peptides," and to prove that his hypothesis was correct he synthesized small chains of amino acids linked to one another by peptide bonds. His interpretations of protein structure were not universally accepted at the time, but gradually criticism thinned and today we accept the fact that proteins are long chains of amino acids linked to one another by peptide bonds.

It is also universally accepted that the chains are unbranched and that despite the extreme diversity of protein types, the twenty amino acids here listed are the nearly exclusive building blocks. This latter fact is noteworthy, for just as the early organic chemists were amazed that living substance should be formed almost entirely from four elements, so should we be amazed that the complexity of living forms is invested primarily in twenty small molecules. Almost any difference between cells, even within the same organism, can be traced to a difference in protein. Every facet of growth is a consequence of protein changes. With the exception of hereditary transmission, every activity of a cell has its roots in one or more proteins. How does this assembly of amino acids come to be a primary substance of life? The question is a deep one and in many respects it is still unanswered, but in just as many respects we now have precise knowledge of what proteins do. This knowledge will be gradually unfolded, but for background purposes we should

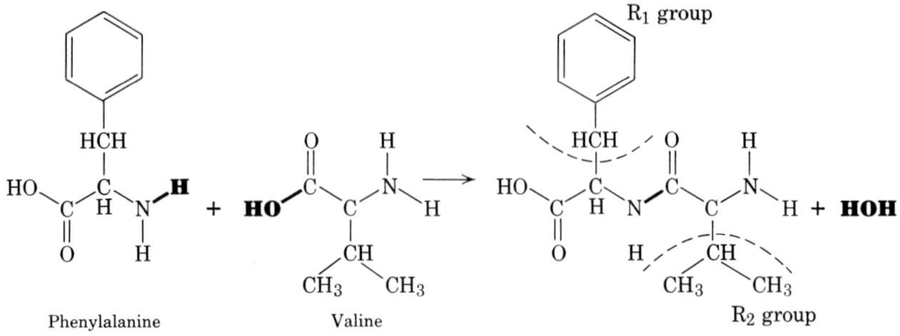

(a) Peptide linkage

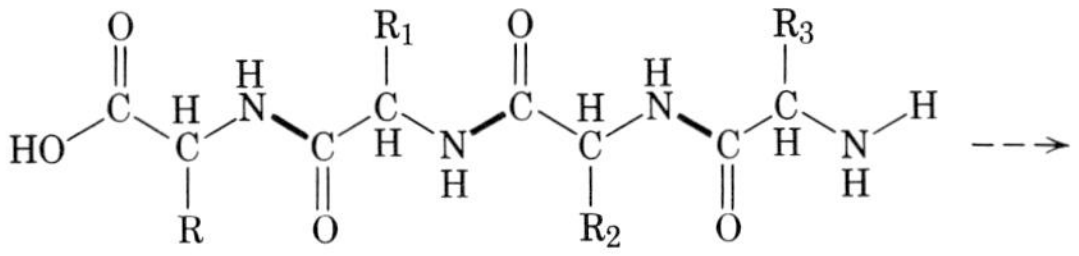

(b) Polypeptide chain

Fig. 4-4. Protein structures.

discuss briefly some primary physicochemical implications of long polypeptide chains.

Properties of Individual Amino Acids in Relation to the Protein Molecule

Once proteins were recognized as a series of amino acids linked to one another by peptide bonds, drawing a structural diagram of the

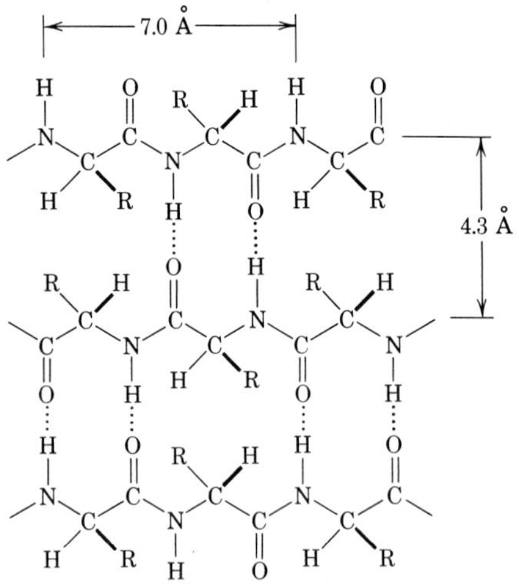

(c) The configuration of polypeptide chain

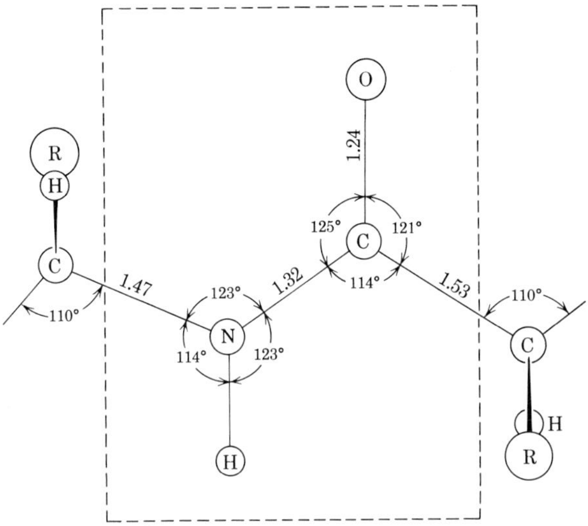

(*d*) Fundamental dimensions of an extended polypeptide chain

polypeptide chain was a relatively simple matter. Since cells are built from small molecules, their characteristic structural orderliness has its origin in the orderliness with which these small molecules are fitted into large ones. That largeness *per se* is not the primary attribute may be seen by contrasting a protein with a starch molecule. In the latter there is only one kind of subunit, glucose; however long the starch molecule becomes it is just more of the same thing. The principal feature of the starch molecule is that glucose molecules are held in a bundle; proteins have this same feature, but it is a minor one compared to the orderly patterns in which these bundles are arranged. This contrast, it should be pointed out, has become fully evident only within the past ten years. Until then our knowledge of the structure of protein molecules rested chiefly on the chemical studies of Fisher which led to the establishment of two chemical characteristics: the identity of the individual amino acids and the peptide linkage.

Some ten years ago the biochemist F. Sanger revolutionized the study of proteins by devising a technique whereby he could determine the *order* in which amino acids were tied together. Success of this technique was made possible by the fact that each protein molecule consisted of a single unbranched polypeptide chain which therefore could have only one free amino group (attached to the α-carbon atom of the amino acid) at one end of the protein molecule, and one free carboxyl (similarly attached) at the other. If a method were available by which amino acids could be sequentially removed from a polypeptide chain, this chemical fact would be of minor importance for analysis. However, hydrolysis of proteins is on the whole a random process irrespective of the reagent used. Acid hydrolysis, for example, yields free amino acids if the reaction is allowed to proceed long enough; if the time is shortened, a mixture of variously sized fragments results. If biological catalysts (See Chapter 5 for "enzymes") are used, some specificity is achieved by way of preferentially cleaving peptide bonds between certain amino acids. This method is generally used when incomplete hydrolysis is desired. Neither method, however, can provide adequate information about sequence.

Sanger's achievement was in the use of a colored reagent (dinitrofluorobenzene) which could interact specifically with the terminal amino group. The bond thus formed is stable under conditions which hydrolyze peptide bonds. An N-terminal amino acid may therefore be identified after the protein has been hydrolyzed and the products separated.

If the colored product contains a single amino acid, identification of the terminal amino acid is direct. If the product contains several amino acids, it is further hydrolyzed after separation from the other polypeptides. Given a variety of colored polypeptide fragments (containing 2, 3, 4, . . . , amino acids), information is obtained not only about the terminal amino acid but also about the sequence of amino acids attached to it. In practice, this procedure cannot be carried out until the entire sequence is elucidated. Instead, the colorless polypeptide fragments obtained in the first hydrolysis are isolated and treated with the reagent to determine their internal sequences. Since all the protein molecules undergoing hydrolysis do not cleave at identical points in the chain, the polypeptide fragments obtained are frequently overlapping. If a sufficient number of analyses are made, these overlapping regions provide clues to the positions of the polypeptides within the original chain. The larger the size of the protein molecule, the greater the number of kinds of polypeptide fragments, and the more arduous is the analysis. Sanger and his colleagues began with the insulin molecule because it is relatively small. Very large protein molecules are yet to be tackled. Nevertheless, strong evidence now exists that differences between protein molecules arise from differences in amino acid sequence within the polypeptide chain. A substitution of one amino acid in a chain of a thousand may be sufficient to produce a distinctive change in the character of the protein. In a sense this discovery brings to a conclusion the task set by Liebig one hundred years ago—to determine the chemical nature of proteins.

In saying that orderly arrangement is a primary characteristic of the polypeptide chain, we imply that the individual amino acids are sufficiently different from one another to display distinctive chemical properties. That there are marked differences between them is apparent from the structural formula shown in Fig. 4-3. Present knowledge of protein behavior leads us to conclude that even those amino acids which have very similar structures have significantly different properties. We would be going far afield, however, if we attempted to earmark the distinctive properties of each amino acid. As an introduction we may simply group the amino acids according to the salient chemical properties. In doing this, the common ends involved in peptide bond formation may be ignored; our attention should be focused on those parts designated R in the diagram of the polypeptide chain, and extending at right angles from it.

The way in which the amino acids are classified depends on the frames of reference. Were we interested in the geometry of the protein chain, we would classify the R groups according to their length and also according to whether or not they were branched. Our primary interest, however, must be in terms of their interactions with water, for although protein constitutes over 70% of the dry weight of cells, water constitutes about 80% of their total weight. No matter how we approach the organization of living material we must account for the arrangement of large molecules in a medium of water. The R groups subtended along the length of the polypeptide chain are surrounded by many times their number of water molecules. A simple calculation will show this. In terms of mass a cell contains five to six times as much water as protein material. But the average amino acid molecule weighs about six times as much as water, and this means that there are approximately 30–35 times as many water molecules in a cell as there are amino acid R groups. To be sure, not all the water in a cell is adjacent to protein, but most of it is, for apart from the special inclusions of a cell (e.g., vacuoles, fat droplets, starch grains, glycogen granules), no sizable region of the cell lacks protein. Thus whatever the R groups of the amino acids may do, their behavior is bound to be influenced by their interactions with water.

In discussing the properties of solutions, we made a broad distinction between substances which were soluble in water (polar) and those which were soluble in fats (apolar). Further along, in discussing the nature of molecules, we pointed out that these wide differences in solubility could be traced to the kinds of atomic groupings present within each molecule. The general principle applied to carbon-containing compounds was that if the bonds formed between carbon and adjacent elements were entirely covalent, the resultant atomic groupings would interact little with water molecules. This relationship was explained by supposing that water molecules have an asymmetric distribution of electrons such that the region close to the oxygen atom is negative whereas that close to the hydrogen atom is positive. On this basis one end or the other of a water molecule would be attracted to those regions of an organic molecule where the constituent atoms were not completely neutral. Furthermore, the electron configurations of carbon and hydrogen are such that they can achieve maximum stability by sharing electrons and hence remaining internally neutral; in contrast, the configurations of nitrogen and oxygen are such that their strong tendencies to attract electrons result in locally charged regions.

Certain of the amino acid R groups are apolar in character. Glycine, alanine, valine, leucine, and isoleucine have the carbons of their R groups saturated with hydrogens. Phenylalanine is in the same category since it contains an apolar benzene ring. Thus whenever these amino acids occur in a protein chain, they will show a marked tendency to interact with other apolar groups and exclude water. Proline, which should have been listed among the apolar types, deserves special mention. It is the only amino acid in which the nitrogen atom at the "common" end of the molecule is bound to the tail end of the R group. Whenever proline occurs, a bend must exist in the polypeptide chain; the use of x-rays in recent years to "photograph" the positions of atoms within a protein molecule has shown this to be the case. It is not, however, for the sake of discussing protein shapes that we have made this point, but only to emphasize the important consequence which the position of a single amino acid may have.

All other amino acid R groups have some degree of polarity, and as with polar substances in general it is convenient to distinguish between them on the basis of their ionizability in water at neutral pH, which is the approximate pH of most cells. Serine and threonine acquire polar characteristics by virtue of a hydroxyl group which, like the hydroxyl groups of carbohydrates, does not dissociate. The hydroxyl group of tyrosine is virtually undissociated at normal intracellular pHs. The nitrogen atom contributes polarity to tryptophan and histidine, although no significant ionization occurs. Cysteine and methionine are the only amino acids containing sulfur, an element that was overlooked in early studies of cell chemistry. Like oxygen and nitrogen, sulfur is an electronegative element, but this property is secondary to the singular chemical role which the sulfur atom of cysteine plays. The —SH ends of two cysteine molecules often combine with one another by eliminating hydrogens and forming a covalent —SS— linkage. Such linkages, frequently found among proteins, stand out because the R group of cysteine appears to be the principal if not the only one among all the amino acids which forms covalent bonds between members of a protein chain. Thus proteins which have many cysteines can form many covalent cross links with one another. Indeed, in many cases the gel-like consistency of cytoplasmic regions is due to these cross links, which have the effect of creating a molecular net.

Physicochemical Properties of Proteins

Of all the physical and chemical properties of proteins, those arising from the ionizable R groups were recognized earliest. Most colloidal experiments were unknowingly directed at these groups alone. Even today most schemes of protein separation and purification take advantage of the fact that each species of protein has a distinctive arrangement of ionizable amino acids. These are of two kinds: those bearing carboxyl groups —COOH, namely, glutamic and aspartic; and those bearing amino groups —NH_2, namely, lysine and arginine. Whenever the R groups of glutamic and aspartic acid dissociate, the protein becomes negatively charged in that region. The —NH_2 groups work in the opposite direction; because of the electronegativity of N these groups combine with H^+ in the presence of water. Both types of reactions are pictured in Table 4-2.

The simplest way to treat the behavior of these ionizable groups is according to Arrhenius' dissociation theory. Although the performance of each particular group within a polypeptide chain cannot be followed, we may presume that the dissociation characteristics of R groups in the protein molecule are similar to those in the free amino acids. Following the same principles applied to solutions of electrolytes, we can predict that the lower the pH the fewer will be the number of dissociated carboxyl groups and the larger the number of charged amino groups; the opposite would occur at high pHs. The proportion of ionized R groups in a solution of amino acids at a given pH may be calculated if the dissociation constants are known. This has been done in Table 4-2. Assuming that the dissociation constants of the R groups of free amino acids also apply to amino acids within a polypeptide chain, the total and net charges carried by a protein at a certain pH may be determined if the composition of the protein is known. Since, as may be seen from the table, neither tyrosine, cysteine, nor histidine is significantly ionized at pH 7, one may infer that within

Table 4-2 Dissociation of Amino Acid R Groups at pH 7.0

Amino acid dissociation form of R group	p*K*	$\text{Log}\left(\frac{\text{dissoc.}}{\text{undissoc.}}\right)$	$\frac{\text{Charged R Groups}}{\text{Uncharged R Groups}}$
Tyrosine 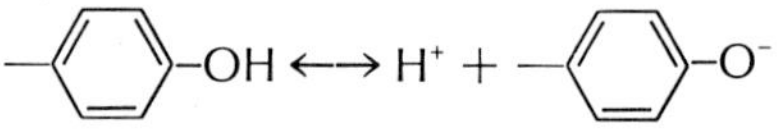	10.97	−3.97	$10^{-3.97}$
Cysteine $-SH \rightleftharpoons H^+ + -S^-$	10.28	−3.28	$10^{-3.28}$
Aspartic $-COOH \rightleftharpoons H^+ + COO^-$	3.65	+3.35	$10^{3.35}$
Glutamic $-COOH \rightleftharpoons H^+ + COO^-$	4.25	+2.75	$10^{2.75}$
Histidine 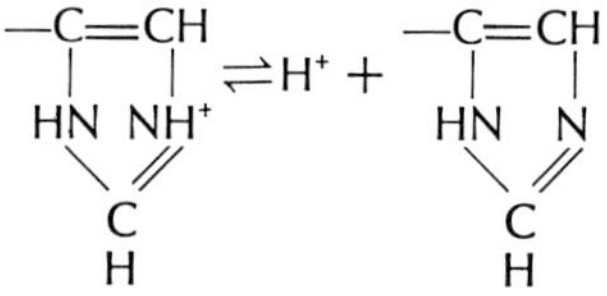	6.1	+3.9	$10^{-3.9}$
Arginine 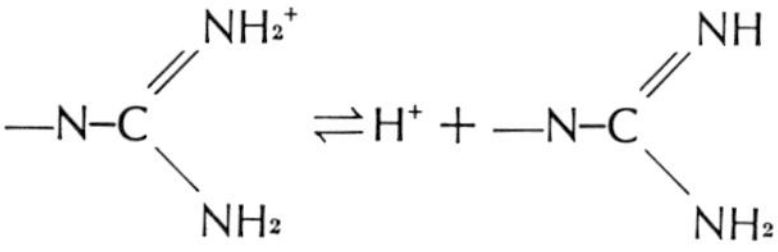	12.48	−5.48	$10^{5.48}$
Lysine $-CH_2 \cdot NH_3^+ \rightleftharpoons H^+ + -CH_2 \cdot NH_2$	10.53	−3.53	$10^{3.53}$

The formula used in determining the ratio of charged to uncharged R groups is identical with that used in connection with buffers:

$$\text{pH (medium)} = \text{p}K\text{ (amino acid)} + \log\left(\frac{\text{dissociated}}{\text{undissociated}}\right)$$

$$\therefore \log\left(\frac{\text{dissociated}}{\text{undissociated}}\right) = \text{pH} - \text{p}K$$

For amino groups the dissociated form is the unionized one, and in computing ratio of charged to uncharged groups the negative value for the logarithm means a preponderance of charged groupings.

living cells the characteristic charges on a protein molecule will be determined by the proportions of lysine, arginine, glutamic and aspartic residues. Although the charges on a protein molecule are by no means the sole determinants of protein behavior, they are extremely important ones. Various consequences of protein charge will be discussed periodically through the remainder of the text, but some of the grosser aspects of the relationship may be mentioned here.

If proteins have fixed charged groups, it should be possible to perform on them the type of experiment which Faraday performed on simple electrolytes: to subject them to an electrical field and demonstrate a directional movement correlated with charge, a procedure known as "electrophoresis." The experiment may be set up in a number of ways. The simplest is to soak a strip of filter paper in a buffer of desired pH, place a small drop of the protein solution at the center of the paper, and attach the ends of the paper to electrodes. Although the protein cannot be seen, its position may be located by a variety of chemical reactions. When current is applied, different proteins migrate in different directions; movement is dependent on the pH of the medium, and for each protein a pH can be found at which it will not move at all. The only satisfactory explanation of this behavior is that the direction of movement is a function of the relative numbers of ionized carboxyl and amino groups on the protein molecule. Since these groups are fixed to the protein molecule they cannot migrate independently, and the force which pulls the molecule as a whole arises from the excess of one group of charges over the other. If the protein does not migrate at all at a particular pH, we must assume that under such conditions the number of dissociated carboxyl groups equals the number of charged amino groups. The pH at which proteins do not migrate is termed the "isoelectric point." Cellular proteins are readily shown to have different isoelectric points. If, for example, all the water-soluble proteins of a cell (obtained by first breaking the cells mechanically and then extracting the mush with water) are spotted on a strip of filter paper and subjected to an electrical field at a given pH, a pattern of distribution is obtained indicating variations both in degree of charge and kind of charge (Fig. 4-5). Most proteins of a cell have isoelectric points in the region 5–6, but certain proteins of the nucleus have isoelectric points in the neighborhood 9–10 because of their high content of arginine or lysine. If most cells have an internal pH of about 7.0, most cytoplasmic proteins would carry a net negative charge; this was once of great interest in colloidal studies inasmuch as the charge on a colloidal particle was considered to be an important factor in its function.

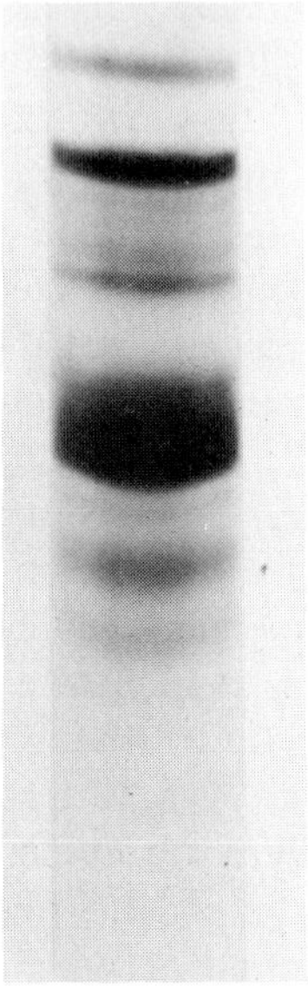

Fig. 4-5. Separation of proteins by electrophoresis. The separation illustrated was achieved by placing a solution of proteins on top of a vertical column consisting of a cylinder of polyacrylamide gel housed in a glass tube. Each end of the column was in contact with a reservoir of buffered solution containing a platinum electrode. On passing direct current between the electrodes, the individual proteins migrated at a rate which was a function of charge and size. After an appropriate interval of time the gel cylinder was removed and placed in a solution of dye which specifically stained the proteins. Each of the bands seen in the illustration represents a different type of protein even though all the proteins were members of the single group called "histones". (Courtesy of William F. Sheridan, Botany Department, University of Illinois.)

Far more important to our understanding of cell behavior than the charge of proteins in general is the charge carried by individual protein molecules. To appreciate this point, however, we need a little understanding of the physical arrangement of protein molecules. What has been said so far represents an incomplete description of proteins. We have described a molecule which appears to be confined to living systems. It is unique not because of its length, although this in itself is a feature of some distinction, but because along its length are displayed the most diversified chemical groupings. If we could see electron densities as we do the contours of a landscape, the protein molecule would assume the majesty of the Himalayas. There would be high peaks of negatively charged regions, deep valleys of positively charged ones, undulating terrain along the nonionizing polar groups, and a contrasting flatness among the apolar ones. The picture may or may not excite the human eye, but in the world of chemical reactions, where electron sharing, attraction, and repulsion are the sole agents of change, the picture must be impressive. No other molecule in the living cell commands such properties, and, it might be added, no other molecule in the cell can perform the functions of protein. With this knowledge alone the emphasis which Liebig and his successors placed on the nitrogenous components of living matter is amply justified.

When the characterization of proteins as long polypeptide chains came to be generally accepted, one of the first questions raised was whether elongated chains existed as such in nature. The question was not raised gratuitously for relatively simple physical methods indicated that this could not always be so. Very many kinds of proteins coagulated when heated in solution to temperatures no higher than 100°C. The term "coagulate" is used to describe an ill-defined type of precipitate, ill-defined because a protein thus precipitated no longer possesses its original physical and chemical properties. It cannot be redissolved under the same conditions as the unheated sample and, as will be later seen, it loses a number of additional properties of biological interest. Yet those who studied the phenomenon discovered that protein did not break down under these conditions; in other words, the polypeptide chain was in no way damaged. To explain such behavior the idea was introduced that the polypeptide chain was somehow folded in the unheated protein molecule; heating caused an unfolding, and the long chains adhered to one another, forming particles too big to be maintained in solution. The term "native" was used to describe protein molecules as they presumably existed in the natural state and "denatured" to describe those whose original folding had been altered. The concept of a long polypeptide chain folding about itself in a very special way was obviously intriguing to those who sought to discover the unique features of protein molecules which made them the dominant molecular species of living systems. In the 1930's much attention was focused on denaturation.

Elaborately folded molecules had never before been seriously considered by chemists, nor were other kinds of molecules known to display the property of coagulation. More specific tests which could provide a better insight into the structure of the protein molecule were sought and found. The simplest test made use of the —SH group on the amino acid cysteine. Several methods were available to measure this group and no other. When these were applied, the number of measurable —SH groups in many proteins appeared to be greater in the denatured than in the native state. To the analyst this meant that when the protein chain was folded, the test reagents could not reach the —SH groups which were tucked inside. Other groups were similarly studied, and increasingly refined methods were applied to determine the kinds and numbers of "hidden" groups in native proteins. Tests were carried out with proteins which had been denatured in a variety of ways—acid, alkali, high concentrations of urea, and the like. The results all pointed in one direction. The majority of cellular proteins were highly folded structures and once caused to unfold they could not, with rare exceptions, be made to return to their original configuration.

We may pause here briefly to get a bird's-eye view of what occurred in this particular channel of biology in the course of 100 years. Around 1850 Liebig's classification of organic substances into carbohydrates, fats, and proteins was an outstanding achievement. The term "protein" introduced by Mulder in 1838 still encompassed a chemical mystery; there was only the strong intuition that this substance held the key to life. To scientists like Liebig and Mulder protein was some kind of giant molecule, although they could not prove it or even adduce evidence for their special belief that all proteins were of one molecular species. To chemists like Graham protein was not a molecule but some unique aggregate—a colloid—which made vital activities possible by virtue of its unusual structure. Within a span of fifty years, close to the turn of the century, Fischer had nailed fast the fundamental proposition that proteins were covalently linked amino acids. This mid-point in our time scale was not the mid-point of progress; compared to what occurred subsequently, it was its beginning. For within the next fifty years not only were purified proteins isolated and crystallized, but the utterly new concept of a highly folded molecular chain was introduced. Yet concepts, as Albert Einstein pointed out, are "free inventions of the human mind"; they are useful to the extent that they invite experiment and valid to the extent that they can be experimentally verified. How do we prove the existence of folded protein chains? Before answering this it is worth reflecting on another of Einstein's observations. He pointed out that if we follow the history of any branch of science, concepts are found to succeed one another in a characteristic way. The earlier the concept, the closer it resembles various facets of one's immediate experiences. Einstein was mainly concerned with physics, but his observation holds true for biology. A newcomer can easily accept the concept of a cell once cells have been seen; so too for nuclei, mitochondria, etc. The concept of large molecules is more difficult for molecules cannot be *directly* visualized, even though their fuzzy outlines may be seen through the electron microscope. Far more difficult, however, are concepts concerning the internal pattern of large molecules. We do not lack ways of obtaining a detailed picture of such molecules; but the ways we have no longer involve the visualization of structures. Superficially, these ways may seem indirect, and their use requires skills which cannot be as casually acquired by most biologists as are techniques of cellular staining. This situation has led some people to maintain that biological knowledge is becoming increasingly specialized. Yet this is true only to the extent that the experimental and theoretical tools required for the understanding of broad relationships have become increasingly diversified. Our principal concepts regarding biological behavior have become fewer rather than more numerous.

Secondary and Tertiary Structure of Proteins

Three lines of experimental study—determination of molecular weights, crystallization, and elucidation of amino acid sequences—pointed to the general conclusion that protein molecules had highly organized structural patterns. If molecular weights were indeterminate, no assurance would be possible that protein molecules were like other, though much smaller, molecules with respect to constancy of size and composition. If crystallization of globular proteins had not been achieved, the purity of preparations would have remained doubtful and studies of intramolecular atomic arrangements could not have been pursued. If amino acid sequences had not been elucidated, the origin of molecular characteristics would have remained in the field of speculation. None of these studies, however, provided any direct information on how the protein molecule is internally organized. Such information came in large part from studies of x-ray diffraction.

In 1912 W. Friedrich, P. Knipping, and Max von Laue reported that if a beam of x-rays passing through a crystal is photographed, the image produced on the plate consists of a dark central spot plus a pattern of fainter spots around it. (See Fig. 4-6.) Laue had originally considered that x-rays, which have a wavelength of the order of 1 Å, would behave on passing through a molecule or group of molecules similarly to light rays passing through extremely narrow slits. His reasoning was based on the fact that interatomic distances are of the order of 1 Å and that such distances would therefore function as did slits in an opaque diaphragm. The discovery that x-rays were actually diffracted on passing through a crystal confirmed Laue's speculations, for the faint spots on the photographic plates obtained by Friedrich and Knipping could only be thus interpreted. This discovery led W. L. Bragg to use x-rays systematically for studies of crystal structure.

Like other tools of analysis developed by physicists and physical chemists, this one was taken up by biologists to meet the challenging problem of elucidating the internal structure of macromolecules. The problem of interpreting x-ray diffraction photographs is by no means simple. But with suitable material and conditions, rigorous interpretation is possible. Any ordered arrangement of molecules fundamentally consists of clusters of atoms which will appear to be arranged in pat-

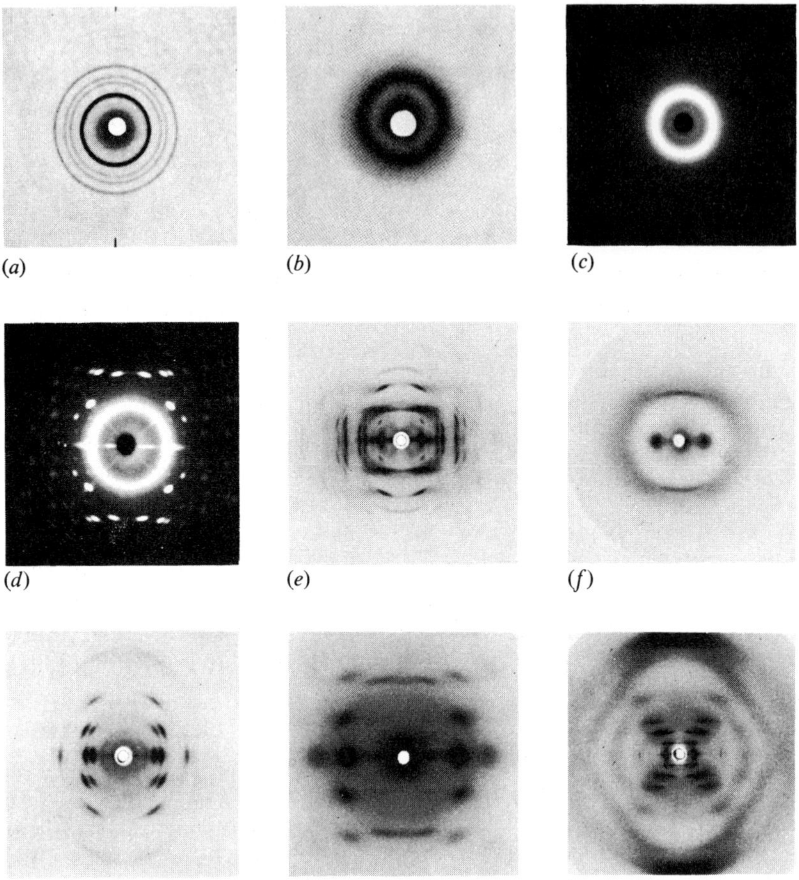

(*a*) (*b*) (*c*)

(*d*) (*e*) (*f*)

(*g*) (*h*) (*i*)

Fig. 4-6. Some x-ray diffraction patterns, all taken on photographic plates mounted perpendicular to the incident beam, behind the sample. (a) Powder diagram, a crystobalite. (*b*) Benzene. (c) Unstretched polyisobutylene. (*d*) Stretched polyisobutylene. (e) Poly-L-alanine, α-form. (*f*) Wool α-keratin. (g) Poly-L-alanine, β-form. (*h*) Silk fibroin. (*i*) DNA. (From Tanford, *Physical Chemistry of Macromolecules,* Wiley, 1961.)

terns of planes, the arrangement of the patterns depending on the angle from which the cluster of molecules is being viewed. X-ray waves passing through a crystal will suffer one of three fates. Some will pass through between the clusters unchanged in direction; others will be scattered if they strike in a direction leading through the nuclei of the atoms; still others will behave as though reflected by the surface of the atomic clusters, the angle of incidence equaling the angle of reflection. However, the spacings between clusters of atoms have an important effect on the direction of the light which emerges from the crystal. The closer the planes are to one another, the greater must be the angle of incident light for the reflected light to appear on the photographic plate. In analyzing atomic arrangements within a crystal, patterns must be recorded for a variety of angles of incidence. From these patterns, however, the relative positions of the atoms may be calculated.

For reasons to be discussed, protein crystals were not the first forms of protein to be analyzed by x-ray diffraction, even though crystal analysis provides for unambiguous interpretation. X-ray crystallography was introduced only about 50 years ago, and the fruitful application of this technique in identifying the position of each atom within a protein molecule is only now beginning.

Fibrous Proteins

When proteins were being mainly studied as colloids, chemists were already aware that they could be grouped into two broad classes. Silk, hair, wool, and horses' hoofs were among the materials identified as virtually pure protein which, apart from chemical composition, appeared to have little in common with the unresolved proteins extracted from the interior of living cells. Proteins of the first group were either insoluble in aqueous media or, if soluble, readily aggregated into gels. The classical gel-forming protein was gelatin which was obtained by heating collagen, a protein abundant in the hoofs of horses as well as in most of the connective tissues of animals. Over the many years that gelatin was being studied as a model protein, it was not understood why its parent protein was insoluble or why heating converted it to a soluble form. However, gelatin met all the chemical requirements of a protein and was justifiably considered to be a useful model substance. Moreover, at least two proteins were known to be "gel-formers" in their native state: myosin of muscle and fibrin, the clotting substance of blood. As physicochemical studies of proteins progressed, these insoluble or gel-forming proteins were called "fibrous," and the other group "globular." The nomenclature was based on the view that fibrous proteins are essentially long polypeptide chains either wound about one another into fibers or interlocked with one another in gels. Globular proteins, on the other hand, were considered to be intramolecularly folded protein chains because of their susceptibility to denaturation and because of their inability to form gels except occasionally after denaturation. A biological parallel could be drawn to the chemical classification; fibrous proteins were usually associated with cell structures, but most globular proteins were linked to the metabolic activities of cells, such as respiration and the synthesis of organic molecules.

For physical reasons fibrous proteins appeared to be the materials of choice for initial studies of the arrangement of long polypeptide chains. The fact that protein fibers such as silk or wool had certain tensile strengths meant that the individual molecules had to be bound to one another by forces whose sum could at least be roughly estimated. Moreover, the presumed absence of any extensive intramolecular folding allowed for the possibility of determining by means of x-ray diffraction whether polypeptide chains were fully or only partly stretched in their natural state. Yet even this apparently simple experimental task is so complicated that an unequivocal result is not obtained by direct analysis of x-ray photographs. X-ray patterns indicate spacings which occur regularly between atoms or groups of atoms; the interpreter must supply an atomic model of the molecule which fits the observed spacings. Sometimes models have been proposed on theoretical grounds and the critical spacings then sought by x-ray analysis.

Two examples will suffice to explain the basic organizational feature of fibrous proteins. In the first we will consider a protein in which the observed spacings are approximately those expected to exist in a fully extended polypeptide; in the other we will consider a protein in which the observed spacings do not match the characteristics of an extended chain. The protein of silk (fibroin) is an almost perfect example of the first. The two characteristic spacings of silk protein at 7.0 and 4.3 Å can be illustrated by drawing two or more parallel polypeptide chains and ignoring the R groups of the amino acids which would be positioned at right angles to the plane of the paper. If such a diagram is compared with a theoretical one for a stretched polypeptide chain constructed on the basis of known interatomic distances and angles, it can be seen that the two are in close agreement. If hair or wool is soaked in water, the fibers can be stretched to almost twice their original length; such stretched fibers have an x-ray diffraction pattern very similar to that of silk (See Fig. 4-4*d*). Hair or wool protein (keratin) when stretched consists of extended polypeptide chains. The pattern is quite different, however, when unstretched fibers are examined. Hence we find a set of atomic groupings spaced at intervals of 1.5 Å. The chain is clearly compressed in some way, and this may be explained without resorting to the impossible device of compressing the atoms or the interatomic bond distances. Solutions to problems of this sort generally depend on a brilliant intuition. In this case we owe a great deal to the chemist Linus Pauling. He made very effective use of mechanical models of atoms with which he built model protein molecules. By making the distances and angles of the mechanical bonds as close as possible to the values obtained from physical studies, he could construct various chain models and analyze each of them for periodic spacings and stability. He and his associates were able to utilize suggestions of predecessors and arrive at a highly satisfactory representation of the compressed polypeptide chain. The principal feature of this representation was a curling of the polypeptide chain into the form of a helix, or spiral, such that adjacent amino acids were displaced from one another by a distance of 1.5 Å along the axis of the molecule (Fig. 4-7). The structure became known as the α-helix structure to correspond with the prefix α which had for some time been used to signify the unstretched form of keratin. Once this model was proposed, investigators sought and found the 1.5 Å periodicity in a number of fibrous proteins and also in the globular protein hemoglobin. Much of the polypeptide chain, even in globular protein, is now believed to assume an α-helix configuration.

The two examples cited pose one common question. What is the force that holds parallel chains together in a fiber and what force maintains the helical structure? It is an axiom of physical science that any arrangement of atoms can be accounted for by the sum of forces acting between atoms. But, the chemical properties of the amino acid R groups cannot account for the forces maintaining this type of structure. Although in many cases —SS— bonds between adjacent cysteines link parallel polypeptide chains (as they do in hair) or even adjacent parts of the same folded chain, and although electrostatic attraction between carboxyl and amino R groups might function similarly, neither of these forces is adequate to explain the helical configuration of proteins or even the forces holding together the molecules in most fibers. Again, we owe to Pauling the application of a bonding concept (hydrogen bonding) which not only solved this particular problem but has found general application in accounting for foldings and associations of large molecules.

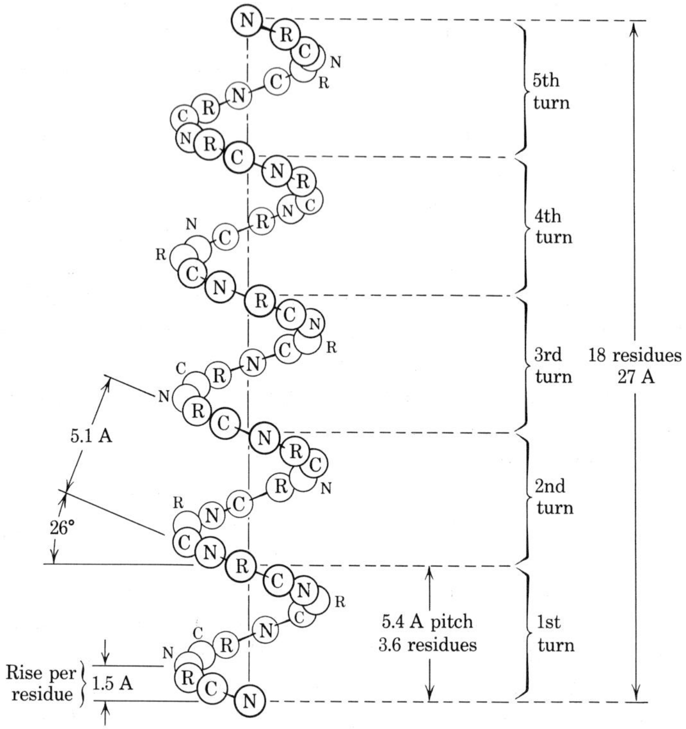

(a)

Fig. 4-7. The alpha helix.

Hydrogen Bonding. Pauling suggested the "hydrogen bond" as the force which holds the polypeptide chain in a helical configuration. It is a unique type of chemical bond which, as the name suggests, involves primarily the hydrogen atom. The idea of a hydrogen bond was first proposed to explain various physical properties of water which seemed anomalous for molecules of that size. The boiling and freezing points of water, for example, were impossible to account for unless the existence of relatively strong forces acting between molecules was assumed. If water molecules existed without intermolecular bonds, the boiling and freezing points should be lower because of the relatively high kinetic energy (energy of movement) associated with small molecules. Continued investigations showed that whenever molecules could so arrange themselves that a hydrogen atom lay between oxygens, nitrogens, or fluorines, some kind of extra bonding occurred, which though weak in comparison with covalent bonds was of sufficient strength to produce detectable physicochemical differences. All such differences were suggestive of multimolecular groupings.

To understand the operation of hydrogen bonds in biological systems, we need to consider the electrical properties of the hydrogen atom, on the

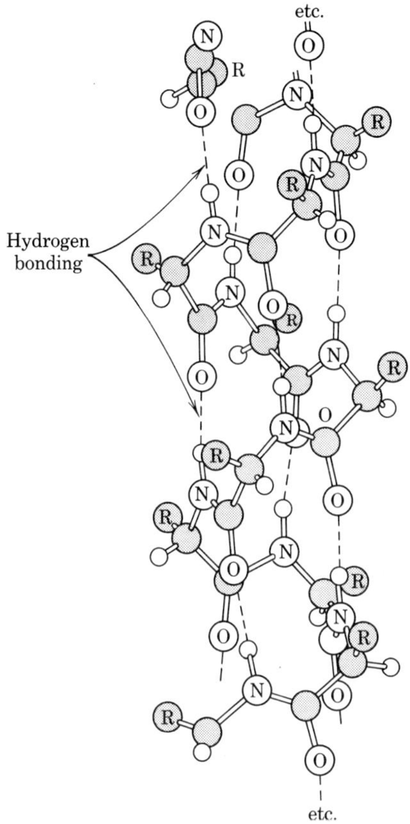

(b)

one hand, and of the oxygen and nitrogen atoms on the other. Since hydrogen has only one orbiting electron, transfer of that electron to the orbit of an electronegative element leaves the hydrogen nucleus (or proton) unshielded. The proton is thus the most densely charged positive body usually encountered in chemistry. By contrast, oxygen and nitrogen are strongly electronegative so that, depending on the electrical properties of their associated atoms, they persist within the molecule as centers of greater or lesser negative charge. The effect of having two such groups close to one another (about 2Å) with one hydrogen between them is illustrated in Fig. 4-8. Whether the electronegative atoms are nitrogen or oxygen, the hydrogen behaves as though it were pulled by two negative centers, and as a result of this the highly positive proton oscillates between the two. Such oscillation is described in hydrogen bonding.

The operation of hydrogen bonding is particularly easy to picture in proteins because the peptide linkage, being partly electrostatic and partly covalent, is able to oscillate between two forms. The conventional covalent

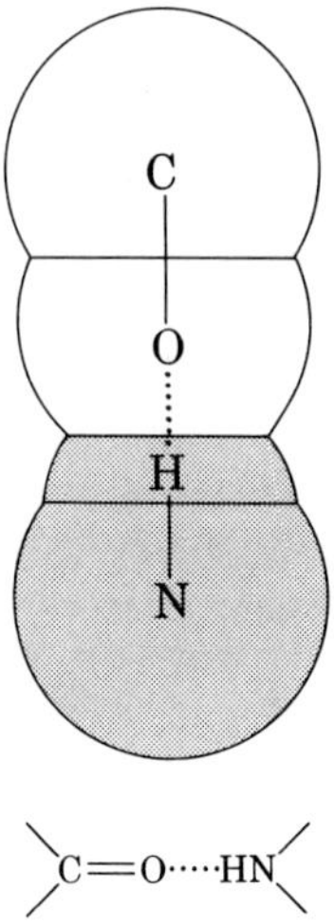

(a)

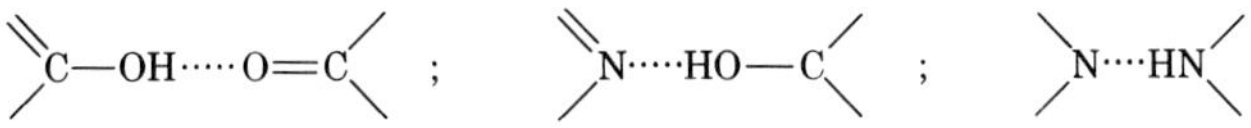

(b)

Fig. 4-8. (*a*) A semidiagrammatic representation of a hydrogen bond between a CO and NH group. The shading indicates that each group is covalently linked to a different molecule (as in nucleic acids) or to different parts of a molecular chain (as in proteins). The bold outlines represent van der Waal radii, the outer limit of the electron cloud surrounding the nucleus of the atom, and into which other atoms do not penetrate unless some linkage is formed between them. If each of the atoms were free, the outlines would be circular. Compared to the diameters of the electron clouds, those of the nuclei are very small. This is especially true for the proton (H) which as a densely charged positive particle occupies an equilibrium position between the two negative clouds. Note that the electron cloud of hydrogen is truncated at both faces indicating an interaction between oxygen and nitrogen. (*b*) Formulas of other types of groupings found commonly in proteins and/or nucleic acids which are capable of hydrogen bonding. (Adapted from R. E. Marsh, R. B. Corey, and L. Pauling, *Biochem. Biophys. Acta.*, **16,** 1 (1955).)

form has already been illustrated and shows no inequality in charge distribution though, for reasons already given, it is presumed to exist. In the alternate form the oxygen atom shares none of its own electrons with carbon, but shares two electrons of the carbon atom to give itself a net negative charge of one; the strongly neutral carbon atom permits this to occur by sharing a pair of electrons from the nitrogen atom, which thus acquires a net positive charge of one.

Such situations are called resonating systems because neither structure appears to persist alone and the bond behaves as though it were constantly alternating between the two forms. As we shall see for nucleic acids, similar resonating systems are commonly found where either nitrogen or oxygen atoms are attached to a carbon atom which is part of a ring structure. If we restrict our attention here to the peptide linkage, one consequence can be clearly seen. At the instant that the oxygen is negatively charged it will have a strong attraction for a proton, whereas the positively charged nitrogen atom will tend to repel its adjacent proton. In the alternative situation the attraction between the oxygen and the proton will no longer exist and that between the nitrogen and the proton will return. Suppose, however, that two parallel peptide chains are close enough so that the C—O of one chain is within 2-3 A of the NH of the other, a distance over which protons are known to resonate. Under these circumstances Pauling conceived that the resonance within the chain became further stabilized by a resonance of protons between the chains as illustrated diagramatically (See Fig. 4-7). In the helical configuration of a protein chain the hydrogen bond acts between peptide linkages of the same chain. As the chain coils, the peptide links come to lie above one another at regular intervals. By hydrogen bonding in a vertical direction the helix is stabilized. Without the implication of such a force, it would be difficult to account for the picture provided by x-ray diffraction.

One important issue concerning the effectiveness of hydrogen bonding in protein molecules exposed to an aqueous environment has not yet been discussed. Water, as pointed out earlier, has a very strong tendency toward hydrogen bond formation. The question has therefore been raised as to why the components of the protein molecule which are potential hydrogen bond formers do not interact with water rather than with other protein molecules. The explanation now favored by physical chemists is that the apolar groups in the protein molecule are an important factor in promoting hydrogen bonding. The reasoning behind the explanation runs as follows: Regions in a polypeptide chain which contain apolar groups tend to exclude water from the vicinity of the chain because the water molecules have very little affinity for the apolar groups and have a great deal of affinity for one another. To the extent that hydrogen bond forming groups are juxtaposed to apolar regions, direct interactions with water are made less likely whereas interactions between the polypeptide chains are increased. The apolar regions of a protein molecule may therefore be viewed as a shield which protects hydrogen bond forming regions from readily interacting with water. Thus, although the helical configuration may be specified by hydrogen bonding, the effectiveness of such bonding is determined by apolar groupings.

Although the concept that long polypeptide chains tend to assume a helical configuration provides a far-reaching generalization on protein structure, it does not provide an adequate account of the forms which specific protein molecules assume. The realization that secondary atomic forces, such as hydrogen bonding and apolar interactions play a key role in helix formation is sufficient to make us pause and reflect on the extent to which such forces may be superimposed on other structural associations. We have considered only peptide bonds with respect to the α-helix, but it would be difficult to suppose that the side chains are without influence on protein structure. Obviously, helices lying parallel to one another are likely to interact through their respective side chains, and this has actually been found for the fibrous protein collagen, where the charged R groups interact strongly. Even more significant, perhaps, is the finding that adjacent helical chains tend to coil about one another. Given the primary molecular coil, it appears as though other coils tend to be superimposed by virtue of molecular associations.

That this point of view is more than mere speculation is made evident through studies of cellular fibers by means of the electron microscope. Some bacteria, for example, have hair-like projections, called "flagella," which are about 180 Å in diameter, too small to be seen in the light microscope. Under the electron microscope, however, some flagella appear to consist of three strands coiled about one another and each measuring 90 Å in diameter. Although the components of the individual strands are difficult to resolve by means of the electron microscope, it is possible to speculate on their nature from other lines of study. X-ray analysis of flagella indicates that they contain protein molecules identical in structure with α-keratin. If the flagellum is assumed to consist entirely of α-helices, an anology may be drawn with the organization of horse hair in which six helical chains of keratin wrap themselves around a seventh one to yield a fiber 30 Å in diameter. By appropriate geometrical calculations it can be shown that seven 30 Å fibers wrapped around one another in helical fashion would give a fiber 90 Å in diameter, which is the dimension observed for the unit strand of the bacterial flagellum under the electron microscope. It is a little early in the history of these studies to draw any sweeping conclusions. Nevertheless, the impressive fact remains that many of the fibrous structures of cells consist of coiled elements and that a reasonable hypothesis links form at the cellular level with the intramolecular arrangement of atoms. Our ability to formulate such a relationship, even if it proves to be incorrect, represents a tremendous leap in thinking from less than fifty years ago when the term "colloid" was vaguely used to explain much of cell behavior.

Fibrous and globular proteins have worlds of their own. Where structural, motile or contractile elements are required, fibrous proteins predominate, but where the complex processes of chemical change are needed, the globular proteins are the unchallenged masters. And to a very large degree the extreme difference in function is mirrored by a corresponding difference in molecular structure. The contrast is all the more remarkable because the basic chemical structures of the two forms of protein are identical. This fact alone argues most eloquently for the extreme physical and chemical flexibility of polypeptide chains.

Globular Proteins

Many experimental approaches to the structure of globular proteins are available. Their solubility in aqueous solutions makes it possible to determine molecular size and shape by a variety of physical techniques. Crystalline preparations of a variety of globular proteins have made possible x-ray diffraction studies which are more precise than those of fibrous forms because crystals represent a highly ordered and repetitive arrangement of individual molecules. The scope of the problem may be illustrated with a few considerations of a specific globular protein. Serum albumin, for example, has a molecular weight of 70,000, a value which has been obtained by osmotic pressure measurements or by diffusion studies. Other techniques indicate that the albumin molecule has the approximate dimensions of 150 × 40 Å, and that it consists of a single polypeptide chain. In evaluating these facts, we may consider first the length of a polypeptide chain with a molecular weight of 70,000. We may estimate roughly that such a chain would have at least 600 amino acids, and since the distance between two adjacent peptide bonds is 3.5 Å, the fully stretched molecule would have to be more than 2000 Å. Assuming an α-helical structure with a spacing of 5 Å for every three amino acids, the chain length would be about 1000 Å. Obviously extended chains could not possibly yield the dimensions of the molecule actually observed, and much the same argument would apply to other globular proteins. These physical facts taken together with the data on denaturation require that the polypeptide chains of globular proteins be extensively folded. A major challenge in modern biology lies in determining not only how they are folded but also why. On top of this there is the all-important question of explaining the relationship between the geometry of a protein molecule and its specific biological function.

Through the remainder of this text globular proteins will be discussed in one connection or another, and the scope of their functions should become increasingly evident. At this stage we wish to draw attention only to some of their general features. Most cells, if disintegrated mechanically in an aqueous medium, release more than 50% of their protein in a soluble form. The majority of such proteins are globular. In their native state they do not form gels or highly viscous solutions as do their fibrous counterparts. If heated, they commonly coagulate for reasons already given; if subjected to extremes of pH, they lose many of their original physicochemical and biological characteristics because of unfolding, or denaturation. Routinely we may test qualitatively for the presence of dissolved protein, either by heating a solution or by adding a variety of acids which precipitate proteins. The chemical uniqueness of proteins is suggested by the fact that these two very simple tests differentiate them from a host of other substances found in cells.

All proteins do not have the same solubility properties, nor do they respond in the same way to treatments by heat or acid. From our earlier discussion of protein composition it should be clear that one factor affecting solubility is the relative number of apolar and polar R groups along the protein chain. One class of proteins, the prolamines, has so high a proportion of apolar R groups that, unlike most other proteins, it is readily soluble in alcohol. Applying the rule that the more a molecule interacts with water, the greater will its solubility be, then another relationship may be predicted. Ionized groups tend to attract more water than nonionized ones. For this reason proteins have their lowest solubility at their isoelectric points. This relationship is not always obvious in native proteins where folding patterns influence water relations within molecules, but it is clearly evident in denatured proteins. Many proteins may be denatured at pHs removed from their isoelectric points and remain in solution, but as the pH of the solution is brought close to the isoelectric point these proteins precipitate. In a qualitative way we explain precipitation by saying that when protein molecules no longer interact strongly enough with water, they interact between themselves, forming bigger and bigger complexes until they become too heavy to remain suspended. This principle may also be applied to the solubility characteristics of native proteins. One of the early techniques used in fractionating proteins was a progressive addition of a salt (e.g., ammonium sulfate) to the mixed protein solution. It was found that individual species of protein precipitated at characteristic concentrations of salt. The phenomenon is known as "salting out," and its effectiveness depends upon the fact that ions of the salt compete with those of the protein for water; the greater the affinity of a particular protein for water, the higher the salt concentration required for its precipitation. The relationship between protein solubility and salt is, however, not consistent through a broad range of salt concentrations. Certain types of proteins (euglobulins) are insoluble in distilled water but are soluble in salt concentrations of the order of 0.02*N*. This "salting-in" effect is not open to a very simple interpretation, and more than one mathematical relationship has been proposed to account for it. Generally, however, low concentrations of ions may interact with polar groups in the R chains in such a way as to enhance their affinity for water.

Since proteins carry both positively and negatively charged groups along their chains, they are susceptible to the formation of insoluble salts for much the same reason that a variety of inorganic salts are insoluble. The explanation given is that the component ions of the salt are held together so strongly that they fail to ionize, have little affinity for water, and hence interact with one another to form a precipitate. Thus cations of the heavy metals (Ag, Zn, Hg, Cu, Pb, and Au) combine with the R-carboxyl groups to render the protein insoluble. A similar situation prevails with respect to a variety of anions. Two widely used protein precipitants, trichloroacetic acid and phosphotungstic acid, are effective largely because their anions form insoluble salts with the positive groups on the protein. Not all salt formations along the R groups lead to precipitation, however. One type of salt formation which has long been of interest in cellular studies is the binding of organic dyes by proteins, one of the principal reactions occurring in the staining of cells. Most such dyes have an ionizable region in the molecule; if the region is negatively charged, the dye is called acidic and if positively charged, it is called basic. Acid dyes thus combine with positively charged R groups, and basic dyes with the negatively charged ones. The relative affinities of different proteins for a particular dye will depend on amino acid composition and, if the protein is in the native state, also on the extent to which the folding of the protein chain leaves dissociable R groups exposed. In staining "fixed" cells, the proteins are, of course, undissolved but the same principle of complementary charge applies.

On the whole, the native configuration of globular proteins depends on a sensitive balance between various interatomic forces: hydrogen bonding, electrostatic attraction, disulfide linkages, and associations between hydrophobic groups. Factors such as temperature and pH tend for various reasons to upset the balance and hence the configuration of the protein. Although we do not yet understand what leads each particular protein type to assume its characteristic folded configuration, we do know that under artificial conditions a protein once uncoiled does not spontaneously return to its native configuration. At one time the "irreversibility of denaturation" was considered to be an absolute characteristic of proteins. But this can no longer be viewed in such extreme terms. Some very small protein molecules have been reversibly denatured, and theoretically at least bigger molecules should behave in the same way if only we could present the appropriate conditions. The conditions obviously exist in the cell, and sooner or later the biologist will come to know precisely what these conditions are.

From the standpoint of future studies, x-ray diffraction holds the principal key to the elucidation of the internal structure of protein molecules. Analyses of fibrous proteins have provided us with the basic picture of "secondary structure," the coiling of the polypeptide chain into a helix. The more complex challenge of explaining how the chain is folded into a globular form ("tertiary structure") has already begun to yield to the crystallographic studies of people like M. F. Perutz and J. C. Kendrew. These rigorous studies make possible the placing of each atom within the native protein molecule. The possibility is close to being fulfilled for myoglobin, a molecule which contains 150 amino acids and which functions in oxygen transport. Cell biology thus has available techniques of structural analysis, which in their resolving power extend from the images seen through the light microscope to the individual atoms in cellular molecules.

NUCLEIC ACIDS

Against the protein molecule with its seemingly unlimited potential for diversity in shape and association, nucleic acid appears to be a rather prosaic mate. The grand span in electron affinities of the amino acid R groups (which makes the electron contour of each protein chain a singular mixture of heights, valleys and plains) is altogether foreign to the nucleic acids. Yet if we were asked to select two types of molecules which are absolutely essential to life, or perhaps even sufficient for it, these are the two molecules one would invariably choose. It might seem strange that nucleic acids which are one-fifth to one-tenth as abundant in living material as are the proteins, and which do not even approach them in respect of structural flexibility, should be given equal footing in the architecture of a cell. Yet once we know that for all its diversity in pattern, each cell comes to life only because it has the ability to reproduce that pattern, the indispensability of nucleic acids becomes clear. It will be some time before this role of nucleic acids is fully unravelled in our study, but for the student led into the strange land of large molecules, it is well to know that this species of molecule has an absolute uniqueness.

The story of nucleic acids begins with the Swiss biochemist Friedrich Miescher, who first discovered them in 1869. Neither he nor anyone else suspected their existence earlier. This was the period when organic chemists felt themselves well on the way to resolving the nature of proteins which, in almost everybody's opinion, represented the key component of life. There was every indication from what we have already said about the proteins that the opinion was a sound one. The discovery of nucleic acids did little to change that opinion. Just as the significance of Mendel's discoveries lay hidden from view, so did that of Miescher's. Some histories of biology written or revised in the late 1950's do not even mention Miescher's name. The historians had good reason. One of the ablest of biological chemists, Albert Matthews, in writing on the chemistry of cells in 1931, expressed the view with a ringing authority that despite all the evidence on the genetic function of chromosomes, the chemical components of chromosomes—nucleic acids and proteins—were in themselves useless items for explaining hereditary behavior. To be sure, Miescher himself had no distinctive insight into the function of nucleic acids. He, his associates, and successors were satisfying the characteristic intellectual curiosity of organic chemists to determine the atomic composition of novel organic molecules.

Miescher happened to be studying the proteins of pus cells. In the course of this he used a familiar technique to render proteins soluble by digesting them with one of the stomach ferments, pepsin. When he examined the pepsin-treated pus cells under the microscope he noticed that the nuclei, though shrunken, were still intact. His curiosity got the better of him and he decided to analyze the chemical composition of the substance remaining in the nuclei which did not behave as proteins should. The substance turned out not to have the composition of protein. Further studies with salmon sperm confirmed the evidence and before long it became clear that the material consisted of large molecules containing carbon, nitrogen, hydrogen, oxygen and phosphorus; that it was acidic in nature, apparently unaffected by heating, and, like some fibrous proteins, was capable of forming gels. The acidity of the substance and its phosphorus content clearly set it apart from proteins, and because it was originally discovered in nuclei, it became known as nucleic acid.

Resolution of the nature of nucleic acids followed much the same procedures used for other large molecules—hydrolysis of the weaker covalent bonds in the molecule. Once this was done and the products analyzed, it became apparent that nucleic acids, whatever their function, were entirely different from proteins except for molecular size. There is little point in sorting out the history of these analyses. As was indicated earlier, the principal motive in these studies was to elucidate the structure of a unique class of organic substances. In searching for sources of nucleic acids, two major classes were found: one predominantly in thymus glands, and therefore called thymus nucleic acid; the other predominantly in yeast and called yeast nucleic acid. These names are no longer used because, after extensive studies, these chemically different nucleic acids were shown not to be distinctively associated with particular organisms or tissues but rather with different regions of the cell. The discovery of nucleic acid localization in cells was primarily due to Robert Feulgen and his students. Initially Feulgen discovered a highly specific color test for thymus nucleic acid which was not given with yeast nucleic acid. When he applied this same color test to intact or sectioned cells he found that the reaction occurred only in the nucleus or on the chromosomes. In later work his student, M. Behrens, disrupted cells mechanically without breaking their nuclei and by analyzing the separated nuclei and cytoplasm found that yeast nucleic acid was characteristically located in the cytoplasm, whereas, as expected, thymus nucleic acid was found solely in the nuclei. On the basis of their respective compositions, the first was called ribonucleic acid and the second, deoxyribonucleic acid. Today they are commonly referred to as RNA and DNA.

Hydrolyzed RNA yields three quite different groups of molecules: phosphoric acid, a sugar called ribose, and four kinds of molecules each comprising a ring structure built out of carbon and nitrogen atoms. Two of these have a single ring and are called pyrimidines; the other two have a double ring and are called purines. Their individual structures are given in Fig. 4-9.

Thus, unlike proteins which are built entirely from one type of molecule, nucleic acids are formed from three. Once the way in which the three types of molecules were linked to one another was understood, a further distinction became clear: the backbone of nucleic acids was formed of phosphoric acid and a sugar, whereas the side chains (analogous to the R groups of the amino acids) were constituted of purines and pyrimidines only. With respect to the structure of the molecule as a whole, the only "interesting" chemical region was that along the backbone where the ionizable phosphoric acid groups provided a strong negative charge. To most biochemists in the 1930's, nucleic acid appeared to be a useful molecule for holding positively charged proteins in place. Hydrolyzed DNA yields products similar to RNA except that deoxyribose replaces ribose and thymine replaces uracil. The arrangements of the molecular subunits are identical in the two types of nucleic acid.

Without trying at this point to discuss the biological functions of nucleic acids, we should, as with proteins, get some general view of what these molecules are. Yet before doing so, it is advisable to sketch in some episodes in the history of the subject following their discovery by Miescher and the subsequent elucidation of their primary chemical structure by organic chemists. Certainly, to the newcomer in biology, structural formulas of large molecules inspire more perplexity than understanding. It is difficult to translate the cold array of atoms which are sketched on this paper into a dynamic scheme that is at once meaningful. The leap from an unseen molecule to the familiar properties of living systems is a long one. But with proteins we could at least follow a relatively simple line of progress beginning with the curiosity of organic chemists in the mid-nineteenth century and ending with the accomplished analyses of physicists and physical chemists in our own time. We cannot do this for the nucleic acids. By the time their chemistry was understood there was so much progress in unravelling the physicochemical properties of cells, especially as they related to proteins, that molecular biologists were no longer simply "looking around". Proteins held so much promise for the biochemists that there appeared to be no need for other vital molecules. The result was that from the 1940's on, nucleic acids pursued the chemists and physicists rather than *vice versa*.

The growing sophistication in knowledge of cell behavior coupled with the widening experimentation in the field of genetics provided increasing evidence of the importance of nucleic acids in cellular function. Most of the evidence came from simple techniques. The Feulgen stain which demonstrated the exclusive localization of DNA on the chromosomes is one example. Another set of stains made it possible to visualize the location of RNA in the cytoplasm of all cells and in all nucleoli. Just as the early chemists discovered the universal presence of proteins in living systems, so did cell biologists discover the universal presence of nucleic acids. All living units down to the simple viruses showed a positive test for one or both of the nucleic acids. Cytologists who relied exclusively on staining techniques for probing cell behavior were increasingly impressed by the fact that RNA, unlike DNA, seemed to vary broadly in quantitative distribution according to the nature of cellular activity. Many cytologists were impressed by the evidence that "active" cells had a lot of RNA whereas "inactive" cells had relatively little. Thus when physicists and chemists were inspired to study the subtle features of nucleic acid molecules, the inspiration came not from a direct curiosity about nucleic acids, but from a desire to account for the overwhelming evidence on the importance of their biological functions.

To express the trend of thinking which led to the elucidation of the molecular structure of nucleic acids, we would first have to cover a great

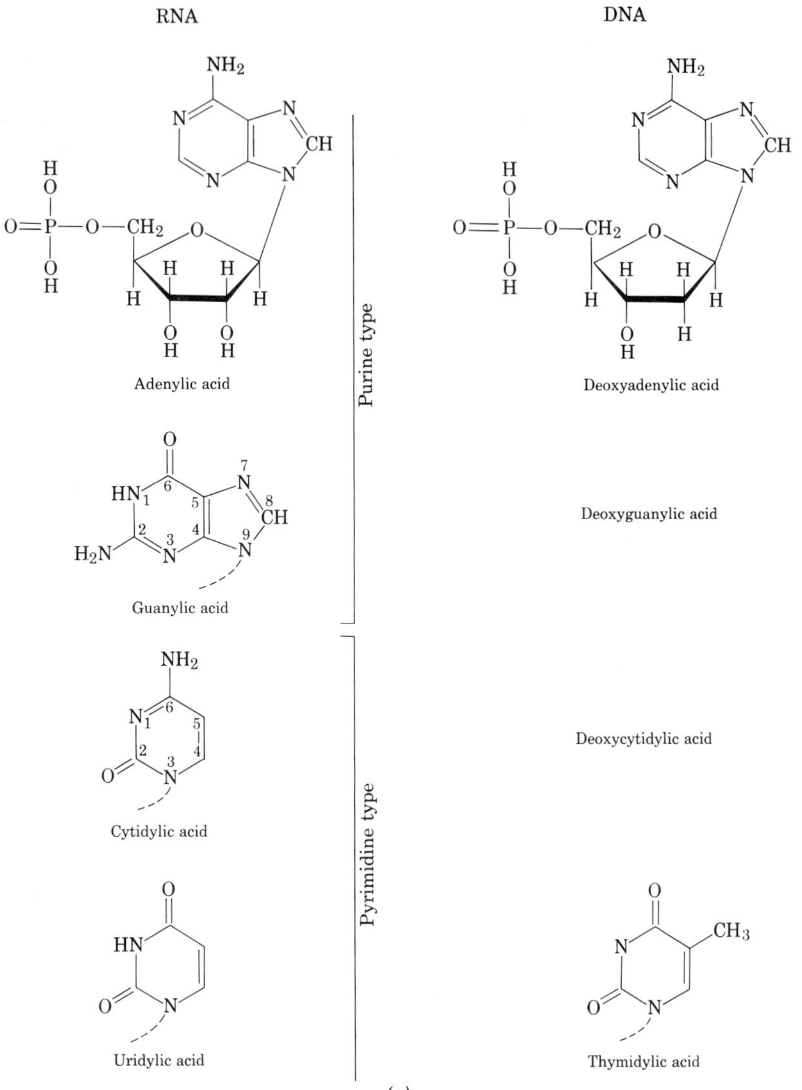

Fig. 4-9. (a) Mononucleotides of nucleic acids. Frequently the keto groups (6 in guanylic, 2 in cytidylic acid) are represented in the enol form (COH). Although

deal in the modern history of cell biology. We would have to formulate and discuss so many basic questions that we would find ourselves immediately forced to complete in one way or another the rest of our familiarizing survey which is the purpose of this section. We have the choice then of leaving the nucleic acids hanging as a queer and some-

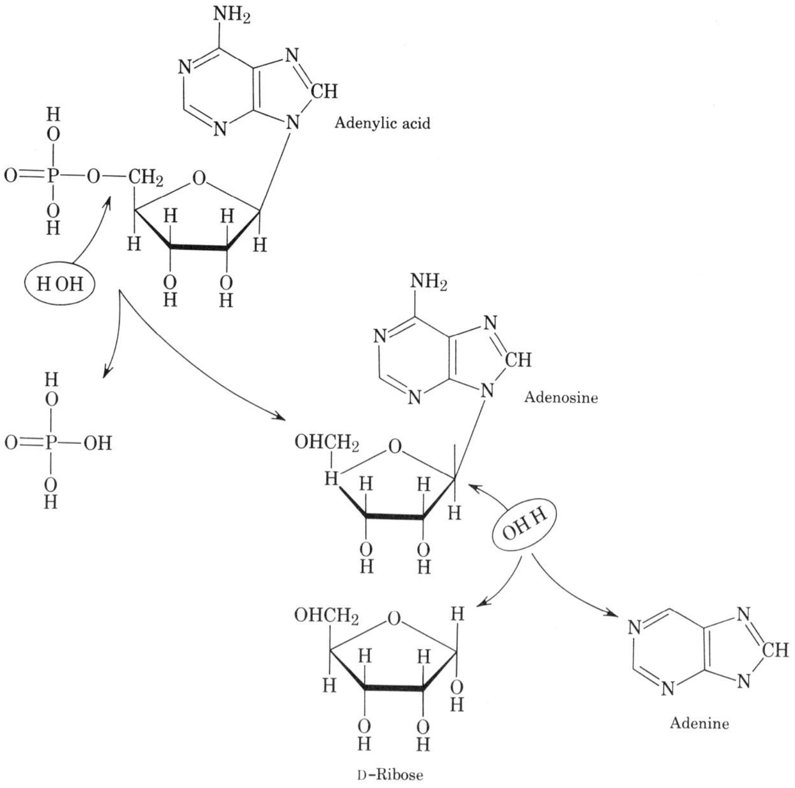

(*b*)

the enol form is assumed by the free bases, this form occurs to a very small extent in the nucleic acid molecules because of the overall electronic configuration. Note the numbering of the purine and pyrimidine rings. The key groups are 2 and 6; these are the critical determinants in pairing. The broken lines indicate the points of attachment of the sugar to the purine or pyrimidine ring. (*b*) Hydrolysis of mononucleotides. In the schemes shown, the first products of hydrolysis are 5′ nucleotides or deoxynucleotides; in these the phosphate groups are attached to carbon atom 5 of the sugar. In the nucleic acid molecule, phosphate forms a bridge between the fifth carbon of one sugar and the third of its neighbor. If the nucleic acid is hydrolyzed chemically, the phosphate linkage at the fifth carbon atom breaks much more readily than at the third. Hence 5′ nucleotides can be obtained only by use of specific hydrolytic enzymes.

what dull molecular arrangement or of by-passing some significant history and providing some finishing touches on the chemistry of these molecules. Since we choose the latter course we must ask the student temporarily to accept on faith the proposition that nucleic acids are the agents of hereditary transmission because it was this proposition which stimulated physicists and chemists to study nucleic acid molecules.

With this background we may return to the physical chemistry of nucleic acids. Our purpose will be served if we restrict ourselves to DNA since the structural principles involved apply equally to both. Secondary properties help to explain why DNA is the principal molecular agent of heredity, but these properties will be considered later. Of interest now are the primary properties which were uncovered after the DNA molecule was identified as the hereditary agent. Once it was recognized that the only possible variables in the DNA molecule were the pyrimidines and purines, a large number of analyses were performed on DNA prepared from different tissues and different organisms in the hope of finding some chemical basis for biological variation. On the whole, the results were disappointing. With a few minor and apparently inconsequential exceptions only two kinds of pyrimidines and two kinds of purines were found in every species of DNA analyzed. Compared to the twenty amino acids in proteins, four seemed to be a very small number. How could we account for the virtually infinite number of genes on a chromosome in terms of different combinations of these four units? Moreover, in the many different animal organisms studied, these units were found to be present in much the same proportions. The picture seemed even darker when the physical properties of DNA were first studied. All DNA solutions were highly viscous, all had the same charge, and all could be precipitated out of solution as fibers by means of alcohol. None of the subtle differences distinguishing protein types could be found among the nucleic acids. On the whole, they appeared to be very long nonfolded molecules with molecular weights of the order of 1–6 million. Results such as these only reinforced the common belief that if any molecule could account for hereditary differences, it had to be protein.

Now and then help comes from the most unexpected quarters, and such help was provided in the case of DNA. The biologist, E. Chargaff, having analyzed a large number of DNA samples, discovered one unusual feature in his results. Regardless of the differences in composition of his DNA samples, small though most of them were, there was one invariant compositional characteristic; the number of purines in a DNA molecule always equalled the number of pyrimidines. Not only that, but adenine always equalled thymine, and guanine always equalled cytosine. This curious relationship became known as "Chargaff's rule," but apart from its demonstrability no sense could be made of it until physicists got to work on the problem with the help of x-ray diffraction studies. These studies were carried out in exactly the same manner as were those on the fibrous proteins, which DNA resembled in various physical respects. As pointed out earlier, interpretation of x-ray diagrams of fibers is very much helped by intuition and this intuition was provided by J. D. Watson and F. H. C. Crick. They did no experiments but culled all the pointers they could from the physicists' data, and in one of those singular displays of creative insight combined Chargaff's rule with x-ray data to propose a structure for the DNA molecule which has undergone no significant correction since its introduction in 1953.

Three important hints were provided by the physicists. The first was that the molecules in the DNA fiber had a highly regular pattern; they had to be long and identical with one another. The second was that even species of DNA with unusual base composition (such as those of some viruses) had x-ray patterns identical with all the others; the structure of the molecules was therefore independent of the relative proportions of the two purine-pyrimidine groupings. This meant that, unlike the proteins, the arrangement of the side groups had no effect on their spacings within the chain. Indeed, the only components which varied at all seemed not to play a role in the structure of the molecule. This, indeed, was the initial opinion. The third and extremely important hint concerned the spacings indicated on the x-ray diagram. A regular clustering of atoms occurred at intervals of about 28 Å. This distance is much greater than the distance between the successive phosphorous atoms shown earlier in the formula for the nucleic acid chain. The chain seemed to be coiled in such a way that several purines and pyrimidines would lie in approximately the same plane at right angles to the long axis of the molecule.

The hints might have remained just that if Watson and Crick had not made mechanical models of atoms and attempted to fit these together in a way which would correspond to the x-ray data. Such fitting is not at all obvious. The rules of the game are strict. Interatomic bonds must have their characteristic distances and angles; rotation of groups with respect to one another must not impose mechanical strain on the model; distances between clusters of atoms in the model must conform with the distances observed in the x-ray diagram. What is a relatively simple matter is to stick all the atoms together so as to form the extended chain; that much is known from chemical analysis, and if the atom models are correctly built for size and bond angles, only one result is possible. But the moment we put such a chain together, we find that it is flexible and can be folded in a variety of ways without straining the model. It is at this stage that intuition operates through a process of trial and error. The success of Watson and Crick lay in their hitting on the idea that if two chains were wound around a cylinder of appropriate diameter, a structure consistent with the x-ray pattern emerged. To make such a structure, however, certain conditions had to be met. The purines and pyrimidines had to face the interior of the cylinder, and most important, to give the structure the symmetry indicated in the x-ray diagram, the two chains had to be wound in *opposite* directions. The latter requirement arose from the fact that the helical model would be meaningless unless some force were present which would preserve the helical shape. That force could be found by supposing that the bases of opposite chains were held together by hydrogen bonds, the same kind of bonds which maintain the helical configuration in proteins. But hydrogen bonding would be effective only if the purines and pyrimidines are brought close enough together to form such bonds (a distance of about 3 A between the atoms involved), and the only way to achieve this was to have a purine in one chain opposite a pyrimidine in the other. Two purines opposite one another occupied too much space in the model; two pyrimidines opposite one another did not come close enough. Furthermore, since only certain atoms within these ring structures were likely to form hydrogen bonds, the only pairs that would fit properly were thymine with adenine and guanine with cytosine. Here then was the meaning of Chargaff's finding that the total number of pyrimidines in a DNA molecule always equaled the total number of purines, and that cytosine was present in equal proportion with guanine while adenine was present in equal proportion with thymine (Fig. 4-10). See Table 4-3.

Table 4-3 Base Composition of DNA from Various Sources

Source	Thymine-Adenine	Guanine-Cytosine
Tetrahymena rostrata	24	76
Mycobacterium tuberculosis	30	70
Chlamydomonas reinhardii	32	68
Triticum aestivum (cultivated wheat) seed embryos	57	43
Bos taurus (bovine liver)	58	42
(bovine sperm)	58	42
Rana pipiens (frog testes)	58	42
Gingko biloba (pollen)	61	39
Bacteriophage T_2	66	34
Micrococcus lysodetikus	72	28

Analyses of DNA composition were originally performed by direct chemical methods. In such studies the molar proportion of thymine was found to be approximately equal to that of adenine, and a similar relationship was found to hold for guanine and cytosine. If allowances were made for experimental error, each pair of bases could be said to consist of equimolar proportions of components. This relationship is essential to the Watson-Crick theory of DNA structure. In recent years chemical analysis has been replaced by physical techniques, principally measuring the density of the molecule. A precise relationship has been found to exist between composition and density; the greater the content of guanine-cytosine, the higher the density. The values listed in Table 4-3 are based on density determinations. The molar equivalence of adenine to thymine, and of guanine to cytosine, is assumed.

One aspect of the distribution of base ratios among different species has provoked a great deal of interest. Vertebrates and higher plants have similar compositions; the guanine-cytosine content varies only between 39–44%. Invertebrates and microbes, on the other hand, show a great deal of variation.

The implications of this model were at once obvious. Like the proteins, the key stabilizing factor was the hydrogen bond; unlike the proteins, the only way in which such a bond could operate was by a combination of two complementary molecules. This requirement for partners was something new and unique in molecular biology. Proteins, by virtue of intramolecular bonding could assume an untold variety of configurations and associations; nucleic acids, on the other hand, appeared to fall into only one type of association—a double helical structure. With this much known it was not at all difficult to speculate that since DNA pairs had to be derived from preexisting DNA pairs, new molecules could be formed by separating the old identical members of a pair and giving each a new partner. How this came to be demonstrated will be discussed in a subsequent section.

LIPIDS

As the nineteenth century drew to a close the remarkable properties of proteins were universally recognized and hints respecting the uniqueness of nucleic acids were already in the air, but "lipids" (fats), like carbohydrates, appeared as prosaic in their chemical properties as they were stubborn in their universal distribution. When organic chemistry got into its swing during the middle of the nineteenth century, lipids proved to be relatively easy targets for analysis. They were obviously a distinctive class of organic compounds inasmuch as they were insoluble in water; Liebig was clearly on solid grounds when he considered them to be one of the three main classes of chemical compounds present in living matter. When the elementary composition of lipids was

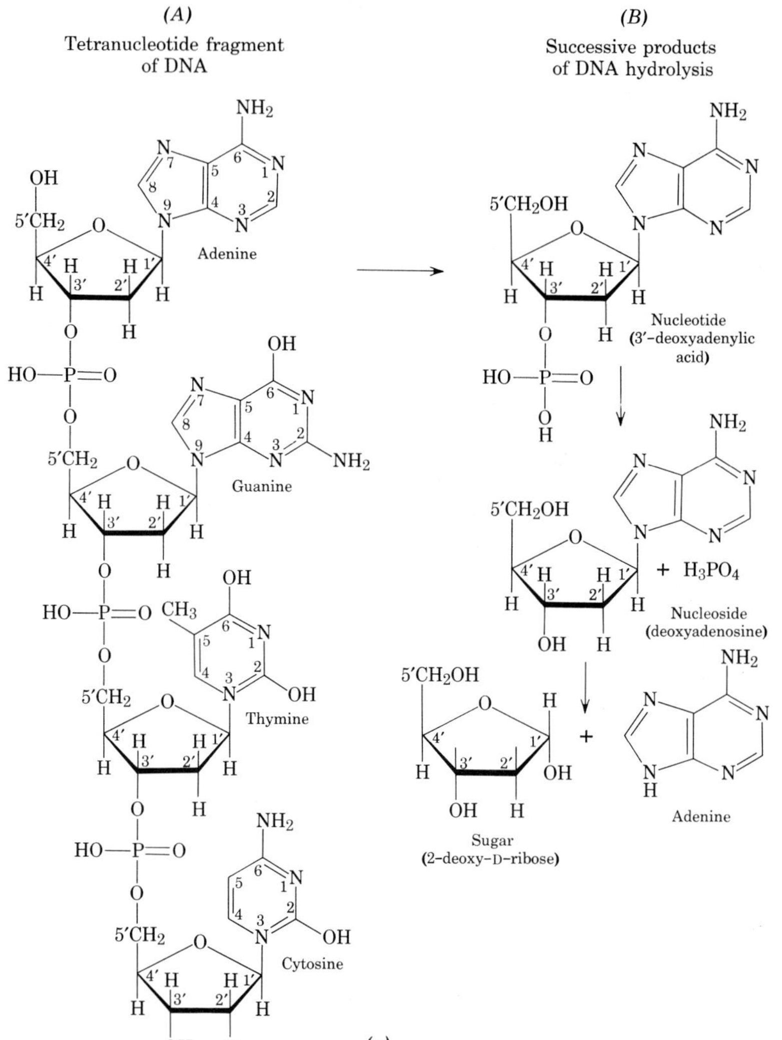

(a)

Fig. 4-10. The structure of the DNA molecule. (*a*) A fragment of the DNA chain. (*b*) Hydrogen bonding in DNA chain. (*c*) A schematic drawing of a DNA molecule. Except for differences noted in Fig. 4-9, RNA and DNA have similar primary structures. The arrangement shown in (*a*) therefore applies to both. The hydrogen bonding which is essential to the maintenance of the double helical structure of DNA is illustrated in (*b*). Note that the number 1 position is the same for all the bases, and hydrogen bonding between these positions is common to both pairs. Differences occur in the 2 and 6 positions. In these, the bonds are between amino and keto groups and therefore, given such a bonding as a requirement for stability, the pairs shown are the only ones possible. Extensive hydrogen bonding of complementary chains or segments is relatively rare in RNA.

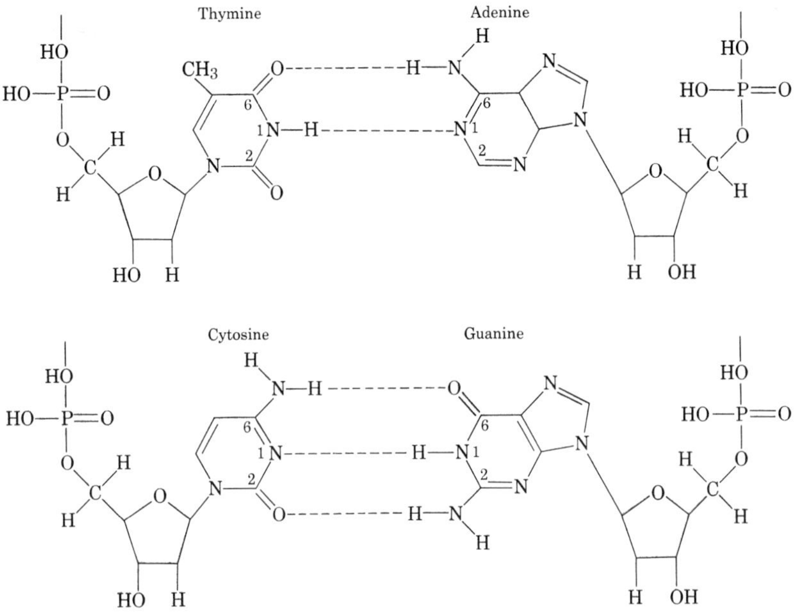
Thymine
Adenine
Cytosine
Guanine

(b)

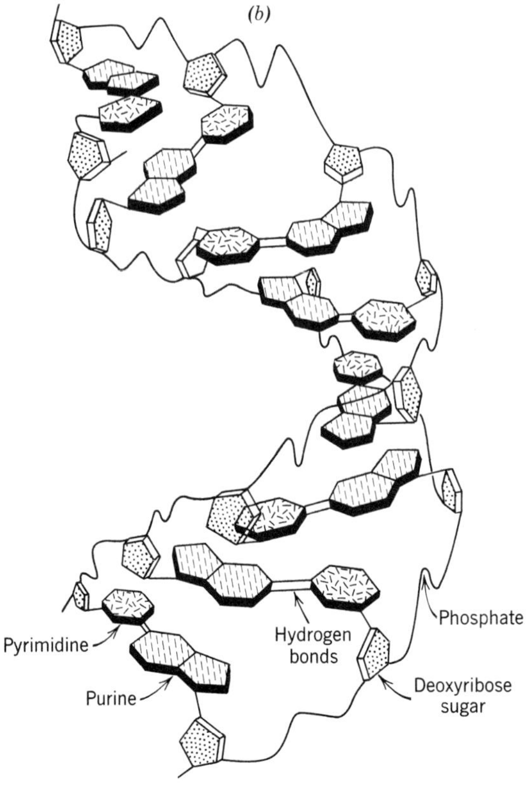
Phosphate
Hydrogen bonds
Pyrimidine
Purine
Deoxyribose sugar

(c)

determined, most of them were found to contain the same elements as carbohydrates but in different proportions. Carbon and hydrogen were major constituents, whereas oxygen was a minor one. When lipids were hydrolyzed into subunits, the majority of them proved to have a simple pattern: they were covalently linked combinations of the carbohydrate glycerol with a variety of "fatty" acids. Glycerol was a three-carbon sugar and the fatty acids were long chains of hydrocarbon groups ending in a single carboxyl. The longer the hydrocarbon chain, the less soluble was the fatty acid in water. A common fatty acid such as palmitic with 15 hydrocarbon groups was obviously highly apolar. No rigid requirements for uniformity in the kinds of fatty acids attached to the glycerol molecule were apparent; in some fats the three attached fatty acids were the same, in others the three were all different. Since the fatty acids tied up their carboxyl groups when combined with glycerol, they came to be known as neutral fats.

Neutral fats are widespread in nature, but if they were the only members of the lipid family one could conveniently reserve the lipids for specialized texts. Certainly the primary function of neutral fats appears to be that of a food reserve. Organisms capable of forming these compounds do so abundantly when they feed in excess, but under conditions of limited food supply they quickly deplete their fat reserves. The phenomenon is particularly well known in our own society where obesity and overeating commonly go hand in hand. Yet the ubiquity of lipids in living systems must have a basis other than food storage if for no other reason than that many types of cells do not accumulate fats for this purpose yet do have fats of some kind in their fabric. Indeed, the only organisms which may not contain lipids are some viruses, and these must multiply within living cells. No living unit that can multiply and grow in an inanimate environment is without lipid materal. By 1900, E. Overton had already concluded on the basis of his permeability studies that all cells must have lipid in their limiting membrane. The case was probably overstated, for it was proposed that the surface membrane was wholly of lipid material, but the idea of the essentiality of lipids was sound. All subsequent studies demonstrate that lipids must play some special role in the organization of cellular membranes, not only at the surface of the cells, but also within the cytoplasm and in organelles such as chloroplasts and mitochondria.

Organic chemists who were curious about the chemical composition of lipids did not approach the subject from the standpoint of cellular organization. Unlike their approach to nucleic acids, biologists took what the organic chemists served and fitted the information into appropriate schemes of cell organization. But there was and still is a great deal more known about the chemical properties of lipids than about their biological functions. This is probably largely due to the fact that the primary activities of living systems, namely, the utilization and conversion of foodstuffs, the reproduction of genetic material, and movement can be traced to proteins and nucleic acids. On the other hand, the details of cell architecture especially as they relate to membranes and their functions have been commonly treated as secondary features of what we have called primary activities. There isn't, however, much to be gained by considering whether or not such treatment is proper; the pertinent fact is that many facets of lipid function are still hidden from view.

There are two groups of lipid molecules which we know to be associated with membranes. Sterols and steroids are one group; except for elemental composition and solubility, they have little in common with the neutral fats. The structure of one of the common sterols, cholesterol, is shown in (Fig. 4-11). Just how this type of molecule associates with proteins is unknown. There is no doubt of its presence in many membranes but the best clue we have to the functional importance of sterols comes from quite a different set of observations. Vitamin D and the male and female sex hormones of animals are all steroids. Certain cancerous growths can be caused by steroids. Yet, for the present these various pieces of information remain unintegrated, and insofar as we are concerned with general properties of cells we can only take note of their universal presence among higher organisms and bear in mind that their localization in membranes coupled with their highly apolar character may have much to do with the kinds of barriers which cells offer to the passage of substances. This just about brings us back to Overton at the turn of the century who accounted for the permeability of different kinds of molecules into living cells by assuming that their surface was fatty in nature.

The second group of membrane-associated lipids is known as phospholipids. In most respects these are identical with the neutral fats, but one difference exercises a deep-seated effect on the physicochemical character of the molecule. The formula for lecithin (a common phospholipid) shows that one arm of the glycerol axis is radically different chemically from the two which are attached to fatty acids (see Fig. 4-11). The arm which carries phosphoric acid and a choline group is highly polar in character. Whereas the first two arms would have a strong affinity for fatty substances, the third would have a strong affinity for polar ones. Such coexistence of apolar and polar groups within a single molecule is common in proteins but phospholipids are much smaller in size and hence are capable of orienting themselves easily

Neutral fats

Basic structure:

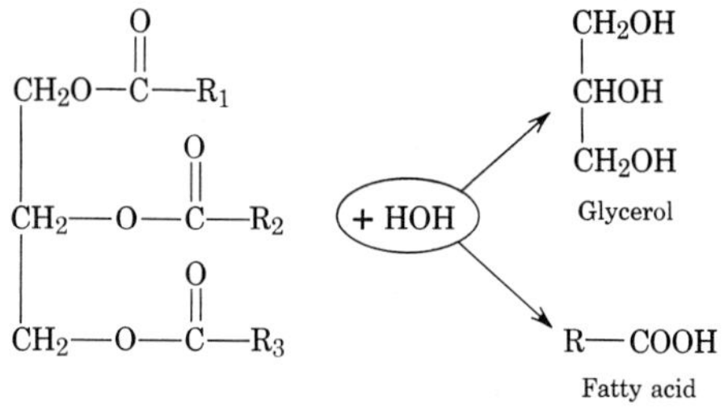

R-groups

Palmitic acid: $CH_3—(CH_2)_{14}—$(saturated)

Oleic acid: $CH_3—(CH_2)_7 \cdot CH{=}CH \cdot (CH_2)_7—$(unsaturated)

(a)

Fig. 4-11. Types of lipid molecules. (*a*) Neutral fats. (*b*) Phospholipids. The molecule glycerol is common to neutral fats and phospholipids. A large number of fatty acids (designated as R) are found in nature, and all the combinations which they form with glycerol are yet to be explored. From the standpoint of cell structure the phospholipids appear to be the more important group of compounds. Neutral fats are common as storage forms of energy, whereas phospholipids are universal components of membrane systems. Note that in phospholipid molecules the highly apolar chains of the fatty acids are oriented in an opposite direction from the polar phosphate complex. Although frequently the fatty acids are represented in straight chains, such chains actually fold in various ways. This feature may be seen in the third group of lipids, the steroids. The upper right part of the formula for cholesterol is essentially a chain of five carbon atoms with a methyl group attached to the first of these and two methyl groups attached to the fifth. This chain may be folded in a variety of ways, and frequently a folding pattern different from the one shown here is drawn in order to illustrate biochemical relationships between cholesterol and other steroids.

within a relatively small distance. The peculiar tendency of phospholipids to orient themselves was long known from the so-called myelin films which they form when suspended in water. A somewhat similar situation occurs when protein, phospholipid, and water are mixed (Fig. 4-12), except that the polar groups of the phospholipid attach to the protein; under the electron microscope the mixture takes on the appearance of a double membrane which looks very much like the natural membranes of cells. It is, in fact, commonly believed that the small phospholipid molecule is a key substance in the structural organization of membrane systems. This belief is reinforced by the fact that phospholipids are universally present in all types of cells, whereas steroids are virtually absent from bacteria and blue-green algae.

With the discussion of lipids we have brought to a conclusion a

Basic structure:

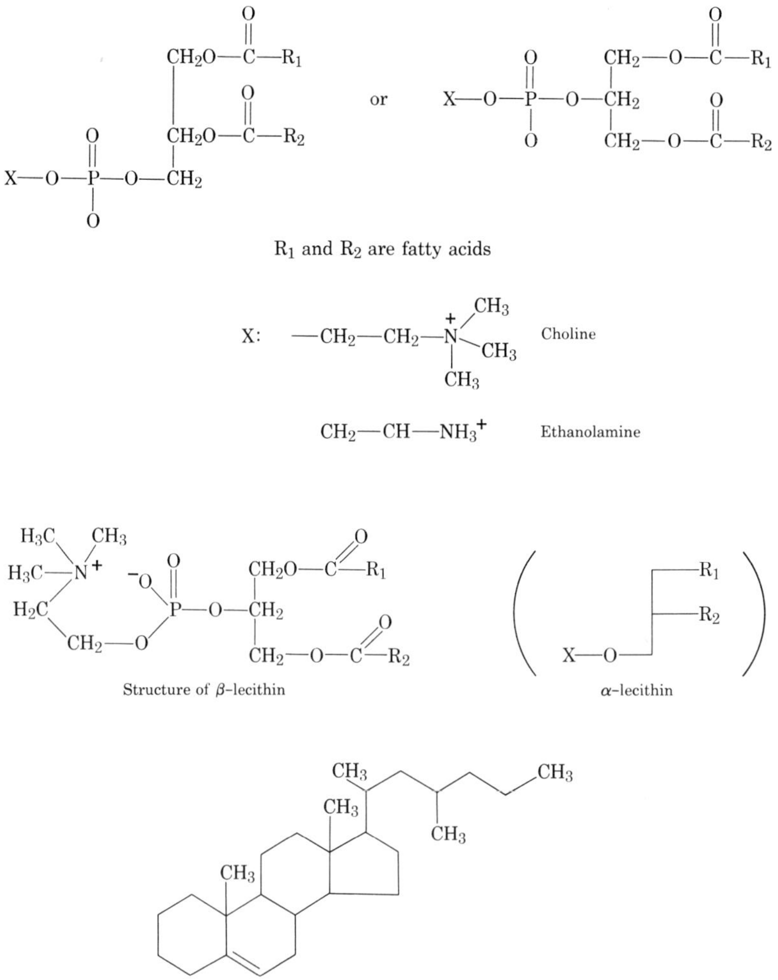

Sterols (cholesterol)

(b)

second broad line of development in cell biology—the identification of organic molecules. In retrospect, it is easy to identify the major advances which have so enriched our background of knowledge. At first the pace was exceedingly slow; until Lavoisier demonstrated that compounds were simply combinations of elements, nothing useful could be achieved in the studies of organic substances. It took almost one hundred years before the foundations which he laid were sufficiently

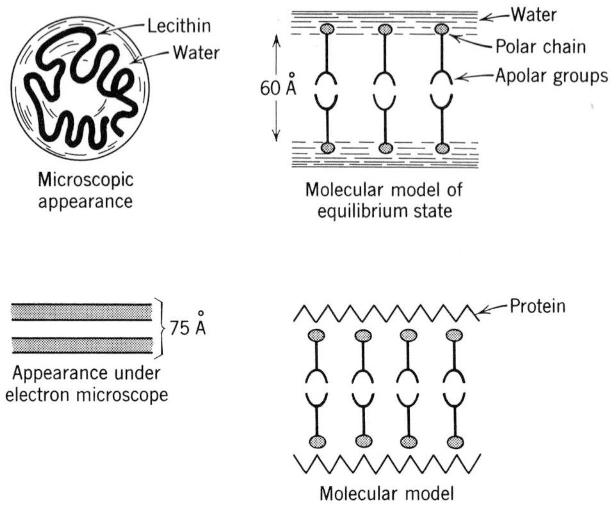

Fig. 4-12. Schematic representation of structure resulting from the mixing of phospholipid with water. (*a*) Myelin forms—a mixture of lecithin and water. (*b*) The unit membrane. The earliest observations of the unique behavior of phospholipids in water were on myelin forms. These were obtained by placing a small amount of lecithin in a droplet of water on a microscope slide. The tendency of lecithin to assume a configuration which would yield a maximum interfacial area between itself and water led to the molecular interpretation shown. Much later, with the advent of electron microscopy, the patterns and dimensions of myelin figures could be determined in the molecular range. Various experiments led to the conclusion that the dense areas represented the regions of the polar groups. Moreover, in the presence of protein the thicknesses found led to the conclusion that the protein had to be in the extended rather that the globular form. This basic type of aggregation in films came to be known as the "unit membrane." Many biologists consider all limiting cytoplasmic membranes to be organized in this way. The question is, nevertheless, still an open one. The molecular model assumes that the hydrophobic regions in the protein molecule do not interact with similar regions in the phospholipids. Such interactions may well exist, and if so, the model would not be valid.

developed to enable organic chemists to begin studying the complex molecules of living matter. And not until the early decades of this century did it become evident that the substance of life, like all other substances, was made up of molecules with a precise atomic composition. For many biologists the concept of extraordinarily large molecules (but molecules nevertheless) was difficult to accept. Colloids, because they were different from most forms of matter, seemed easier to reconcile with the unique properties of life. Only in our own time have we come to realize how much bigness in itself can alter the properties of a molecule. Looked at in historical perspective, modern cell biology is logically developing a subject initiated by nineteenth century organic chemists who were curious about the molecular composition of living matter. Large molecules as such are not, however, life itself. They are structures subject to other natural forces which give living systems their dynamic qualities. It is to these other forces that we now turn, and their elucidation too, can be traced back to the scientific studies of the late eighteenth and nineteenth centuries.

5

The Study of Chemical Change

Just as diversity was apparent from the dawn of human consciousness, so was the persistence of change. Shelley pithily expressed the feeling of all men of all ages when he wrote "Worlds on worlds are rolling ever From creation to decay, Like bubbles on a river Sparkling, bursting, borne away." No scientist could escape this sense of change, and if many ignored it in their studies, just as many pursued it with unbounded enthusiasm. The alchemists are often looked on as dedicated to uncovering the secret of change, but they hold little more than historical interest for us. Goals far less spectacular than the transmutation of elements had to be met if understanding of chemical changes was to make any progress. And without such understanding no realistic interpretation of living processes was possible. Theories about life abounded, but at best they represented comforting explanations concocted in the shadow of almost complete ignorance.

Skilled observers were not lacking. Useful anatomical descriptions of plants and animals date back to the Greeks, notably Aristotle and Theophrastus. The studies of human anatomy published by Vesalius around 1543 are without peer. In 1735 the preliminary draft of *Systema Naturae* by Carolus Linnaeus provided, apart from genetics, most of the basic concepts of taxonomy. Although without these studies biology would have remained a chaotic display of forms, with them alone there was little possibility for deeper understanding. Chemical descriptions too, however acutely they touched on functions, could not master the process of change until some general conceptions of physicochemical change came into being. Two principal barriers to the understanding of change existed when Antoine Lavoisier was conducting his researches at the close of the eighteenth century. It is impossible to comprehend a process of change unless one can also identify in the same process the components which remain unchanged. The historic lift which Lavoisier gave to chemistry was precisely in the elucidation of this paradox. He demonstrated that matter was neither created nor destroyed, and that the persistence of matter was to be explained by the immutability of the elements. Once Lavoisier's views had become established, the chemist had a solid footing, for he could take it as axiomatic that chemical changes were merely recombinations of identifiable and immutable chemical elements.

Difficult as it was for scientists to achieve this simple characterization of chemical behavior, there was a second barrier which proved even more formidable. Material description of change was but a half-description; some explanation of the mechanisms of change was also required. Lavoisier may have shocked a prejudiced world, but he did not, at least in his chemical studies, strain the intellects of his fellowmen. The idea that matter consisted of unchanging elements was a familiar one; the novelty of Lavoisier's ideas lay in the proposition that these elements were measurable entities free of the somewhat mystical attributes inherent in the phlogiston theory. But to explain change called on the powers of human abstraction, and abstractions arouse prejudiced suspicions. Galileo and Newton gave the world of science its first major abstraction—the attractive force of gravitation. It was clearly set forth by Newton in 1687. The conception of energy, which was essential to explain the nature of change, was not adequately elaborated until the middle of the nineteenth century, about 150 years after Huygens, a contemporary of Newton, hazily perceived it. The history of biology clearly mirrors the disparity in developments of the concrete facts of chemistry and the abstract conceptions of physics. Liebig, the early master of biochemistry, could find no flaws in his belief that decay and fermentation were simple chemical processes unrelated to the vital properties of organisms. Yet this view was expressed almost 200 years after Newton's publication of the *Principia* and less than 100 years before our own time.

THE CONCEPT OF ENERGY

Theories of matter based on concrete entities, the atoms, for example, may seem complex or strange but they do not strain the intellect. Abstract propositions are not always amenable to pictorial representation, and at the extreme their understanding requires talents which are better displayed by the philosopher than by the empirical scientist. Even today an uninformed biologist will gracefully accept the existence of the protein molecule but he is frequently dismayed by a term such as "free energy." In almost everybody's mind there is a level of abstraction at which an uneasy truce exists between the eagerness to understand and the willingness to ignore. Our own age, like any other, manifests the consequence of this trait.

Nothing could have been more apparent to man than that a moving body, say, a stone, was no different in composition from the same body at rest. Explanation of motion was one of the earliest challenges to the powers of human abstraction and the Greeks had already responded to this challenge in sophisticated ways; Aristotelian interpretation of motion remained dominant for about 2000 years. The idea that a "force" is essential to motion was commonplace, but the abstraction as such did not specify a predictable relationship between the two. Obviously, force could not be tangibly isolated; it could only be measured. The great seventeenth century revolution in science began with the attempt to resolve in a quantitative way the phenomenon of motion. It reached its peak through the work of Galileo and ultimately of Newton. The laws of motion which Newton formulated did not explain the instrinsic nature of force any more than did the speculations of Aristotle, but they did show that a predictable relationship existed between the unseen force and directly measurable properties such as mass, velocity, and acceleration. In this achievement lay one of the principal keys to all future developments in science: the intangible properties of matter—motion, heat, electricity, magnetism, and light—were shown to be accessible through measurement. To this day it is clear that however we interrelate these properties, and whatever the analogies we may choose to draw for purposes of didaction, we cannot escape from the realm of abstractions.

The philosophic implications of Newton's mechanical universe had an immediate impact on biological speculations. But the impact was diffuse, for Newtonian mechanics could be aimed at only a few specific targets. The relevant developments in physical and organic chemistry were yet to come. More important, a number of relationships had yet to be established between the laws of motion and the different properties of matter. This string of relationships leads to the concept of energy, without which biology could never have emerged as a dynamic science.

Newton's first law states that a moving body maintains its state of motion unless acted on by some outside force. This means that the moving body has "something" derived from the application of force which a resting one does not. The "something" was called momentum and was defined as mass × velocity (mv). Newton did not try to explain force, though he considered it essential to define force in a way which would permit its measurement. His solution was simple. Since a moving body of constant mass does not alter its velocity unless force is applied, then force could be measured as a function of the acceleration it produces. A force (F) is therefore equal to mass × acceleration (ma). Nothing was yet said about energy; nor was there any need to. A hint for such a need lay in the relationship pointed out by Huygens, among others, that when moving objects collide the total momentum of these objects remains the same. The simplest illustration of this is that of a moving ball striking a stationary one under conditions where the moving ball stops and the stationary one begins to move. The principle of conservation of momentum states that $m_a v_a + m_b v_b = m_a v_{a1} + m_b v_{b1}$. If $m_b v_b$ *and* $m_a v_{a1}$ are zero, the total momentum of ball *a* has been transferred to ball *b*. What happens, however, as is evident in common experience, when ball *a* comes to rest due to frictional forces alone? Momentum is not conserved; has it then simply vanished? The explanation was contained in experience almost as old as history itself. The use of friction to generate heat and produce fire is among the earliest of human inventions. How do we link the production of heat with loss of momentum? Here again we must operate among abstractions, for there is no way to show a tangible conversion of intangible momentum into intangible heat. Christian Huygens perceived the broad issue perhaps a little more clearly than his contemporaries, but he did not concern himself with heat if only because accurate thermometers had not yet been invented and heat remained an ill-defined property.

Huygens introduced a new equation to characterize the interaction between idealized elastic spheres. Alongside the equation for conservation of momentum, he introduced:

$$m_a v_a^2 + m_b v_b^2 = m_a v_{a1}^2 + m_b v_{b1}^2$$

Similar though the equations appear to be, one cannot be derived algebraically from the other. The new expression mv^2 implied different relations. Velocity has direction and must be represented algebraically as plus or minus. Velocity squared is always positive, and to a mathematician this meant that there must be some property of matter which is distinct from velocity. Baron von Leibniz called it "*Vis visa*" (vital force); about 100 years later it was called "kinetic energy." The name itself imparted little by way of understanding, and the equation as given did not extend comparisons beyond those concerning elastic bodies in motion.

One extension of this equation, however, set the stage for the full development of an energy concept. It was obvious that force could be applied to an object for different lengths of time; in terms of motion this meant that force could be applied over different distances. The product force $\times$ distance thus had a distinctive meaning and was called "work." The practical usefulness of this simple equation became fully evident during the Industrial Revolution when objective criteria were necessary to compare the efficiencies of different machines. The term "horsepower" serves as a reminder of the times when the work power of machines was being compared to that of horses. The utility of the term is minor compared to an equation which could be algebraically deduced and which specified the amount of work which had to be done to arrest a moving body. The equation read simply

$$F \times \text{distance} = \tfrac{1}{2}\, mv^2$$

In this equation lay the experimental solution to the relationship of loss of momentum and production of heat. A direct test of the equation was easily accomplished by measuring the distance required for a known force to arrest a moving body of known mass and velocity. All that was needed to extend the application of this equation was to devise a way of producing heat by mechanical work in an arrangement where both parameters could be measured. If heat could be equated with work, it could also be equated with mv^2. And if mechanical work could be used to produce heat, could it not also be used in measurements of other intangibles such as electricity?

Four men stand out in the history of the subject—Count Rumford, Julius Mayer, James Joule, and Hermann Helmholtz. Together they unequivocally established that for a given amount of mechanical work a definite amount of heat or other form of energy was produced. Count Rumford initiated the sequence by measuring the heat produced during the boring of a cannon. Mayer dealt lucidly with the abstract proposition that energy was indestructible. Joule brought to virtual completion the demonstration of a strict equivalence between work and heat. He measured the amount of mechanical work necessary to operate a generator in order to produce electrical current yielding a defined amount of heat, the amount of work necessary to cause water to flow through narrow pipes in order to yield a given quantity of frictional heat, the amount of work required to compress a gas so as to produce a given amount of heat, and the amount of heat produced for a given amount of work in stirring a paddle wheel in an insulated water container. The reference points were always work done and heat produced, because in each case precise measurements were possible. It soon became clear that regardless of the experimental system used, the equivalence was the same. If the operation of an electrical generator yielded the same result as the operation of a paddle wheel, then a common denominator must underlie all these changes and the denominator must be distinct from the physical and chemical properties of the components of the system.

It was the physiologist and physicist Hermann von Helmholtz who perceived that the capacity of any system to do work was a measure of the energy available. His law of conservation of energy showed, with mathematical precision, that in all natural phenomena energy could neither be created nor destroyed. And if chemical, mechanical, electrical, or magnetic systems could directly or indirectly be caused to produce heat, it was clear that the common denominator, energy, was present in all natural systems. The impact of this declaration was probably even greater than that of the conservation of mass, for it made plain that in all processes involving change two properties remained constant—mass and energy. With Einstein's equation $E = mc^2$ (where c = velocity of light) the two laws became one. Table 5-1 serves mainly to indicate the arbitrary nature of the energy units, but it also provides a source of reference for interpreting various equations.

None of these arbitrarily defined and interrelated values made energy less abstract a term, for energy *is* an abstraction. Generally, we speak of forms of energy—light, heat, motion—as though energy were something we first knew and came to identify in a number of forms. To represent our thought processes more accurately we should use the phrase "energy of forms"; yet the fact that we habitually speak otherwise, testifies to the secure position of this concept in our thinking. It should not be too difficult to see how establishment of the energy concept provided for a new turn in biological thinking.

Any number of simple examples could be chosen to illustrate this. Living organisms produce heat, a fact which greatly intrigued Lavoisier; but heat is a form of energy and since it cannot be created the organism must have acquired a source of energy from the outside world. The wood of a tree, if burnt, yields a great deal of heat, but this tree began from seed and could not have released so much energy unless it was acquired during growth. Indeed, whereas Liebig could argue that fermentation was unrelated to the vital properties of organisms, his successors could no longer do so. For just as the organic chemist was bound by the proposition that the elements of compounds were in-

Table 5-1 Energy Conversions

Mechanics

1 dyne of force produces an acceleration of 1 cm/sec^2 on a mass of 1 g
1 erg of work equals 1 dyne acting over 1 cm
1 joule of work equals 10^7 erg

Heat

1 calorie equals amount of heat required to raise temperature of 1 g of water from 15.5° to 16.5°C

Electricity

10 amperes of electrical current passing through a conductor of 1 cm length shaped as an arc of 1 cm radius exerts a force of 1 dyne on a magnet of fixed strength (1 unit magnetic pole) placed at the center
3600 coulombs of electrical current equal 1 ampere × 1 hr
1 volt of electrical potential between two points or systems is capable of causing a flow of 1 ampere over 1 ohm of resistance

Equivalences between energy forms

Mechanical equivalent of heat: 4.184×10^7 ergs of work = 1 calorie
Electrical equivalent of heat: 1 coulomb passing between 1 volt difference = 10^7 erg
Chemical equivalent of electricity: 1 faraday (96,500 coulombs) causes a chemical change in 1 gram equivalent of substance

Constants used in thermodynamic and other equations

R (the "gas constant"). The appropriate number is indicated by the units used in a particular equation:

$$8.3 \times 10^7 \text{ erg}; \ 1.987 \text{ cal}/1 \text{ degree} \times 1 \text{ mole}; \ \frac{0.0821 \text{ liter} \times 1 \text{ atmosphere}}{1 \text{ degree} \times 1 \text{ mole}}$$

G, Force of acceleration due to gravity: 981 cm/sec^2

destructible, so were the new physiologists bound by the principle that every change in an organism had to be balanced not only in terms of weight, but also in terms of energy. It was not at once apparent how and in what forms energy was transmitted, but it became axiomatic that an explanation which did not account for energy was no explanation at all.

Wherever there is matter, there is some form of energy. This is the simplest statement of the Einstein equation, the truth of which has become painfully apparent in the production of nuclear weapons. As technology grows, so does the ability of man to transform one kind of energy into another. Solar batteries transform light into electrical energy, and the chemical energy in a flashlight battery is transformed into electrical energy which in turn is transformed into light. The momentum of falling water is used to generate electricity, and electrical energy is used to pump water to various heights. However familiar these processes have become, energy remains the unseen and intangible component. If we no longer attribute to energy the mysterious powers that were commonly assigned to it even after the age of Newton, it is not because we have come to understand its intrinsic qualities. We have mastered the mystery only in the sense that we are able to write equations which predict the course of chemical and physical change. But energy remains the unseen merchant, the friendly ghost who trades the properties of matter in so utterly predictable a fashion that his invisibility has come to be regarded as an unimportant idiosyncracy of character.

CHEMICAL EXCHANGES BETWEEN CELL AND ENVIRONMENT

The concept of energy could be meaningfully applied to biological systems only after certain basic chemical interactions between living forms and their environment were understood. Historically, however, the sequence was not simple. Ideas related to theories of energy were being developed at the same time that chemical theory was acquiring durable foundations. And both fields of inquiry were having an impact on biological thinking. In retrospect, we can see that a concept of energy was a primary requirement for understanding the chemical dynamics of cells. Yet long before this requirement became clear, the dependence of organisms on their environment for growth and maintenance was a well documented fact. There are two aspects to this dependence: the acquisition of chemicals for the building of organic molecules and the acquisition of energy for the performance of various kinds of work. The first aspect presented itself primarily as a problem of elucidating specific chemical reactions, the second of interpreting chemical reactions in terms of concepts of energy. Respiration, fermentation, and photosynthesis were the three processes of chemical exchange which needed explanation and a full explanation is yet to be achieved. Details of these processes will be considered in Part II; this chapter provides the outline to serve as background for future discussions.

One facet of change must have impressed itself early on human consciousness: that breathing ceased with death. Some 2000 years after the flourishing of Greek civilization the meaning of this was still entirely unclear. In the seventeenth century no one had any chemical understanding of air. There were some fuzzy ideas such as those of

Hooke, who between 1660 and 1680 suggested that air might have two parts. If nothing else, the extreme interest in gunpowder forced speculations such as Hooke's. Some of his contemporaries even conceived of a basic identity between inorganic combustion and respiration. Yet however perceptive some of these suggestions, they held little sway in directing chemistry and biology until the mysticism surrounding the nature of air gave way to concrete demonstrations of its actual composition. This is one item in the history of science in which biology and chemistry are so intertwined that it is impossible to discuss the two disciplines individually. When Robert Boyle developed his vacuum pump he clearly observed, as did some of his predecessors, that removal of air extinguished flames and made life impossible. And when Joseph Priestley discovered oxygen in 1774, he did so by demonstrating that on heating the red oxide of mercury, a gas was given off which made life possible. Lavoisier drew heavily on the findings of his contemporaries, and they were all interested in the nature of gas exchanges between organisms and their environment. That interest, as we have said, dates back to the unseen dawn of human history.

It is a simple matter to describe Lavoisier's contribution to this basic biological problem. He succeeded more clearly than his contemporaries because he made precise measurements. In doing so he was able to demonstrate that during combustion the material combusted combined with oxygen. By 1782 he had more or less convinced the scientific world that respiration was a process in which oxygen was consumed and carbon dioxide produced. This was in no sense either a complete description or an explanation of the process. Lavoisier had the chemical description; he did not, as we shall see, understand its implications. Yet once the description was at hand, its universality was soon recognized. The biologist Lazzaro Spallanzani, his contemporary, made plain before the close of the eighteenth century that all forms of life absorb oxygen and give off CO_2. Moreover, he demonstrated that not only did organisms respire in this way, but that living tissues within an organism also absorbed oxygen and evolved carbon dioxide. Since tissues are assemblies of cells, we may attribute to Spallanzani the discovery that living cells respire. By the early nineteenth century one of the major categories of chemical exchange between cells and their environment could be written in precise chemical terms:

$$C_6H_{12}O_6 + 6O_2 \rightarrow 6CO_2 + 6H_2O$$

Except for the energy component, this equation retains its validity today. Carbohydrate oxidation is the principal source of energy for all respiring cells.

Against this generalization must be placed other observations which indicated that respiration was not a universal characteristic of life. Spallanzani himself had observed that snails would produce CO_2 in an atmosphere of nitrogen or hydrogen. Although the exchange of oxygen for CO_2 appeared as a fairly clean and clearly definable process, the array of other chemical changes which had been observed to occur in living systems was neither clear nor readily definable. A number of these changes related to technological arts such as alcoholic fermentations and bread leavening, which had been practiced for thousands of years. Together with the extremely old observation that tissues of dead organisms undergo decay ("putrefaction") they provided a target for scientific curiosity which, at a time of growth of chemical knowledge, was bound to attract many experimentalists. Fermentation, leavening, and putrefaction were all manifestations of chemical change and all involved organic materials. Gas evolution and heat production were common to all three. These superficial similarities were recognized, but throughout the eighteenth century and much of the nineteenth the common factors were lost in metaphysical speculations. Justus von Liebig, who rightly saw that these processes had much in common with food breakdown in living organisms, refused to acknowledge any relationship between such chemical breakdown and vital functions. It may not be at once obvious why these special instances of organic activity should count significantly in the development of biology. However, the deeply held prejudices which perceived only an unbridgeable chasm between organic decomposition as a chemical force and organic synthesis as a vital force made a fruitful study of vital functions impossible. This chasm was widened rather than narrowed by organic chemists such as Friedrich Wöhler who regarded the chemical synthesis of urea as evidence that chemical transformations of organic compounds proceeded independently of vital activity. Two of the very great achievements in nineteenth century biology were the demonstrations that both processes were due to living organisms, and that only living organisms possessed the machinery which made these processes possible.

Studies of putrefaction and fermentation provide one of the liveliest chapters in the history of biology, for it is here that the great battle concerning spontaneous generation was waged. Fundamentally, however, the issue of spontaneous generation, though related to the problem of energetics, was not part of it. The point which had to be clarified was how respiration and fermentation were equivalent insofar as both processes were essential to the maintenance of life. Louis Pasteur played a major role in this clarification. His original interest was in clarifying the chemistry of well known and widely utilized fermentative processes; his principal achievement lay in demonstrating that these processes were as essential to the viability of microorganisms as was respiration to the viability of animals.

Spallanzani had inferred from his various observations that both fermentation and putrefaction were causal factors in growth of microorganisms, but the inference did not convince his contemporaries. Even when Schwann and others had discovered that yeast cells were associated with alcoholic fermentation, the response of chemists such as Liebig and Wohler was to attack the idea that the association had any real significance. In the midst of this controversy Pasteur provided the clear and convincing demonstration that the growth of a population of yeast cells was correlated with the extent of alcoholic fermentation. Yeast cells were a material of choice for these studies for one very important reason. As will be discussed in Part II, such cells are capable of both fermenting and respiring. If a suspension of yeast is aerated, fermentation does not occur. Pasteur showed both that in the presence of air sugar was principally converted to CO_2 and that in the absence of air it was converted to CO_2 plus alcohol, and that the amount of yeast growth per unit mass of sugar consumed was much greater under aerobic conditions. The thesis that carbohydrate breakdown was essential to life could no longer be questioned in face of this demonstration.

Not only did Pasteur clarify the relationship between yeast cells and fermentation, but he also demonstrated that the type of fermentation (e.g., lactic acid vs. ethanol production) depended on the type of microorganism present. Unlike yeast, some microorganisms could not even survive in the presence of air. For such organisms, fermentation was obligatory. From these studies Pasteur generalized that the maintenance of life required either respiration or fermentation, the latter occurring in the absence of air. What Pasteur did not perceive, however, were the processes within the cell which led to the formation of the different products. The manner in which cells satisfied their universal need for energy was nevertheless basically demonstrated even though the significance of energy transformations and the specific mechanisms of transformations were yet to be elucidated.

The basic chemical equations for fermentation were more difficult to establish than those for respiration or photosynthesis. Fermentations involved a variety of organic by-products, whereas the other processes could be summarized largely in terms of gas exchanges. The elucidation of photosynthesis had the same meteoric history as that of respiration. In 1772 Priestly discovered that green plants could restore air which had become spoiled by animal respiration. About seven years later

Jan Ingenhousz demonstrated that such restoration was impossible unless light was also present. In 1782 Jean Senebier showed that "fixed air" (carbon dioxide) was essential to the formation of "dephlogisticated air" (oxygen), and following this Ingenhousz showed that the only possible source of carbon for plants was that found in carbon dioxide. When Nicolas de Saussure at the beginning of nineteenth century demonstrated that water as well as carbon dioxide had to be used in photosynthesis in order to account for the weight of carbohydrate synthesized, the chemical description was completed.

The inadequacy of the chemical description was perceived by Mayer, one of the four individuals cited in the development of the concept of energy. He pointed out that the light which was essential to photosynthesis was the source of energy for the process. The green plant was the instrument which converted light energy into a chemical form. In photosynthesis, more than in any of the other processes, a clear distinction could be drawn between the need for energy and the mechanisms by which energy was transformed. Mayer was concerned with the first of these relationships, and not until recently has any progress been made with respect to the problem of energy transformations.

THERMODYNAMICS AND CELL BEHAVIOR

Helmholtz's principle of conservation of energy became known as the first law of thermodynamics (dynamics of heat), for it signaled a new field of study concerned exclusively with the laws governing the transfer and transformation of energy. By itself the first law provided only a broad outline. The biologist could confidently say that respiration or fermentation must be the source of energy for organisms, because this process represented the chemical course of ingested food. Or he could assert that sunlight must be the source of energy for photosynthesis because if the breakdown of carbohydrate plus oxygen into CO_2 and water yielded energy, then the formation of carbohydrate and oxygen from CO_2 and water required energy. Similar reasoning could be applied to the formation of substances in the course of growth and to mechanical work which organisms perform. But once these relationships were appreciated, they were no more than rough guides. The new limitations opened new possibilities, for the biologist was now challenged to answer the question of how these processes made energy available and how the available energy was utilized. This basic question (which will be systematically considered in Part II) would have offered little resolution were it not for additional rules of behavior provided by thermodynamics and modern physics.

From what has been said about the nature of those properties once called "imponderables," it should be clear that thermodynamics was bound to be concerned with abstract mathematical propositions yielding equations that could be experimentally tested. It would take us too far afield to consider even the major systems subjected to thermodynamic analysis, even though they provided additional concepts which we routinely apply to biological studies. For our purpose we will discuss briefly two principal concepts which underly all the molecular changes occurring within a cell.

One of the questions which presented itself to the new science was whether any rules could be formulated concerning the capacity of systems to do work, regardless of the particular nature of the system or of the form of energy made available for the performance of work. In the pure world of Newtonian mechanics this question did not arise. A mechanical force applied under ideal conditions could obviously be totally transformed into work, but in a world where steam engines were being used, where chemical batteries were generating electricity, and where electricity was producing heat or performing mechanical work, the answer was by no means obvious. Heat was a form of energy and yet heat could be harnessed to perform work only where differences in temperature existed. How did temperature difference relate to the work capacity of a system? And was there any over-all rule determining which chemical reactions could occur and which could not? Thermodynamics was looking in effect for a universal law which governed the possibilities of change.

As a result of this search a new concept was introduced—free energy. This was defined as the energy available in a system for doing work. The word "work" is not to be taken in its literal sense. In the thermodynamic sense "work" is $F \times s$ (distance); Fs is equal to $\frac{1}{2} mv^2$, which in turn is equal to defined amounts of all the other forms of energy we have considered. Thus the free energy of a system is energy which is removable from that system, but whether the energy removed is used to perform mechanical work or is locked in some other form is immaterial to the definition. Of primary interest to the biologist is the sweeping rule which states that a chemical reaction will occur only if it results in a decrease of free energy by the reactants. This at once set down a rigid requirement for all chemical changes in living things. Every reaction system in a cell must lead to a set of products in which the free energy is decreased. Thus a reaction which is impossible in the inanimate world is also impossible in the living world. The biologists had now to explain the chemical miracle of life not in terms of transgressing universal physical laws, but in terms of their application.

Before turning to the question of how free energy is measured, one general feature of the concept is worth noting. The definition does not specify the channels through which the energy is extracted. If, for example, glucose plus oxygen has a higher free energy relative to CO_2 plus water, then according to the free energy concept the reaction can take place regardless of the kinds of intermediate chemical changes in the process. Without knowing the details of a biochemical reaction, we have a rule which enables us to predict whether the components assigned to a reaction mixture make the reaction possible. We can say at once, for example, that the chemical reactions occurring in photosynthesis are by themselves impossible. The chemical description is therefore incomplete and can only be made complete by adding to the system some source of energy which will satisfy the rule; for photosynthesis this energy is light. In this way we may examine every constituent of the living cell and predict whether the ingredients we suspect to be involved in a particular synthesis are by themselves adequate to accomplish it. It will later be seen, for example, that none of the large molecules in a cell can be formed solely out of their subunits.

The free energy rule enables us to go beyond this qualitative description. By calculating the free energy in a particular reaction we can predict the total work possible. The oxidation of one mole of glucose, for example, yields the equivalent of 673 kilocalories of heat. If we want to know how a muscle can perform a given amount of work, we immediately set down the proposition that at least 1 mole of glucose must be completely oxidized for the equivalent of 673 kcal of work performed. Precisely the same kind of reasoning can be applied to every physiological process, whether it involves mechanical, electrical, or chemical work. It is worth reflecting on how this game of rules with an abstract property, which we can never hope to see or isolate, can dictate so unequivocally the events that occur in living systems. No exceptions whatsoever have been permitted; the rules apply as much to the simple virus as to complex man.

How is free energy determined? We must first clarify "reaction system" since it has implications not only for studies of present day life, but also for speculations concerning life's origin. Of the various types of reaction systems it is best to consider those which have the same characteristics as living organisms. Although basic thermodynamic relationships may be derived from the operation of a steam engine, this mechanical system involves changes in temperature, pressure and volume, whereas organisms generally function under conditions of constant temperature, pressure, and volume. The world of *chemical* changes is the one into which living systems fall, and it is this world which we will subject to thermodynamic considerations in clarifying reaction system.

It is well known that fats become rancid on exposure to air and that the rancidity is largely due to chemical oxidation rather than to the action of microorganisms. From this observation we would correctly conclude that fat *oxidation* is an energy-yielding process. If, however, these same fats are maintained in an atmosphere of hydrogen, they become reduced and we would therefore have to conclude that fat *reduction* is an energy-yielding process.

The conclusions are obviously incompatible, but since the facts are established beyond doubt our conclusions must be misrepresenting the situations. The misrepresentation is easily located. In each case we failed to identify the total reaction system. In the first our system was fat plus air; in the second it was fat plus hydrogen. The biologist is accustomed to think of biochemical reactions in terms of air and so usually omits its mention. Suppose, however, as was probably the case when life originated, that the atmosphere contained no oxygen and was even partly composed of free hydrogen. Under such conditions the statement that respiration of carbohydrates is a source of energy would be meaningless and in error. What must therefore be appreciated is that the thermodynamic rule can only be applied in a comparative way. We must always compare initial with final products. There is no free energy of a compound completely isolated from every thing else. The definition of free energy as "the capacity to do work" means that the compound must change when releasing that energy. The extent of change and any surrounding conditions influencing the change will affect the free energy value. The free energy of the reaction glucose + air yields CO_2 + water is very different in magnitude and sign from that of glucose + air yields the elements C, H, and O.

In order to illustrate how free energy is measured, we must consider "equilibrium reactions." Using the classical example of a system containing ethyl acetate, water, acetic acid, and ethyl alcohol, we may write the equation for the reaction as:

$$C_2H_5OH + CH_3COOH \longleftrightarrow C_2H_5OOCH_3 + H_2O$$

This reaction occurs when we mix equal amounts of ethyl alcohol and acetic acid and allow the mixture to stand; ethyl acetate and water are formed. From this we would conclude that as the reaction moves from left to right there is a decrease in free energy of the system. But if ethyl acetate and water are the starting materials, the reaction proceeds from right to left and we are now obliged to reach an opposite conclusion. Here again we must have omitted an essential factor from our conclusion. The omission becomes apparent if the reaction, whether started from right or left, is allowed to continue until no further change occurs. At that point it will be found that a definite proportion of the four components persists, and these amounts can be characterized in the same way as was done for the dissociation of weak electrolytes. Thus

$$\frac{[\text{Ethyl acetate}][H_2O]}{[\text{Ethyl alcohol}][\text{Acetic acid}]} = K$$

Since we already know from previous considerations that the value K remains constant and that changes in concentration of any one component will induce shifts in all, we may draw two conclusions about free energy relations. First, the concentrations of reactants have a bearing on the free energy value of the reaction. For this reason all tables of free energy values refer to specific states, usually the "standard state" which indicates molar concentrations of reactants and products. Second, the *direction* of the free energy change is determined by the relative concentrations of reactants and products. This second conclusion in effect states that a quantitative relationship exists between the equilibrium constant of a reaction mixture and the free energy of the reaction. Once the system is in equilibrium, no energy is available for work. Superficially, the statement appears to be a rather simple one. But if we try to explain it in molecular terms, the apparent simplicity at once disappears. It is easy enough to picture a situation in which two molecules A and B react to form C and to explain it by supposing that when the two molecules join to form a new compound they release a certain amount of energy. But why, then, should changing the relative concentrations of A, B, and C have any bearing on the energy relationship? If C is the more stable combination, how would increasing its concentration relative to A and B alter this comparative stability?

The substance of the question had its origins before the development of thermodynamics. It was implicit in the Arrhenius equation governing the partial dissociation of weak electrolytes. And yet we can only reiterate what has been said earlier. The rules are given by Nature; the scientist verbalizes them in mathematical language. Some rules are readily visualized whereas others are not. In this particular instance we are required to visualize a situation in which a particular ratio of molecules A, B, and C represents conditions of maximum chemical stability such that there is no tendency for changes to occur. The task is not impossible but it will suffice here to recall the phenomenon of mixed electrovalent and covalent bonds; the situation is not merely analogous but identical in principle. The mixture exists because it is the most stable compromise between the individual tendencies of the two processes. There are many reactions in which equilibrium never occurs in practice, as, for example, the reaction of metallic sodium with water. In such cases the equilibrium point is so far to one side that derivation of free energy values from equilibrium constants is impossible.

Consider the reaction A + B = C for which an equilibrium constant can be experimentally determined. By convention we calculate K_{eq} as the product of the equilibrium concentrations on the right side of the equation divided by that of the concentrations on the left side. Thus for the reaction as we have written it

$$K_{eq} = \frac{[C]_e}{[A]_e\,[B]_e}$$

Subscript *e* signifies concentrations at equilibrium; it is not generally used but we have included it for subsequent calculations. We may now frame the following question. If we combine the three substances A, B and C, each at a concentration of *1M*, in pure water, will they react and if so what would be the free energy of the reaction? We already know from our definition of free energy that any reaction which does take place must be accompanied by a release of free energy. Qualitatively, we could predict the course of the reaction from the equilibrium constant. If it is greater than 1, the reaction would be expected to proceed from left to right; if it is less than 1, it would proceed in the other direction. Our problem is finding some way to measure the magnitude of the free energy change. The reason for choosing 1*M* concentrations is that it simplifies the derivation and is also arbitrarily defined as "standard conditions." Reference tables on free energy values of chemical reactions apply to these conditions.

In order to derive a formula, we must enter a world of fantasy. We are required to imagine that one mole (i.e., 6×10^{23} molecules) each of A, B, and C interact without at any time changing the standard conditions specified initially. If concentrations are permitted to change as the reaction proceeds, then so do the conditions. It is important to bear in mind that a *1M* concentration does not imply the presence of one mole of substance. This would be true only if the volume of the reaction mixture were 1 liter; if it were 1000 liters, the last mole of substance to interact would do so under the concentration conditions approaching equilibrium. But as equilibrium is approached, the free energy of the reaction approaches zero. In a 1000 liter solution the last mole converted would yield little free energy, and yet the purpose of defining standard conditions is to provide a single value for the free energy released by the interaction of one mole of each of the substances. The paradox of interacting compounds altering the concentrations while at the same time leaving the concentrations unaltered is solved by pretending the impossible to be possible. An infinitely large volume of solution containing compounds

A, B, and C at 1*M* concentrations is placed in a vessel with a molecular trapdoor. Each time a very small number of molecules interact to reach equilibrium they are carried out through the trap door until one mole of the three components have thus interacted. Since thermodynamics is concerned solely with initial and final states, the physical chemist can take advantage of his ingenious device by ignoring the operation of the trapdoor and considering only the change in concentration which has occurred among the compounds removed from the reaction mixture. It is worth noting that although our interest is in the chemical reaction, calculation of the free energy value may be based entirely on changes in concentrations. From the standpoint of thermodynamics any change involves work. The amount of work done by a system is a function of the energy released; it is irrelevant to thermodynamic calculations whether the free energy is used to change concentrations or to generate electricity. For any particular system the total work done is a constant. It so happens that concentration changes may be equated with free energy changes in a simple way. When a substance changes from a concentration C_1 to C_2, the change in free energy is given by the equation

$$\Delta F = RT \log_n \frac{C_2}{C_1}$$

We will not attempt to derive this equation, but its form will gradually become familiar, as it applies equally to changes in volume, pressure, and electrical potential. Note that if C_2 is less than C_1, the logarithm will have a negative value, as will the free energy change. A negative ΔF means a decrease in free energy, and therefore dilution occurs spontaneously. How the energy released on dilution may be used is another matter. A simple way to illustrate this is by returning to the phenomenon of osmotic pressure. Put into distilled water a semipermeable bag containing a concentrated solution of a substance to which the bag is impermeable, and the penetration of water into the bag may be used to generate a pressure which the skillful technologist can convert to work.

We may list the changes which have occurred in each of the reactants A, B, C in the following way (where subscript 1 = 1*M* concentration and *e* equilibrium concentration):

$$(1)\ [A]_1 \rightarrow [A]_e \qquad \Delta F = RT \log_n \frac{[A]_e}{[A]} = RT \log_n [A]_e\ (A = 1)$$

$$(2)\ [B]_1 \rightarrow [B]_e \qquad \Delta F = RT \log_n \frac{[B]_e}{[B]} = RT \log_n [B]_e$$

$$(3)\ [C]_e \rightarrow [C]_1 \qquad \Delta F = RT \log_n \frac{[C]_1}{[C]_e} = -\ RT \log [C]_e$$

We have reversed the order of component *C* because it is on the opposite side of the original equation. This states that if we sum up the changes listed and subtract them from the original equation, we should be left with the final equation. Thus

$$\begin{array}{rrcl} & [A]_1 + [B]_1 & \longleftrightarrow & [C]_1 \\ - & [A]_1 + [B]_1 + [C]_e & \longleftrightarrow & [A]_e + [B]_e + [C]_1 \\ \hline & -[C]_e & \longleftrightarrow & -[A]_e - [B]_e \end{array}$$

Or
$$[A]_e + [B]_e \longleftrightarrow [C]_e$$

The sum of the free energy changes just listed is

$$\Delta F = RT \log_n [A]_e + RT \log_n [B]_e - RT \log_n [C]_e = RT \log_n \frac{[A]_e [B]_e}{[C]_e}$$

Because

$$K_{eq} = \frac{[C]_e}{[A]_e [B]_e},$$

therefore

$$\Delta F = -RT \log_n K_{eq}$$

The latter is the fundamental equation linking free energy change with the equilibrium constant and, because it applies to the standard conditions set, the free energy change is designated as $\Delta F°$. By an algebraic process identical with the one just used, it is possible to show that for any other set of concentrations

$$\Delta F = \Delta F° + RT \log_n \frac{[C]}{[A][B]}$$

For this transformation the concentrations chosen need not be the same for all. If, for example, we were concerned with the same reaction in a cell and found that the following concentrations were maintained in the cell—A, 0.2M; B, 0.2M and C, 0.05M—our first question would be whether this represents an equilibrium relationship. If it did, we would conclude that the cell cannot be gaining any free energy from the reaction because at equilibrium there is no tendency to change and $\Delta F = 0$. Use of the formula would in any event indicate this. Thus at the concentrations found in the cell ΔF would be

$$\Delta F = \Delta F° + RT \log_n \frac{[0.05]}{[0.2]\,[0.2]}$$

It is obvious that as we make C smaller the added expression becomes more negative, which would further increase the negative value we supposed the $\Delta F°$ to have. We would therefore conclude that under the conditions existing in the cell, that particular reaction was yielding more energy than the standard one.

This particular game of numbers has certain limitations. The fact that the concentrations we find in the cell are not equilibrium ones means that other reactions must be occurring which prevent the achievement of equilibrium. If the equilibrium constant were 4, then the very low concentration of C must be attributed either to a very rapid and continuous production of A and B or to a very rapid removal of C. In either case we are implicating other reactions, and since these must also involve free energy changes, we have a decidedly incomplete picture of how compounds A, B, C function in providing energy. Compounds A and B would be involved in the free energy changes associated with the reactions producing them, and C would be similarly involved in removal reactions.

In 1885 Le Châtelier stated the principle that if a change occurs in one of the factors affecting a system in equilibrium, the system will tend to adjust itself to annul the effect of that change as far as possible. This principle is, of course, a verbalization of thermodynamic relationships bearing on equilibria. It had a great deal of impact on the thinking of many biologists who considered this principle to characterize much of living behavior. The responses of cells to a variety of stimuli —heat, cold, light, electric and mechanical shock—lent themselves easily to that interpretation. This view was extended to molecular reactions. The cell was pictured as containing a variety of reaction systems in which the components were at equilibrium. When a muscle was active, for example, it was presumed to increase the rate of respiration by drawing off the products of respiration and thus upsetting the equilibrium. The argument was extended to all the chemical reactions occurring in a cell, and this interpretation made it appear as though the cell acted by continuously readjusting to the requirements of thermodynamic equilibria. Although current evidence does not rule out that certain adjustments may occur in this way, it definitely does rule out that all adjustments occur in that way. The most forcible argument against the view is based on a very obvious property of cells; they grow and they age. They do not remain the same. In the total cycle of life there cannot be a continuous readjustment to maintain a state of equilibrium.

There is, however, a more specific consideration bearing on thermodynamic equilibria. Let us note first that in determining the free energy of a reaction we have equipped ourselves with essential basic information about the possibility of the reaction occurring and about the maximum possible yield of energy which such a reaction permits. Moreover, we must also take into account the very many reactions for which the equilibrium point is hypothetical because experimentally the reactions move entirely in one direction. We cannot set up an equilibrium reaction system between proteins and amino acids, nor between nucleic acids and their precursors. Other methods must be used to calculate free energy changes, the basis for which will be discussed in the concluding paragraphs. But granted that a reaction is thermodynamically possible, its occurrence, as we shall see, is not at all assured. This remark takes us to the components of cells called "enzymes."

ORGANIC CATALYSIS

The principle of energy conservation, like that of mass conservation, provided a theoretical foundation on which chemical change could be rationalized. The concept of free energy provided the specific rule for predicting the course of chemical reactions. The principles and the concept were nevertheless of little use to the biologist without concrete instances of actual chemical reactions occurring in living cells. Respiration, fermentation, and photosynthesis were three such instances. What was known about them in the nineteenth century amounted to no more than a balanced chemical summary of exchanges between the cell interior and its environment. They were ultimately recognized as sources of energy because thermodynamic theory clarified the indispensability of energy to chemical and physical change. A deep chasm of unknowns nevertheless prevailed between the apparently rigorous explanation of these processes as energy sources and the actual nature of the component events. In what way, for example, did sugar utilization provide energy? What basis existed for the breakdown of sugar into CO_2 and water by aerobic organisms and for the breakdown into CO_2 and ethanol by some anaerobic ones? In what way did such breakdowns furnish energy? Answers could not be provided by the concept of free energy alone. The biologist first had to explain why organic reactions which released free energy occurred only in living organisms, and not in an inanimate environment. If free energy change were a sufficient condition for a chemical reaction, Liebig's distinction between organic change and vital activity would have been correct even though Liebig's reasoning was without the benefits of thermodynamics.

Long before the issues could be phrased in this manner, biologists had made observations which pointed to the presence of some cell constituents capable of promoting chemical reactions. As far back as 1752 René Reamur was curious about the problem of food digestion, and to satisfy his curiosity he experimented with a bird (a kite) which had the distinctive habit of regurgitating anything indigestible. He induced it to swallow some small laboratory vessels and collected the stomach or "gastric" juices which the bird returned along with the vessels; he found that the juices would dissolve foods. This was the first clear demonstration that the stomach produced a substance which was independently capable of breaking down foods. The rapidity of the process distinguished it from putrefaction and fermentation. Many studies of this kind were carried out over the next hundred years, the most outstanding of which was by the physiologist Claude Bernard. He had learned about the many interesting observations which an American army surgeon, William Beaumont, had made on a patient with a hole in his stomach caused by a gun-shot wound. By creating an artificial passage directly from stomach to exterior, Bernard hoped to make rather precise studies of the digestive process. He discovered that the principal phase of digestion occurred not in the highly acid stomach but in the duodenum just beyond, and that this digestion was made possible by substances secreted by the pancreas. He did not describe digestion in a loose way, but specified that pancreatic juice was capable of breaking down fats into glycerol and fatty acids, starch into sugar, and proteins into small soluble components. This was the clearest demonstration yet that living cells could produce and secrete substances which had highly specific properties. Bernard further believed that if cells could produce these so-called "ferments" they could also produce other kinds of ferments which were capable of chemical syntheses. In this matter he was a little ahead of his time. But he pushed in that direction and, by showing that liver was capable of synthesizing a starch-like substance which he called "glycogen," he disproved the long-held belief that plants manufactured carbohydrates whereas animals could only break them down.

Alongside the observations on fermenting agents produced by higher organisms, a great deal of attention was focused on microorganisms, especially yeast. Although the successes of organic chemistry put blinders on many scientists which required the genius of a Pasteur to remove, these successes also led to a number of discoveries on yeast ferments which eventually provided the basis for modern biochemistry. By 1830 A. P. Dubrunfaut was able to show that yeast cells converted sucrose to a mixture of sugars. Thirty years later Marcellin Berthelot isolated the active agent of this conversion by precipitating a yeast extract with alcohol. He named the agent *ferment inversif*, a name which was later changed to "invertase," the inverting enzyme. The inversion was the hydrolysis of sucrose into its molecular subunits, glucose and fructose. The discovery of invertase was clearly a historic step in that it provided the first demonstration of a normal intracellular component which functioned in the transformation of a carbohydrate utilized by the cell in fermentation. From this discovery the word "enzyme" emerged which was used by Willy Kühne to designate all organic catalysts. The demonstration of a cell component which could cause the hydrolysis of sucrose naturally led to speculation concerning the possibility that the so-called "vital processes" such as respiration and fermentation were due to similar components. Bernard sought some experimental way of demonstrating the relationship during the last few months of his life. The posthumous publication of his working notes led Pasteur to rather strong criticisms of the view that vital processes could be effected by extracted cellular components.

A triangle of controversies thus dominated the polemics of biologists and organic chemists during the latter half of the nineteenth century. Against those who believed that organic chemical changes were incidental to life, were those who argued that such changes were inseparably bound with life. And in mutual conflict with both views were those who rejected the association as either incidental or inseparable. Pasteur demonstrated that the incidental relationship was untenable; an accidental experimental observation brought about the final resolution of the conflict. Edward Buchner was grinding yeast cells with sand in order to obtain extracts which might be used for therapeutic purposes. To preserve these extracts he added large quantities of sugar, but in doing so he observed an evolution of gas from the liquid. On further examination of the phenomenon, he discovered that the gas evolved was CO_2 and that alcohol was being formed in the extract. This momentous discovery laid the foundations of modern biochemistry, for it provided an unambiguous demonstration that a vital process such as fermentation could occur outside the cell if the necessary cellular components were properly extracted. These components were clearly the catalysts which Bernard had sought in the twilight of his life to account for the vital activities of living systems.

What were these ferments? The basis for an answer to this question was already being laid some twenty-five years before Bernard conducted his experiments. In 1812 Gustav Kirchhoff discovered that starch could be converted into sugars in the presence of acid, but that the acid remained unchanged throughout the reaction. A little later Sir Humphry Davy found that ethyl alcohol could be oxidized in air if platinum were present, although the platinum remained unchanged. In 1822 J. W. Dobereiner performed a classic experiment showing that hydrogen and oxygen could combine to form water at room temperature if finely divided platinum were present in the mixture. Such studies multiplied. By 1838 J. J. Berzelius used the word "catalysis" to describe the action of those substances which affected the rates of chemical reactions but themselves remained unchanged. To Berzelius it was obvious that living systems must have catalysts to effect chemical changes. When Kühne coined the word enzymes in 1878 to designate organic catalysts, the evidence for their presence inside cells was beginning to accumulate rapidly.

When catalysts, especially of inorganic reactions, began to be studied, it became clear that the presence in a reaction mixture of an otherwise inert substance with large surface area was a major source of catalytic power. Dobereiner's use of finely divided platinum to catalyze the formation of water from hydrogen and oxygen was a striking example, and by 1900 there were very many examples to choose from. But given this fact, how is it to be explained? We must assume that there is something in the nature of a chemical reaction which is distinct from free energy requirements. The assumption was not made after the first elaboration of thermodynamic principles; it was already perceived by Svante Arrhenius who contributed so much to our understanding of electrolytes. The effect of temperature in accelerating reaction rates was well recognized. A simple explanation of this might have been that increased temperature caused an increased frequency in molecular collisions making interactions between opposing molecular species more likely. Stated in this way, however, the explanation is wanting. If collisions alone were a sufficient condition for molecules to react, almost all chemical reactions would go to completion at room temperature in less than a second. It was possible to calculate on other grounds that collisions occur with a very high frequency; the approximate doubling of many chemical reaction rates for every 10° rise in temperature could not possibly be explained by assuming that every time two potentially reactive molecules collide a reaction ensues. It was nevertheless taken as axiomatic that increasing the temperature of a system increased the rate of molecular movement. This follows from the definition of "kinetic energy" (energy of motion) as equal to $\frac{1}{2}\,mv^2$ and also the demonstration that kinetic energy is equal to $\frac{3}{2}\,RT$. Thus as the temperature increases, so does the velocity. Since collision itself, although increasing in frequency with increasing temperature, cannot account for the acceleration of reaction rates, the alternative is to suppose that the increase in velocity, and hence of kinetic energy, has additional consequences. Arrhenius developed the idea that in order for two molecules to interact they must have a certain minimal energy value, and he called this value the "activation energy" E_a. From other relationships he deduced the equation:

$$\ln k = -\frac{E_a}{RT} + \ln A$$

where k is the reaction rate, $\ln A$ is a constant, R is the gas constant. If E_a is also a constant, a plot of ln k against 1/T should yield a straight line. This relationship is found experimentally.

The mechanism of this relationship cannot be simply explained, although an analogy should help to clarify it. Let us suppose that two molecules with adequately high velocities collide head-on. They might behave in much the same way as Huygen's perfectly elastic balls and simply recoil from one another, the sum total of their kinetic energies remaining the same. Or, we might imagine that some of the energy of the impact is transferred to some other form of energy within the molecule, say, the vibration of the electrons. Accordingly, the electrons, having somehow acquired additional energy, are now less stable in their arrangement. Such instability could not be permanently maintained, and the atom would therefore tend to readjust the vibrations (if that were the initial effect) of its electrons so as to return to a normal and stable energetic level. But two potentially interacting molecules which are similarly affected and in close proximity would likely respond to the situation by interacting with one another and thus achieve even stabler configurations than the previous ones. The picture is an analogy, and it can serve at best to underline the experimental evidence that increasing the kinetic energy of molecules increases their frequency of interaction in a precise and predictable manner. From the first law of thermodynamics we take it as a certainty that the kinetic energy involved in collisions can be translated into other forms. But exactly how this increased kinetic energy so affects the peripheral electrons as to facilitate their rearrangement would require us to probe deeply into intraatomic properties.

The popular diagram of the activation energy principle is illustrated in Fig. 5-1. To this ought to be added the fact that in any population of molecules, kinetic energies are not uniformly distributed. The numbers of molecules with energy values much above or below the mean will be small. Thus the effectiveness of temperature with respect to activation energy does not imply that all molecules in a system achieve the necessary level, but that by increasing the mean kinetic energy the numbers of molecules which acquire an energy of activation is increased.

Granted the correctness of the above explanation, how does this apply to the phenomenon initially cited that catalysis can be effected by provision of large surface areas? Clearly, in neither surface catalysis nor enzyme catalysis does temperature enter as a decisive factor. Yet the point of beginning with the catalytic effect of temperature was to take advantage of the wealth of physical understanding of its operation. Temperature effects led to a new concept, activation energy. With our present physical picture of the universe, it is inconceivable that a principle, if correct in one situation, should not be correct in another; otherwise it could not be regarded as a fundamental statement of behavior. If activation energy is adequate to account for the extremely slow rates

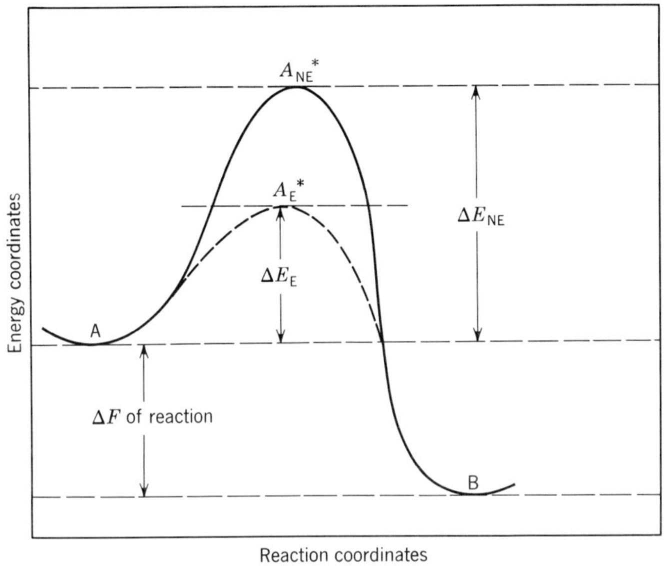

Fig. 5-1. Diagram showing the energy barriers of a reaction $A \rightarrow B$. A_{NE}* indicates the activated complex in a nonenzymic reaction and A_E* the activated complex in an enzyme-catalyzed reaction. A is the initial substrate and B the product. ΔE_{NE} is the energy of activation for nonenzymic and ΔE_E for the enzymic reaction. ΔF is the difference in free energy in $A \rightarrow B$. (From Conn and Stumpf, *Outlines of Biochemistry*, Wiley, 1963.)

of reaction in systems which are potentially very reactive, surface catalysis must have some way of raising interacting molecules to the necessary level without raising their kinetic energy. There appears to be one alternative: if the kinetic energy is not raised, the activation energy must be lowered. This explanation is applied not only to surface catalysts but also to enzymes. We might have drawn this conclusion directly for enzymes without implicating the inorganic surface catalysts, but there is good reason for not having done so. Electron micrographs of almost any type of cell show, from the numerous membranes within the cell, that the ratio of surface to volume is large. The potential effects of surfaces on the progress of reactions within a cell need to be discussed.

One concrete statement can be made about surface catalysts or enzymes. Inasmuch as the catalyst must exercise some force on the molecules concerned, such that they become more susceptible to interaction, the distance between reacting molecules and catalyst can be no greater than the known distances over which interatomic forces operate. Since we know such distances to be of the order of 1–3 Å, it means that some type of chemical bond must be formed between the two. This consideration is of fundamental importance. Although a number of attempts have been made to prove that long–range forces (acting over distances much greater than 3 Å) operate in biological systems, there has been no satisfactory proof and only questionable theoretical logic. Since the first step in catalysis would be the formation of some new chemical bond between molecule and catalyst, we must suppose that here too an activation energy exists, but that its magnitude must be much lower than that required for the interaction between the molecules themselves. This first step can be explained in terms of the behavior of substances which provide large surface areas. Finely divided solids, such as charcoal, have a high tendency to "adsorb" a variety of dissolved molecules when dispersed in the solution. The use of charcoal to remove colored materials from sugar extracts is an old practice. The extensive studies of adsorption phenomena led to the generalization that many kinds of molecules tend to concentrate at interfaces (water-air, water-oil, water-solids). According to thermodynamic law such concentration could occur only if accompanied by a decrease in free energy. The decrease is commonly explained by supposing that molecules at a surface are subject to forces different from those acting on molecules at the interior, whether in a solid, liquid, or gaseous system. Interior molecules are surrounded on all sides by similar molecules, whereas those at the surface can only be so surrounded on three sides. The response of solutes to the "different" side will very much depend on the nature of the interface. Mercury molecules, for example, interact less with glass or air than between themselves; a drop of mercury on a glass plate tends to assume a spherical form so as to reduce the mercury-glass and mercury-air interfaces. A water drop spreads over a nongreasy glass surface because water molecules interact more strongly with glass than between themselves. Water does not spread over a greasy surface because apolar fat molecules have little affinity for the polar groups of water.

The tendency for liquid-solid or liquid-air interfaces to assume maximal or minimal areas is a relatively simple example of how physicochemical forces act at interfaces. These same forces also act in three-component systems such as water dispersed in an aqueous solution. The extent to which solute molecules bind to the surface of the solid depends on the relative degrees of interaction between the molecules and water, and between the molecules and the solid. The combination of these interactions provides the unique physicochemical situation characteristic of interfaces. Although the explanation of how this combination lowered activation energies was not obvious, the fact that the provision of adsorbing surfaces catalyzed many chemical reactions impressed both chemist and biologist.

The formation of a bond between a molecule and the surface of a catalyst is insufficient to explain the nature of catalysis. As stated earlier in discussing temperature of activation, a factor which increases the rate of a chemical reaction must somehow affect the behavior of the outer electrons so as to render them less stable than in their normal state and hence more susceptible to adopting other more stable configurations. If only one molecule is involved (in simple decomposition, for example), we could explain surface catalysis by supposing that the bond formed between the molecule adsorbed and the adsorbent so alters the electron pattern of the adsorbed molecule as to increase the probability of decomposition. If, however, two or more kinds of molecules are involved, the catalyst can only be effective if the interacting molecules are located adjacently on the surface. To the extent that reactants concentrate at the surface of the catalyst, the probability of such a juxtaposition is correspondingly increased. A second factor must also be considered, that of molecular orientation. Presumably a certain group or groups within a molecule will have a preferential affinity for the surface of the catalyst, and as a consequence adsorbed molecules will become oriented. Such orientation may position a molecule for more effective interaction either with its neighbor or a dissolved molecule which randomly collides with it. Whatever the details of the catalytic mechanism, we may write a general equation for catalysis in which the catalyst is treated as though it formed an unstable complex with the reactants or "substrates." Letting C denote catalyst, and S substrate molecules, then

$$(1)\ C + S \rightleftharpoons CS \quad (2)\ CS \rightleftharpoons C + \text{products, or } CS \rightarrow C + \text{products.}$$

The reaction may be reversible or irreversible, but this property is not a function of the catalyst used. The total free energy change which determines the direction of the reaction is a thermodynamic function. The catalyst does not affect the equilibrium constant of the total reaction.

Studies of inorganic catalysis made available a principle by which possible chemical reactions could be realized. The demonstration of enzymes within cells provided proof that the same or a similar principle governed chemical transformations in living systems. What remained entirely unclear was the nature of the intracellular components which behaved as catalysts. The colloidal theory of cell organization which was so ably developed by Thomas Graham offered an easy interpretation. If adsorbing surfaces catalyzed a variety of inorganic reactions, colloidal particles should behave similarly with respect to organic transformations. The problem of identifying the chemical nature of enzymes was difficult. Individual enzymes are present in very small amount, and to isolate specific enzymes when their chemical nature was uncertain and subject to a great deal of dispute was an almost impossible task. Many attempts were made to purify and identify enzymes, and even in the 1920's as great a chemist as Richard Willstatter reached the negative conclusion that proteins could not possibly be enzymes. Another outstanding biochemist, Otto Warburg, concluded that proteins merely functioned to provide large surface areas in oxidative processes, for he was impressed by the fact that inorganic iron adsorbed on charcoal was an effective catalyst for the oxidation of organic compounds.

As in other controversies the issue was finally resolved by direct experimental demonstration. Two individuals, first J. B. Sumner and then J. H. Northrop, independently succeeded by 1930 in purifying and crystallizing two enzymes; both of these were shown to be proteins. This marked the beginning of intensive experimentation in the purification, crystallization, and physicochemical characterization of the very many enzymes which underlie cell metabolism. Within thirty years cell biochemistry moved from the stage at which enzymes were identified as proteins to the stage at which the spatial arrangement of their atoms was being investigated by x-ray diffraction techniques.

A background question may be raised at this point. What properties of protein molecules contributed to their catalytic capacity? The explanation of catalysis as a property of the surface area of cell colloids had to be discarded even if such colloids were assumed to be proteins. For the crystalline enzymes prepared by Northrop and Sumner had highly specific properties. One catalyzed the breakdown of urea, the other catalyzed the hydrolysis of protein. And as the number of purified enzymes increased, the evidence that each type of enzyme protein catalyzed a specific type of chemical reaction became overwhelming. Interest therefore turned from the general property of catalysis to the specificity of enzyme catalysis. The evidence that once enzymes were denatured (by heat, acid, or alkali) catalytic activity was lost clearly suggested that the tertiary structure of the protein molecule was an important factor in the process.

Even today our understanding of the mechanism of enzyme catalysis remains incomplete. Some features of the mechanism are nevertheless reasonably clear. The fact that the R groups of amino acids provide a wide variety of physicochemical properties makes it possible for proteins to interact with a wide variety of substrate molecules. The virtually unlimited number of ways in which amino acids can be arranged along a polypeptide chain makes possible a correspondingly unlimited number of combinations of adjacent R groups. One or more regions in a protein molecule could have a highly specific affinity for a particular substrate molecule because of a complementary arrangement in the reactive groups of the substrate and the combining region in the protein molecule. Such specificity is further increased by the fact that the polypeptide chain in an enzyme is a folded structure, the geometry of the folds either facilitating or hindering the substrate-enzyme interaction. The folds could affect the positioning of adjacent interacting molecules, thereby increasing catalytic efficiency. How enzyme substrate complexes lower activation energies is a question we will not attempt to answer. One of the very important facts concerning the behavior of cells is that most, if not all, chemical reactions within a cell are catalyzed by enzymes, and that for each type of reaction a specific enzyme can be found.

THE CONCEPT OF ENTROPY

To bring this section to its proper close we should consider one grand concept which underlies all that has hitherto been discussed. We have been criss-crossing between a world of abstractions and a living world filled with a variety of concrete chemical changes. In doing so, we have attempted to explain how these abstractions originated and how they have proved indispensable to the understanding of biological phenomena. A little recapitulation may prove helpful. Terms such as momentum, force, energy, and work have been introduced; all of these have a familiar ring, but none can be visualized. In one way or another philosophers had sensed their existence from the earliest of times. Beginning with Newton's age, these terms were given precise meaning, but only in the sense that each could be defined in a formula equating the term with certain measured parameters. And by a variety of mathematical manipulations they could all be equated to one another.

All of these definitions involved mechanical measurements. A giant advance was made when James Joule applied a known amount of mechanical work and measured the heat produced. Heat, it should be observed, is no more tangible a component than those above. It may be defined in terms of specific heat, which is the amount of heat required to raise the temperature of a gram of substance by 1°. A thermometer does not measure heat; usually it measures volume change in a mercury column. The definitions of heat, specific heat, and temperature are in themselves circular. Joule made it possible to equate heat with mechanical parameters. Much the same applies to electricity and magnetism. Electric current could be defined in terms of magnetic deflection (the basis of most meters), and also by the amount of heat produced in passing through a conductor. Static electricity could be measured in terms of the force exerted in keeping two similarly charged bodies (pith balls, gold leaves) apart. The discovery by Alessandro Volta that chemical reactions could generate electrical current, the demonstrations by Faraday that current could be used to generate chemical change, and the long-standing observations that chemical reactions could produce or even remove heat, eventually led to the tying up of all the diverse phenomena of nature into sets of equations known as thermodynamics. The strength of the system was that it allowed for absolutely no exceptions. This, more than any demonstration of specific chemical reactions, spelled the termination of eighteenth and nineteenth century preoccupations with vital forces.

To discover a rule of life was one thing; to assert that life was completely subject to the rules governing all of nature was a challenge to which vitalists could respond effectively only by proving the contrary. The proof was not forthcoming. The predictability of thermodynamics was indeed frightening. Some comfort came later from modern physics when men like Werner Heisenberg enunciated the "uncertainty principle," which laid bare a critical weakness of thermodynamics—that it could account for the behavior of any large population of atoms or molecules but it could not predict the course of an individual one. In sophisticated language the criticism read that it was impossible to measure simultaneously the position and velocity of an elementary particle. Throughout his life Einstein objected to the implications accorded this principle. He claimed the uncertainty to arise from the shortcomings of theory and experimental technique; colorfully, he commented that the Creator did not play dice with the Universe. The last chapter is thus by no means written.

"Uncertainty" is not our final stop, although it is a view which biologists should be aware of. Evolution itself embodies an uncertainty principle. But leaving this point aside, there still remains the great synthetic concept of entropy, which was sneaked by in discussing the concept of free energy. We set forth dogmatically the proposition that in chemical reactions the free energy of the system had to be decreased in order for the reaction to be possible. We then indicated the manner in which such free energy could be calculated. But what is the source of this concept? We must return to the problem of heat for an answer. Consider first an experiment in which measured amounts of heat are supplied to a container of water in so ingenious a fashion that the container is kept perfectly insulated and none of the added heat is lost. If the temperature of the water is recorded at each addition of heat, a relationship will be found between the amount of heat added and the increase in temperature. The relationship may be quantitatively formulated, and a new parameter, the specific heat of water, determined. The temperature will increase more or less steadily as heat is added until 100°C is reached (assuming that pressure is maintained at 1 atmosphere). At this point, a certain amount of additional heat will produce no change in temperature, and after this addition has been made the liquid will change to vapor. The apparently ineffective heat is known as the latent heat of evaporation. We need carry the experiment no further to pose the problem. Temperature is presumed to be a measure of the kinetic energy of the water molecules, and yet in the above experiment we have identical masses of water at identical temperatures but containing different amounts of heat. The same paradox could be illustrated by dissolving KCl in water. Again, assuming complete isolation of the system, the temperature of the solution would be found to drop although no heat has been lost. In both cases a curious effect has been observed: The ratio of heat contained to temperature has fallen. What is the significance and meaning of this effect?

The second law of thermodynamics was formulated to account for a universal phenomenon related to this curiosity. It stated that only those reactions were possible in which the ratio of heat content to temperature (Q/T) increased. As in the first law, no exceptions were admitted. It applied to every kind of phenomenon, not merely the chemical one. To this ratio the term "entropy" was assigned. If every change had to occur along with an increase in entropy, it followed that in a universe undergoing change entropy was steadily increasing. Accordingly, we might be gloomy about the future, about the time when Q/T would be uniform throughout the universe and all change would cease, life obviously being included.

There is no question that this second law has proved to be inviolate. What, then, is its meaning? In the example of adding heat to water, the "ineffective" heat was associated with a change from a liquid to a gaseous state. In dissolving KCl, the adsorbed heat was associated with a dispersion of the ions making up the crystals of salt. In both cases heat energy was consumed in causing an increase of "disorder" in the systems. In the first, heat was added from the outside and the temperature remained constant while the association of water molecules in the liquid was transformed into a dissociated and less orderly gaseous assembly; in the second, heat not being supplied, the temperature dropped while the orderly crystals of KCl were being randomized as ions in solution.

Thus although entropy has been usually defined as Q/T, it is equally definable as the measure of randomness of a system. The more disorderly a system, the greater its entropy. And according to thermodynamics whenever reactions occur in the universe, the resulting effect is an increased disorder.

It may be a little difficult to digest the idea that adding heat to a system need not increase its temperature, or that decreasing the temperature of a system need not involve a loss of heat. These are nevertheless facts. But what gives the second law its universal interest is often not even stated in terms of heat. It is the concept of relating the orderliness of things to this heat phenomenon that is so pregnant with meaning. Until this last section we have not mentioned "order" even though it must have been recognized by man as early as any other natural property. Not only are we aware of orderliness in organisms, but we are constantly aware of it with respect to our own activities It is a commonplace that keeping things in order requires effort and that disorder can result from neglect. But even if we move from the familiar level to the distant molecular one, we quickly appreciate the fact that a precisely folded protein molecule is far more orderly than an unorganised assembly of amino acids. We readily acknowledge that the probability of obtaining an organized grouping of playing cards by tossing a pack in the air is infinitesimal. But what is almost unbelievable is that we can take this abstract property and link it quantitatively to two seemingly irrelevant properties—heat and temperature. This point should engage the attention of every scientist; not the vague notion that order requires energy, but the specific fact that order is quantitatively related to heat and temperature.

When we discussed the free energy of chemical reactions, we were really considering one of three terms in an equation. We may see this by setting forth the following proposition. Let us suppose that we are interested in determining the energy available from oxidizing glucose into CO_2 and water. This would be rather difficult to do by determining the equilibrium constant since the equilibrium is much too far over in favor of the products. We could, however, determine its value by arranging an oxidation experiment so as to convert the free energy of the reaction into heat. This, in fact, is the course taken when the energy is not trapped in some other form. If we measure the heat of reaction in this way, we find that the heat produced does not correspond to the amount of work which the reaction system should yield. In fact, if heat produced were a sufficient measure of energy available, dissolving KCl in water should be impossible because the solution gets cold; using heat production as the only criterion, this would mean that adsorption rather than a production of energy has occurred. There are, indeed, many such "endothermic" reactions. We therefore cannot write the equation $\Delta F = \Delta H$ where ΔH is by convention the heat absorbed by the system. What we have failed to consider in the reaction is the fact that any disorder created will be accompanied by an absorption of heat, so that the heat we measure is the algebraic sum of that arising from the free energy change as well as the entropy change. To take account of this relationship the formula $\Delta F = \Delta H - T \Delta S$ was devised. The term S represents entropy, or Q/T.

This equation has a number of interesting features. We have already noted that a negative ΔF means a decrease in the free energy of the system and is therefore a measure of the energy available. We have also noted that a negative ΔH represents heat evolved so that the evolution of heat in a reaction tends to make ΔF negative. An increase in entropy would leave the $T \Delta S$ term negative so that it too would contribute to a $-\Delta F$. Taking the equation as a whole, we at once see that the free energy could be negative even if heat were absorbed or entropy decreased, provided that the algebraic sum of the two were negative. Thus if entropy were to decrease, the term $T \Delta S$ would become positive; but if the heat loss were great enough and ΔH were negative, the sum of the two expressions would remain negative. In effect, we are saying that a particular chemical reaction could increase orderliness. This would appear to contradict the initial statement, that reactions must go in the direction of disorder. The relationship is of fundamental importance to living processes since growth obviously results in an increased ordering of substances. If this relationship did not exist, the formation of specific protein molecules would be impossible.

Does this relationship transgress the second law? A physicist who knew nothing more about life than that it was an exceedingly complex system would say that life must continually pay a price for the order it preserves. He would not need to, and might not even care to, learn the intricacies of molecular arrangement in living cells. He would make one unequivocal statement. Nature tends to disorder, and, since life is ordered, such order is maintained by creating even greater disorder elsewhere in the universe. This, in fact, is the case for any chemical reaction we may select. To keep the free energy negative, the heat energy lost by the system must exceed the energy required to increase order. We have not concerned ourselves with the fate of that heat because it is of no particular interest to the chemical reaction. If we include the surroundings of the reaction system in our equation and then ask whether the Q/T ratio was greater before or after, the latter would turn out to be correct. It may be said that the evolution of life has purchased order at the price of disorder. The steady flow of energy from sun to earth and its associated randomizing of their differences in energy is the price paid for the maintenance of order in living systems.

SUMMARY

We began this section of chemical change after having discussed two broad topics: the applicability of physicochemical laws governing the behavior of solutions to the activities of cells, and a resolution of the chemical composition of cells in terms of molecular constituents. In the course of these discussions a variety of specific cellular characteristics were touched on, some of which the student will find useful to commit to memory, and many of which will serve to stimulate further readings. Moreover, the development of each of the topics had ideological repercussions which far outrun the ground they cover. Looking back into the history of cell biology, studies of solutions yielded the first clear demonstrations that physicochemical laws could be directly applied to vital behavior. Organic chemical studies provided the counterpart: that the constituents of living cells were no more than molecules although molecules of distinctive composition.

The next logical step was to find some relationship between physicochemical laws, the organic molecules of life, and the total behavior of living cells. And we indicated how abstract concepts were indispensable here. The law of conservation of matter led rather quickly to a chemical resolution of characteristic vital activities such as fermentation, respiration, and photosynthesis. When that level of understanding was achieved, deeper abstractions were called for. It was not at all obvious how the pumping of blood, the contraction of muscle, the movements of microorganisms, and the synthesis of large organic molecules could be related to the chemical properties of vital constituents or to the physical laws of solutions. Minds as perceptive as those of Lavoisier and Liebig were defeated by the challenge. Not until the commonplace term "energy" had been given a precise and quantitative meaning could the challenge be met. It is well to remember that prominent figures such as Mayer and Helmholtz, who led physics in giving "energy" a scientifically useful meaning, were biologists in the full sense of the word. They were equally at home in both fields. The presentation of physicochemical laws as an integral part of cell biology is thus hardly a novel departure; on the contrary, it is a return to an older period of history when intellectual curiosity had little taste for unreal subdivisions of Nature.

PART TWO

THE FORMULATION OF PROBLEMS

Problems about cells may be formulated in a variety of ways, and no single way is superior to all others. The choice of formulation is, of course, influenced by the pattern of current biological interests, but it is also a matter of personal preference. A very common approach to cell biology —and one that is highly successful for teaching purposes —is to formulate problems in terms of individual cell structures and their interactions. Once the student is acquainted with the general microscopic and submicroscopic organization of cells, he can readily follow discussions which center about the question of what each subcellular organelle does. Questions such as "What does the nucleus do?" or "What does the nucleolus do?" have a specific meaning and their objective is readily comprehended. This approach, nevertheless, has not been used. Instead, problems have been formulated with respect to generalized but abstract propositions. "How do cells maintain order?"

"How do cells transmit order?" "How do cells regulate their behavior?" These are the principal questions on which the formulation of problems was based. From the didactic standpoint the approach has a number of drawbacks. The student is plunged into an abstract realm. He is being led from universals to particulars, a process which is contrary to common mental habit. Yet despite these drawbacks the approach has one decided advantage. It makes the process of abstraction and the handling of abstractions more familiar to the student. As the science of biology becomes increasingly sophisticated, it also becomes increasingly filled with abstract propositions. To make the organization of the text consistent with projected trends in cell biology, we decided to turn to this second type of approach.

SECTION A

The Maintenance of Order

Order, like energy, is an abstract term. Although difficult to define precisely, the quality of orderliness is easily recognized, be it in the arrangement of a house, the design of a machine, the plan of an essay, or the behavior of a crowd. That cells are highly ordered systems compared with their surroundings needs no demonstration. And, on the basis of thermodynamic principles discussed in Part I, no special arguments are required on behalf of the proposition that the creation or maintenance of order requires an expenditure of energy. The problem is to explain the relationship between energy and order in terms of concrete mechanisms. Without thermodynamics, biologists would find themselves lost in the mists of metaphysical speculations; with thermodynamics, they have a set of "ground rules" which set sharp limits to speculation. The possible and impossible are given by thermodynamics; the actual has to be discovered through imaginative experimentation. The proposition that energy is essential to order has to be matched by demonstrations of the forms of energy which living systems use and how each of these forms contributes to specific types of order. Such demonstrations are products of twentieth century biology, and out of them has grown a new body of generalizations—based not on abstract speculations but on concrete mechanisms. It is to these mechanisms that this section is devoted.

The evidence that all living cells must either respire, ferment, or photosynthesize was discussed in Part I. So too was the general interpretation that these processes are the sources of energy for cellular activities. Although a somewhat detailed account of the mechanisms by which cells acquire and transform energy will be given in this section, the account will not take the form of a separate analysis of each of the processes. A number of generalized questions will first be posed, questions which do not reflect the history of the topic but which do reflect the logic of modern concepts of energy relationships. The purpose of these questions is to define the kinds of universal compounds cells must produce in order to make energy available for cellular syntheses. The section therefore begins by discussing the essential properties of different forms of chemical energy, and only later are the actual processes which give rise to these forms fully described. Occasional reference to the charts on metabolic transformations may prove helpful in reading the first few chapters.

A very important question related to the general subject of energy transformations is considered in various chapters. To what extent is each of the specific mechanisms in such transformations universal? The problem is important because of the fact of metabolic interdependence within the world of organisms. In considering biological diversity a distinction must be made between fundamental processes which have been acquired by only certain types of cells and fundamental processes which evolved in various forms among different organisms. Photosynthesis is an example of the first source of diversity; different types of fermentation are an example of the second. In both cases, however, the apparently extensive differences between metabolic patterns can be expressed in terms of a few fundamental mechanisms.

6

Sources of Cellular Substance

PRIMARY PROCESSES

Ultimately, living matter is synthesized from inorganic substances. One obvious aspect of ordering by cells is their capacity to select certain elements from their environment and to convert them into organic compounds. The kinds of conversions and their corresponding energetic requirements are not the same for all essential elements. A broad distinction can be made between metallic and nonmetallic elements with respect to the chemical modifications essential to their incorporations into living matter. Although metallic ions perform critical functions in the metabolism of cells, they commonly constitute only 1–2% of cellular substance. The energy drain which metallic ions impose on cells is almost entirely concerned with their accumulation or secretion, a process which is to be considered in a later chapter. The nonmetallic elements, on the other hand, account for almost all the mass of living substance, and out of them are formed all the macromolecules. The conversion of these elements—carbon, hydrogen, oxygen, nitrogen, sulfur, and phosphorus—from their inorganic into their organic associations imposes a major drain on the energy resources of cells, a drain which is distinct from that required for their accumulation.

Properties of Metallic Elements. The chemical differences between metallic and nonmetallic elements have a direct bearing on their respective roles in cell functions. The nonmetals are primarily, though not entirely, characterized by formation of stable covalent bonds. The covalent bonding, as discussed in Part I, confers a relatively high degree of chemical stability on organic molecules. Metallic ions, on the other hand, have two distinctive properties of paramount importance in biological functions. (1) Their capacity (especially copper and iron) to undergo oxidation-reduction with relatively small changes in free energy. Since, as will be discussed later, the basic mechanisms in respiratory oxidations and photosynthetic reductions involve a series of electron transfers, metallic ions appropriately located in protein molecules are extremely effective in mediating such transfers. Even the free metallic ions are often good oxidative catalysts, though not nearly as effective as when linked to a protein molecule. (2) Metals possess two types of valence, "primary" and "secondary." The primary valence is designated by the number of charges on the cation (which may be one or more) and such valences are responsible for the type of intramolecular bonds discussed in Part I. Secondary valences are designated by "coordination numbers" and these permit secondary bonds to be formed between metal ions and various groupings of nonmetallic elements (e.g., $—NH_2$, CHO, CO) found in organic molecules. The chlorophylls, cytochromes, and hemoglobins are examples of metal coordination complexes. Metals like Fe, Cu, Mo and Co, which have strong coordination tendencies, can alter their primary valences by virtue of oxidation-reduction and yet remain linked to the organic molecule catalyzing oxidation-reduction reactions. The metals Mg and Ca, which have weak coordination tendencies, can be highly effective in providing transient links between enzyme and substrate, a property which accounts for their role in various enzyme systems. The monovalent ion K^+ neither undergoes oxidation-reduction nor forms coordinate linkages, but is the major cation of cells and is absolutely essential for life. Why this is so, and why Na^+ cannot replace it, is still poorly understood.

To explain the fact that metallic elements do not constitute the molecular skeletons of living matter would require a formidable degree of chemical rationalization along the lines just discussed. A much simpler, though less profound argument could be offered as explanation. The nonmetallic elements are far more abundant. The relative scarcity of metals is probably the major cause of energy expenditure for their incorporation into living matter. A striking example of this is provided by the tunicates which utilize vanadium ion in their oxygen transport system and must therefore concentrate it from the extremely dilute condition occurring in seawater. Although vanadium is an extreme example, all of the metals listed in Table 6-1 are usually concentrated by cells from their environment in the course of growth.

Since the objective of this chapter is to explain how energy is utilized to form organic compounds from inorganic ones, the first question which should be explored is whether the transformations can be expressed in general terms. Must the various types of organic compounds be considered separately, or can they be considered as a single group in which the constituent elements differ from their inorganic associations by certain universal characteristics? This type of question is implicit in every formulation of a biological problem. If generalization is impossible, so is simplification. If we had to account for energy expenditure by cells in terms of each kind of organic molecule, the energy problem could not be formulated, but only cataloged. Such cataloging is unnecessary for common chemical differences can be found between the organic and inorganic associations of the nonmetallic elements.

Table 6-1 Some Characteristic Bondings of Nonmetallic Elements in Cells and in the Inanimate Environment

	Bond Types	
Element	Environmental	Cellular
Carbon	CO_2, CO_3^{2-}	$—CH_3$, $—CH_2OH$, —CHO, —COOH
Oxygen	O_2	
Hydrogen	H_2O	
Nitrogen	N_2, NO_3^-	$—C—NH_2$
Sulfur	S, SO_4^{2-}	—SH
Phosphorus	PO_4^{3-}	$—O—(PO_3)^{2-}$

Extracellular oxygen is listed as O_2, but it may be seen from the table that all other elements occur partly or entirely as oxides in the environment. Oxygen constitutes nearly half of the weight of the earth's crust in which it exists in the form of oxides. Using oxygen as a point of reference, the simple yet striking feature of the table is that, excepting phosphorus, the elements are combined with less oxygen in living matter than in the inorganic forms. The stores of CH_4, NH_3, and H_2S present in some localities are products of organisms although CH_4 may represent remains from a time when the earth's atmosphere was highly reducing.

The characteristic associations of the elements are shown in Tables 6-1 and 6-2. The major and most important feature of this listing is the fact that the elements carbon, nitrogen, and sulfur are all associated with fewer oxygens and more hydrogens when present in organic forms. Commonly, the organic forms of these elements are described as the reduced condition, whereas the inorganic forms are described as the oxidized condition. Although this description will be seen to be technically incorrect, it adequately describes the major difference between organic and inorganic associations. This difference holds for all the major constituents of cells. Therefore, regardless of the kinds of compounds cells form from these elements, a common and primary requirement is their reduction.

Table 6-2 The Principal Metallic Elements of Living Matter

Element	Valences in the Environment	Intra-cellular Valences	Functions
Iron (Fe)	2-, 3-	2-, 3-	A constituent of the cytochrome enzymes which function in electron transport; and of the hemoglobins which transport oxygen.
Copper (Cu)	1-, 2-	1-, 2-	Present in cytochrome a which transfers electrons to molecular oxygen. Also present in a variety of oxidizing enzymes, especially in those acting on phenolic-type compounds. In various invertebrates, the oxygen-transporting proteins (hemocyanin) contain copper rather than iron.
Manganese (Mn)	Mainly 2-, some 3-	2- (3- ?)	A co-factor in some oxidative systems, especially those involving the splitting of peroxide. Functions analogously to magnesium ion in some enzyme systems (e.g., peptidases)
Cobalt (Co)	2-, 3-	(2- ?), 3-	Present in the structure of vitamin B_{12} which is among a group of coenzymes associated with catalyzing various molecular transformations. One of these is the conversion of ribosyl compounds to deoxyribosyl ones.
Molybdenum (Mo)	Several valences but mainly 6-	6-	Present in some oxidizing enzymes (xanthine oxidase). This element also functions in the very important enzyme system for nitrate reduction.
Zinc (Zn)	2-	2-	Present in various enzymes (alcohol dehydrogenase, carbonic anhydrase)
Magnesium (Mg)	2-	2-	The principal metallic ion in enzyme systems governs the transfer or release of phosphate groups (phosphatases, phosphorylases). This role is related to the fact that Mg ions have a high affinity for complexing with phosphate groups.
Calcium (Ca)	2-	2-	A cofactor in some enzyme systems but plays an important, though still unexplained, role in the surface membranes of cells. Also serves as a binding agent in the pectic layer of plant cell walls and in the bones of vertebrates.

Potassium (K)	1-	1-	The principal monovalent cation of the cell interior. Cells cannot survive in the absence of K^+, and thus far no single function has been found for it which would entirely explain its critical role. A number of enzyme systems including those synthesizing protein require K^+ for activity. Cells usually expend appreciable energy in maintaining high levels of K^+ and low levels of Na^+.

This table has been drawn up to serve as a source of reference and also to illustrate the fact that, unlike nonmetallic elements, the organic and inorganic associations of metals do not differ with respect to states of oxidation. Metals like Fe and Cu which have two stable valences are of major importance in the mechanisms which cells use to oxidize some substances at the expense of reducing others. Of equal importance to cellular systems is the fact discussed in the text that metals also have the capacity to form "coordination complexes," a property which is utilized in the attachment of substrates to enzymes.

Generally, conditions prevailing in the "biosphere" (that portion of the earth containing living forms) favor oxidation. Presumably at one time in the history of the earth, the atmosphere was highly reducing and even the small organic molecules which must now be synthesized by cells formed spontaneously. In a reducing atmosphere free energy is released on reduction; in an oxidizing atmosphere energy is required for reduction, whereas oxidation releases energy. The ability of cells to synthesize organic components thus depends on their ability to transform part of their available energy into compounds which are capable of reducing the oxidized forms of essential elements.

The one exception to this generalization is the element phosphorus which exists as the trivalent phosphate ion both intracellularly and extracellularly. In organic compounds, however, phosphate is linked through oxygen to a carbon atom, or in some cases to a nitrogen atom. Phosphates so linked are said to be "esterified," and (for reasons to be discussed later) esterification is an energy-requiring process. Phosphate esters may degrade spontaneously and release energy in the reaction, but they do not form without an energy-donating mechanism. The exceptional behavior of phosphate compounds is matched by their exceptional function. Without phosphates, organic syntheses by cells would be impossible.

Although we may characterize reduction and esterification as energy-requiring mechanisms in the primary steps of forming cellular substance, one question which ought to be considered is whether all cells are capable of performing these steps. That this is clearly not the case is known from the restricted distribution of photosynthesizing cells. Sulfur and nitrogen reduction are also restricted in occurrence. Only phosphate esterification is a universal property. We might therefore say that whereas phosphate esterification is essential to the life of every cell, reduction of other nonmetallic elements is essential to the living world as a whole but not to cells in particular. This sweeping statement must nevertheless be qualified, and the qualifications will be made most apparent by considering the nonmetallic elements individually.

The Carbon Cycle. The distinctive feature of photosynthesis is the reduction of carbon compounds by means of light energy. Neither the utilization of CO_2 as a carbon source nor the release of oxygen is essential to the reaction even though both processes occur in most photosynthetic organisms. These points will be fully considered later. The uniqueness of photosynthesis lies in the reduction of carbon compounds, such as CO_2, by means of light energy. All cells have mechanisms to reduce oxidized carbon groups and they generally have mechanisms for incorporating CO_2 into carbon compounds. In formulating the problem of how energy is utilized by cells for the reduction of carbon compounds we must divide the problem into two parts—the mechanisms which cells use to effect reduction, and the sources of energy which cells use to drive the reduction. Basically, the mechanisms are common to all cells and only with respect to the utilization of light as a source of energy do we find a restricted distribution. Since generally, though not exclusively, nonphotosynthetic cells derive their energy from carbohydrate oxidation, the carbon cycle in the living world may be represented as a production of carbohydrate in photosynthetic organisms and an oxidation of carbohydrate in nonphotosynthetic forms (Fig. 6-1).

The Nitrogen Cycle. Although elemental nitrogen constitutes close to 80% of the atmosphere, exceedingly few organisms are capable of incorporating it in this form. Green plants are capable of utilizing either nitrates or ammonia as sources of nitrogen; a few bacterial species can utilize nitrate but most can metabolize ammonia; animals cannot survive with ammonia as the principal nitrogen source. Thus three distinct metabolic processes are associated with the utilization of inorganic nitrogen by cells: the reduction of elemental nitrogen, the reduction of nitrates, and the incorporation of reduced nitrogen into organic molecules. All three processes require energy, but unlike

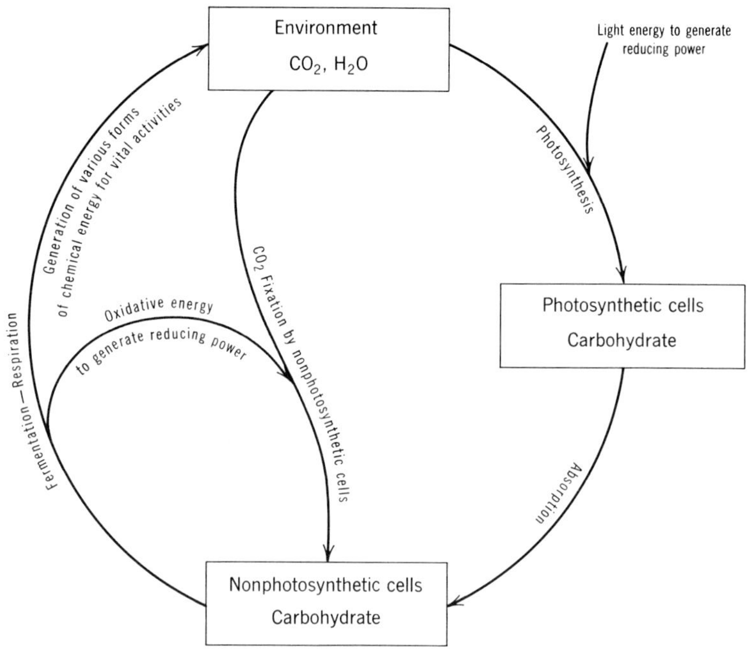

Fig. 6-1. The carbon cycle. A simplified representation of the relationship between photosynthetic and nonphotosynthetic cells with respect to sources of energy for reduction of carbon compounds.

the formation of organic carbon, it is the capacity to utilize energy for nitrogen reduction rather than the source of that energy which defines the limited types of cells capable of surviving on inorganic nitrogen. Most cells can incorporate ammonia into at least one organic acid and thus convert it to an amino acid; in animals, however, some of the carbon skeletons of the amino acids cannot be synthesized, and this deficiency makes the utilization of ammonia as the sole nitrogen source impossible.

Graphic representations of the nitrogen cycle, such as the one shown in Fig. 6-2, illustrate the utilization and conversion of various forms of nitrogen by different groups of organisms. The differences between organisms with respect to their patterns of nitrogen metabolism indicate that cells vary considerably in their capacity to convert inorganic nitrogen to organic forms. As pointed out earlier, this variation cannot be formulated as a problem in energetics; such variation is due to the kinds of enzymes present in a cell which permit or prohibit the utilization of trapped energy for nitrogen reduction and utilization. Universal patterns of nitrogen utilization begin at a biochemical

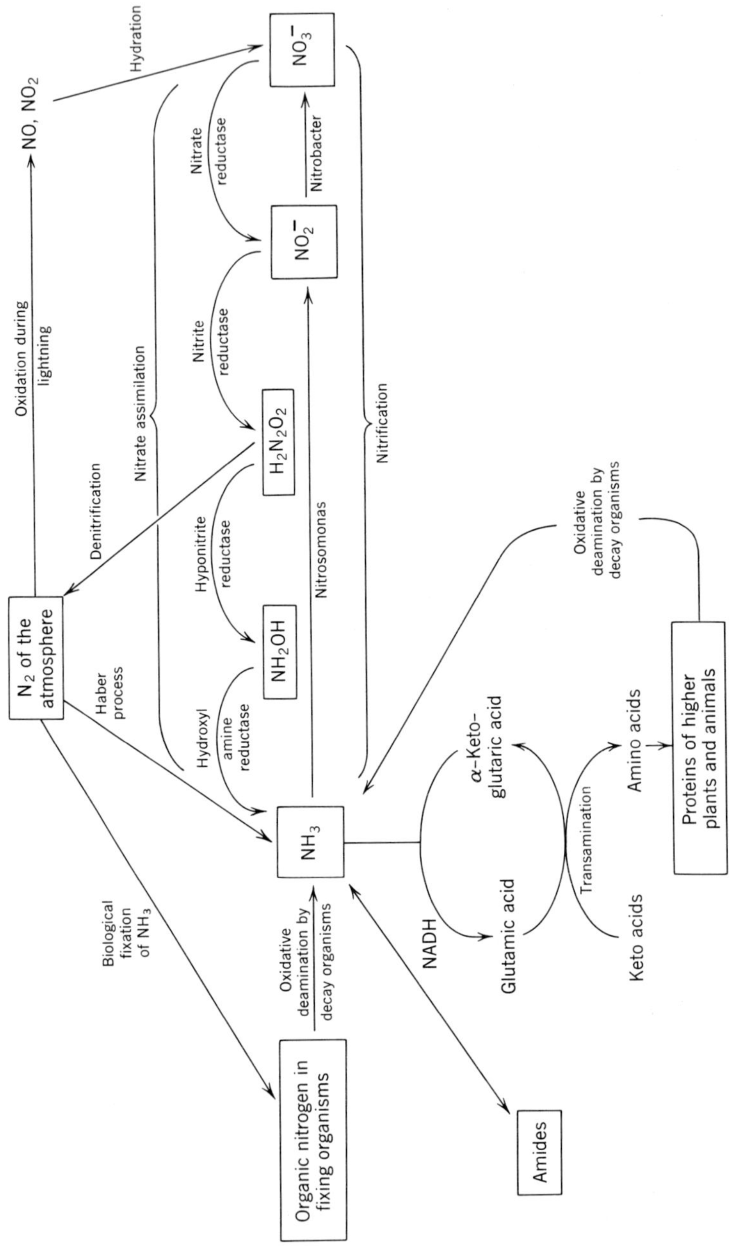

Fig. 6-2. The nitrogen cycle. (From Conn and Stumpf, *Outlines of Biochemistry,* Wiley, 1963.)

stage coincident with or beyond the formation of small nitrogenous organic molecules. These patterns are evident in the synthesis of proteins from amino acids, and of nucleic acids from nucleotides.

The Sulfur Cycle. The reduction of inorganic sulfur compounds, like that of nitrogen compounds, occurs only in certain groups of organisms. And, as in nitrogen metabolism, differences in sulfur metabolism between groups of organisms are attributable to the kinds of enzymes present. The characteristics of the sulfur cycle are shown in Fig. 6-3. Except for the fact that most microorganisms can reduce sulfate, whereas comparatively few can reduce nitrate, the two cycles are quite similar.

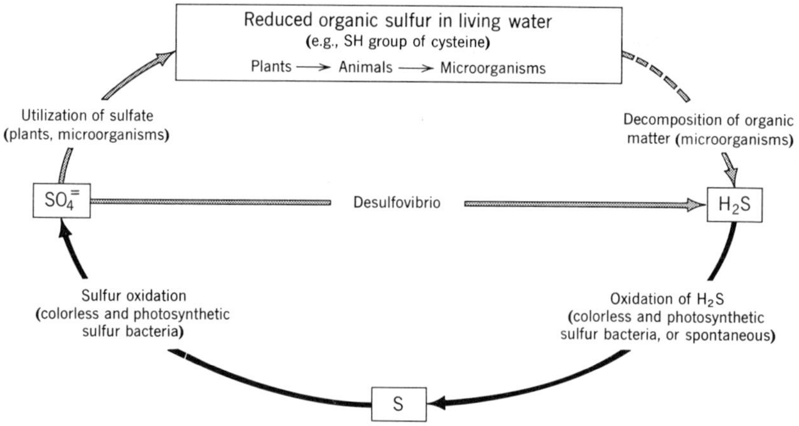

(*a*)

(1) Anaerobic oxidation (*Desulfovibrio*) of sulfate

$$2CH_3-\underset{\displaystyle H}{\overset{\displaystyle \overset{\displaystyle H}{O}}{\overset{|}{C}}}-COOH + SO_4^{2-} \rightarrow$$

Lactic acid

$$2CH_3-COOH + 2CO_2 + H_2S + 2(OH)^- + \text{energy}$$

Acetic acid

(2) Sulfide and sulfur oxidation (*Beggiatoa*)

$$2H_2S + O_2 \rightarrow 2S + 2H_2O + \text{energy}$$

$$2S + 2H_2O + 3O_2 \rightarrow 2SO_4^{2-} + 4H^+ + \text{energy}$$

(*b*)

Fig. 6-3. (*a*) The sulfur cycle. The oxidations of the sulfur atom are shown as solid black arrows. The reductions are shown as gray arrows. Reactions involving no valence change are shown as broken lines. (*b*) Oxidation and reduction of sulfur as energy-releasing processes. (Part a from R. Y. Stanier, M. Doudoroff, and E. A. Adelberg, *The Microbial World,* Prentice-Hall, 1963.)

Oxidation-Reduction of Nonmetallic Elements: Formulation of the Problem

The preceding discussion about the relationships between energy and the primary steps in acquisition of essential elements may be summarized as follows (1) Insofar as cells expend energy in acquiring metallic elements, such expenditure is mainly directed at overcoming the forces of diffusion. The uptake of metallic ions is thus part of the broader problem of how cells control the movement of substances. (2) Among the nonmetallic elements phosphorus is distinctive inasmuch as its incorporation into organic matter involves no oxidation or reduction but only esterification. Since phosphate esterification is one of the principal mechanisms by which cells make energy available for their different activities, this process becomes central to any consideration of energy utilization. (3) The remainder of the nonmetallic elements undergo reduction as a condition of incorporation into organic molecules.

A few simplifying generalizations were made about oxidation-reduction processes. Light is the ultimate source of energy for the reduction of carbon compounds into carbohydrate; subsequent oxidation of carbohydrate provides the necessary energy for most cell functions. The reduction of nitrogen and sulfur is a secondary process dependent on the normal energy reserves of the cell and the presence of appropriate enzymes to utilize that energy. These generalizations are not entirely valid because of certain exceptions. A brief discussion of the exceptions should enable us to formulate the problems surrounding oxidation-reduction in more general terms even though a more exacting discussion of the topic is reserved for a later chapter.

The sulfur cycle serves conveniently to exemplify the "exceptions." The fact that one group of microorganisms (Desulfovibrio) converts sulfate to a by-product H_2S (see Fig. 6-3*a*), and that another group (colorless sulfur bacteria) converts H_2S to the by-product sulfate (see Fig. 6-3*b*) contradict thermodynamic theory. For if sulfide oxidation releases energy, then sulfate reduction should absorb energy. And the juxtaposition of these two processes in nature makes no more thermodynamic sense than the juxtaposition of carbohydrate oxidation and carbohydrate synthesis without the intervention of light energy. If nothing were known about the sources of energy available to these two groups of microorganisms, we might resolve the contradiction by supposing that Desulfovibrio simply throws away a lot of its energy acquired through carbohydrate oxidation in order to reduce sulfate to H_2S. But this is not the case.

A key to explaining the apparent contradiction lies in the fact that Desulfovibrio organisms live in anaerobic environments (mud or bogs), whereas the colorless sulfur bacteria or algae live in aerobic ones. One group cannot survive in the presence of oxygen, the other group cannot survive without it. Desulfovibrio utilizes the oxygen of sulfate to oxidize its carbohydrates; the oxidation yields energy, and H_2S is a by-product just as H_2O is a by-product of respiration in most aerobic forms. Beggiatoa (a colorless sulfur bacterium) which lives in an H_2S-rich environment uses the hydrogens of the sulfide as reducing agents of carbohydrates, sulfur or sulfate being the by-product. In the absence of free oxygen—a characteristic of the Desulfovibrio environment—sulfate reduction at the expense of carbohydrate oxidation yields energy. In an oxygenated environment, however, the free energy of the reaction involving sulfide and sulfate would favor sulfate formation; the oxidation of sulfide to sulfate would release energy. Beggiatoa, having the appropriate enzymes, can therefore take advantage of a sulfide-rich environment.

Despite the fact that these two groups of organisms catalyze diametrically opposite reactions, and also despite the fact that an intense metabolism of sulfur compounds is alien to other organisms, the macromolecules of these organisms are fundamentally the same as those of all others. Carbon and nitrogen reduction and phosphate esterification are as much a primary requirement for sulfur-metabolizing cells as they are for other cell forms. All cells have the same specialized reducing molecules and the same phosphate esters which are used to build macromolecules. The differences between various groups of cells lie in the particular set of reactions which furnish a supply of these two basic forms of chemical energy. The preponderance of aerobically respiring cells often leads us to forget that carbohydrate oxidation by molecular oxygen is but one of several sources of energy. The universal feature of energy transactions is the provision of reducing power.

From the standpoint of organisms or populations, the specific nature of energy yielding reactions is highly relevant to physiological function and ecological interactions. The dependence of animals on plants for a source of carbohydrate, or of plants on nitrogen-fixing organisms for a source of nitrate, are pertinent illustrations. But from the standpoint of cellular functions, the different modes of energy acquisition should be regarded as special expressions of a single underlying mechanism. The fact that one mode of energy acquisition predominates over all others does not make it either more important or more fundamental as a cellular mechanism. The important question is whether several basic mechanisms exist or whether one universal mechanism exists which is adaptable to various environmental conditions. The correct answer is no longer in doubt, and the problems surrounding the acquisition of nonmetallic elements and their oxidation-reduction will now be briefly stated.

The occurrence of the nonmetallic elements (O, N, S, C) in a relatively reduced state in the major molecular constituents of cells is universal. So too is the energy requirement for effecting an over-all reduction. But each of these elements is capable of playing a role in meeting the energy requirement which is distinct from, even if related to, the role which each plays in the structure of biological molecules. All cells derive their energy from oxidation-reduction reactions. And all such oxidation-reduction reactions involve two or more of these nonmetallic elements. Hence nonmetallic elements are used both as fuel and as units of biological structure. When used as fuel, they appear as by-products in compounds of relatively low free energy. Clearly, such elements can be used for fuel only when they are present in amounts which far exceed those needed for the building of cellular molecules. An organism living in an anerobic environment could not possibly use free oxygen as part of its fueling system. What all cells must do is to provide from this group of elements two categories of compounds essential to every oxidation-reduction process: an oxidant and a reductant. One element can be reduced only if another becomes oxidized. In plant photosynthesis the oxygen in water becomes oxidized while the carbon in carbon dioxide becomes reduced; in respiration of higher organisms the carbon in carbohydrate becomes oxidized while free oxygen becomes reduced as water. To put the relationship in more general terms, which will be fully discussed under "energy acquisition," all cells require an electron donor (the reductant) and an electron acceptor (the oxidant). And common to all cells is a mechanism for transferring electrons in such a way as to permit the free energy released in the course of transfer to be converted into a chemical form utilizable for a variety of physiological needs.

SECONDARY PROCESSES

Cells consist of two size classes of molecules, small and large. The chemical evidence, discussed to some extent in Part I, clearly indicates that the large molecules (macromolecules)—protein, nucleic acid, carbohydrate—are formed by chemical linking of small ones. All cells must therefore convert the nutrients they absorb into a variety of small molecules and link them into large ones. As pointed out in Chapter 5, many cells lack the capacity to synthesize all the small molecules essential to their function, but all cells have the capacity to synthesize a large variety of these. Secondary processes, or "intermediary metabolism," thus refer to the mechanisms by which cells convert the essential elements absorbed as nutrients into small and large organic molecules.

Even a cursory inspection of the formulas of cellular organic compounds is sufficient to indicate that they have a common structural characteristic. All such molecules are basically chains of carbon atoms. In some compounds—purines and pyrimidines, for example—the chain may be closed to form a ring and a few nitrogen atoms may interrupt the otherwise pure carbon atom skeleton. In others, chains of 30 or more carbon atoms may be linked without interruption—in the fatty acids, for example. But with few exceptions, which are restricted to very small molecules such as urea, the major feature of the atomic skeletons of all cellular substances is a direct linking of carbon atoms. And even nitrogen-containing molecules are formed from precursors, notably amino acids, in which the nitrogen atom is attached laterally to the long axis of a carbon skeleton. If, therefore, cells must provide energy to reduce their essential nonmetallic elements, they must also provide energy for linking carbon atoms. All cells have the capacity to link carbon atoms and hence they also have the mechanisms to draw energy for that purpose.

To suggest that intermediary metabolism is primarily a problem of linking carbon atoms might seem far removed from the actualities of cell behavior. For cells link carbon atoms in many ways and the ingredients necessary for synthesizing one type of molecule are quite different from those required for synthesizing another. Compounds are generally not built by sequential additions of carbon atoms. Frequently parts of different molecules may be interchanged or entire molecules combined with one another. Yet all these additions, exchanges, and removals, though their number is still undetermined, may be attributed to one type of mechanism—enzyme catalysis. For each type of group linked, interchanged, or removed there is a specific enzyme. Thirty years ago it would have been impossible to do more than illustrate the nature of intermediary metabolism by means of well-studies examples. Today we may go beyond the specific characteristics of each enzyme system and address ourselves to one of the main questions in this section. Are there any universal mechanisms by which cells use energy to assemble small organic molecules? Since the basic skeleton of such molecules is a serial linkage of carbon atoms, the question may be set in the specific terms of the forms of energy essential to effecting C-C linkages.

The enzymatic reactions in Fig. 6-4 have been selected to illustrate a basic and common mechanism underlying the synthesis of carbon chains. The first point to be noted is that although carbon dioxide is used for extending a preexisting carbon chain, no equation is included for the serial linking of carbon dioxide molecules. Carbon dioxide is, in fact, an essential nutrient for all types of cells, but regardless of whether cells draw their energy from photosynthesis or respiration, carbon dioxide molecules are used to lengthen chains by attaching them to preexisting organic molecules of three or more carbon atoms. Usually carbon dioxide is considered essential to photosynthetic cells because of its role as a carbon source, but this

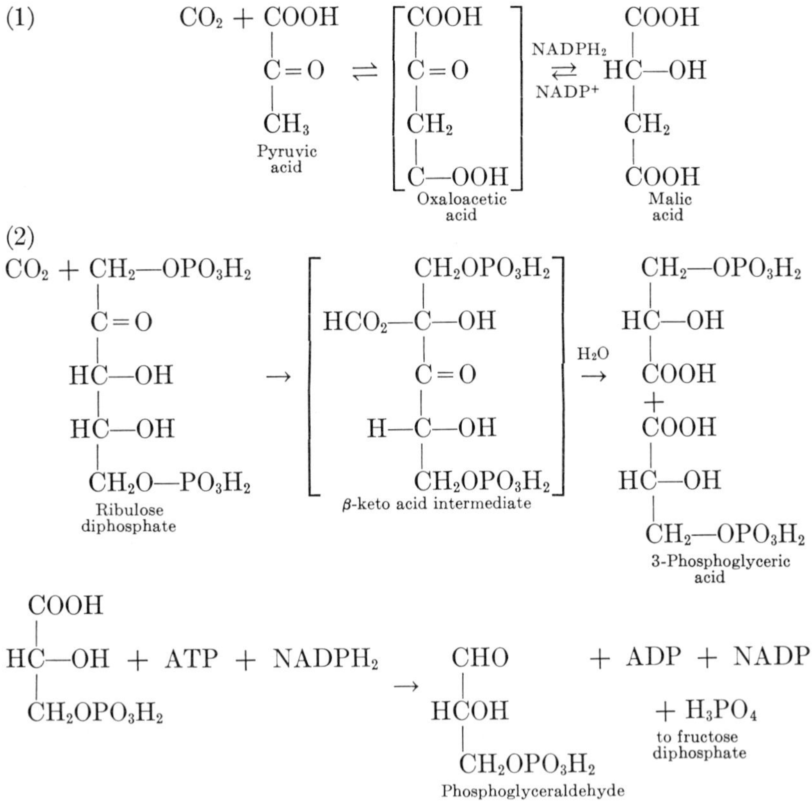

Fig. 6-4. Two systems for building carbon chains by CO_2 addition. The nature of the reducing compounds NAD and NADP are discussed in the text. (For details of reaction, see glycolysis chart, Fig. 7-6.)

is a particular instance of its utilization, not the sole one. The general need for carbon dioxide has been difficult to demonstrate because this compound is the most common product of respiration, and not until the introduction of isotopic techniques (see end of chapter for brief discussion of techniques) was a clear demonstration of this need possible. Enough experiments have now been done to establish that the level of carbon dioxide, or of its counterpart, the bicarbonate ion, may have a critical influence on the course of cell development. Some fungi will not develop their reproductive cells at low levels of carbon dioxide, and cell division in certain algae has been shown to be inhibited or accelerated according to the availability of carbon dioxide or bicarbonate in their medium. The full significance of carbon dioxide in physiological processes is by no means understood, but the fact that cells must build some of their carbon chains with CO_2 is clearly established.

The first two equations shown in Fig. 6-4 illustrate a universal type of energy requirement. In one reaction carbon dioxide is coupled with pyruvate, a common intermediate of carbohydrate breakdown in all types of cells. In the second reaction it is coupled with ribulose diphosphate, a compound used for the purpose in photosynthetic systems. The ways in which these acceptor compounds are formed need not concern us here, for the important point is that in neither case has any requirement for energy been indicated in the first two reactions. Thermodynamically this must mean that carbon chains can be extended spontaneously by adding CO_2 to the type of compounds shown in the equations (a keto group adjacent to the carbon atom linking with the CO_2). Why, then, do we speak of an energy need for carbon-carbon linkages? The answer lies in the nature of the products, which in Fig. 6-4 are shown bracketed to indicate that they are unstable compounds. In the absence of secondary changes such compounds would not accumulate to any appreciable extent; if oxaloacetate, for example, were dissolved in water, it would gradually decompose to CO_2 and pyruvate, and equilibrium would be reached (assuming that CO_2 was not allowed to escape) only when a very small amount of oxaloacetate was left. The secondary changes required for stabilization are the energy-consuming ones, and, regardless of the specific type of coupling, the form of energy is the same—a source of reducing compound. The action of the reducing compound is shown by the two equations in Fig. 6-4. In equation (1) the action is direct; pyruvate $+CO_2$ is converted to malate, a compound which enters into a number of other cellular reactions. In equation (2) the action is indirect. The combination of ribulose diphosphate with CO_2

yields a compound which splits into two three-carbon molecules. The first effect of the combination is therefore to decrease the length of the carbon chain. However, by utilizing a source of reducing power and a compound which will be discussed soon (ATP) cells are able to restore the two three-carbon molecules into one carbohydrate molecule consisting of six carbon atoms.

Coenzymes

One of the universal features of carbon chain building, not only in these reactions but in all others, is the need for reducing the complexes initially formed. Not only is such reduction essential to stabilization, but in many cases the compounds required by cells (e.g., fatty acids and amino acids) consist of chains in which the carbon atoms are fully reduced. It is with respect to the agents of reduction that the building of carbon chains may be described in universal terms. These agents are called coenzymes and are of two kinds—the nicotinamide group and the flavin group. Coenzymes are present in all cells, even though all cells cannot synthesize them; for those cells which cannot, "vitamins" are required as subunits for synthesis of the coenzyme compounds.

The term coenzyme was introduced when evidence was first obtained that certain enzymes would not act unless these relatively low molecular weight substances were present (Fig. 6-5). Conversely, coenzymes cannot be oxidized or reduced unless specific enzymes are present to catalyze the necessary reactions. To explain why their complex structure is essential to the role they play would be impossible without profound chemical considerations. We may more easily cite the fact that these compounds are universally present. Inasmuch as they become oxidized in the course of reducing linked carbon atoms, they must become reduced through some other channel in order to be reused. This channel is the one of energy acquisition which will be discussed in subsequent chapters.

Both the nicotinamide and flavin coenzymes contain "nucleotides," a term designating all those compounds which are serial linkages of a purine or pyrimidine ring, a five-carbon sugar, and a phosphate. Nucleotides are the subunits of nucleic acids. This fact may have great evolutionary significance. We may reasonably presume that nucleotides were abundant during the early evolution of life and that while some were used to build nucleic acids, others were modified for serving as oxidation-reduction intermediates. The modified portion of the molecule, not the nucleotide component, is the oxidation-reduction site.

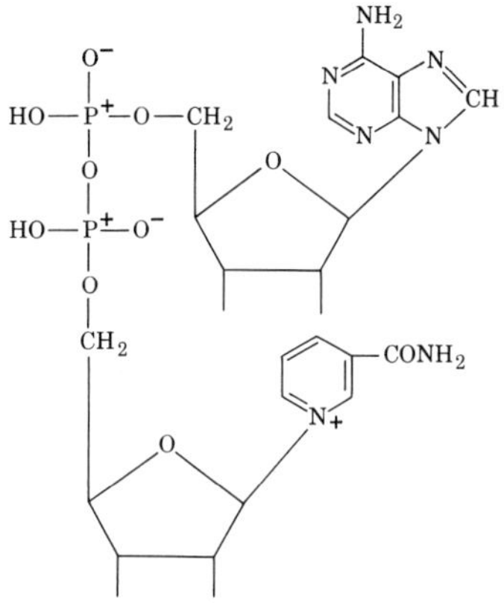

Nicotinamide-adenine dinucleotide (NAD^+)
DPN^+ or coenzyme I

(a)

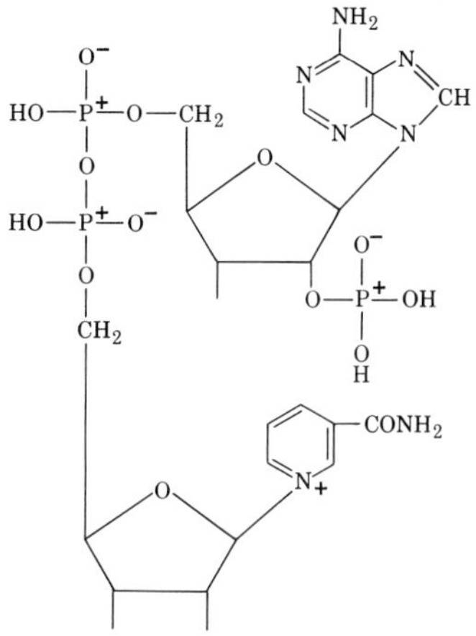

Nicotinamide-adenine dinucleotide phosphate
($NADP^+$)
TPN^+ or coenzyme II

(b)

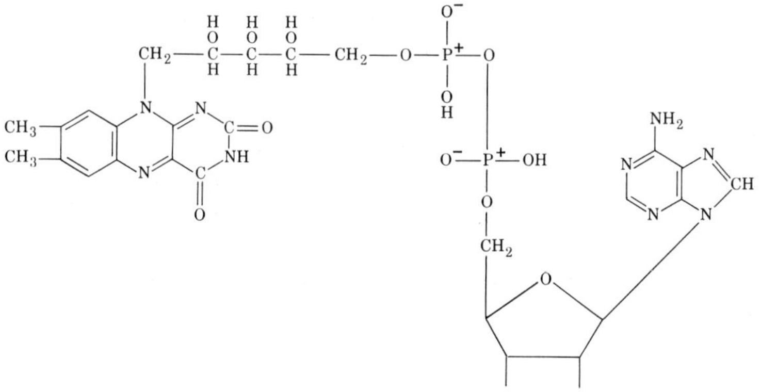

Flavin adenine dinucleotide (FAD)

(c)

Fig. 6-5. Coenzymes used in reduction of carbon compounds. Structural changes on reduction shown in Fig. 8-2.

Of the two categories of oxidizing-reducing coenzymes, the nicotinamide-containing group is the most abundant in cells. The formulas are shown in Fig. 6-5*a*, as are the alterations produced in the nicotinamide ring on oxidation or reduction. By the rules of chemical nomenclature, the coenzymes are called nicotinamide adenine dinucleotide (NAD) and nicotinamide adenine dinucleotide phosphate (NADP); they are commonly referred to by the initials shown in parentheses. From the standpoint of oxidation-reduction activities the two coenzymes are virtually identical. However, certain enzymes will catalyze reactions only with NAD, others only with NADP. The difference between these two otherwise identical coenzymes has little significance in terms of specific biochemical reactions, but it matters greatly in terms of the organization of reactions within cells. One example may be cited here to illustrate this feature of cell organization. Two enzymes in cells can oxidize malate to pyruvate or vice versa. One of these, requiring NADP, has already been discussed in connection with carbon linking. The other requires NAD, but it is associated with mitochondria, which are the sites for converting carbohydrate into other forms of chemical energy.

The second category of oxidizing-reducing coenzymes are flavin dinucleotides (Fig. 6-5*c*). Not only does their reducible component differ markedly from nicotinamide, but their association with proteins is different. The nicotinamide adenine dinucleotides are generally rather loosely associated with their enzymes, and a large proportion of them are usually recovered in unattached form when cells are broken. The flavin dinucleotides are tightly bound to the enzymes with which they are associated, and the kinds of situations in which the flavin dinucleotides serve to reduce the carbons of organic molecules are generally limited; their role is far more prominent in energy conversion.

Carbon chain building by addition of CO_2 to preexisting carbon chains is a universal process, as is the stabilization of such chains by reduced coenzymes. Whatever the ultimate source of energy—photosynthesis, respiration, or varieties of sulfate and nitrate metabolism—an essential and universal form of energy is the reduced coenzyme. Reducing power, however, is used not only to stabilize CO_2 additions, but also to stabilize the union of larger carbon chain fragments. An example of such a union is given in Fig. 6-6, but biochemistry texts should be consulted for a more extensive discussion of these reactions.

Most carbon chain building in cells occurs by the addition of molecules containing two or more carbon atoms to another carbon chain. The long carbon chains of the fatty acids, those of many amino acids, and the complex patterns of ring structures in the steroids are built in this way. Such chain building, like the lengthening of carbon chains with CO_2, requires effective reductions. Unlike CO_2 additions, however, multicarbon unit additions cannot be effected unless the units exist in an activated form. The activation of these units is closely tied to the utilization of carbohydrates for energy purposes. Although such utilization will be discussed in detail later, the immediate objective here is to describe a second universal form of chemical energy for the building of organic molecules—the activated carbon chain complex.

(*a*)

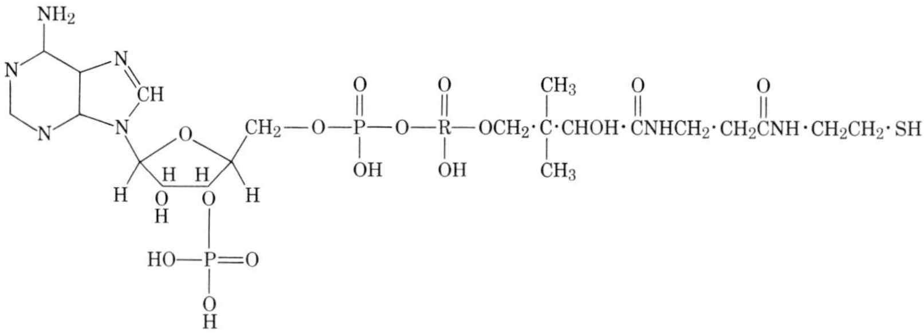

(*b*) Net reaction in the formation of a thioacyl linkage

$$R{-}C(=O){-}OH + HS{-}CoA \rightarrow R{-}C(=O){-}S{-}CoA + H_2O$$

(*c*) A condensation reaction

$$CH_3\cdot C(=O){-}S{-}CoA + CH_3\cdot CH_2\cdot CH_2\cdot C(=O){-}S{-}CoA \rightarrow$$
$$CH_3\cdot C(=O){-}CH_2\cdot CH_2\cdot CH_2\cdot C(=O){-}S{-}CoA + HS{-}CoA$$

(*d*) Reduction of carbonyl group

$$CH_3{-}C(=O){-}CH_2\cdot CH_2\cdot CH_2\cdot C(=O){-}S{-}CoA + NADP\cdot H_2 \rightarrow$$
$$CH_3\cdot CH(OH){-}CH_2\cdot CH_2{-}CH_2\cdot C(=O){-}S{-}CoA + NADP$$

Fig. 6-6. Carbon chain building by condensation of multicarbon units. Units are described as activated when combined with coenzyme A. Note that such activation consists of a bond between a partially oxidized carbon atom (—C(OH)=O) and the —SH group of the coenzyme. The mechanism by which this bond is formed will be discussed in a later chapter. Activated carbon units may consist of 2 or many carbon atoms, but they are all characterized by the thioacyl linkage. For stabilization of the chains formed, the —CO group must be reduced.

Molecules of two or more carbon atoms are ultimately derived from a breakdown of carbohydrates. Paradoxical though it may seem, cells build long carbon chains by first breaking down preexisting ones. Since ultimately the "preexisting" chains are carbohydrates produced in photosynthesis, the provision of multicarbon fragments is essentially a phenomenon of carbohydrate breakdown. All cells are capable of fragmenting carbohydrates and of linking these fragments into carbon chains which are much longer than the substrates from which they were derived. Fragmentation of carbohydrates nevertheless involves the rupture of carbon-carbon linkages. If cells were to lose all the energy thus released, the rebuilding of carbon chains would be relatively inefficient. Cells, however, do have a way of conserving the energy released on bond rupture, and the mechanism of conservation involves another universally distributed coenzyme generally known as coenzyme A (or CoA).

As the carbon atoms in a carbohydrate molecule become progressively oxidized, a stage is reached at which the linkage between adjacent carbon atoms becomes highly unstable, leading to a loss of the most oxidized carbon in the form of CO_2. (Details of this are given in Chapter 7.) Thermodynamically, this instability implies a release of energy on "decarboxylation." Cells conserve the energy otherwise lost as heat during decarboxylation, by means of an enzyme system which links the decarboxylated fragment to coenzyme A. The number of carbon atoms in the decarboxylated portion is irrelevant to the functioning of the linkage with coenzyme A. Given the appropriate condensing enzyme, coenzyme A-carbon chain complexes of different sizes may be put together as illustrated in Fig. 6-6. The universal feature of these complexes is that a terminal and partly oxidized carbon atom is linked to a sulfur atom in coenzyme A by a "thioacyl bond." Thus in order to "activate" multicarbon molecules, the cell must first oxidize a terminal carbon atom, and once the activated fragment is attached to another chain, the oxidized carbon atom must again be reduced. The principal energy forms which all cells must provide to synthesize their essential organic molecules are reduced coenzymes and acyl-coenzyme A complexes. How cells provide these is the subject of Chapter 7.

SYNTHESIS OF MACROMOLECULES

The macromolecules of cells—proteins, nucleic acids, polysaccharides—are synthesized by linking small molecules. Amino acids are linked to form protein, nucleotides to form nucleic acids, and monosaccharides to form polysaccharides. Synthesis of macromolecules by sequential addition of single carbon atoms is unknown. The linkages between the molecular subunits are basically identical inasmuch as their formation involves the elimination of water. When macromolecules are split into their subunits, one molecule of water is consumed for each linkage split, and hence the term "hydrolysis" for the procedure used in analyses of macromolecular composition. Although each type of molecular subunit requires a specific enzyme for attaching it to a macromolecular chain, all enzymes effecting intermolecular linkages do so by eliminating one molecule of water for each linkage formed.

If water elimination is the common chemical feature of macromolecular syntheses, is it also the common energy drain? The answer is in the affirmative, and the reason may be seen in the two model reactions shown in Fig. 6-7. Water elimination is not spontaneous but requires an external source of energy; in macromolecular syntheses this particular process is the common energy-requiring mechanism. In order to appreciate the factors which make this type of reaction a drain on the energy reservoir of cells, let us review briefly the characteristics of equilibrium systems from the standpoint of free energy change.

One feature of chemical reactions discussed in Part I is the tendency for many systems to achieve a stable condition in which both reactants and products remain in characteristic proportions. In such systems we cannot correctly speak of free energy change without referring to the initial concentrations of all components in the system. Thus a mixture of ethyl acetate and water will react to form acetic acid and alcohol, but the reverse will occur if the initial mixture is alcohol and acetic acid. Theoretically, an equilibrium point exists for almost every chemical reaction such that the addition of components to an equilibrium mixture will cause a change in the direction of removing some of the added components. In practice, equilibrium mixtures are rarely obtained for reactions involving a large free energy change, as, for example, the addition of metallic sodium to water, but such mixtures are the rule rather than the exception for reactions involving free energy changes of the order common in biochemical systems. Even though equilibrium may occur at a point at which components on one side of the equation are many times more concentrated than those on the other, a basic factor in determining whether or not a particular reaction can occur is the relative concentration of reactants in the system.

(*a*) Protein synthesis

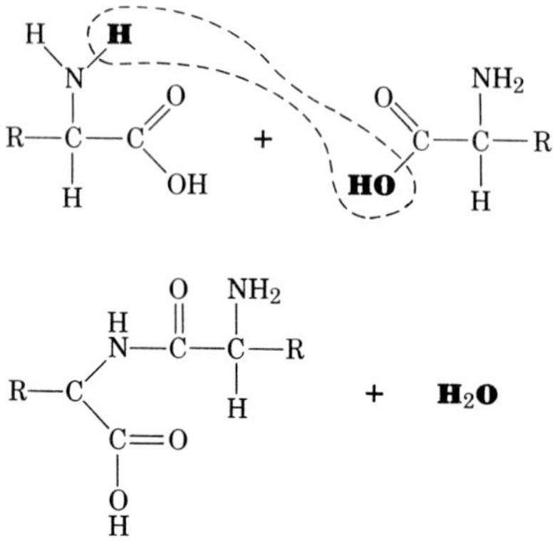

(*b*) Polysaccharide synthesis

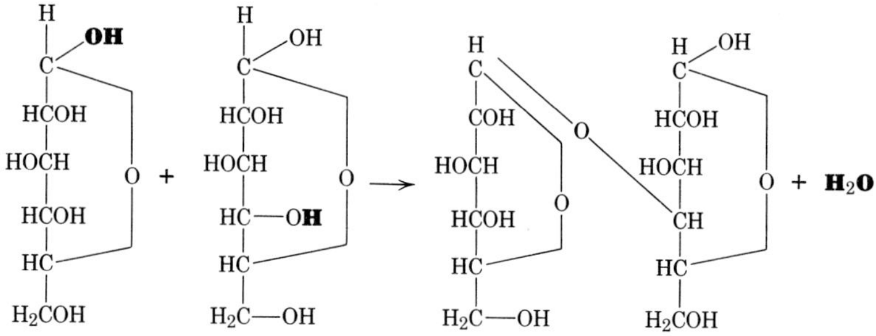

Fig. 6-7. Macromolecular synthesis by water elimination—a simplified representation to indicate the net chemical changes which a cell must effect in such synthesis.

An outstanding feature of living systems is the abundance of water in the molecular framework. As discussed in the section on pH determinations, the concentration of water is approximately 56*M*. By contrast, the concentrations of most small molecules which serve as precursors for macromolecules are rarely greater than about 0.001*M*. Clearly, the major thermodynamic obstacle which cells must overcome in assembling macromolecules is the presence of water. There are a few cellular reactions in which elimination of water does proceed without the addition of energy (for example, see glycolysis, discussed in Chapter 7, in which phosphoglyceric acid is converted to phosphoenol pyruvate), but macromolecular syntheses are not among them. If no other proof were available, the rapidity with which a variety of cellular enzymes hydrolyze macromolecules once cells are irreversibly injured would be sufficient. Just as an aerobic environment favors oxidations, an aqueous environment favors hydrolyses. All cells must therefore have a mechanism whereby energy can be used to overcome the adverse conditions created by the abundance of water.

Phosphate Linkages

Phosphorus plays its outstanding role in furnishing this mechanism. Phosphate ions are unique among the nonmetallic elements in that they do not undergo oxidation-reduction; as pointed out in Chapter 5, their distinctive combination in organic matter is in the form of phosphate esters. The esters are of several types; by virtue of the chemical properties of certain types the cells are able to overcome the adverse conditions surrounding the synthesis of macromolecules by water elimination. The formation of phosphate esters by energy-yielding processes will be taken for granted here, and only the mechanism by which they make syntheses possible will be considered.

The types of phosphate bonds found in small cellular molecules are illustrated in Fig. 6-8. Of particular interest are those labeled "high energy bonds." Chemically, the term is a misnomer, but biologically it is appropriate. For, chemically, these bonds release relatively little energy when disrupted, but because they are sufficiently unstable in water to render them highly reactive with a variety of organic molecules, they represent a source of energy for biochemical processes. This instability provides a key to understanding the basic mechanism by which the high energy phosphate bonds make possible macromolecular syntheses. The equation shown in Fig. 6-9 indicates that disruption of a phosphate ester bond leads to the consumption of one molecule of water. Stated thermodynamically, hydrolysis of such bonds releases free energy. With respect to this property, the biochemical strategy of the cell may be simply described. By means of appropriate enzymes the water equivalents essential for hydrolysis are derived from the subunits to be joined in macromolecular synthesis rather than from water itself. The mechanism is illustrated in Fig. 6-9. Whereas in hydrolysis the free energy released is lost as heat, in synthetic processes part of the energy is conserved in the formation of peptide or other bonds.

The commonest form of high energy bond is found in the molecule adenosine triphosphate (ATP). All of the known nucleotides, however, may also occur in the triphosphate form. In nucleic acid synthesis all four nucleotide triphosphates must be present; in other syntheses specific nucleotide types are essential to correspondingly specific reactions. Uridine triphosphate (UTP), for example, is a common en-

(*a*) Chemical groups

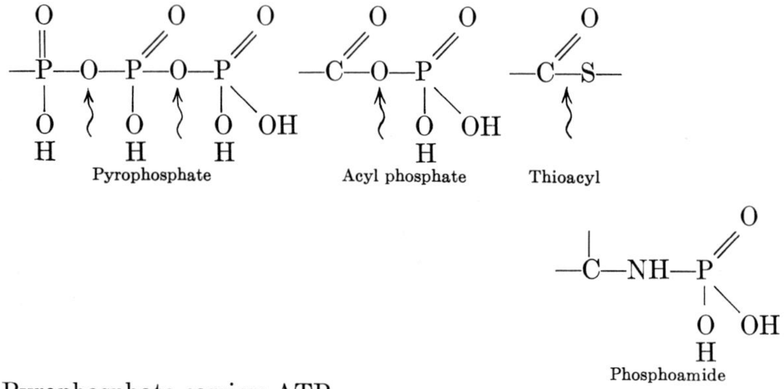

(*b*) Pyrophosphate carrier: ATP

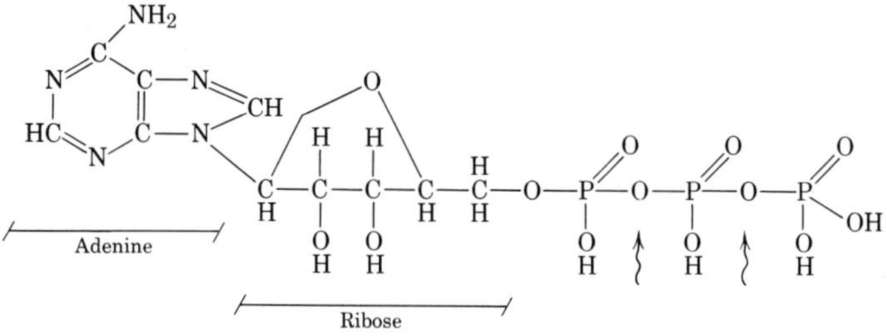

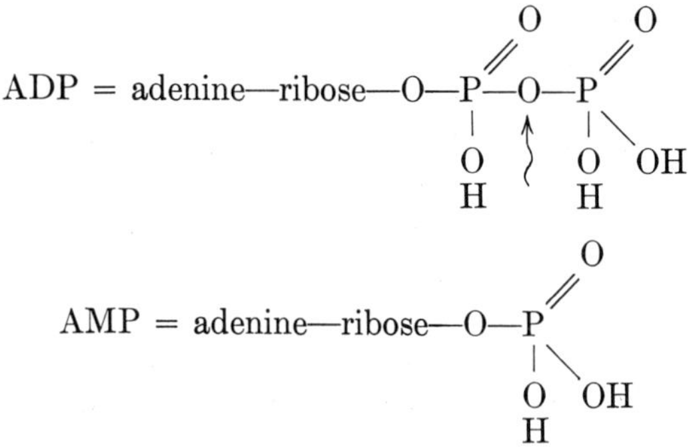

Adenosine = adenine—ribose

Fig. 6-8. "High energy" bonds.

Fig. 6-9. Reactions of ATP.

(1) Hydrolysis

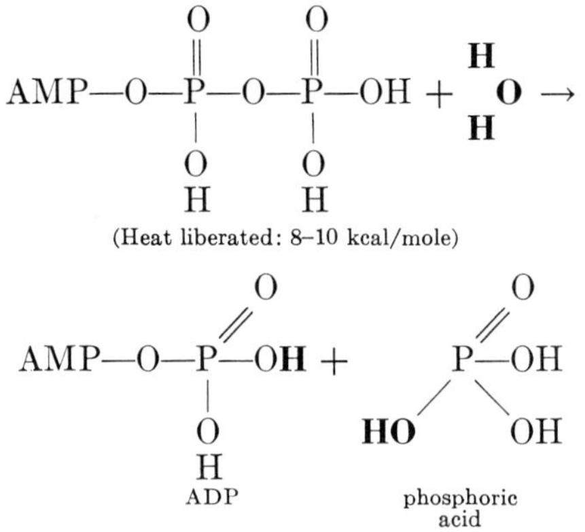

(2) Simplified model of polysaccharide synthesis

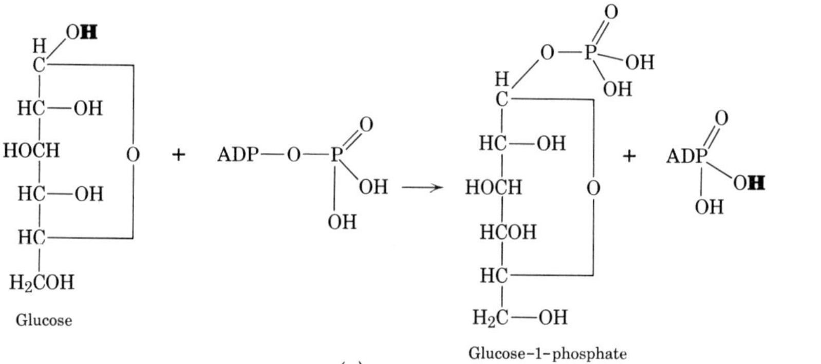

(*a*)

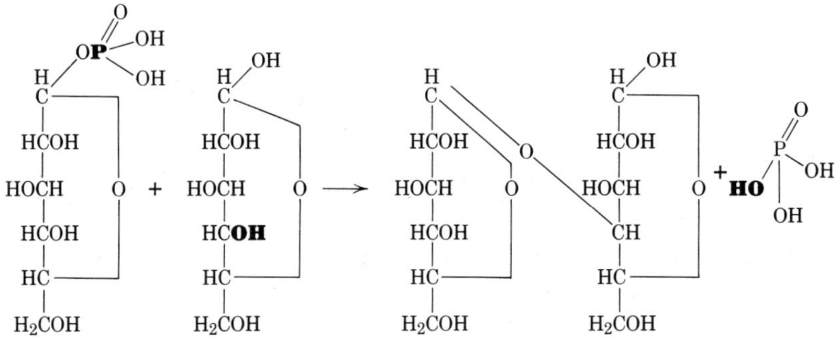

(*b*)

Fig. 6-9 (continued).

(3) Pyrophosphate cleavage in nucleic acid synthesis (X represents any purine or pyrimidine ring)

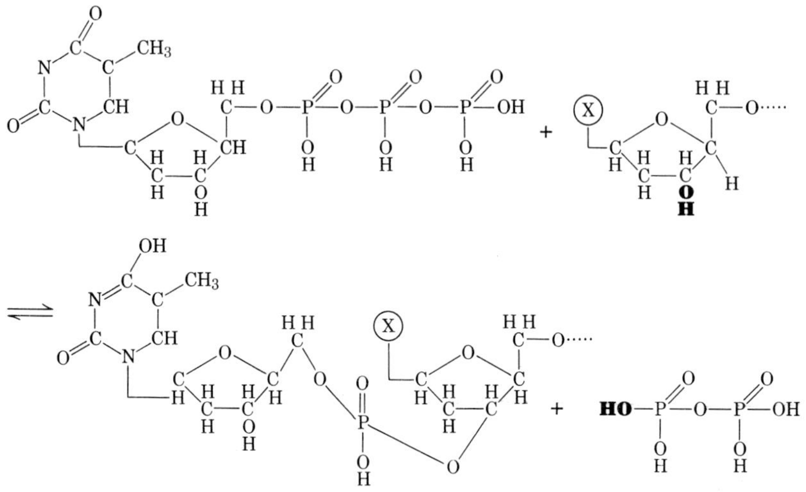

(4) Phosphorylation of thymidine (thymidine kinase)

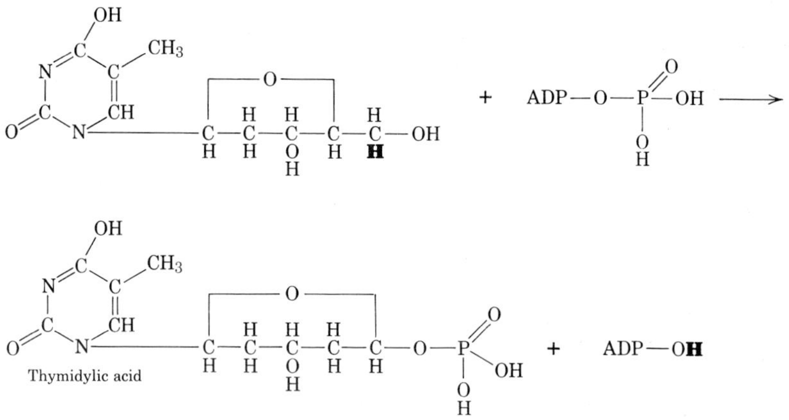

Thymidylic acid

ergy source for the linking of sugars, and guanosine triphosphate (GTP) is essential to certain steps in protein synthesis. The source of chemical energy lies in the phosphate bond, but the specific process by which the energy is transformed depends on the rest of the molecule for an effective coupling between the enzyme catalyst and the substrate to be altered. Cells also have the enzymes to transfer high energy bonds from one nucleotide carrier to another; in terms of a

Fig. 6-9 (continued).

(5) Exchanges of pyrophosphate bonds (kinases) (Y or X represent any nitrogen base + sugar)

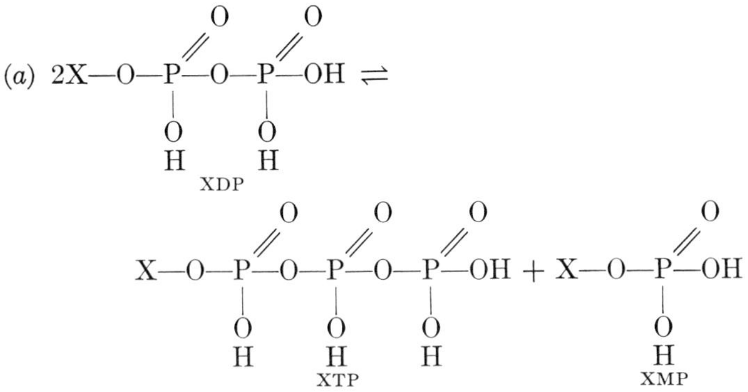

(*b*) XTP + YMP ⇌ XDP + YDP
2YDP ⇌ YMP + YTP

source for these various forms of chemical energy, the formation of one type assures a supply of the others.

To summarize, the energy essential to the synthesis of cellular constituents is required in three forms: a reduced coenzyme, a coenzyme carrier for small carbon chains, and a high energy phosphate linkage. Certain specialized types of coupling (e.g., the attachment of a formate group to a carbon chain) have not been considered, for the major purpose of this chapter has been to reduce the diversity of energy transactions to the fewest possible number of basic mechanisms. The next objective is to elucidate the mechanism by which these basic energy carriers are provided.

7

Sources of Energy: Carbon Compounds

The provision of cellular substances would appear to have contradictory requirements. On the one hand, oxidative destruction of carbon-carbon linkages is necessary for the production of generally utilizable forms of energy such as ATP or NADH, while a formation of carbon-carbon bonds is essential in the synthesis of cell constituents. The principal channel by which all cells acquire energy for chemical and mechanical work is the partial or complete oxidation of carbohydrates. This is true whether the carbohydrate is endogenously formed through photosynthesis or whether it is acquired as a nutrient. Carbohydrates, however, are also the source of carbon skeletons, which by enzymatic transformation furnish the molecules for macromolecular syntheses. The picture is oversimplified since many cells do have the capacity to utilize noncarbohydrates for energetic purposes; such capacity, however, is derived from the ability of cells to transform a large variety of molecules into intermediates of carbohydrate breakdown. Basically the apparent contradiction is universal. Carbohydrates must be broken down to provide flexible forms of chemical energy, but at least parts of their molecules must be conserved to furnish carbon-carbon linkages.

The contradiction has been labeled "apparent" for very good reasons. The effective utilization of energy released in the oxidation of carbohydrates is governed by certain thermodynamic principles. The basic equation applied to chemical reactions, $\Delta F = \Delta H - T\Delta S$, describes the total process but does not specify the extent to which the free energy released can be converted into usable energy. The oxidation of glucose to CO_2 and water releases about 670 kcal of energy per mole, but the conditions under which the energy is released determine how much of the energy is lost as heat. In order to make possible efficient conversion of energy, the degradation of a glucose molecule must proceed in step-wise fashion such that in any single reaction the ΔF value is small. It is a characteristic of chemical reactions that the smaller the ΔF, the smaller the irreversible loss of free energy as heat. The thermodynamic ideal is a reaction in which the ΔF approaches zero and the reversibility of the reaction approaches completeness. In practice ideal reactions are not very useful for effecting chemical conversions; enough free energy must be made available in any single reaction to permit the formation of at least one chemical bond. Thus the actual pattern of carbohydrate breakdown in a cell, though not approaching the thermodynamic ideal, is step-wise. Since long carbon chains are built by enzymes which condense carbon units smaller than the original carbohydrate molecule oxidized, the conversion of carbohydrate into flexible forms of energy or into novel carbon chains has at least one common requirement.

Not only is there a general requirement for relatively small carbon-containing molecules in the building of long carbon chains, but different kinds of such small molecules are necessary as precursors for different kinds of syntheses. In this respect, too, carbohydrate oxidation and carbon chain synthesis follow a common pathway. The intermediate breakdown products of carbohydrate oxidation furnish the necessary kinds of molecules for the synthesis of essential constituents. Inspection of Figs. 7-1, 7-4 in which the course of carbohydrate oxidation is diagramed, will reveal a funneling of various intermediates into different synthetic routes. Glycerol, for example, is used for lipid formation, ribose for nucleotides, acetyl-coenzyme A for hydrocarbon chains, α-ketoglutarate for amino acids, and succinate for porphyrins. Thus although cells may be superficially described as either oxidizing carbohydrates for energetic purposes or conserving carbohydrate for carbon chain building, they may be more accurately described as transforming carbohydrate along a pathway which ultimately leads either to the production of general energy donors or to the provision of essential carbon skeletons.

The partitioning of carbohydrate along this pathway is a process which cells must regulate. All cells must convert part of their carbohydrate supply into compounds like ATP or NADH, for such energy donors are essential to the joining of carbon skeletons and to their modification. But cells vary a great deal in their relative requirements for common energy donors and specific carbon skeletons. Growing cells have a large need for carbon skeletons because they are synthesizing new constituents. Muscle cells, by contrast, require few carbon skeletons for synthetic purposes, but require abundant amounts of ATP as fuel for mechanical work. An elongating plant cell channels most of its carbohydrate into the formation of cell wall material, using relatively little of its energy for synthesis of cytoplasmic material; cells of a storage organ, such as a potato tuber, convert most of their energy supply into starch. A zoospore utilizes its energy supply for flagellar movement during its motile stage, but shifts its metabolism radically once it begins to grow. Thus not only do different cell types vary in their pattern of carbohydrate utilization, but the same cell undergoes marked changes in pattern during the course of its development. How such shifts occur is a point of major interest to the cell biologist, but an understanding of this phenomenon is impossible without first understanding the mechanisms by which carbohydrate is metabolized.

Underlying the complex charts of metabolic pathways, which today are commonplace, is a formidable body of experimental techniques. The transformations which a glucose molecule undergoes in a cell are difficult to demonstrate directly. In the early development of biochemistry very many guesses had to be made to explain observations and to plan new experiments. For although the characterization of the terminal products of glucose metabolism was a relatively simple matter, identification of the intermediates was difficult not only because such intermediates occur in very low concentrations but also because the sequence in which intermediates are formed cannot be tracked directly. The earliest technique to be applied in biochemical studies, and one which remains indispensable, is that of skillful organic chemistry. Molecular structure had to be determined for each of the compounds found in cells, whether such compounds occurred in high or low concentration. Even so, the biochemist had to exercise a great deal of selection. The separation of small and large molecules was a comparatively straightforward task, because macromolecules are generally precipitated by acids or alcohol. But acidic or alcoholic extracts of cells probably contain almost as many compounds as are found listed in the organic section of a chemistry handbook. The less that was known about metabolic pathways, the more difficult the choice of compound to purify. The task was helped considerably by the discovery of metabolic inhibitors, substances which poisoned specific enzyme reactions. Such poisoning caused the accumulation of intermediates which could then be isolated and their structure determined. Few, if any, inhibitors are entirely specific in their effects, and their use is therefore limited. A far more fruitful approach was provided by the discovery that enzymes could be studied in test tubes. As a result of this, one principle became firmly established in biochemical studies. Any reaction presumed to occur in a living cell could be demonstrated by isolating and purifying the appropriate enzyme or enzymes. As the number of enzymes isolated and characterized grew, so did our knowledge of metabolic pathways. The introduction of isotopes into biochemistry opened a new era in metabolic studies. The most commonly used isotopes are the radioactive isotopes because their presence can be detected with a very high degree of sensitivity. By feeding cells compounds containing radioactive elements (for example, C^{14} in place of C^{12} atoms in glucose) the natural fate of these compounds can be followed. Cells do not discriminate between radioactive and non-radioactive forms of the element, so that normal transformations are not affected even though the products of transformation can be identified by virtue of their radioactive content.

GLYCOLYSIS

If cells are suspended in water or a solution of sucrose, disrupted mechanically by means of one of several devices (e.g., a Waring Blendor, a mortar and pestle with an abrasive such as sand), and the suspension then centrifuged, a clear solution of protein and various other organic molecules is obtained. All the enzymes essential to the first phase of carbohydrate metabolism (glycolysis) are found in this solution. The use of sucrose in this procedure has particular relevance to the relationship between these enzymes and the structural organization of cells. Disruption of cells in sucrose media preserves the structural integrity of subcellular components—nuclei, chloroplasts, mitochondria, membranes, and ribosomes. These structures are sedimented during centrifugation, and since only a very small fraction of the enzymes in question is found in the sediment, the inference generally made is that these enzymes are distributed in the "ground substance" of the cell—that part of the cell in which the subcellular components are embedded. The initial metabolism of carbohydrate, as well as many other molecular transformations, appears to be catalyzed by enzymes which are not organized into specific structures. The situation is comparable to a solution of enzymes in a test tube in which the activity of each enzyme depends on a random collision between enzyme and substrate. Whether this is in fact the case is not at all clear, but, as will be seen, there is a marked difference with respect to the organization of enzymes governing the first phase of carbohydrate metabolism (glycolysis), and those governing the second phase which completes the conversion of the first products to CO_2 and water.

The term glycolysis was originally used to designate the process in which the polysaccharide glycogen was converted to a three-carbon product, pyruvate. Interest in glycogen as a starting point arises from the fact that many types of animal cells convert excess carbohydrates into glycogen, which becomes available as a source of energy when extracellular carbohydrate is insufficient. The essential features of glycolysis are much the same whether the carbohydrate used is glycogen, starch, sucrose, or other storage forms. Most, if not all, cells have the capacity to store food reserves. These features are the conversion of polysaccharide into hexoses, the splitting of hexoses into three-carbon molecules, and the oxidation of such molecules to yield pyruvate. A detailed representation of the various enzymatic steps occurring in glycolysis is shown in Fig. 7-1.

Glycolysis is characterized by four principal types of reactions: phosphorylation, intramolecular rearrangements, molecular cleavage, and partial oxidation. All such reactions are catalyzed by specific enzymes.

Phosphorylation

Phosphorylation is a universal feature of glycolysis. Although enzymes have been found which oxidize hexoses directly, such oxidations do not lead to a conservation of energy for cellular activities. Glycolytic enzymes have a very high affinity for phosphorylated compounds, and all cells have the capacity to attach phosphate radicals to certain hexoses. A number of explanations may be offered to account for the fact that the initial stages of carbohydrate metabolism involve phosphorylated rather than free sugars, but these will not be discussed here. The important point is that unless hexoses are converted to phosphate compounds, they are more or less ignored by the enzymes in the cell. Since, however, the formation of such compounds consumes energy, cells must initially expend energy for phosphorylation in order to gain energy in later steps of oxidation. If the starting compound is a polysaccharide, the first phosphorylation reaction utilizes the energy already present in the glycosidic bonds; an enzyme (phosphorylase) combines with one of the hexoses in the polysaccharide chain and transfers that hexose to a phosphate ion. The enzyme thus conserves the energy originally used to form the glycosidic bond in the polysaccharide and uses it to effect a phosphate ester bond. Either type of bond is cleaved by hydrolysis, and hence each type of bond is formed by an elimination of water. Water elimination, as pointed out earlier, requires energy, and this requirement is avoided by enabling the reaction to take place without the intervention of water.

If a hexose is the initial carbohydrate source it is phosphorylated by ATP; the second phosphate, which must be attached to the other end of the hexose molecule, is in all cases supplied by ATP. Since the equivalent of one molecule of ATP is used in the formation of each glycosidic bond in the polysaccharide, initiation of glycolysis requires an expenditure of energy equivalent to that used in forming two molecules of ATP from two molecules of ADP.

One general feature of how cells control the direction of reactions is made apparent by examining the enzymes involved in the synthesis of a polysaccharide such as starch or glycogen. In eq. (7-1) the phosphorylase reaction

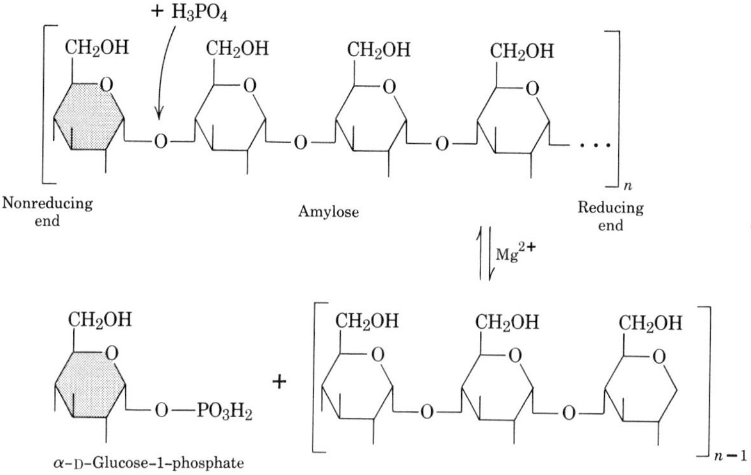

(7-1)

is shown to be reversible. Indeed, if the reaction is carried out in a test tube with the isolated enzyme, polysaccharide may be readily obtained by adding a relatively large amount of glucose phosphate and a relatively small amount of phosphate ion. The equilibrium point, however, favors the glucose phosphate so that only a small amount of polysaccharide is formed. On the other hand, polysaccharide plus inorganic phosphate in the presence of the enzyme yield a high proportion of glucose phosphate. How do cells manage to store a great deal of polysaccharide if supplied with amounts of glucose which are even slightly in excess of their carbon requirements? We might explain such synthesis by assuming that a lot of glucose phosphate remains unused, thereby pushing the reaction in the direction of polysaccharide synthesis. This, however, is not the case. Cells have a different enzyme system which couples glucose with uridine triphosphate (UTP), and the product, uridine diphosphoglucose, is acted on by another enzyme which transfers the glucose to polysaccharide as shown in eq. (7-2). From the energetic standpoint, the cost of forming polysaccharide in this way is twice that of forming it by the phosphorylase system, two pyrophosphate bonds being used instead of one. In the uridyl system, however, equilibrium favors polysaccharide synthesis, and we may generalize this particular example by stating that in most macromolecular syntheses cells expend extra energy to assure formation of polymer. The same removal of two pyrophosphate bonds is found in nucleic acid and polypeptide syntheses. Cells are not always parsimonious about energy expenditure.

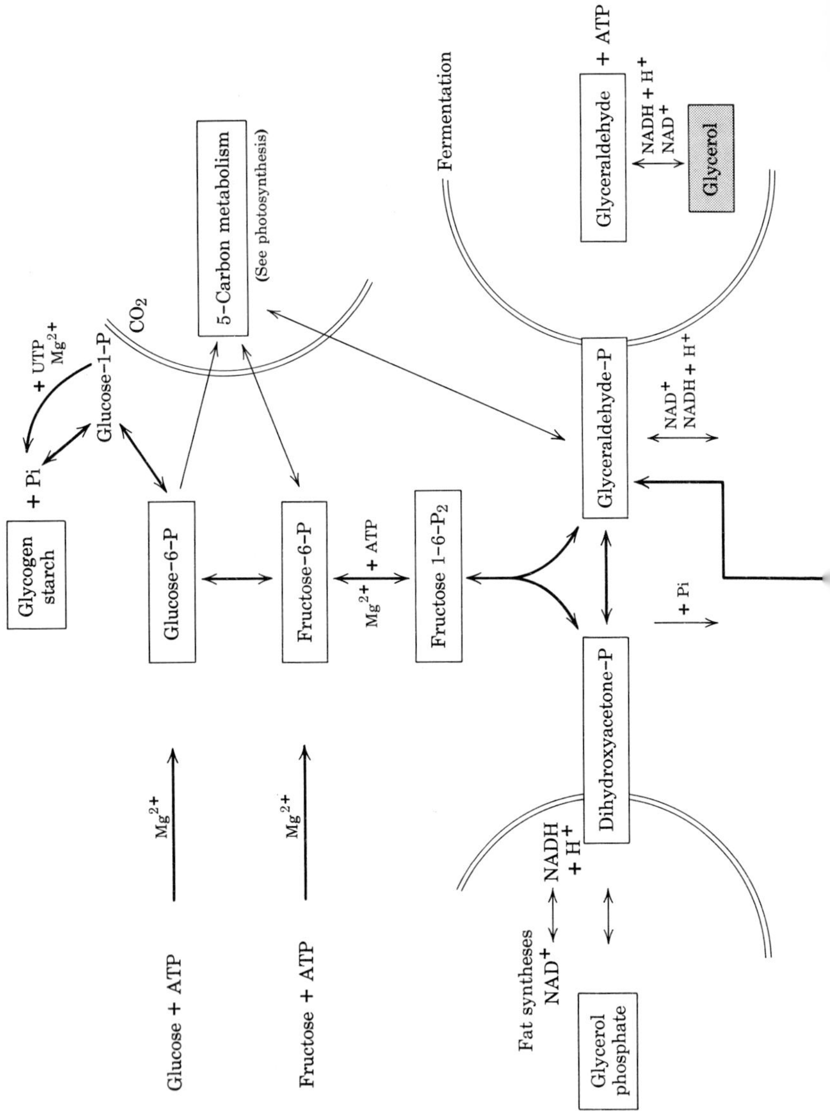
Glycogen starch
+ Pi
+ UTP
Mg^{2+}
Glucose-1-P
CO_2
5-Carbon metabolism
(See photosynthesis)
Glucose-6-P
Fructose-6-P
Mg^{2+}
+ ATP
Fructose 1-6-P_2
Glucose + ATP
Mg^{2+}
Fructose + ATP
Mg^{2+}
Fermentation
Glyceraldehyde
+ ATP
NADH + H^+
NAD^+
Glycerol
Glyceraldehyde-P
NAD^+
NADH + H^+
+ Pi
Dihydroxyacetone-P
NADH
+ H^+
Fat syntheses
NAD^+
Glycerol phosphate

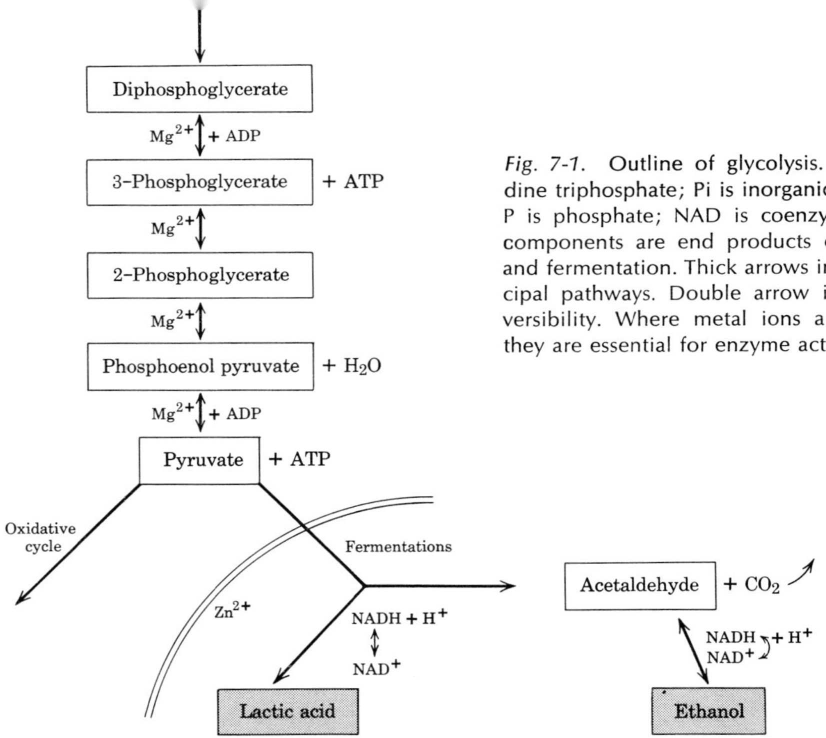

Fig. 7-1. Outline of glycolysis. UTP is uridine triphosphate; Pi is inorganic phosphate; P is phosphate; NAD is coenzyme. Shaded components are end products of glycolysis and fermentation. Thick arrows indicate principal pathways. Double arrow indicates reversibility. Where metal ions are indicated they are essential for enzyme activity.

(7-2)

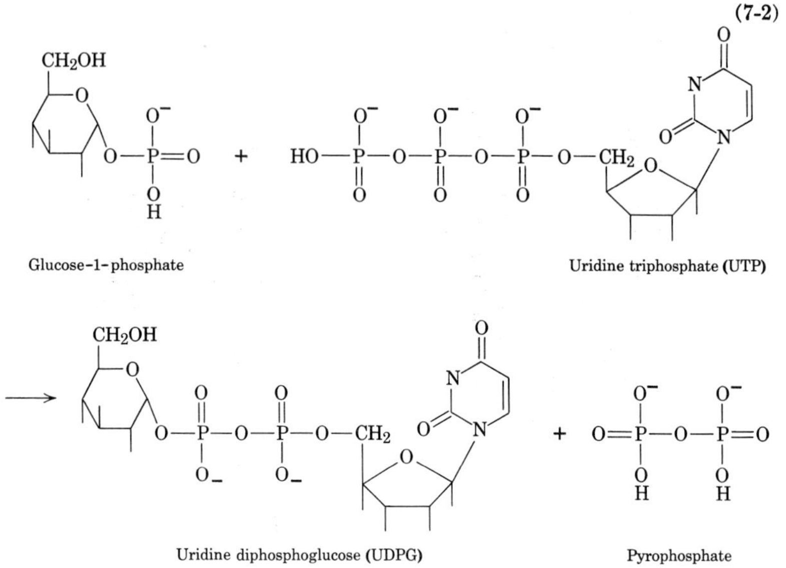

(a)

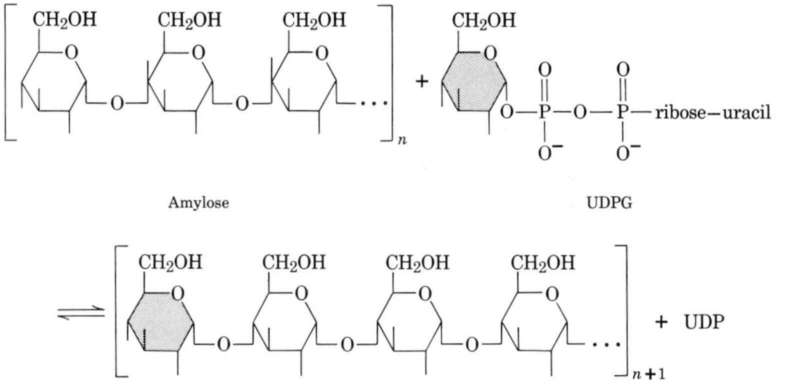

(b)

Intramolecular Rearrangements

Various types of intramolecular rearrangements are evident in Fig. 7-1 and need not be discussed in detail. Equations (7-3), (7-4), and (7-5) show these rearrangements; they are reversible reactions requiring no energy input. Indeed, the list of cellular enzymes capable of catalyzing intramolecular transformations is rather long, and they constitute the simplest example of reactions which are thermodynamically possible; yet, they do not occur to any appreciable extent outside of cells because of the absence of a specific catalyst. Two rearrange-

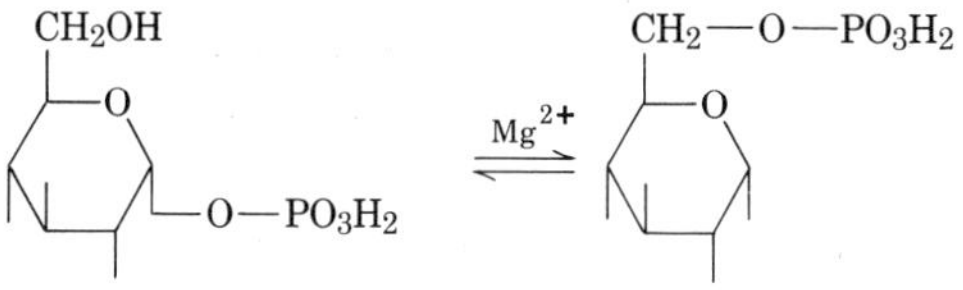

(7-3)

Glucose-1-phosphate Glucose-6-phosphate

ments in glycolysis may be briefly considered. The transfer of a phosphate group within a molecule (e.g., glucose-1-phosphate to glucose-6-phosphate) requires a trace of glucose diphosphate. This reaction again illustrates the principle that whenever a cell transfers a bond which has been formed by the elimination of water, it uses enzymatic mechanisms which keep water out of the trans-

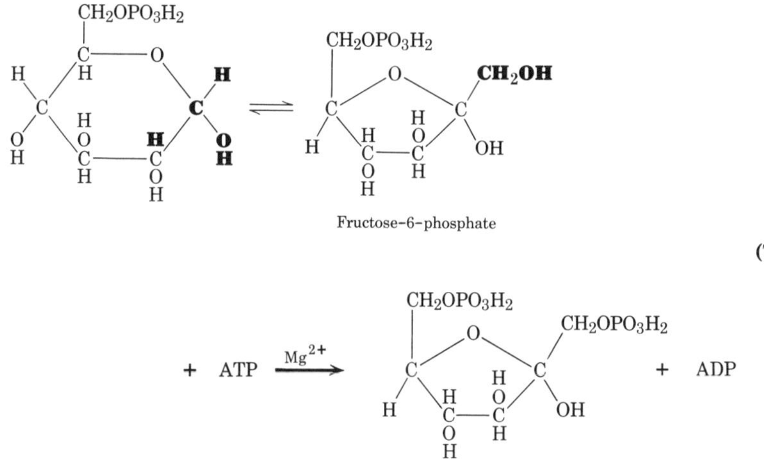

Fructose-6-phosphate

(7-4)

Fructose-1-6-phosphate

fer reaction. In phosphate transfer the enzyme brings the monophosphate and the diphosphate molecules together, and, with the two substrates in close proximity, the diphosphate form donates one of its phosphates to the desired position in the monophosphate and retrieves a phosphate from the undesired position in the original monophosphate

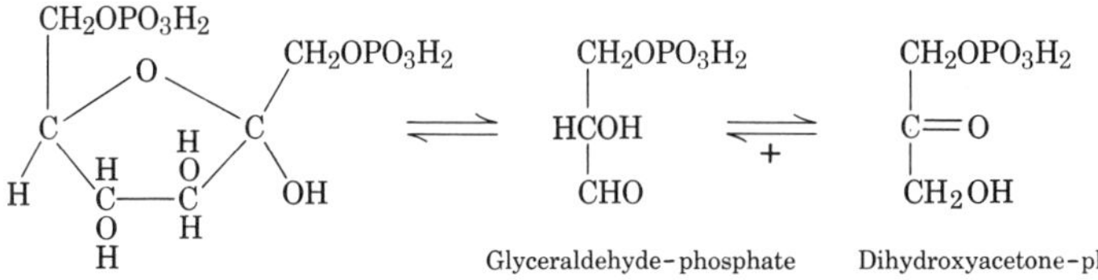

(7-5)

Glyceraldehyde-phosphate Dihydroxyacetone-phosphate

form. Exactly how the reactants are positioned at the enzyme surface is still unknown, but the existence of some mechanism which permits phosphate transfer while excluding interaction with water may be confidently inferred. Indeed, all phosphate transfers, whether intra- or intermolecular, proceed by the same type of mechanism.

A second type of intramolecular transformation which merits special comment is the conversion of glucose phosphate to fructose phosphate. The keto sugar splits readily in the presence of the enzyme aldolase to yield two triose molecules (phosphoglyceraldehyde and dihydroxyacetone phosphate) which are interconvertible. In effect, the splitting has yielded two identical halves so that both halves of what was originally an asymmetric molecule may be acted on by the same enzyme. The cleavage of fructose diphosphate into triose phosphates is the only cleavage occurring in glycolysis. Since this cleavage is reversible and requires no input of energy for either direction of the reaction, the energy contained in the carbon-carbon linkage has been conserved in the products. Thus up to the step of glyceraldehyde phosphate production, the glycolytic enzymes may be considered as having effected a variety of molecular transformations without either dissipating or yielding energy to the cell.

Glyceraldehyde Oxidation

The oxidation of phosphoglyceraldehyde is the first and only step in the main glycolytic pathway which involves a release of free energy and a conversion of the energy thus released to two of the generalized forms of chemical energy—NADH and ATP. In principle the reaction may be simply illustrated by omitting the phosphate group of the glyceraldehyde. Oxidation occurs in converting the aldehyde group to a carboxyl one. The oxygen is derived from water, which transfers its hydrogen to a hydrogen acceptor. This basic reaction occurs through the whole process of carbohydrate oxidation, cells using an external source of oxygen only in the final stages to reoxidize the hydrogen acceptor. When cells cannot utilize such oxygen, the hydrogens are transferred to some other organic compounds.

The actual process of glyceraldehyde oxidation is, however, far more complex than that illustrated in eq. (7-6). The total reaction in the cell involves phosphoglyceraldehyde, phosphate ion, NAD, and enzyme. A semidiagramatic representation of the reaction is given in Fig. 7-2. The representation is hypothetical and may prove to be incorrect, but it is worth examining in detail because it illustrates several basic principles of metabolism.

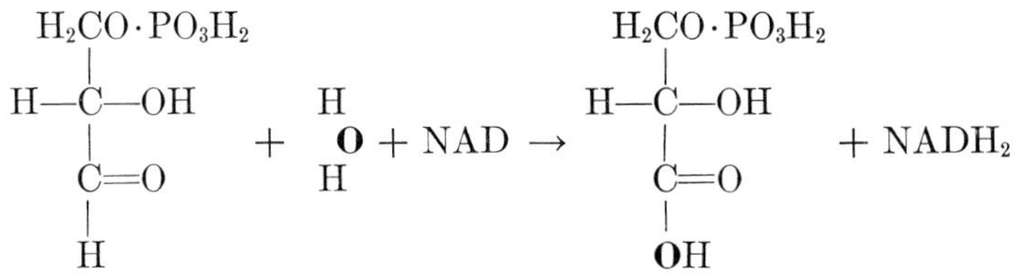

(7–6)

The first feature to be noted is the attachment of coenzyme to the protein. In all enzyme reactions requiring a coenzyme this type of arrangement prevails. A second feature which may not be apparent in the figure is that the region where the enzyme is attached represents a small part of the protein surface. This region is generally termed the "active site," and a great deal of evidence is now being accumulated on the functions of those amino acids which are located in the region. In this particular enzyme the —SH group of cysteine plays a critical role; such —SH groups function in many enzymes, and we may presume that the mechanism of their action is general. The third feature of the reaction scheme is the nature of the changes which occur when the enzyme acts as catalyst. All such changes are interchanges between different parts of the enzyme with different parts of the substrates. When the substrate glyceraldehyde phosphate collides with the enzyme, the S atom of cysteine exchanges partners. The hydrogen of the aldehyde group which is to be oxidized combines with nicotinamide at the point where the sulfur was previously bonded, and the sulfur atom combines with the carbon atom of the adlehyde group at the point where the hydrogen was bonded. The result is the transfer of a hydrogen atom from substrate to coenzyme with the simultaneous formation of a "thioacyl" bond, the same type of high energy bond discussed for coenzyme A condensations (Chapter 6). When inorganic phosphate ion collides with the enzyme substrate complex, a second interchange occurs. A hydrogen atom from the phosphate attaches to the sulfur, displacing the phosphoglyceryl group, which simultaneously attaches to the oxygen atom of phosphate to which the hydrogen was originally bonded. This interchange results in the addition of a second hydrogen to the enzyme-coenzyme complex and a release of the oxidation product, diphosphoglyceric acid. The enzyme is returned to its original state by transferring the acquired hydrogens to free NAD.

From the standpoint of illustrating the strategy of cellular enzymes in avoiding the adverse effects of water, this is a model reaction. The two bonds which were formed—thioacyl and acyl phosphate—hydrolyze readily with water, and if the structure of the enzyme were not designed to exclude interactions between free hydroxyl and hydrogen ions, the

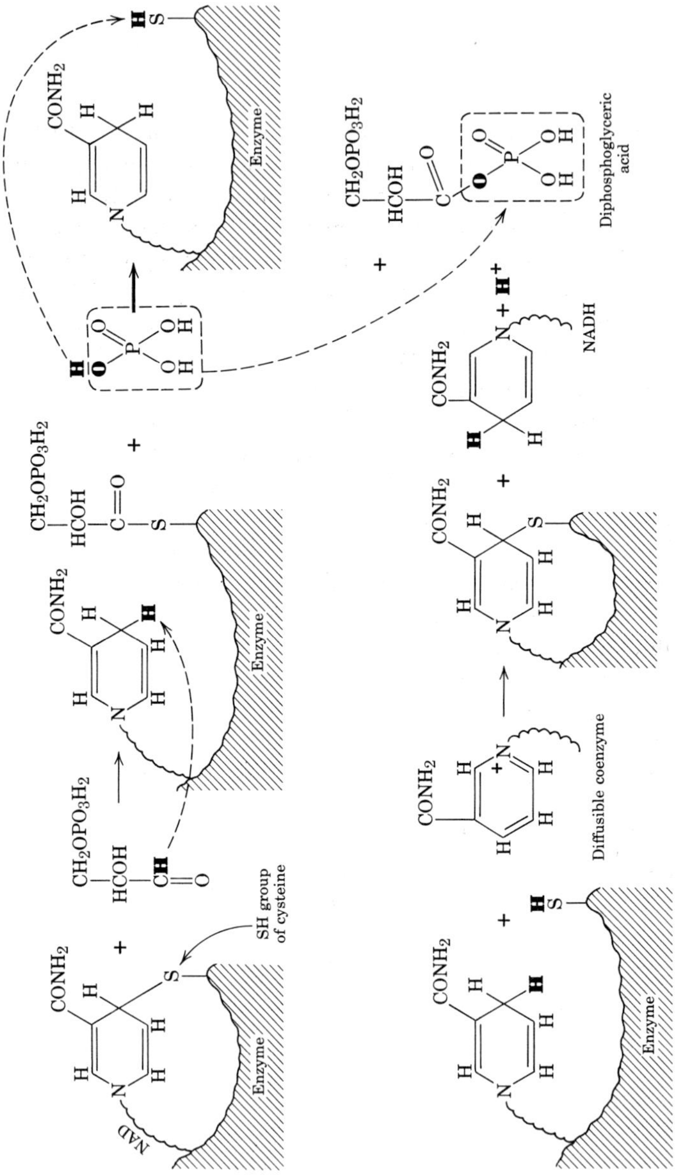

Fig. 7-2. A proposed scheme for interaction of enzyme, substrate, and coenzyme in glyceraldehyde-P oxidation.

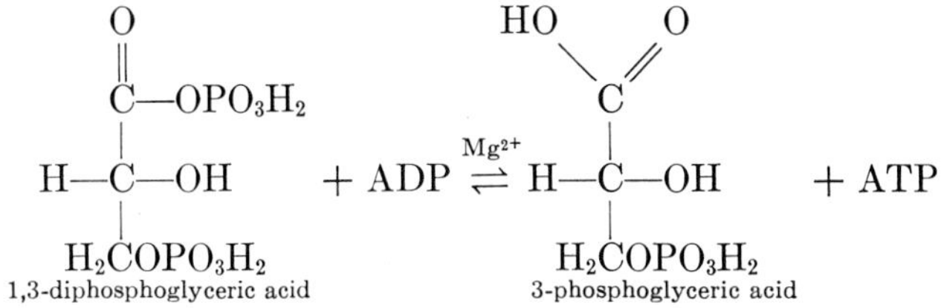

(7-7)

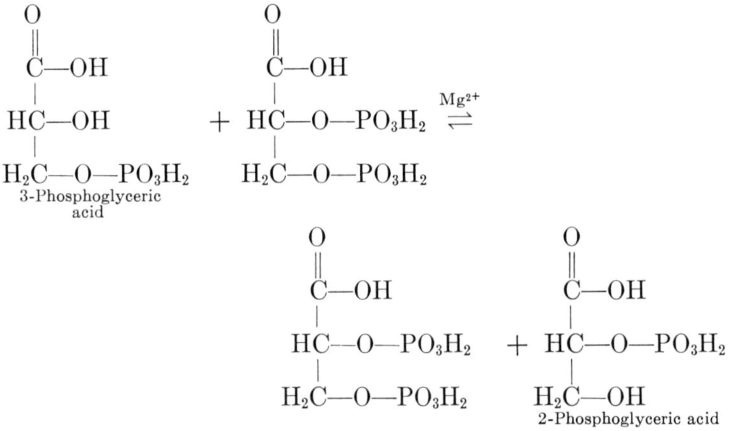

(7-8)

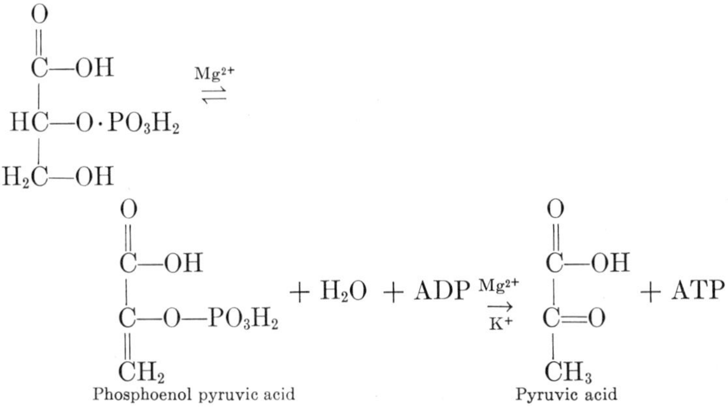

(7-9)

energy of oxidation would have been lost as heat. Instead, the enzyme is so constructed as to react preferentially with phosphate ions, and this preference provides the fundamental mechanism whereby the energy released on oxidation is transferred to a chemical bond which can be used for water elimination reactions. The acyl phosphate bond in diphosphoglyceric acid is not used directly for this purpose; but in the presence of an appropriate enzyme the phosphate is transferred to ADP to form ATP, thereby leaving phosphoglyceric acid as shown in eq. (7-7). By a series of conversions shown in eqs. (7-8) and (7-9) the second phosphate is used to regenerate another molecule of ATP, thereby returning the original energy investment made to initiate carbohydrate oxidation. A most instructive feature of this whole oxidative step is that despite the specificity of the enzyme for its substrate, the kinds of chemical bonds used to effect the oxidation are identical with those used in other energetic reactions. The variety of enzyme reactions masks the economy of enzyme mechanisms.

FERMENTATION

The formation of pyruvic acid represents a pivotal metabolic point for most cells. What happens beyond this point serves to classify the world of organisms into two groups—those which ferment and those which respire. The basic difference between these two groups lies in the nature of the substance which becomes reduced as a result of carbohydrate oxidation. Regardless of mechanism, oxidation of a substance must be accompanied by reduction of another substance. Glycolysis in both groups of organisms leads to a reduction of NAD, and, as will be seen, such reduction also occurs in many other oxidative reactions. Thus NADH is a reduced product common to both fermenting and respiring cells. The amount of NAD in a cell is limited, however. This compound must be synthesized by the cell, and no cell could survive if it did not possess some means to reoxidize NADH, for the energy required in its synthesis far exceeds that which is obtainable from its reduction. In respiring organisms the ultimate acceptor of hydrogens from NADH is oxygen; in fermenting ones the acceptor may be one or more of a variety of organic compounds. For this reason fermenting organisms produce only partially degraded carbohydrates as a by-product of their metabolism.

The by-products of fermentation are generally listed as lactic acid or ethyl alcohol. The enzyme reactions which lead to their formation are shown in Fig. 7-3 along with others leading to different products. Superficially these reactions would appear to contradict the general principle that carbohydrate oxidation releases energy whereas carbohydrate reduction requires an input of energy. Without close inspection of the reactions we might well ask how it is thermodynamically possible for pyruvic acid, which is the product of an oxidative process releasing energy, to become reduced without the expenditure of energy, at least in an amount equivalent to that produced during oxidation (one mole-

Fig. 7-3. Oxidation of NADH without free oxygen.

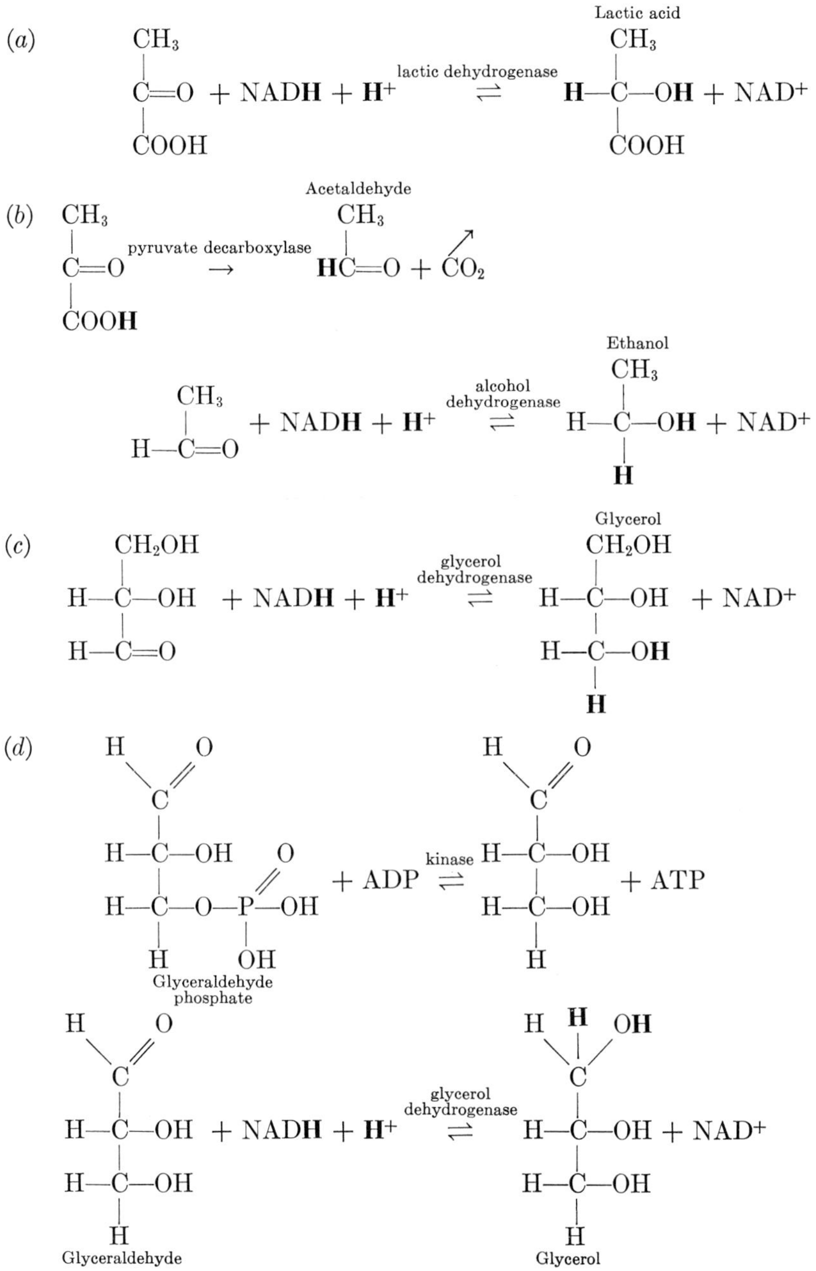

Fig. 7-3 (continued).

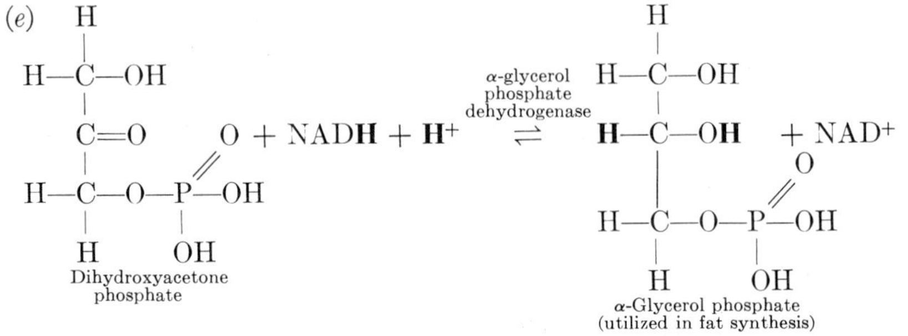

cule of ATP and one of NADH). The answer lies in the nature of the reduced product which is different from the reduced precursor. The net energy expenditure in reducing pyruvate to glyceraldehyde is one molecule of ATP plus one molecule of NADH; that of reducing pyruvate to lactate is only one molecule of NADH. The difference between the two represents the gain to the cell. Both types of reactions are readily reversible. None of the reactions, if carried out in a test tube with isolated enzyme, go to completion. Were it not for the fact that NADH is being continuously generated in glycolysis and that the by-product of fermentation is lost to the medium, the reaction would come to a halt within the cell. Even so, in all fermenting systems the gradual accumulation of by-product ultimately leads to a cessation of growth. This fact is well recognized, and perhaps deplored, in commercial alcoholic fermentations.

Fermentation thus represents an inefficient way of producing ATP from carbohydrate. The glucose molecule, if totally oxidized to CO_2 and water, releases about 670 kcal/mole of glucose oxidized, whereas the production of two moles of ATP from two moles of ADP (the net gain in fermentation) represents about 20 kcal. Were ATP the sole utilizable source of energy, it is doubtful whether fermenting organisms would survive. We must recall, however, that reducing power and carbon chains are equally essential forms of energy in the formation of cell components. All of the reduced coenzymes in a fermenting cell are not reoxidized by pyruvic acid; some are reoxidized in the processes leading to the hydrocarbon chains in fats or other small molecules like those of some amino acids. Neither are all the glucose molecules metabolized to the characteristic by-product of the fermenting organism. A fraction of the carbon chains formed along the glycolytic pathway are acted on by enzymes similar to those in respiring organisms, and these reactions lead to similar essential products. Just as a respiring cell must regulate its metabolism so as to balance its needs for ATP (which is produced by destruction of carbon chains) against its need for carbon chains (which limit the production of ATP), so must a fermenting cell regulate its metabolism to balance the discarding of carbon chains (which leads to NADH reoxidation) against the preservation of carbon chains (which in effect deprives the cell of ATP).

The world of fermenting cells is diversified almost beyond the reach of the imagination. One common feature holds them together: the need to extract energy from metabolism without having recourse to molecular oxygen. Bacteriologists are well aware of this unusual world for it is largely populated by bacteria, but for many a biologist they remain a casually recognized subgroup of cells. It is impossible in a text of this size to describe even partly the intricate patterns of chemical changes which such cells display. Merely to list the kinds of compounds which they can use as energy sources, let alone the kinds of compounds they produce, would take several pages. There is hardly an organic compound which one species or another cannot use; proteins, amino acids, organic acids, fats, and carbohydrates are targets of destruction. "Industrial fermentations" is a familiar phrase because of the diversity of organic by-products these organisms furnish at an efficiency which the organic chemist has yet to attain. For the organism they are by-products simply because the lack of oxygen (or the lack of necessary enzymes to utilize oxygen) prevents it from converting its organic sources to carbon dioxide and water. Such organisms do not resort to unusual principles of energy conversion; they effect distinctive chemical rearrangements because of the distinctive conditions under which they live. Their ability to do so enables them to prosper where other organisms fail, and this, of course, is the reason for their long survival.

Some microorganisms can grow under both aerobic and anaerobic conditions. They are known as "facultative anaerobes" to distinguish them from "obligate anaerobes" which are poisoned in the presence of oxygen. The common baker's yeast is a good example of a facultative anaerobe. In the absence of oxygen yeast cells produce CO_2 and ethyl alcohol; in its presence they produce CO_2 and water. Under aerobic conditions yeast cells grow and divide more rapidly than under anaerobic ones, a contrast which serves to illustrate the fact that the ability to utilize molecular oxygen as a hydrogen acceptor permits a more efficient metabolism of carbohydrate as a source of energy.

Many types of cells in higher organisms can ferment under anerobic conditions although they vary in their ability to withstand the lack of oxygen. Most such cells die when deprived of oxygen, but some may survive for extended periods of time. Cells in the roots of rice grow in poorly aerated muds of the rice fields; those of wheat would quickly perish under similar circumstances. Even among animals unusual situations may be found. Cancerous cells, for example, ferment vigorously, a property which has claimed the curiosity of investigators in the field of cancer. Yet whether or not cells of higher organisms actually ferment, most have the enzymes to do so. The enzymes which make possible fermentation may be readily isolated from the tissues of respiring organisms. Why cells thus equipped do not ferment is a question which still puzzles biologists; all that is definitely known is that some regulatory device suppresses fermentation when respiration is possible.

OXIDATION OF PYRUVIC ACID

One of the earliest pieces of evidence for the existence of a difference in subcellular organization between enzymes associated with pyruvate oxidation and those associated with glycolysis and fermentation was found in studies of disrupted cells. As far back as 1930, the noted biochemist Otto Warburg observed that the capacity of cells to utilize oxygen was localized in particles which did not disperse when cells were broken in aqueous media. Some twenty years later, when techniques were improved for the fractionation of subcellular components, the observation was clearly confirmed and explained. The capacity of cells to oxidize pyruvic acid to CO_2 and water was found to be restricted to the mitochondria. Even in microorganisms which lack typical mitochondria pyruvic acid oxidation has been found to depend on enzymes attached to mitochondria-like membranes. Thus one conclusion could be clearly drawn, especially in relation to the cells of higher organisms: The total complement of enzymes necessary to oxidize pyruvic acid to CO_2 and water is restricted to a specific organelle of the cell—the mitochondrion.

From this conclusion alone a second very important conclusion follows. The free energy available from the oxidation of glucose is about 670 kcal/mole, but the energy derived from the single oxidative step in glycolysis is the equivalent of two moles of ATP plus two moles of NADH (1 NADH will later be shown to equal 3 ATP), which adds up to about 80 kcal/mole. Most of the free energy available from glucose oxidation must therefore remain locked up in pyruvic acid, and since pyruvic acid is oxidized by mitochondria, these subcellular structures must be the major source of energy for aerobic cells. This conclusion is supported by electron microscope observations. Cells with high rates of respiration have a high concentration of mitochondria. In cells with specialized functions which have high energy requirements mitochondria are concentrated in the regions exercising those functions. In muscle cells mitochondria are lined up adjacent to the contractile fibers; in kidney cells, which expend a great deal of energy in regulating the flow of solutes, the mitochondria are concentrated in the vicinity of the surface membrane.

One of the aims of the biochemist is to isolate individual enzymes in order to determine their specific properties. In studying pyruvate oxidation, biochemists fulfilled this aim to the extent that all the enzymes responsible for the breakdown of pyruvate have been isolated and characterized. The history of the subject is a long one, involving not only enzyme characterization but also the development of techniques for disrupting the oxidative particles and bringing the constituent enzymes into solution. From such studies, along with those which led to a characterization of the intermediates in pyruvate oxidation, a picture eventually emerged which is summarized in the next paragraph (see also Fig. 7-4).

The first step in pyruvate oxidation is a decarboxylation of pyruvate similar to that in alcoholic fermentation, but differing in so far as the product is acetyl-coenzyme A rather than free acetaldehyde. In fermentation the energy released by cleavage of pyruvate is lost as heat; in respiration it is conserved in the form of a thioacyl bond. The second step in the process is the condensation of the acetyl group with a four-carbon molecule (oxaloacetate), using the energy of the thioacyl bond to effect the condensation. The six-carbon molecule thus formed is then progressively degraded to oxaloacetic acid which, in effect, plays the role of a catalyst. In the course of oxidative degradation each step, with the exception of one, leads to a reduction of NAD. The consumption of oxygen is therefore not directly related to substrate oxidation, but is due to a very different enzymatic system which transfers hydrogens from NADH to molecular oxygen. This system will be discussed in Chapter 8.

Pyruvate decarboxylation: The enzymatic machinery necessary to preserve the energy of the C-C bond in the course of decarboxylation is extremely complex compared with that essential to decarboxylations during fermentation. Several enzymes and several coenzymes are involved. The sequence of enzymatic events may be summarized thus:

$$\underset{\displaystyle \overset{\|}{O}}{CH_3—C—COOH} + CoA—SH + NAD^+ \xrightarrow[\text{TPP}]{\text{lipoic acid, } Mg^{2+}}$$

$$\underset{\displaystyle \overset{\|}{O}}{CH_3—C—S—CoA} + NADH + H^+ + CO_2 \qquad (7\text{-}10)$$

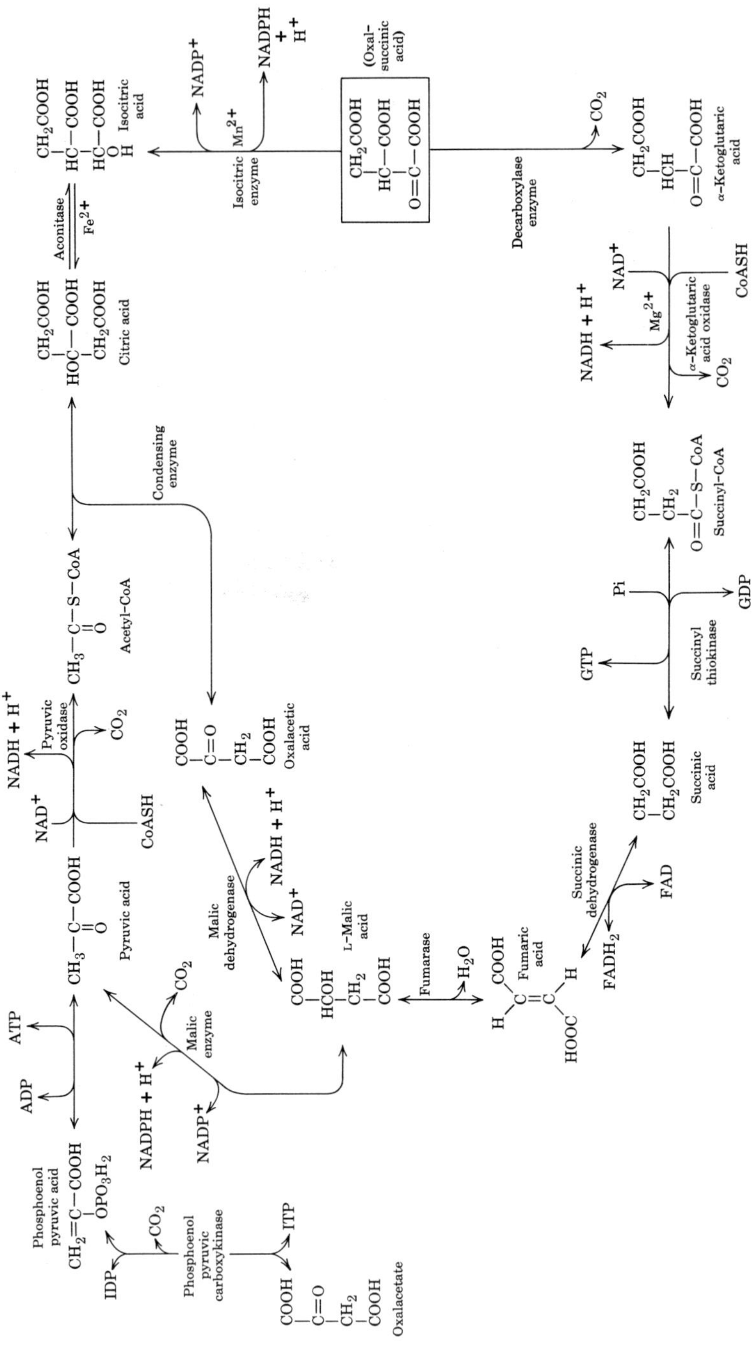

Fig. 7-4. The Krebs cycle.

The formulae for those coenzymes not previously illustrated are shown in Fig. 7-5 and some of the main reactions detailed in Fig 7-6. Again, the principle of maintaining the substrate molecule attached to the enzyme-coenzyme complex that was noted in previous reactions occurs here. Although the enzymes are not illustrated, the coenzyme substrate complexes occur in association with enzyme protein. A new type of bond is postulated for the attachment of the acetyl group to the coenzyme thiamine pyrophosphate (7-6*a*) but the thioacyl bond (7-6*b* and 7-6*c*) should now be familiar from the several other reactions discussed. Thiamine pyrophosphate is the universal coenzyme for decarboxylations, including those which occur in fermentation; the conservation of energy in the oxidative system appears to be tied to the presence of a second coenzyme, lipoic acid, which makes possible the formation of a thioacyl bond. In the absence of lipoic acid, acetaldehyde detaches from the coenzyme and the free energy is liberated as heat. In the course of oxidative decarboxylation lipoic acid is progressively reduced (Steps 2 and 3). Its reoxidation occurs via NAD^+, and the product NADH is oxidized through the electron transport system.

As pointed out in Chapter 7, acetyl-CoA is the principal mechanism which cells utilize for building carbon chains. Other enzyme systems forming acetyl-CoA must therefore be present in cells which do not possess the oxidative decarboxylation mechanism. A series of enzyme reactions leading to acetyl-CoA from acetaldehyde is shown here:

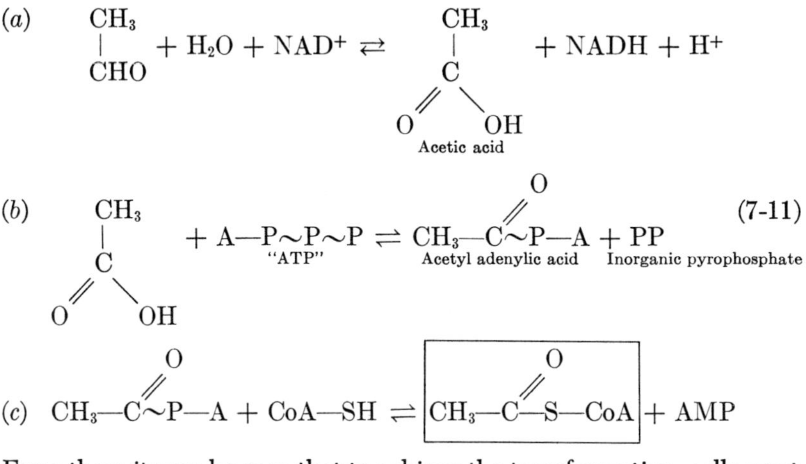

(7-11)

From these it may be seen that to achieve the transformation, cells must expend two pyrophosphate linkages and this is a measure of the energy conserved by cells which have an oxidative decarboxylation mechanism.

(*a*) Thiamine pyrophosphate

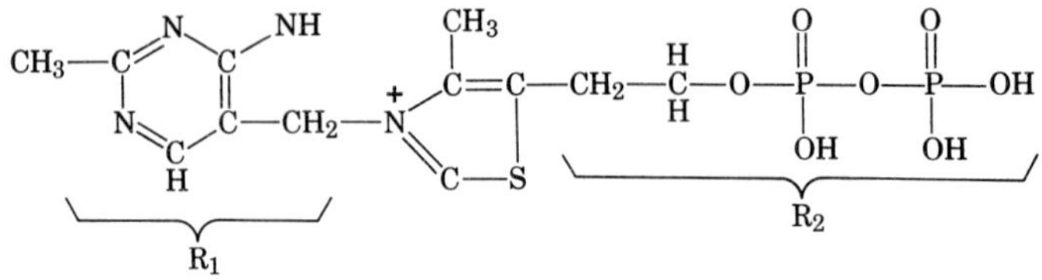

(*b*) Lipoic acid

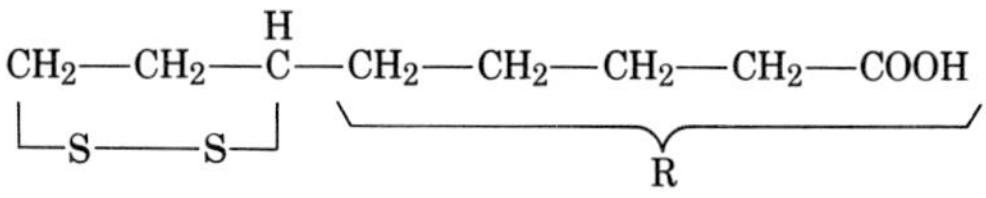

Fig. 7-5. Structure of coenzymes associated with oxidative decarboxylation. (The structure of coenzyme A is given in Fig. 6-6.)

THE CITRIC ACID (OR KREBS') CYCLE

The complete sequence of enzyme reactions is given in Figs. 7-4 and 7-6. One of the major features of these reactions is the operation of a cycle. Its discovery by Hans Krebs represented an outstanding achievement in the field of carbohydrate metabolism. Elucidation of the glycolytic pathway provided no pointers to the existence of a cyclical system for the oxidation of pyruvate. Once the cycle was discovered, it became apparent that in order to degrade short carbon chains and conserve the energy of degradation, attachment of the short chain to a longer one was the mechanism of choice. In this cycle oxaloacetate serves as the carrier and is recovered after two carbons of the six-carbon compound have been degraded to CO_2. The sequence of enzymatic events is basically similar to that observed in glycolysis. A number of intramolecular rearrangements are followed by an oxidation, leading to an unstable molecule, oxalosuccinate, which is decarboxylated without gain in energy. The next step is an oxidative decarboxylation of α-ketoglutarate by a mechanism identical in principle with that described for pyruvic acid. The energy of the thioacyl bond in succinyl-CoA is transferred to guanosine diphosphate to form guanosine triphosphate (GTP) unless the succinyl-CoA is used as such in other syntheses. The specificity of the enzyme system which transfers the energy of the thioacyl bond to GTP is unusual; generally the recipient is ATP. The significance of this reaction is not clear even though GTP is a specific energy donor for a number of syntheses including one step in the synthesis of protein. Cells generally have enzymes capable of transferring pyrophosphate linkages from one nucleotide to another. Succinate oxi-

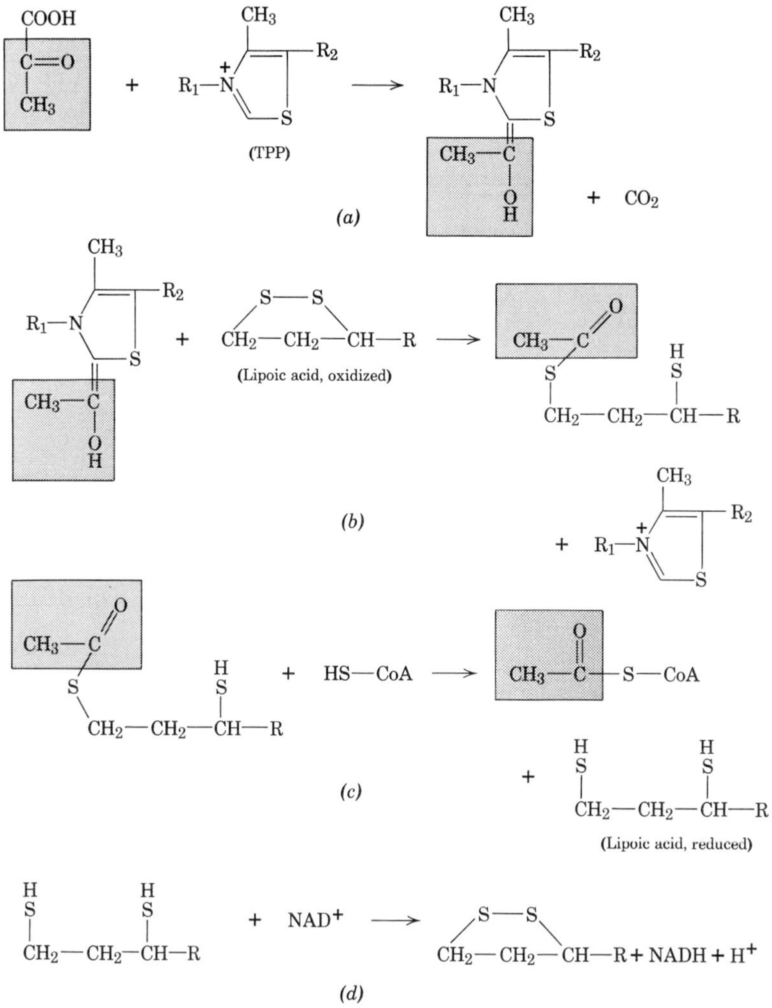

Fig. 7-6. Principal reactions in oxidative decarboxylation of pyruvate.

dation is also distinctive in its mechanism. Unlike the other oxidations which occur in the cycle, hydrogens are transferred to a flavin which does not, as in oxidative decarboxylation, transfer them to a nicotinamide nucleotide, but interacts directly with the electron transport system.

The net result of pyruvate oxidation is the production of one pyrophosphate linkage in GTP, four molecules of NADH, and one molecule of $FADH_2$. Twice those numbers are produced for every molecule of glucose degraded, so that if only from the standpoint of reducing power oxidation of pyruvic acid is about five times as effective a source as glycolysis. How this reducing power can be transformed into ATP will be considered in Chapter 8, but at this stage it is more important to consider the cycle in relation to the third requirement for cellular syntheses—the provision of carbon chains.

The principal carbon skeletons formed in the course of pyruvate oxidation which are used in the building of distinctive cellular constituents are indicated in Fig. 7-4. The chemical aspects of these relationships may be briefly summarized. Acetyl-CoA provides the units for building long hydrocarbon chains such as are found in lipids, carotenoids, and rubber. It also provides units for the synthesis of cyclic hydrocarbons such as are found in the steroids (e.g., cholesterol, steroid hormones). The details of the various enzymatic processes which ultimately yield these compounds will not be discussed in this text. The important feature is that the energy provided for condensation in the acyl-CoA units plus the energy provided by NADPH (NADP may also be reduced by NADH) is sufficient for all these syntheses. We may note incidentally that these various products, different though they appear to be in their final structure, have a common and simple derivation.

α-Ketoglutarate is the principal, if not the sole, point at which ammonia is attached to the carbon skeleton. A number of enzymes exist which can transfer the amino group from the product glutamate to other carbon skeletons, especially those of organic acids, to yield amino acids. Ultimately, therefore, α-ketoglutarate is the source of amino acids for proteins, and conversely it is the point through which amino acids may be converted into other forms of chemical energy. Since the amino acids aspartate and glutamate are used for the building of purines and pyrimidines, ketoglutarate and the other organic acids in the cycle which may be converted to amino acids are carbon sources for the building of nucleic acids.

Succinate is the principal carbon skeleton used in synthesizing porphyrins, which are the backbone of the pigments used in photosynthesis and respiration (the chlorophylls and cytochromes). In all, therefore, the compounds which appear as intermediates in the oxidation of pyruvic acid are sources of carbon skeletons for each of the principal cell constituents. The significance of the Krebs' cycle to cell function cannot therefore be limited to its efficient production of NADH or ATP; the pathway of pyruvate oxidation is also the pathway from which most of the principal skeletons are provided for the macromolecules of cells.

THE ROLE OF MITOCHONDRIA IN CARBOHYDRATE TRANSFORMATIONS

There are two extremes of interest in the metabolic organization of cells for carbohydrate transformations. A biochemist (hypothetical, of course!) may be entirely concerned with the kinds of enzymes essential to the complete oxidation of pyruvate and his overwhelming concern for molecular transformations may lead him to be indifferent to the question of where these transformations are occurring in the cell. A biologist, on the other hand, may be entirely preoccupied with the problem of assigning specific biochemical roles to the different subcellular structures and show little concern for the details of molecular mechanisms. Yet neither interest is in itself sufficient to deepen our understanding of cell behavior. The (hypothetical) biochemist must begin to ask whether the occurrence of certain enzymes in organized aggregates has a bearing on their molecular properties, and the (equally hypothetical) biologist must begin to ask whether the special properties of aggregated enzyme systems have a bearing on subcellular function. Thus the topic of carbohydrate transformations lends itself to two interrelated considerations: the nature of the enzyme systems present in mitochondria; and the role of mitochondria in carbohydrate transformations.

Studies of the localization of enzymes in isolated fractions of disrupted cells lead to one unambiguous conclusion. Mitochondria are not the exclusive sites of all the enzymes associated with pyruvate oxidation. Some such conclusion might have been reached from the fact that fermenting cells, which do not oxidize pyruvic acid, are able to furnish the various carbon skeletons formed in the Krebs' cycle. More direct evidence is, however, available. If any of the intermediates found in the Krebs' cycle is added to a suspension of mitochondria, oxygen is consumed and CO_2 produced. This emphasizes the point previously made that mitochondria contain the full complement of Krebs' cycle enzymes plus the enzymatic machinery for terminal oxidation. If, however, the activities of individual enzymes are studied in each of the cell fractions, the pattern of exclusive localization disappears. All the enzymes of the Krebs' cycle have not been fully studied in this way, but it has already been established that the enzyme rearranging citrate, "aconitase," and enzymes oxidizing isocitrate and malate are also found in the soluble proteins of the cell. Conversely, one of the fermenting enzymes, lactic dehydrogenase, also appears to be present in mitochondria. A similar lack of exclusive localization with respect to electron transport will be considered in the next chapter. The presence of similar enzymes in mitochondria and other subcellular fractions does not imply an identity of the enzymes in question. In fact, the opposite is certainly true for at least some enzymes. Thus the oxidation of citrate and malate occurs in mitochondria with NAD as hydrogen acceptor, whereas NADP is the hydrogen acceptor outside the mitochondria.

The role of mitochondria in carbohydrate transformations cannot therefore be defined as the exclusive seat of all enzymes essential to pyruvate oxidation. Rather, one must consider this role in terms of the fact that mitochondria are uniquely equipped to utilize molecular oxygen. The presence of the full complement of pyruvate oxidizing enzymes in mitochondria would imply that the proximity of the dehydrogenases yielding reduced coenzymes to the enzyme system consuming molecular oxygen has a distinct functional advantage. On the other hand, the presence of at least some enzymes essential to pyruvate oxidation outside the mitochondria would imply that certain carbohydrate transformations are required in other regions of the cell. Moreover, mitochondria contain a number of other enzymes which are not directly involved in pyruvate oxidation. Thus although we may correctly describe mitochondria as the "powerhouses" of cells because of their ability to utilize molecular oxygen in pyruvate oxidation, the description is inadequate. Were the enzymes for pyruvate oxidation the sole and unique components of mitochondria, we might suppose that mitochondria were structures formed by the cell for the express purpose of pyruvate oxidation. Using this supposition as a starting point, we might investigate how the enzymes are put together or why they constitute so efficient an organization. Many biologists, however, have asked very different questions: How did mitochondria originate? Could it be that they were once separate organisms which, in the early stages of evolution, invaded cells? Was the eventual adaptation of mitochondria to a completely intracellular life the consequence of mutual advantages conferred on host and invader? We may easily visualize a situation in which the capacity of mitochondria to utilize molecular oxygen conferred a distinct selective advantage on cells which contained them, and in which the provision of various substrates and enzymes by the host provided a far more favorable environment for mitochondrial multiplication. Speculations such as these cut into the question of biochemical function, and leave us prepared for the possibility that these subcellular organelles may have carried with them functions other than those associated with terminal oxidation. For the present we can only say that mitochondria are very important sites of carbohydrate transformations but that the biological functions of these structures have yet to be completely elucidated.

The problem of biological functions is distinct from, even if related to, that of biochemical mechanisms. We should still ask how the presence of the oxidative enzymes within the mictochondria is related to mitochondrial structure. In describing the appearance of cells in Part I, it was pointed out that mitochondria are enclosed by a membrane and contain an inner membrane with many infoldings. Mitochondria have been ruptured by mechanical procedures, and by this means it has been possible to estimate that about 50% of the protein remains as part of the membrane system on disruption and that the remainder is recovered in solution. The dissolved proteins have not been completely characterized, but among them are many of the dehydrogenases composing the Krebs' cycle. As a rough approximation it may be said that the enzymes responsible for the oxidation of NADH are attached to mitochondrial membranes whereas those immediately responsible for substrate oxidation (with the exception of succinate) are either loosely attached to the membranes or are randomly dispersed in the interior of the mitochondrion. Since the significance of such enzyme partitioning requires an understanding of the electron transport system, further discussion of the topic will be deferred to Chapter 8.

8

The Acquisition of Oxidative Energy: Electron Transport

A cell which has not funneled off any carbon fragments for organic syntheses, or used NADH for reducing carbon chains, has extracted relatively little energy in the course of degrading a glucose molecule through glycolysis and the Krebs' cycle. For each molecule of glucose degraded two molecules of ATP were formed during glycolysis and two molecules of GTP were formed during ketoglutarate oxidation. The four pyrophosphate bonds formed are equivalent to an acquisition of approximately 40 kcal/mole of glucose metabolized. Since the free energy of glucose oxidation to CO_2 and water is about 670 kcal/mole, the conclusion must be drawn that the sole remaining product of glucose degradation, NADH, is a potential source of the major portion of energy available from such degradation. Indeed, regardless of the particular enzyme reaction from which it was generated, NADH (or NADPH) provides the major source of ATP in all respiring cells.

The conversion of the available energy in reduced coenzymes into pyrophosphate bonds of ATP has two major characteristics. (1) Molecular oxygen is the ultimate agent of oxidation (except for certain microorganisms which may use the oxygen in sulfate or nitrate for the same purpose; see Fig. 6-6). (2) ATP is concomitantly produced during oxidation. The coupling of ATP formation to oxygen utilization is generally called "oxidative phosphorylation." The fact that the capacity of many types of cells to utilize molecular oxygen is restricted to the mitochondria at once suggests that oxidative phosphorylation is also restricted to these subcellular organelles. In other types of cells, especially those of plants, where oxidations also occur outside of the mitochondria, the energy yields from nonmitochondrial oxidations are negligible, and such oxidations are usually directed at compounds other than the reduced nicotinamide coenzymes. Thus for virtually all aerobic cells mitochondria are the major source of ATP.

The ability of mitochondria to convert NADH into pyrophosphate bonds of ATP may be demonstrated experimentally. Isolated mitochondria supplied with NADH, ADP, and inorganic phosphate consume oxygen and inorganic phosphate while producing NAD and ATP. For each atom of oxygen consumed, 3 molecules of ATP are produced and 1 molecule of NADH is oxidized. Thus the energy released on the oxidation of NADH by molecular oxygen is sufficient to form at least three pyrophosphate linkages. (To account for this observation we must clarify the mechanism by which oxidation occurs and by which the oxidation is coupled to the esterification of inorganic phosphate into ATP.

THE CHEMICAL NATURE OF OXIDATION

Oxidation is generally defined as a loss of electrons; the reverse holds for reduction. Combinations of substances with either oxygen or hydrogen constitute special instances of oxidation-reduction; neither of these elements is essential to the process. See Fig. 8-1. Three features of the process are perhaps self-evident: (1) No free pool of electrons exists, and the gain of an electron by one atom must be accompanied by a corresponding loss from another. (2) Electrons do not move over appreciable distances between atoms in the course of their transfer, and oxidation-reductions can occur only under the same conditions as other chemical reactions, namely, by close juxtaposition of the interacting atoms. An exception to this, which will be discussed later, is the move-

(*a*) Dehydrogenation

$$RCH_2 + X \rightarrow RC + XH_2$$

(*b*) Hydrogen oxidation

$$XH_2 + Y \rightarrow X + Y^{2-} + 2H^+$$

(*c*) Oxygen reduction

$$Y^{2-} + O \rightarrow [O^{2-}]$$
$$[O^{2-}] + 2H^+ \rightarrow H_2O$$

Fig. 8-1. Schematic representation of hydrogen oxidation and oxygen reduction in biological systems. X is a reducible coenzyme capable of accepting hydrogens from a substrate. Y is a coenzyme capable of accepting or donating electrons.

ment of electrons along metallic conductors. (3) Since changes in electron number produce changes in atomic charge, oxidation-reduction products can exist in the presence of ions of opposite charge in solution or as compounds in which the net charge is zero.

A simple illustration of oxidation-reduction is the conversion of hydrogen and chlorine into hydrochloric acid. In this process an

$$H_2 + Cl_2 \rightarrow 2H^+ + 2Cl^-$$

electron is transferred from the hydrogen to the chlorine atom, hydrogen being the reductant, chlorine the oxidant. A slightly different reaction is the oxidation of zinc by silver chloride in which zinc is converted from an elemental to an ionic form while the reverse happens for silver.

$$Zn + 2AgCl \rightarrow ZnCl_2 + 2Ag \; (Zn \rightarrow Zn^{2+};\; Ag^+ \rightarrow Ag)$$

Oxidation-reductions cannot be so simply illustrated for organic compounds. At the beginning of this section, the oxidized and reduced states of the carbon atom were defined in terms of the relative numbers of associated hydrogens and oxygens. Yet throughout the cycle of glucose oxidation, carbon retains a valence of 4. In terms of loss or gain of electrons carbon atoms cannot be described as undergoing oxidation during respiration. This statement may be somewhat qualified by taking into account the electronegative nature of oxygen, which would tend to pull electrons from the carbon atom into the orbits of the oxygen atom. Such a qualification cannot be applied to combinations between hydrogen and carbon, since the hydrogen atom is equally stabilized by sharing an electron with carbon as by losing one to become a proton. The carbon atom in CH_4 is hydrogenated; by definition it is not reduced.

Oxidation-reductions in carbohydrate metabolism are different from the examples given above in which electrons are directly transferred from one substance to another. The key to understanding these is the fact that the hydrogens bonded to carbon atoms are potential sources of reducing power. Oxidation of carbohydrate molecules involves a sequence of hydrogen removals by enzymes appropriately called "dehydrogenases." The action of these enzymes has already been described in discussing glycolysis and pyruvate oxidation. Ultimately, and prior to their combination with oxygen, hydrogens thus removed are converted into protons, the electrons being transferred to one of the oxidation-reduction coenzymes which are thereby reduced. (Fig. 8-1). Dehydrogenation of carbohydrates does not therefore represent a direct oxidation of the carbon atom. The common reference to carbon oxidation is an indirect one since the hydrogen atom, and not the carbon, is the substance oxidized, and such oxidation is identical with that illustrated for metallic ions. In cellular systems carbohydrate oxidation does not proceed without oxidation of hydrogen atoms. Conversely, in carbohydrate synthesis hydrogenation of carbon does not proceed without the reduction of protons, a process to be considered under photosynthesis.

The principal role of oxygen as an oxidant in carbohydrate oxidation is also indirect. As pointed out earlier, molecular oxygen does not enter into the process until after all the carbons of the glucose molecule have been converted into CO_2. Oxidation by molecular oxygen occurs as the terminal step of respiration in which the electrons transferred from hydrogen to the coenzymes are in turn transferred to oxygen, and as a result the negatively charged oxygen atoms combine with protons to form water. Oxygen is therefore an oxidant inasmuch as it is an electron acceptor. The oxidation-reduction aspect of carbohydrate metabolism may therefore be described as the formation of protons from hydrogen and the acceptance of electrons by oxygen. Common to all respiring cells is the possession of enzymes capable of transferring electrons from an electron donor, usually hydrogen, to an electron acceptor, usually oxygen. The actual mechanisms for oxidation-reduction in cells correspond much more closely to the basic definition of oxidation as electron loss than to the limited description of oxidation as a process of oxygen addition.

The essential pattern of electron transfer is illustrated in Fig. 8-2. Glyceraldehyde is used as a simple model to illustrate the initial event of dehydrogenation. The net reaction is the substitution of a hydroxyl group for a hydrogen atom in the aldehyde portion of the molecule. However, if two important details of the reaction are noted, it will at once be recognized that the conversion of glyceraldehyde to glyceric acid is not effected by direct substitution of hydroxyl for hydrogen. First, it may be seen that the coenzyme essential to the reaction is converted from a cation in the oxidized state to a neutral molecule in the reduced state. An electron must have therefore been transferred to NAD to cause the charge neutralization. Second, the proton appearing on reduction is derived from the hydroxyl group attached to the carbon atom in the hydration step; the electron released in the formation of this proton neutralizes NAD^+. Thus the net result of glyceraldehyde oxidation has been the transfer of an electron and one hydrogen atom to NAD, the net charge in the mixture remaining the same because of the simultaneous release of a proton on neutralization of NAD^+. The NADH molecule is a potential source of two electrons, one which

(*a*) Dehydrogenation of glyceraldehyde

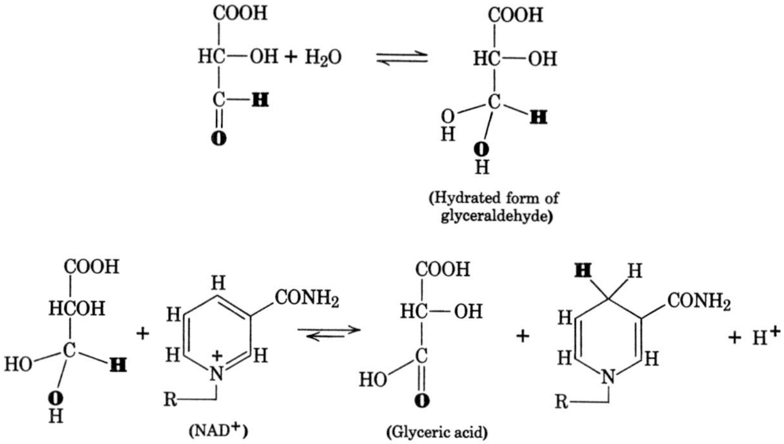

(*b*) Hydrogen transfer

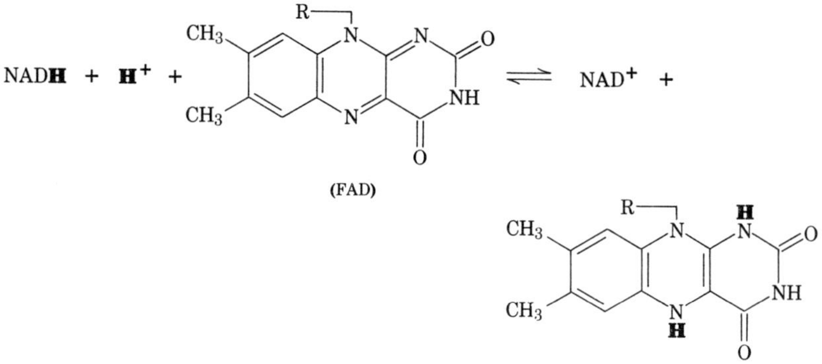

(*c*) Electron transfer

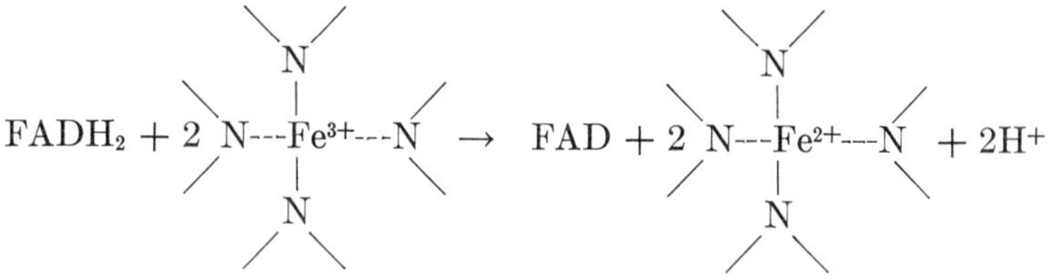

(*d*) Oxygen reduction

$$2CytFe^{2+} + O + 2H^+ \rightarrow 2CytFe^{3+} + H_2O$$

Fig. 8-2. Basic pattern of oxidation mediated by electron transport systems.

corresponds to the proton in solution, the other which is contained in the covalently bonded hydrogen atom. In the simplified sequence illustrated in Fig. 8-2 NADH is oxidized by FAD, in the course of which both electrons and protons are incorporated into the reduced FAD molecule. Oxidation of $FADH_2$, however, results in the release of two protons and the transfer of two electrons to the cytochrome system. These electrons continue to be transferred through a series of cytochromes until they are transferred to oxygen and water is formed.

OXIDATIONS AND FREE ENERGY

In Part I a basic thermodynamic generalization was made about the nature of chemical reactions, namely, that reactions will only occur if accompanied by a decrease in free energy (ΔF), and that the total energy available from any reaction could not exceed the ΔF value. This generalization also covers oxidation-reductions since they represent a particular type of chemical reaction. To leave the subject of oxidation-reduction merely by saying that it is a special instance of chemical reactions in general would be to overlook the overwhelming importance of this process in cell energetics. All cellular energy is ultimately derived from oxidation-reduction processes, and for this reason a quantitative expression of the tendency of a cellular constituent to donate or accept electrons is of fundamental interest. In this respect the specific nature of the compound is of secondary interest. For if compounds A and B have the same electron-donating tendencies, then however different they might be, the transformation of one into the other cannot be a source of cellular energy. What is therefore needed is a system for comparing the electron-donating tendencies of different compounds. From such a system it should be possible to predict the direction of oxidations within cells and also the amount of energy which may be released by any particular oxidation-reduction reaction.

A system of this kind has been erected because the tendencies of substances to yield or absorb electrons can be measured by electrochemical techniques. A detailed account of these techniques is given in the special section below, but for the general reader a few of the important principles are sketched here.

The basic principle of electrochemical measurement is that a substance capable of yielding electrons will do so to a limited extent if in contact with a metallic conductor. Two such substances maintained in separate solutions but each in contact with one end of the metallic conductor will generate a voltage if their electron-donating tendencies differ. The electrode in contact with the substance having the greater electron-donating tendency will appear as the negative pole. By using one particular electrode system as a standard of reference (the hydrogen electrode), the electron-donating tendency of any substance can be expressed as a voltage. The more negative the voltage, the greater the electron-donating tendency. Thus of two substances, the one with the more negative voltage can be expected to oxidize the other and release energy in the process.

The conditions under which a voltage is measured affect the value obtained. For comparative purposes these conditions must be standardized. Temperature, concentration, and pH are all important factors. One other variable merits brief discussion: the relative concentrations of oxidized and reduced forms of the substance measured. If, for example, substance A donates an electron to a metallic wire, it becomes oxidized to A^+. No matter what substance is chosen, a solution of that substance alone will ultimately contain two substances (the reduced and oxidized forms) if any electron transfer occurs. From previous discussions of equilibrium systems it should be clear that the tendency of a substance to become oxidized is influenced by the amount of oxidized substance already present. A solution, for example, in which initially A is the only component would have a greater tendency to donate electrons than one in which A^+ is present in twice the concentration of A. The standard oxidation-reduction potential (E_0) is obtained by measuring the voltage generated when an electrode is immersed in an equimolar mixture of the reduced and oxidized forms of a substance and connected at the free end to a hydrogen electrode. The free energy of any oxidation-reduction reaction between two sets of substances with a potential difference of E may be calculated from the equation $\Delta F = n\mathfrak{F}E$, where n is the number of electrons transferred per mole of substance and $\mathfrak{F}$ is the faraday constant. In this way it has been possible to calculate the free energy made available on oxidation of NADH to NAD with molecular oxygen (see following section).

Electrochemical Measurements of Oxidation-Reduction Systems. The core of electrochemistry lies in the fact that electrons can move through metals. Not only can the flow of electrons be measured as electrical current but the *tendency to flow* can be measured as voltage. This property opens the possibility for a kind of measurement otherwise impossible in chemical reactions. Oxidation-reduction has been defined as the gain and loss of electrons between interacting atoms. If a system can be physically arranged such that transfer of electrons is effected through a metallic conductor, we could, by measuring voltage, measure the *tendency* for an atom in one species of molecule to donate electrons to an atom in another.

The simplest way to explain the arrangement for measuring electrical potentials is to begin with systems which, by themselves, have little biological interest—metallic elements exposed to water or to a solution of their ions. In Fig. 8-3*a* a piece of copper is shown immersed in a solution of copper sulfate. Under such conditions, or even if the metal is immersed in pure water, metallic copper releases copper ions into solution, leaving the body of the metal with an excess of electrons. The strong electrostatic attraction between the negatively charged metal and the surrounding ions prohibits a separation of negative metal and ions, and hence a direct verification of this picture. However, by assuming that this does occur we are able to explain virtually all electrochemical phenomena. The equilibrium equations for the copper-copper sulfate system are shown in Fig. 8-3*a*; corresponding ones may be written for other metals. The source of electrons (solid copper in the illustration) is designated as the electrode, and the whole system is usually referred to as a "half-cell" although electrode is often used for both. It is important to note that the tendency of the metallic element to release electrons is affected by the concentration of metallic ions originating from the salt. A more elaborate type of half-cell, and one commonly used for reference, is the calomel electrode. Since mercury is liquid, an unreactive solid metal must be in contact with it to act as a conductor if measurements of electron flow are to be made. Mercury is the source of electrons; mercurous chloride affects electron release in the same way as $CuSO_4$ does in the copper electrode. This particular mixture was developed for reasons of stability and reproducibility; it has no special significance for the fundamental reaction.

Fig. 8-3. Special section on redox potentials.

(*a*) Single electrode or "half-cell"

(1) Copper-copper sulfate electrode

$$CuSO_4 \rightleftarrows Cu^{2+} + SO_4^{2-}$$

$$Cu \rightleftarrows Cu^{2+} + 2e^-$$

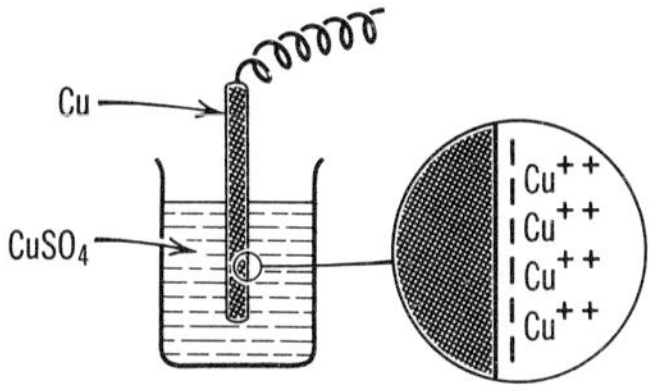

(2) Calomel electrode

$$HgCl \rightleftarrows Hg^+ + Cl^-$$

$$Hg \rightleftarrows Hg^+ + e^-$$

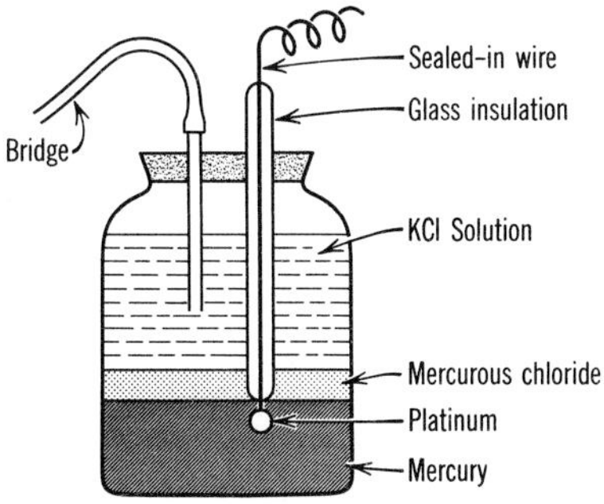

Fig. 8-3 (continued).

(3) Hydrogen electrode

$HCl \rightarrow H^+ + Cl^-$

(H^+ adsorbed on Pt)

$H \rightleftarrows H^+ + e^-$

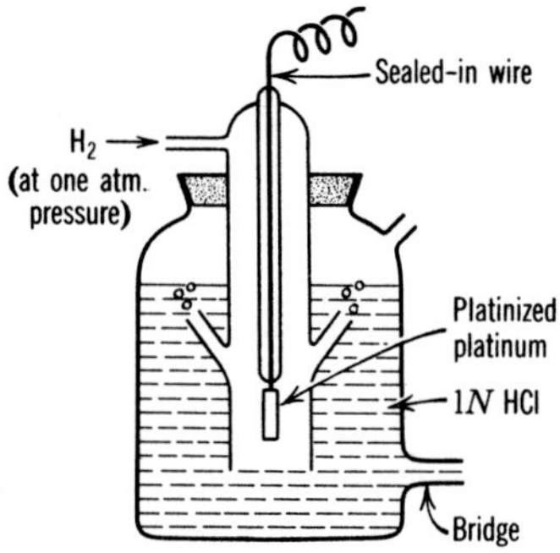

(*b*) Measurement of potential (standard potential of electrodes at 25°C = E_0)

(1) Reactions at hydrogen electrode

$$H \rightarrow H^+ + e^-$$

(tendency for excess H^+/Cl^-)

(2) Reactions at Cu-$CuSO_4$ electrode

$$Cu \rightarrow Cu^{2+} + 2e^-$$

(tendency for excess SO_4^{2-}/Cu^{2+})

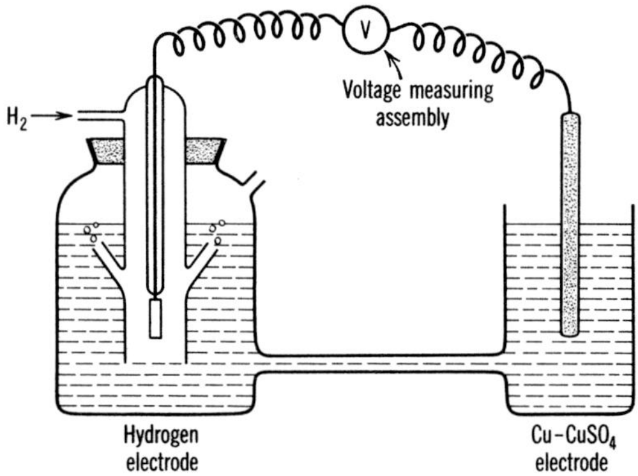

(3) Arbitrary standard (hydrogen electrode = 0)
Metals immersed in solution of corresponding ion (concentration of ion = $1N$)

$Li \rightarrow Li^+ + e^-$	−2.96 v
$Zn \rightarrow Zn^{2+} + 2e^-$	−0.76 v
$Cu \rightarrow Cu^{2+} + 2e^-$	0.334 v
$Hg \rightarrow Hg^{2+} + 2e^-$	0.78 v
$Ag \rightarrow Ag^+ + e^-$	0.80 v

Fig. 8-3 (continued).

(*c*) Derivation of electrochemical equations

(1) Equivalents

The Faraday ($\mathfrak{F}$) is equal to approximately 96,500 coulombs of current. This is equivalent to the passage of Avogadro's number of electrons (6.06×10^{23}). Thus 1 $\mathfrak{F}$ is able to convert 1 mole of Ag^+ to Ag.

The greater the difference in potential between points in which electrons pass, the greater must be the energy of each electron. Therefore the energy equivalent of 1 $\mathfrak{F}$ of current is a function of the voltage. It has been experimentally determined that 1 $\mathfrak{F}$ passing over a difference of 1 volt = 23,000 calories.

(2) Electrical work

Defined as the amount of current × voltage (i.e., difference in electrical potential) and expressed as

$$\text{Work done} = n\mathfrak{F}E$$

where n is the number of altered charges per ion and E the voltage. In $Ag^+ \rightarrow Ag$, $n = 1$; in $Cu^{2+} \rightarrow Cu$, $n = 2$.

From (*b*3)	Ag:Ag	=	0.80 v
	Cu:Cu	=	.34
Net voltage for this cell		=	0.46 v

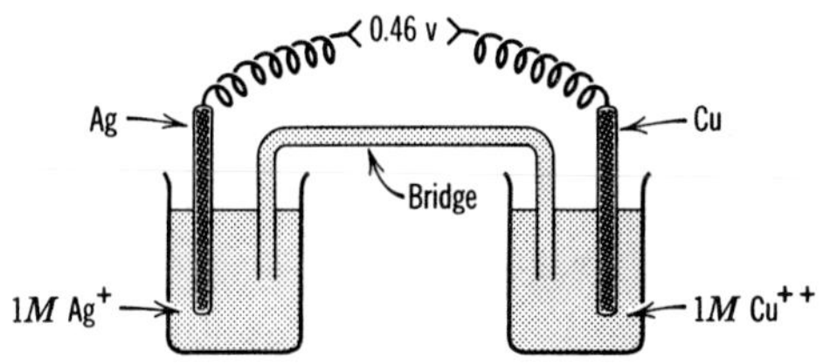

The net reaction if cell were permitted to reach equilibrium would be $2Ag^+ + Cu \rightarrow 2Ag + Cu^{2+}$.

We may calculate the amount of work done if two moles of electrons are transported from the copper electrode to form two moles of silver, or vice versa. The signs will be opposite, but the absolute values will be the same. In general, except when comparing potentials, it is convenient to forget about signs in setting up equations because the definition of positive and negative poles is purely arbitrary and may be indicated after the identities are written.

Work done in transforming 1 mole of Cu = $2 \times \mathfrak{F} \times 0.46$ ($n = 2$)

Work done in transforming 2 moles of Ag = $2(1 \times \mathfrak{F} \times 0.46)$ ($n = 1$)

$2 \times 23{,}000 \times .46 = 21{,}160$ cal/mole Cu or per 2 moles Ag^+.

Fig. 8-3 (continued).

(3) Equivalence between electrical and chemical work

Use E_0 to symbolize the voltage of this cell since the ions are at a concentration of 1M (the concentration used in calculating standard free energies $\Delta F°$). Identical derivations could be made for any electrode system under standard conditions. It is to be understood that ordinarily the symbol E_0 is applied to the potential of a standard electrode as measured against the hydrogen electrode.

The equilibrium constant for the equation $2Ag^+ + Cu \rightarrow 2Ag + Cu^{2+}$ may be written as

$$\frac{[Ag]^2[Cu^{2+}]}{[Ag^+]^2[Cu]} = K_{eq}$$

The powers must be inserted to be consistent with the stoichiometry of the equation; i.e., if 2 moles of one component are involved, the concentration of that component must be squared. The equation may be simplified because the metallic elements do not dissolve and their concentration is arbitrarily set as 1 under all conditions. Thus,

$$\frac{[Cu^{2+}]}{[Ag^+]^2} = K_{eq}$$

From an earlier derivation it is known that $\Delta F°$ per mole of Cu oxidized would be $RT \ln K_{eq}$. But under reversible conditions the work done and the free energy change are equal. The work done electrically must be the same as the work done chemically

$$n\mathfrak{F}E_0 = RT \ln K_{eq} \quad \text{and} \quad E_0 = RT/n\mathfrak{F} \times \ln K_{eq}$$

To determine the ΔF per mole for any other set of concentrations we do the following:

$$\Delta F = \Delta F_0 + RT \ln \frac{[Cu^{2+}]}{[Ag^+]^2}$$

The concentrations of Cu^{++} and Ag^+ are those in the new set of conditions. By substitution we may write a parallel equation:

$$E = E_0 + RT/n\mathfrak{F} \ln \frac{[Cu^{2+}]}{[Ag^+]^2}$$

In this equation copper is oxidized and silver ion reduced. Much the same relationship would exist for any two metals in an oxidation-reduction reaction. A more general form of the equation may therefore be written:

$$E = E_0 + RT/n\mathfrak{F} \times \ln \frac{[\text{oxid}]^a}{[\text{red}]^b}$$

E_0 represents the potential when the two components are present at $1M$ concentrations; [oxid] and [red] are the concentrations under any other set of conditions; and the superscripts a and b are to match the stoichiometry of the equation.

(*d*) Oxidation-reduction potentials

(1) Arrangement for measuring potential of a substance which can be present in solution in a reduced or oxidized state, e.g., ferric and ferrous ions.

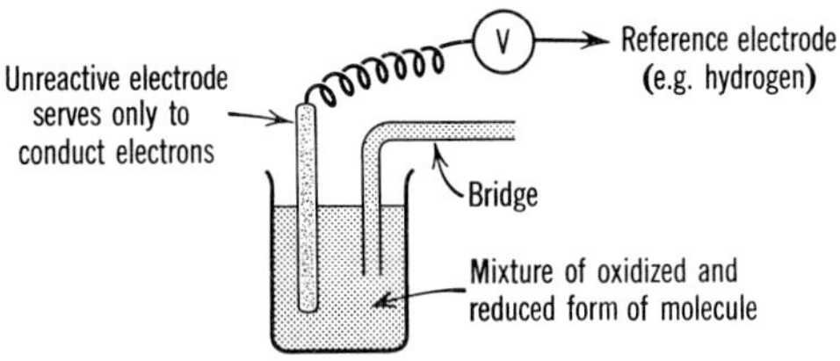

Equation for mixture is $Fe^{2+} \longrightarrow Fe^{3+} + e^-$. If half-cell shown is connected to hydrogen electrode, electrons may travel through wire and the direction of flow will be determined by the relative tendencies of half-cells to donate electrons. For this class of substances define E_0 as the "redox" potential, which is the potential recorded against the hydrogen electrode when the reduced and oxidized forms of the solute are present in equal concentrations. Since many cellular reduction systems (and others as well) are dependent upon the concentration of hydrogen ion, use term E_0' to designate E_0 at pH 7. If E_0 is known, then the potential of other mixtures containing different concentrations of the solutes may be calculated from the formula already given for transforming E_0 to E.

(2) E_0' for some biological systems (measured at 25–30°C)

NAD^+:NADH + H^+	−0.32 v
FAD:$FADH_2$	−0.12 v
Cyt c^{3+}:Cyt a^{2+}	+0.26 v
O:H_2O	+0.82 v

Total energy change in reaction: $NADH + H^+ + O \longrightarrow NAD^+ + H_2O$

$$E_0'(O/H_2O) - E_0'(NAD^+/NADH) = 1.14 \text{ volts}$$

$$\Delta F = -n\mathfrak{F}E = -2 \times 23{,}000 \times 1.14 = -52{,}400 \text{ calories}$$

(*e*) Effect of ion concentrations on electrode potentials

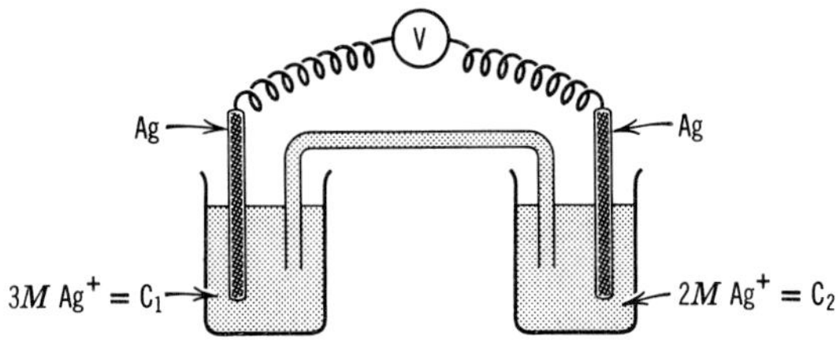

Fig. 8-3 (continued).

The expression is derived by equating the work done chemically in transporting 1 mole of Ag^+ from C_1 to C_2 with that done electrically by transporting 1 mole of electrons from C_2 to C_1.

$$\text{Chemical work} = RT \ln (C_2/C_1) \qquad \text{Electrical work} = n\mathfrak{F}E$$
$$E = RT/n\mathfrak{F} \times \ln (C_2/C_1)$$
$$\therefore n\mathfrak{F}E = RT \ln (C_2/C_1)$$

The greater the concentration of cations surrounding an electrode, the lower its tendency to donate electrons and the more positive its potential.

The following values may be substituted in the equation:

$$R = 1.987 \text{ cal/degree} \times \text{mole}; \; T = 303° \text{ Abs.}$$
$$n = 1; \; \mathfrak{F} = 23{,}000 \text{ cal/v}; \; \ln (C_2/C_1) = 2.3 \times \log_{10} (C_2/C_1)$$

The expected voltage in the cell illustrated would be approximately:

$$E = \frac{2 \times 300}{23{,}000} \times 2.3 \log(C_2/C_1)$$
$$= \frac{600}{10{,}000} \times \log (C_2/C_1)$$
$$= 0.06 \log (C_2/C_1) \text{ in volts}$$
$$\log (2/3) = -\log 1.5 = -.18 \qquad \text{cell voltage} = 0.011 \text{ v}$$

It may be seen that for a tenfold difference in concentration there will be 0.06 v difference regardless of the electrode system used.

The hydrogen electrode which is used as the ultimate standard of reference is not metallic. If "spongy" platinum (so treated as to have a large surface area) is immersed in a solution of HCl saturated with hydrogen gas, the hydrogen tends to deposit electrons on the platinum surface and to release hydrogen ions. The situation is otherwise much the same as in the calomel electrode. Platinum does not enter the reaction but simply serves as an electron conductor. The equilibrium as shown involves hydrogen ions, hydrogen atoms, and electrons. By so constructing the electrode vessel that the HCl solution is continuously exposed to a defined pressure of hydrogen, the tendency for electrons to be deposited on the platinum can be rigidly controlled. It was arbitrarily established that a gas pressure of 1 atmosphere and a concentration of 1*N* HCl should represent the standard reference condition.

The absolute electron-producing tendency of a single electrode cannot be measured; instead, the relative tendencies of two electrode systems are compared. Two connections between the half-cells are required for the comparison; the reason for this may be discussed in relation to Fig. 8-3*b*. Let us suppose, as is actually the case, that when a hydrogen electrode is connected to a copper-copper sulfate one, there is a flow of electrons from hydrogen to copper. At the hydrogen electrode hydrogen ions will be produced as the electrons move away, and copper will be deposited at the copper electrode by the interaction of incoming electrons with copper ions. If nothing else happened, there would be an excess of cations in the hydrogen cell and an excess of anions in the copper cell. From what we know about the extremely strong forces of electrostatic attraction, virtually no movement of the electrons would occur under these circumstances because the excess cations in the hydrogen cell would exert a strong pull on the migrating electrons, and the excess anions in the copper cell would repel them. If, however, the ions were free to move between the two cells no such inhibition would exist. This requirement, unfortunately, is the Achilles' Heel in all measurements of potential. The ideal connection would be one which permitted only compensating ions to pass through, but this is impossible to achieve. A number of problems arise when any type of connection is made. Since, however, an exceedingly small number of ions need to pass through when a potential as measured (enough to compensate for the electrons which deflect the voltmeter needle), the passage may be small and so arranged as to minimize diffusion. Frequently a tube filled with agar gel saturated with KCl is used. The passage is referred to as a "bridge"; in the absence of a bridge no voltage is recorded even though the metallic portions are connected.

Using the type of arrangement illustrated in Fig. 8-3*b*, the tendencies of a variety of metals to emit electrons have been measured when bathed in a solution of their own ions. A few such values are listed. The potential of the hydrogen electrode is arbitrarily defined as zero, and the potentials recorded for the metals are called "standard potentials" (E_0). The convention generally favored by biologists (though not by physical chemists) is to designate those electrodes which tend to transfer electrons to the hydrogen electrode as negative, and those which tend to accept electrons as positive. Thus the more negative a potential, the greater the reducing power. By comparing the potentials of zinc and copper electrodes, we can say at once that zinc will reduce copper ions. This is precisely the same type of answer we have sought for all chemical reactions using the free energy sign as indicator. The distinctive feature of such potential measurements is that we obtain values under nearly ideal thermodynamic conditions.

Only a few algebraic manipulations are necessary to relate the potential measured with other thermodynamic properties. The equivalence between amount of current and chemical change is known from the work of Michael Faraday. (Fig. 8-3*c*). The formula for electrical work is also indicated in the figure. Since ΔF equals the work done under ideal reversible conditions, an equation between free energy, potential, and amount of current can be set up. Since we already have an equation between ΔF and the equilibrium constant for any chemical system in which the components are present in unit molar concentrations, we may therefore relate equilibrium constant to electrical work. In discussing the relationship between free energy changes for any arbitrary set of concentration of chemical reactants and the equilibrium constant of the reaction, it was shown that this could be simply derived by adding a factor to the standard free energy change ($\Delta F°$). The factor was $RT \times \ln (C_1/C_2)$ (RT multiplied by the natural logarithm of concentrations of products/concentrations of reactants). Only those concentrations need to be included which are not in the standard state (that is, at a concentration of $1M$). The only alterable component, however, is the concentration of ion; the metal is not in solution and the electrons cannot be added or subtracted except insofar as they adjust themselves to an equilibrium position. In effect, therefore, the change in free energy between two electrodes immersed in solutions other than those specified by the standard state depends only on the concentration of the ions (Fig. 8-3*e*). This relationship is made particularly evident when two half-cells containing the same metal but different concentrations of metallic ions are connected with one another. With such an arrangement a voltage difference will be recorded. The magnitude of the voltage may be predicted from an equation derived by calculating the amount of work which would be done if one mole of silver ion at concentration C_1 were diluted to a concentration C_2 under "imaginary reversible conditions." This should be identical with the amount of work done electrically if one mole of electrons were transported from C_2 to C_1, resulting in the removal of one mole of silver ions in cell 1 and the production of one mole of silver ions in cell 2. It will be later seen that this equation has widespread importance in biological systems.

RESPIRATORY PIGMENTS

From the type of data shown in Fig. 8-3 we may calculate that the energy made available on oxidation of the hydrogens associated with NAD to water is about 54 kcals/mole of NADH. This amount of energy is more than sufficient to account for the experimental demonstration that such oxidation yields three moles of ATP. On thermodynamic grounds we might also predict that the oxidation would be step-wise in order to maximize the conversion of energy released into ATP. This is, in fact, the case, and the discussion which follows concerns the nature of the cellular substances which mediate the oxidative reactions.

Oxidation, like all other metabolic processes in cells, is catalyzed by enzyme proteins. The fact that the oxidation-reduction potential of NADH is low enough to release free energy upon oxidation does not imply that the reaction will occur spontaneously. NADH may be exposed to air for relatively long periods of time without the formation of NAD. An enzyme is clearly necessary, but what makes the oxidative enzyme system unique is that substrates in the usual sense are not present for enzymes to act on. Hydrogen atoms or electrons are the sole substrates for catalysis. The enzyme proteins themselves do not interact directly with these substrates, and in all oxidative systems each of the enzymes is associated with a relatively small organic molecule which, like NAD, may be reversibly oxidized or reduced either by transfer of a hydrogen atom or an electron. If the organic molecule is loosely attached to the protein, it is generally called a coenzyme; if it is tightly bound to the protein molecule, it is called a "prosthetic group."

The organic molecules which act as oxidation-reduction sites fall into three groups: nicotinamides, flavins, and cytochromes. Each of these groups spans a circumscribed range of oxidation-reduction potentials, the nicotinamides being at the lowest and the cytochromes at the highest end of the scale. The structure of nicotinamides and flavins has already been illustrated in Fig. 6-5; that of the cytochromes is shown in Fig. 8-4. Only a few features of their chemistry can be discussed here, for the mechanisms by which these substances become oxidized and reduced have been the objects of profound physicochemical studies.

The nicotinamides and the flavins have certain common subunits. In both, the constituent undergoing oxidation-reduction (the nicotinamide ring or the isoalloxazine configuration) is attached to the nucleotide adenylic acid. In other respects, however, their respective properties are dissimilar. The nicotinamides are loosely bound to their corresponding enzymes. When cells are disrupted, a large proportion of NAD and NADP is found unassociated in the medium. How much of this apparent lack of association is due to a weak bonding of coenzyme to protein within the cell is not clear, but the conclusion generally drawn is that a very large fraction of the nicotinamides is mobile and broadly distributed through the cell. The flavins, by contrast, are more tightly bound to their associated proteins. This property is reflected to some extent in the fact that flavoproteins have a much broader range of oxidation-reduction potentials than do the nicotinamides. The latter have the same oxidation-reduction potential irrespective of the enzymes with which they are associated. Flavoproteins, on the other hand, have some members which can reduce the nicotinamides (as in oxidative decarboxylation), others which can oxidize the nicotinamides, and still others which can interact directly with molecular oxygen. Indeed, the nicotinamides may be considered as anaerobic coenzymes. They can be reduced by substrates and reduce substrates in turn, or reduce flavoproteins; they cannot interact with molecular oxygen. The flavoproteins can be reduced by nicotinamides or by substrates (as in succinic dehydrogenase) and reduced flavoproteins can be oxidized either by the cytochromes or by molecular oxygen. They may therefore be considered as either aerobic or anaerobic coenzymes. From an

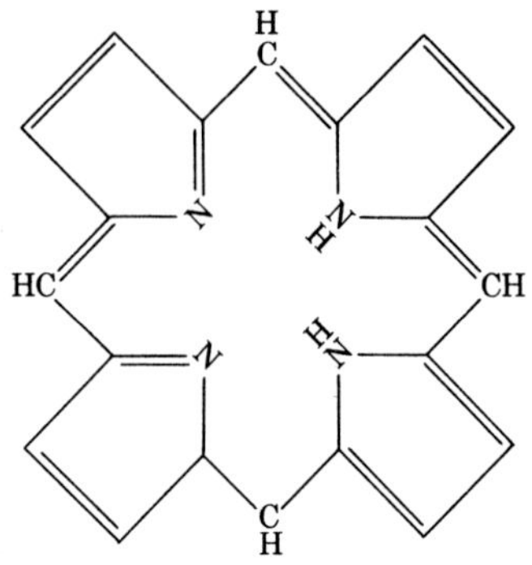

(a)

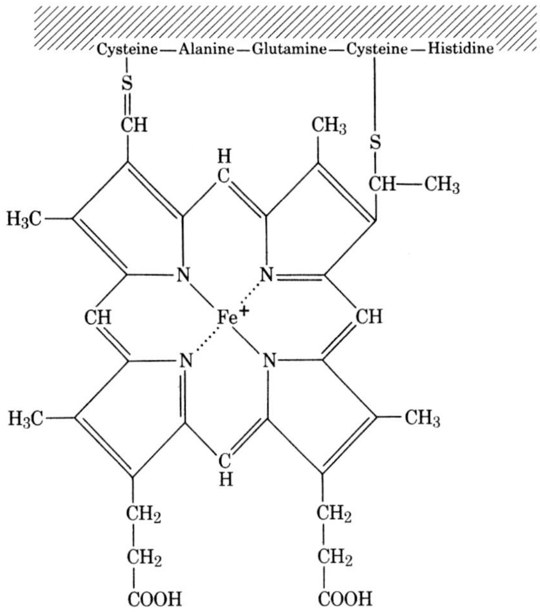

(b)

Fig. 8-4. The porphyrins. *(a)* Skeleton of porphyrin molecule. *(b)* Cytochrome C. Note the covalent linkage between porphyrin and cysteine-SH groups of protein. The attachment of porphyrin to protein is presumably effected by an enzyme.

evolutionary standpoint the nicotinamides would probably be regarded as the earliest of the coenzymes, since it is generally supposed that in the early stages of life oxygen was lacking in the atmosphere. The fact that virtually all the primary oxidation-reduction steps in carbohydrate metabolism are mediated by the nicotinamides supports this view. Fermentation without nicotinamides is unknown.

The range of oxidation-reduction potentials in flavoproteins is due to the bonding between flavin and protein moieties of the enzyme; this characteristic also applies to the cytochromes. Interaction between the two moieties alters the chemical characteristics of the prosthetic group so that, depending on the type of protein, the oxidation-reduction potential of the whole enzyme may differ to a greater or lesser degree from that of the free flavin. Whereas in the nicotinamide-coupled dehydrogenases the protein acts purely as catalyst, in the flavin dehydrogenases it acts both as catalyst and modifier of the oxidation-reduction potential. The spread in flavoprotein potentials is associated with the fact that these enzymes catalyze a variety of oxidative reactions, only few of which are involved in the principal channel of carbohydrate oxidation. Flavoproteins, like nicotinamide enzymes, are found in many parts of the cell, but unlike the nicotinamides, flavin prosthetic groups are not found in free form. Moreover, a number of the flavoproteins other than those present in mitochondria have been found to be associated with cytoplasmic membranes; by contrast, the nicotinamide dehydrogenases, as pointed out earlier, are commonly found in the soluble fraction of disrupted cells.

One major functional distinction between flavoproteins and cytochromes lies in their respective reactions with molecular oxygen. Flavoproteins yield hydrogen peroxide; cytochromes yield water. Hydrogen peroxide is toxic to cells; and were it not for the fact that the enzyme catalase which converts hydrogen peroxide to water is a common constituent of cells, oxygenations by flavoproteins would be likely to prove lethal. This relationship itself raises an important question as to whether the flavoproteins could ever have functioned as the sole terminal oxidases at one stage in cell evolution. The question is made even more significant because the prosthetic group of catalase is a porphyrin, which is the generic name for the class of organic compounds encompassing the cytochromes. A striking example in support of the argument that porphyrins are essential to survival in an oxygen-rich environment is the behavior of a species of anaerobic bacteria. Such bacteria die when exposed to oxygen, but survive if a porphyrin compound is added to the medium. The porphyrin appears to be taken up by the cells and acts as a catalyst to destroy the hydrogen peroxide formed by the flavoproteins when the bacteria are exposed to oxygen.

The world of porphyrins is indeed distinctive both structurally and functionally. Porphyrins are synthesized by a very different route from that leading to nicotinamides and flavins. Their chemical properties are very much tied to the incorporation of a metal into their structure—notably, iron and copper for the cytochromes and magnesium for chlorophylls. Although metallic ions are also associated with the activities of flavoproteins, the tight relationship between metal and prosthetic group function clearly evident in the porphyrins has not been demonstrated to operate in the flavoproteins. The cytochromes are exclusively transporters of electrons. In none of the cytochromes do we find combinations with hydrogen in the course of reduction. Although the distinctive property of oxidation by molecular oxygen is frequently attributed to the cytochromes, the actual distinctiveness lies in the capacity to transfer electrons. A number of bacterial species can oxidize in the absence of oxygen. Two examples are given in Fig. 8-5. From these it may be seen that given appropriate electron acceptors, the cytochromes can perform their distinctive function even in the absence of free oxygen.

The full variety of porphyrins in general and cytochromes in particular is yet to be disclosed. Cytochromes vary not only with respect to their specific proteins but also with respect to the subgroupings in the porphyrin skeleton. How much of biological survival in an oxygen-free world is dependent on porphyrins is manifest from their variety of functional activities. The transport of oxygen by hemoglobins, the destruction of peroxides by catalases and peroxidases, the release of free oxygen in photosynthesis, and the utilization of free oxygen in respiration are all mediated by porphyrin compounds. Leaving aside the

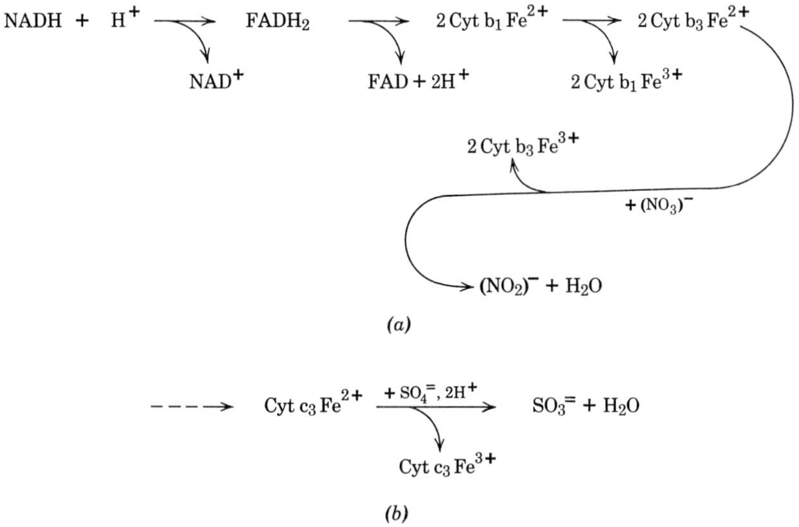

Fig. 8-5. Electron transport in the absence of free oxygen. *(a)* Respiration of *E. coli* under anaerobic conditions but in the presence of KNO_3, which acts as electron acceptor. *(b)* Electron transport in the obligate anaerobe *Desulfovibrio*, using sulfate as electron acceptor.

special case of the chlorophylls, the diversity of functions is attributable to the combined effects of modifying the porphyrin skeleton and the protein structure to which the porphyrins are attached. Even within the special sequence of electron transport from NADH to molecular oxygen, five different cytochromes have been found to be operative, each with a characteristic and different oxidation-reduction potential. Moreover, it is doubtful that all aerobic cells have identical complexes of cytochromes; the vastness of this particular field of molecular morphology has yet to be measured.

Spectrophotometric Assays. All three categories of oxidative cofactors can be classified as pigments. This property may seem to be incidental to their function, but from the standpoint of experimental analysis it is of major importance. The nicotinamides are not pigments in the conventional sense, for they are colorless in visible light. In ultraviolet light they manifest the characteristics of a pigment, that is, they absorb certain wave lengths of light and transmit others. Flavins, like nicotinamides, absorb in the ultraviolet, the principal agent of absorption being adenylic acid. However, flavins also absorb in the visible range and are occasionally referred to as the "yellow enzymes." Cytochromes are red, and the prosthetic group is commonly referred to as the "heme pigment." Experimentally, the significant feature is not the color as such, but the fact that the absorption characteristics of these pigments depend on their oxidation-reduction state. In Fig. 8-6 the absorption spectra of the reduced and oxidized forms of NAD and cytochrome C are illustrated. These differences in absorption spectra are widely used to measure oxidative processes. With suitable arrangements it is even possible to make such measurements using either intact cells or intact mitochondria. Since each of the cytochromes has distinctive absorption bands, the behavior of the individual enzymes in the electron transport system can be followed. Spectrophotometric techniques have thus made available to biologists an immense fund of information concerning the operation of the cytochrome systems. These techniques have also been successfully exploited in studies of photosynthesis.

MITOCHONDRIAL STRUCTURE AND OXIDATIVE PHOSPHORYLATION

To discuss mitochondria as subcellular structures which are solely specialized for oxidative phosphorylation would contract our perspective of mitochondrial function. For mitochondria are not only the sites of pyruvic acid oxidation; they are known to catalyze other chemical syntheses and to be capable of accumulating inorganic ions. However, to consider mitochondria as unique sites of oxidative phosphorylation is entirely appropriate. We have already observed that mitochondria

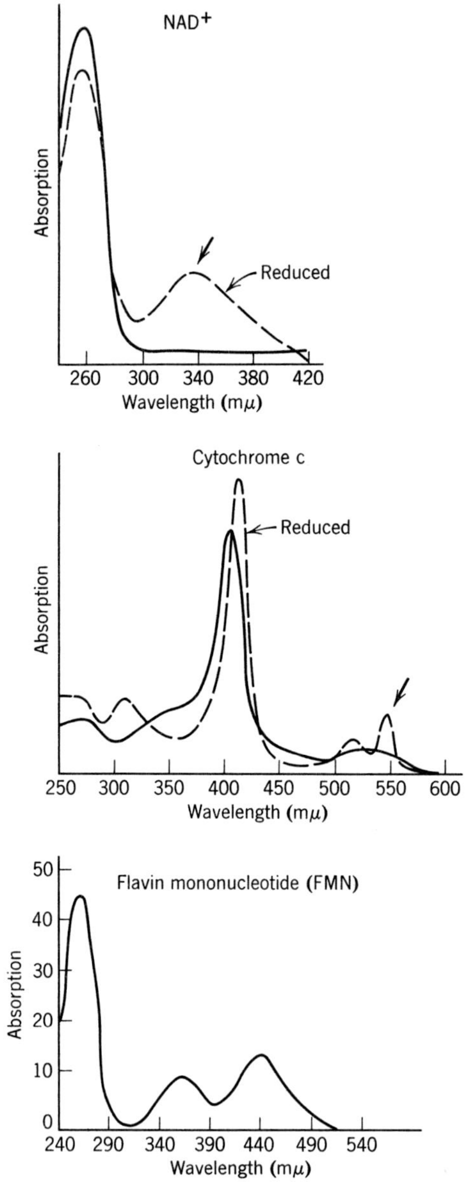

Fig. 8-6. Absorption spectra of some hydrogen or electron carriers. For NAD and FMN (or FAD), the high absorption at 260 mμ is due to adenine. Reduction of the nicotinamide ring produces an increased absorption at 340 mμ; reduction of alloxazine nucleus in flavins causes an increased absorption at about 465 mμ. There is no 260 mμ peak in cytochromes, adenine not being present. The 400 mμ peak is characteristic of the porphyrins. Heavy arrow shows wavelength commonly used to measure reduction.

are often found located in regions of cells known to require large amounts of energy, for example, in muscle fibers and kidney membranes. Histochemical techniques—those utilizing colored reagents which accumulate at those sites in the cell where the enzyme tested for is present—demonstrate the exclusive localization of certain oxidative enzymes in the mitochondria. Observations such as these coupled with the biochemical studies of cellular fractions all point to the single conclusion that wherever mitochondria are present, they constitute the major, if not the sole, site of oxidative phosphorylation. Whereas the dehydrogenases essential to pyruvate oxidation have counterparts elsewhere in the cell, the capacity to transfer electrons to molecular oxygen has a unique localization.

From a purely chemical standpoint the presence of the various hydrogen and electron transporting enzymes might appear to be sufficient to explain mitochondrial behavior. This, however, is not the case. Not only must we explain how the conversion of ADP to ATP occurs simultaneously with oxidation, but we must also explain the efficiency with which the whole process is executed. Broken mitochondria do not perform like intact ones; even injured mitochondria lose the capacity to catalyze the full oxidative sequence from pyruvate to CO_2, water, and ATP. Furthermore, injured mitochondria display enzyme activities, such as the hydrolysis of ATP, which are undetectable in normal ones. The evidence is overwhelming that the organization of mitochondria is a fundamental factor in their oxidative function.

The most important feature of mitochondrial organization is its membranous character. Shape and size of mitochondria are not known to have any influence on their metabolic activities. In some cells mitochondria may be nearly spherical; in others they are extremely elongated. Mitochondria have been frequently observed to pinch in two during cell growth and division; unlike chromosomes, they are not differentiated along their longitudinal axis. Mitochondria would appear to consist of many functional oxidative units, all of which are components of the membranes. Electron microscope studies of mitochondrial membranes reveal that the internal areas have an asymmetric pattern (Fig. 8-7). The interior face of the membrane is distinctly different from its outer one. Since biochemical studies of disrupted mitochondria show that the electron transport system from NADH to molecular oxygen is localized in the membranes, the argument for a very close relationship between membrane structure and oxidative function is strong indeed.

This relationship has been studied extensively in recent years through combined use of electron microscopic and biochemical techniques. The major conclusion drawn from these studies is that the

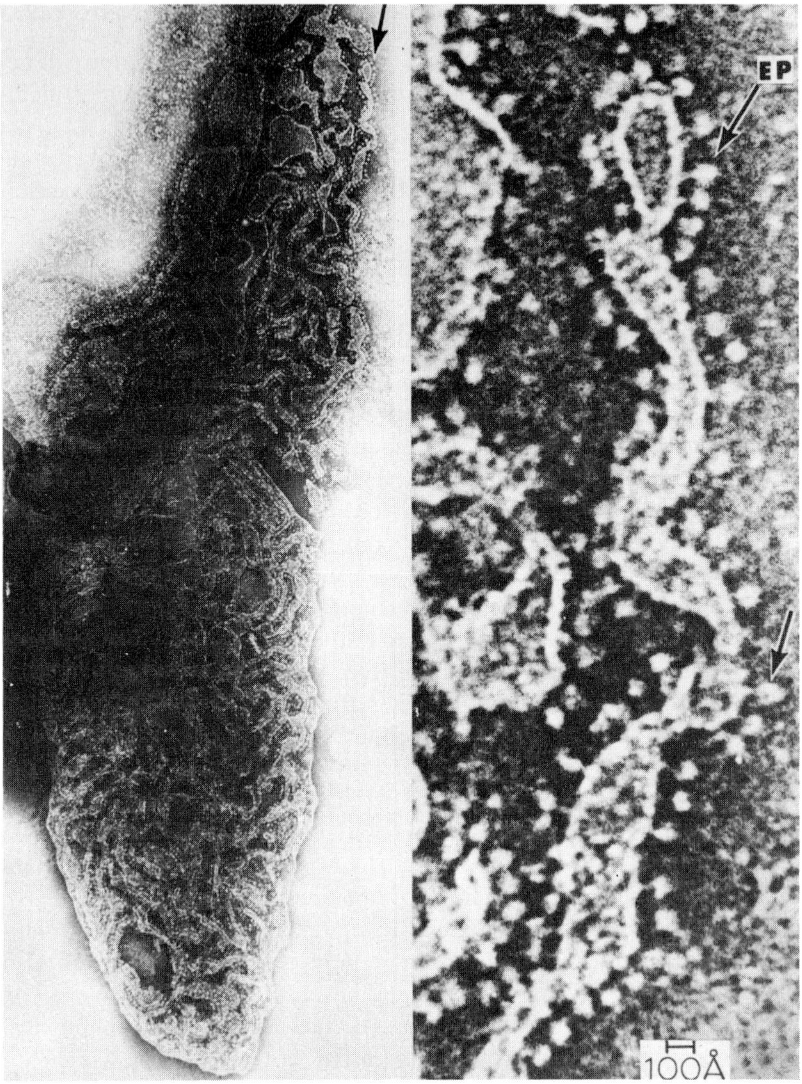

Fig. 8-7. Flattened membrane of beef heart mitochondrion prepared for viewing under the electron microscope by "negative staining." In this procedure the background rather than the tissue is coated with an electron scattering substance (e.g., phosphotungstic acid). The picture on the left (43,500 ×) shows numerous particles attached to the folded surface membrane. The picture on the right (285,000 ×) provides a more detailed image of the arrangement. The "elementary particles" (EP) have been considered to be aggregates of the electron transport enzymes. Although not apparent in these micrographs, the particles have been shown to face the interior of the mitochondrion and to be attached to the inner but not the outer membrane. (From H. Fernandez-Moran, T. Oda, P. V. Blair, and D. E. Green, "A Macromolecular Repeating Unit of Mitochondrial Structure and Function," *J. Cell Biol.*, **22**, No. 1, 1964, p. 71.)

electron transport system is localized in the particles subtended from the inner membrane. The details of structural organization are, however, far from being completely understood. Under the electron microscope the "particle" appears as a complex body consisting of a head, stalk, and a base through which attachment to the membrane occurs. This complex body has not been isolated as such "in vitro." The isolated electron-transport particles are more or less spherical. How much distortion occurs either in the electron micrographs or in the isolated material is still an open question. We do not therefore know whether stalk, head and base are each associated with different sets of enzymes or even whether only a segment of the electron transport chain is found in a simple particle. Despite these limitations in understanding, the studies as a whole have clarified the fact that membranes performing energy conversions have a distinctive type of molecular organization and we should expect such membranes to differ from other types of membranes which function primarily to regulate diffusion. This still leaves the problem of elucidating the relationship between membrane organization and the individual steps in electron transport.

Explanation of this relationship has not been satisfactory. The established facts underlying this relationship may be described as follows. Mitochondrial membranes contain the complete sequence of enzymes—nicotinamides, flavins, and cytochromes—essential for the oxidation of NADH or of the flavin directly reduced by succinic dehydrogenase. Free nicotinamides are also present in the interior of the mitochondrion and presumably serve to transfer hydrogens and electrons produced by the dehydrogenases to the protein-associated coenzymes by a mechanism similar to the one described earlier for phosphoglyceraldehyde dehydrogenase. The sequence of electron carriers has been determined by several experimental techniques, including the use of poisons which specifically inhibit one particular transfer in the chain. This sequence and also the oxidation-reduction potentials of the individual components are shown in Fig. 8-8. From the data on potential differences, the points at which electron transfer could release sufficient energy for the formation of ATP from ADP can be calculated. The calculations are consistent with experimental demonstrations of the points at which ATP is formed.

The importance of membrane organization to this coordinated picture of oxidative phosphorylation is indicated by other kinds of experiments. Normal mitochondria do not oxidize NADH if they are deprived of inorganic phosphate or ADP. This property clearly indicates that phosphorylation is coupled to oxidation, for if either a source of inorganic phosphate or an acceptor for it are absent the enzymes halt

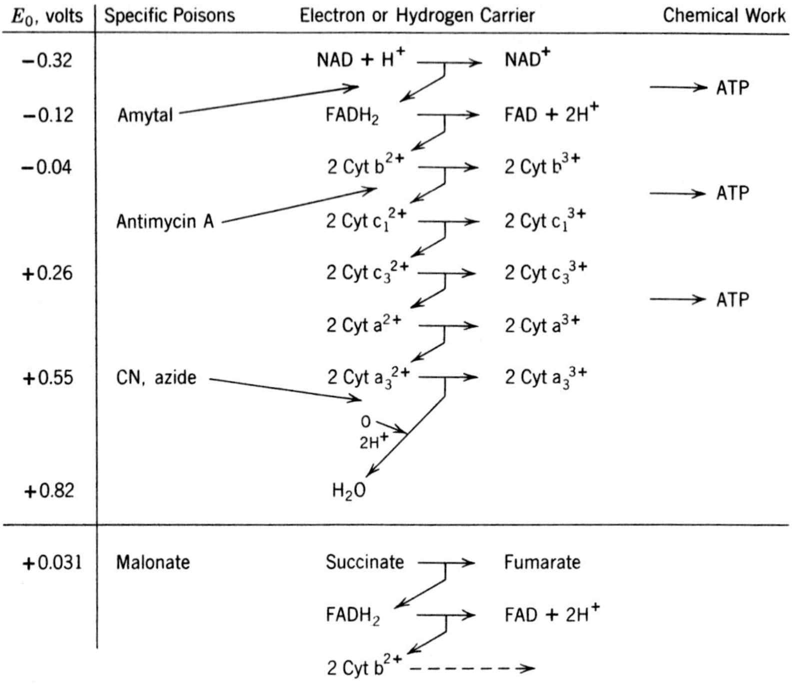

Fig. 8-8. Electron transport in aerobic organisms.

the transport of electrons. If mitochondria are injured (by exposure to distilled water, for example, or by allowing them to remain in suspension for several hours), this tight coupling is lost. Not only does oxidation proceed in the absence of phosphorylation, but it proceeds at a faster rate. The organization of the mitochondrion thus imposes a restraint on oxidations which is lost on even slight injury. The capacity to phosphorylate is not lost by this injury because mitochondrial fragments can form ATP as a result of oxidation. What is lost is the high efficiency of the process. There observations lead to the conclusion that the intact membrane facilitates a coordination between the enzyme components which is otherwise impossible.

The importance of the membrane in coordinating enzyme activities is also indicated by experiments in which some of the flavoproteins and cytochromes have been isolated in relatively pure form. Such pure forms of the enzymes, although completely separated from their normal membrane association, still retain their specific catalytic properties. The purified flavoprotein succinic dehydrogenase specifically dehydrogenates succinate; the purified flavoprotein cytochrome C reductase still transfers electrons from NADH to cytochrome C; purified cytochrome C is not oxidized by air, but is oxidized by purified cytochrome oxidase. Membrane organization does not therefore confer specificity on the different enzymes; the basic molecular properties remain the same within and apart from the membrane. Membrane organization would appear to modify and coordinate the activities of these enzymes. How modification is achieved remains poorly understood. Even the simple comparison of reaction rates between free and attached enzymes shows no uniform relationship; some enzymes show greater activity when dislodged from membranes, whereas others show a decrease.

One process which begs explanation and which probably limits our understanding of membrane function is the nature of electron transport. In metabolic processes such as glycolysis the coordinated activity of the different enzymes is relatively easy to explain even if the interpretation is oversimplified. We need only postulate that an enzyme modifies a particular substrate molecule when the two collide. Given this principle of the effectiveness of random collisions, we may picture a substrate molecule being gradually transformed as a result of a sequence of collisions. Such a picture is impossible for electron transport because electrons are not set free in solution. The alternative is to suppose that electrons are transferred when the enzymes themselves collide. If the electron-transporting enzymes are fixed to the membrane, as they appear to be, random collision is unlikely. Various explanations have been proposed, but none adequately clarifies the question of how electrons are transported along a series of oxidation-reduction enzymes. For this particular problem our understanding is not limited by a knowledge of the individual molecular components, but rather by our poor understanding of the ways in which their organization into membranes affects their activities.

A most important but poorly understood activity of mitochondrial membranes is the coupling of phosphorylation to oxidation. The evidence for such coupling is unambiguous, but its explanation remains a subject of much study. There is general agreement that during electron transport some membranous component must become modified so as to couple with inorganic phosphate in a high energy bond after which the phosphate is transferred to ADP. The mechanism of such coupling might be similar to that described for glyceraldehyde phosphate oxidation. However, unlike this oxidation, a substrate molecule cannot be implicated. If the membranous component combines with a phosphate as a result of reduction, it must become reoxidized on transferring the phosphate molecule. Some evidence points to the component being a protein but the problem remains unresolved.

Apart from their role in coordinating the activities of component enzymes, mitochondrial membranes also function as diffusion barriers. Isolated mitochondria show osmotic responses typical of intact cells. They swell in distilled water (and thus lose various soluble components such as NAD) and contract in concentrated solutions of sucrose. NADH is not oxidized by normal mitochondria, presumably because the coenzyme does not penetrate, but is readily oxidized if the mitochondria are exposed to distilled water. In cells treated with fluorocitrate, a compound which inhibits citrate oxidation, the mitochondria accumulate a great deal of citrate. This indicates that substrates do not readily diffuse through mitochondrial membranes. Mitochondrial membranes may therefore be said to furnish a "microenvironment" within cells. By restricting the flow of solutes, the intermediates in pyruvate oxidation remain accessible to the dehydrogenases within the mitochondria. Such restrictions raise other problems, however, for mitochondria must be able to absorb pyruvate and ADP as well as to secrete ATP. In ADP-ATP interchange an alternative is possible; conceivably, an enzyme at the surface of the mitochondrion could transfer high energy phosphate from internal ATP to external ADP, thereby regenerating ADP inside and replenishing the supply of ATP outside. The problem of solute transport across mitochondrial membranes is basically the same as that of cellular membranes in general.

In summary, electron transport from NADH to molecular oxygen is the major source of ATP for cells. Glycolysis alone yields 2 molecules of ATP and 2 of NADH per molecule of glucose oxidized. NADH produced in glycolysis can be oxidized by the mitochondria to yield 6 more molecules of ATP. During pyruvate oxidation the equivalent of 5 molecules of NADH are generated, which yield 15 molecules of ATP. Since each molecule of glucose gives rise to 2 molecules of pyruvate, 30 molecules of ATP are generated per molecule of glucose during pyruvate oxidation. The sum total of glucose oxidation is therefore 38 molecules of ATP. The most accurate determinations of the free energy of a pyrophosphate bond give a value of about 8 kcal/mole. Thus if all the energy made available by glucose oxidation were transformed into ATP, the yield would be 304 kcal/mole. The total energy possible from glucose oxidation is 670 kcal/mole. The efficiency of transformation is therefore less than 50%; the remainder of the energy must be lost as heat. Respiration without heat production is impossible.

9

The Acquisition of Light Energy: Photosynthesis

The abundance of terrestrial life flows from the abundance of solar radiation. Without radiation there would be no photosynthesis; without photosynthesis respiring organisms would quickly fail for want of oxygen, and all organisms would ultimately perish for want of carbohydrate. The dependence of the living world on green cells is by itself impressive enough to justify intensive studies of photosynthesis. Less sweeping but more intellectual reasons could be given for such studies, the most obvious one being the challenging nature of the photosynthetic mechanism. All the preceding discussions in this section have dealt with energy transactions which represent chemical transformations. For photosynthesis a totally new form of energy has to be introduced, that of radiation.

Plant physiologists had long recognized that photosynthesis could be resolved into two distinct types of reactions—a "light reaction" and a "dark reaction." The first reaction referred to a set of reactions for which light was essential, and the second referred to a set of chemical reactions whereby the products of light transformations were utilized to form carbohydrate. Today the component processes of photosynthesis may be distinguished in more specific terms. Light energy is utilized to produce two familiar forms of energy—reduced coenzyme (NADPH) and ATP; in the dark reaction these two forms of energy are used in conventional biochemical reactions to form carbohydrate from CO_2 and water. The dark reactions of photosynthesis are essentially resolved and will be briefly described at the end of this chapter. The light reactions remain a problem. Even though the products of this reaction have been identified, the fundamental question of how radiant energy is harnessed by chloroplasts to yield chemical reducing power is unresolved.

The neatness of metabolic patterns is so apparent in this simple description of photosynthesis that it deserves some comment. In carbohydrate oxidation NADH (the equivalent of NADPH) and ATP are produced; in photosynthesis carbohydrate is produced from NADPH and ATP. The enzymes which synthesize the carbohydrate are similar to, when not identical with, those which mediate carbohydrate breakdown. The light reaction does not give rise to any distinctive reducing agent; on the contrary, it produces the most common reducing coenzyme in the cell, and the one with the highest reducing power. Furthermore, the reduction occurs by the same process which occurs in oxidation, namely, the transfer of electrons and hydrogens to the coenzyme. The principal problem can therefore be simply defined. How does light energy make available hydrogens and/or electrons for reducing NADP?

This problem cannot be meaningfully discussed until certain basic properties of radiation are understood. First, the nature of light energy must be explained. Second, the mechanism by which a molecule can absorb the energy of light must be clarified. Third, the behavior of a molecule which has absorbed such energy must be described. The first and second requirements are general issues in photochemistry. The third requirement can be meaningfully met only if the kinds of cellular molecules absorbing light and the organization of these molecules in subcellular structures are known.

THE NATURE OF LIGHT ENERGY

From the time of Newton physicists sought to develop a satisfactory concept of radiation. To enter into a detailed development of this concept would take us too far afield; instead, three basic features of radiation will be summarized, and these should serve as background for discussing the photosynthetic mechanism.

(1) The energy of every form of radiation may be measured. (2) Velocity is an intrinsic attribute of radiation, and this attribute distinguishes it from all other energy forms. Paradoxically, Albert Einstein, the founder of relativity, was the first to propose that the velocity of light was an absolute. On theoretical grounds he used the absolute value as a constant in equating mass with energy. It was shown experimentally that radiation emitted at one point takes a finite time to

reach another. But how to explain the physical movement of radiation which travels through a vacuum as readily as through air is a challenge that has never been met on familiar terms. A simple analogy can be drawn between radiation and waves on the surface of water inasmuch as all radiation is considered to be a form of pulsation having characteristic frequencies. Each pulse is in the shape of a wave; the greater the number of pulses per second, the shorter the length of the wave. The relationship between the two is easiest to visualize in terms of a particle which describes a wave as it moves along. Since the speed of that particle is absolute (the velocity of light), it must form larger waves if it pulsates less frequently as it moves. *Every kind* of radiation may be described in terms of wavelength and frequency, and differences between kinds of radiation can all be reduced to these same terms. (3) Although radiation does not consist of wave-like motions of the medium (passage through a vacuum precludes this simple picture), its physical behavior can best be explained by assuming that waves do consist of discrete packets of energy analogous to pulsating particles. The packets are called "quanta"; for visible radiation a specific term "photon" is employed. The greater the frequency of pulsation, the greater is the energy content of the quantum; thus the shorter the wavelength, the greater the energy content.

The quantitative measurement of radiant energy began with the studies of Gustav Kirchhoff who demonstrated that an object which absorbs all incident radiation converts it into heat. He established the relationship by constructing a hollow black sphere perforated by a pinhole through which light of known intensity was shone. By measuring the temperature change of the sphere as light was absorbed, Kirchhoff was able to demonstrate an energetic equivalence between heat and light. The relationship is by no means obvious, and its full implications took many years to be understood. Names as well known as Max Planck and Albert Einstein were prominent in the history of the subject. Today it is still not fully clear how radiation in general, and light in particular, effect energy transfers, but the quantitative features of the basic relationships are established beyond doubt.

The relationship between the energy contained in a photon and the light in question is quantitatively given by the simple equation $E = h\nu$, where h is Planck's constant, and ν is the frequency of radiation. The frequency is equal to the velocity of light divided by its wavelength. Hence the equation may also be written $E = hc/\lambda$ where c is the velocity of light, and λ is the wavelength. For those interested in the relationship of light energy to chemical reactions, the equation may be written in a particularly useful way. The form of the equation derives from a law of Albert Einstein and Johannes Stark that when a molecule is activated by light it absorbs only one photon. Since all chemical reactions are referred to in terms of moles (or their equivalents), the unit Einstein was coined to represent 6.06 x 10^{23} photons. If one Einstein of photons is absorbed by a substance, we infer that one mole of substance is involved in the absorption. We may also express the equation in terms of the number of calories which one mole of substance would absorb in interacting with one Einstein of light:

$$E_{(einstein)} = \frac{2.86 \times 10^8 \text{ calories}}{\gamma \text{ (in Å)}}$$

An example of green cells absorbing light of 6500 Å may be used to illustrate how this equation is applied to relate light energy with chemical energy. If the cells absorbed 1 Einstein of light, it follows from the equation that 1 mole of chlorophyll molecules absorbed 45,000 Cal. (or 45 Kcal). This amount of energy is sufficient to reduce one mole of NADP or to produce three moles of ATP. Whether this does in fact occur will be discussed later. The question which must first be discussed is how light energy alters a molecule so as to enable it to perform otherwise impossible chemical reactions.

The component of a molecule which absorbs light energy is the electron (the only other component present is the atomic nucleus). Very high energy radiation is necessary to penetrate the atom, and such energy is several orders of magnitude higher than that found in visible radiation. In all cases the energy value of each quantum of radiation is inversely proportional to the wavelength. To choose familiar examples, ultraviolet radiation of 2000 Å can cause more severe molecular disturbances than visible radiation, whereas radio waves about (10^6 Å) have virtually no effect on molecular stability. By contrast, cosmic radiation (0.001 Å) may be sufficient to damage atomic nuclei. Since a single molecule absorbs only one quantum of light in each reaction, the energy of the quantum absorbed indicates the category of change we might expect. Thus the value of 45-50 Kcal/mole previously calculated for visible light is in the range of ordinary chemical reactions, which are the ones essential to photosynthesis. The wavelengths of different qualities of radiation are given in Fig. 9-1, and the energy content of their respective quanta may be calculated from the equation given earlier.

The absorption of a light quantum, or photon, is not a direct function of its energy level. If, for example, a solution of chlorophyll is placed in the path of a beam of white light, certain wavelengths are absorbed by the chlorophyll molecules and others are not. Chlorophyll solutions

Light Quality	Wavelength (Å)	Sample Values of Quantum Energy Wavelength	(kcal/einstein)
Ultraviolet	100–3900	(2860 Å)	100
Violet	3900–4300	(4000 Å)	72
Blue	4300–4700		
Blue-green	4700–5000	(5000 Å)	57
Green	5000–5600		
Yellow	5600–6000	(6000 Å)	48
Orange	6000–6500		
Red	6500–7600	(7000 Å)	41

Fig. 9-1. Wavelengths of ultraviolet and visible light.

are green because they strongly absorb red light, but do not absorb green (Fig. 9-2). Some property of a molecule must therefore determine what quality of light, if any, one of its electrons will absorb. Since the electron is the agent of absorption, we may safely predict that the property in question relates to it.

In order to present the modern conception of the selective absorption of photons by electrons, it would be necessary to discuss in full the modern theory of atomic structure. Such discussion is beyond the scope of this text, but in order to appreciate the implications of the phenomenon one essential condition of photon absorption will be briefly discussed. An electron which absorbs a photon must obviously increase its energy, for energy cannot be lost. One of the characteristic ways in which an electron expresses its increased energy is by changing the orbit of its spin about the atomic nucleus. This change is such that the difference in energy levels between old and new orbits exactly equals the energy of the photon absorbed. The orbits which any particular electron may occupy are limited in number, and unless the photon can raise the energy of an electron to a defined orbital level that photon will not be absorbed. The relatively low energy levels of light quanta severely limit the kinds of electrons and the kinds of photons which can meet this requirement. The chlorophylls are remarkable among the organic molecules for the extent to which they absorb visible light. The cytochromes, even though very similar in structure, absorb visible light with a much lower efficiency. The unique capacity of chlorophyll to initiate photosynthesis begins with its capacity to absorb a high proportion of the energy in visible light.

An electron which has absorbed a photon is said to be in an "excited state." The term is meant to describe the fact that such an electron created an unstable condition within the molecule. The instability is of short duration. Excited electrons return to their normal energy level

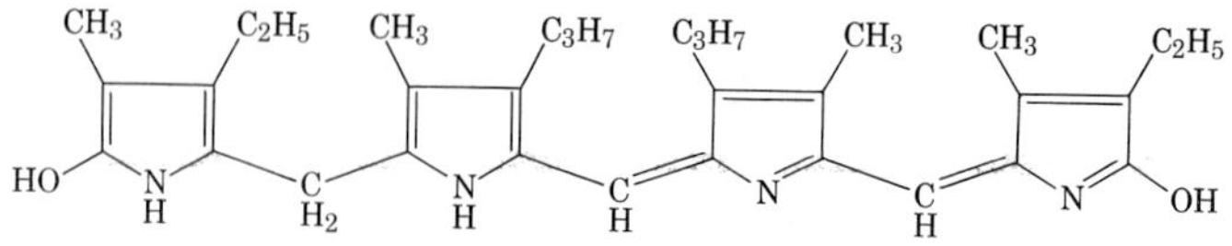

(a)

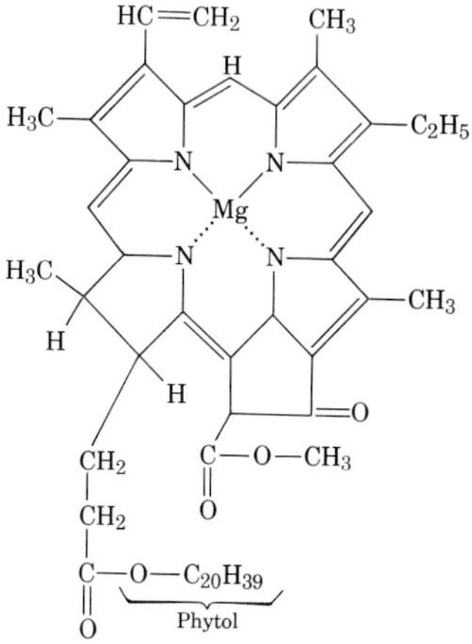

(1)

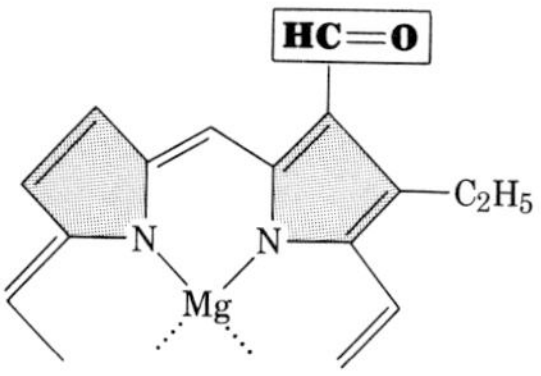

(2)

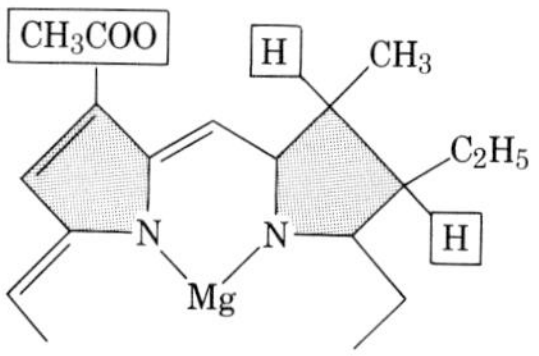

(3)

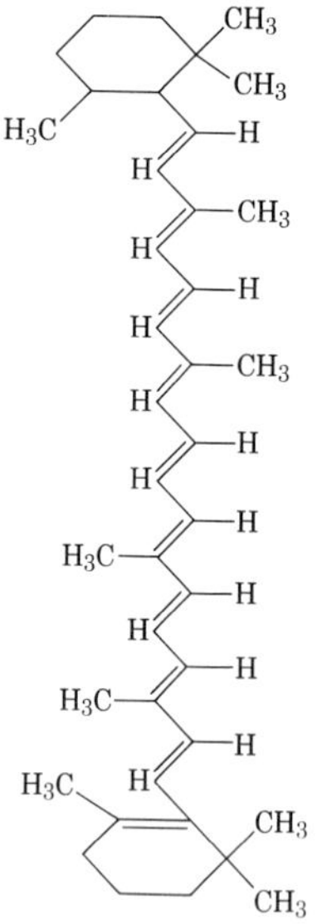

(4)

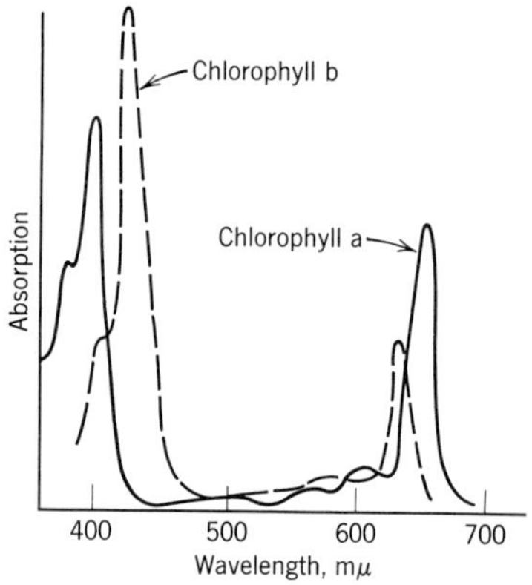

(5)

(b)

Fig. 9-2. Pigments in photosynthesis. *(a)* Phycobilins. *(b)* Chlorophylls: (1) chlorophyll a, (2) chlorophyll b, (3) bacteriochlorophyll, (4) β-carotene, (5) absorption spectra of chlorophylls a and b. Note structural differences between chlorophyll b and bacteriochlorophyll; the groups enclosed by boxed lines differ from corresponding groups in chlorophyll a.

within about 10^{-8} seconds, and during that interval the acquired energy must somehow be transferred. The transfer commonly occurs in one of four ways. (1) The excited electron may reemit a photon of lower energy; in such emission the quality of light emitted has a longer wavelength than the light absorbed. Phenomena of this type are known as fluorescence. (2) The molecule containing the excited electron may acquire increased kinetic energy and dissipate the energy as heat by colliding with other molecules; this is a common event when a colored solution absorbs light without the occurrence of any of the other changes listed here. (3) The energy acquired by the electron may be so great as to cleave a chemical bond and thus decompose the molecule. (4) The unstable molecule may interact with another molecule and produce some chemical change. This is basically what happens in photosynthesis.

Photosynthesis is thus made possible by the absorption of photons by certain electrons in the chlorophyll molecule. The structure of the chlorophyll molecule is such that the light is absorbed with high efficiency and is also such that on absorption, a return to a normal energy level is achieved by channeling the released energy into reactions which lead to the production of NADPH and ATP.

THE BIOLOGICAL PROCESS

In order to distinguish between incidental and essential processes in photosynthesis, a brief survey of the characteristics of photosynthetic systems should prove helpful. The survey is rather simple for the world of green plants. Water molecules are split, yielding free oxygen; the hydrogen released is used to reduce CO_2 to carbohydrate. The world of bacteria, however, compels us to question this simple representation of photosynthesis. There are two groups of bacteria, the "purple" and the "green," which must or can use light energy for growth but do not split water, evolve no oxygen, and do not necessarily use carbon dioxide as a source of carbon. Without entering into the detailed biochemistry of such photosyntheses, we can make the following generalizations.

(1) In all photosynthetic processes so far examined cells require some source of hydrogen which is released by light energy for the reduction of relatively oxidized carbon compounds. In green plants the source is water; in some bacteria H_2S, which is plentiful in their environment, serves the purpose; and, for other species a variety of organic compounds which are more reduced than carbohydrate serve as

well. Without specifying how the chemical reaction takes place, we may say that part of the over-all photosynthetic process involves the splitting of some molecule to release hydrogen. It follows from this that oxygen release is an incidental result associated with the use of water as a source of hydrogen.

(2) Although carbon dioxide is the most common substance reduced in photosynthesis, partially reduced compounds such as acetate may replace it.

(3) Several forms of chlorophyll are found in nature (Fig. 9-2). All plants contain chlorophyll a, and this pigment is generally associated with the capacity to split water and release oxygen. Bacteria which do not split water contain other forms of chlorophyll. Photosynthesizing cells generally contain more than one type of light absorbing pigment. In most plants chlorophyll b is present along with chlorophyll a, but in some algae and also in some bacteria accessory pigments may be present which are not chlorophylls. These pigments are the phycobilins (Fig. 9-2*a*). Phycoerythrin, which belongs to this latter group, is present in the red algae. The yellow carotenes which are generally found along with chlorophylls do not appear to play any direct role in photosynthesis, but serve to protect chlorophyll from destruction by surplus light irradiation.

(4) The type of pigments present in a photosynthesizing cell determines the quality of light which can possibly be effective in promoting photosynthesis. If, for example, the chart of wavelength vs. color (Fig. 9-1) is compared with the absorption spectrum of chlorophyll (Fig. 9-2*b*), it may be seen that chlorophyll absorbs very little light in the green region but absorbs a great deal on either side of it. This property should lead us to conclude that green light supports very little photosynthesis in green plants, and this is exactly what experimental studies show. It is possible to test the relationship in a much more precise manner by shining light of different wavelengths on green cells and comparing the yield of photosynthesis for each of the wavelengths used. When this is done a curve ("action spectrum") comparable to the absorption spectrum is obtained, but modified in a very important way (Fig. 9-3*a*). Since pigments other than chlorophyll A are also present which can utilize light energy for photosynthesis, the action spectrum obtained is a composite of the absorption spectra of chlorophyll A and the other pigments. Since red algae, for example, contain phycoerythrin, they can use green light effectively for photosynthesis (see Fig. 9-3*b*).

The energetic relationships between the different pigments in a particular photosynthetic system are not entirely clear. A commonly held view is that the pigment absorbing at the longest wavelengths is primarily responsible for photochemical activities, and that the others which absorb at shorter wavelengths transfer their absorbed energy to the primary pigment (chlorophyll a, in green plants). Such a transfer is thermodynamically possible, since the shorter wavelengths have higher energy and the quanta absorbed by the pigments in this region could be reemitted at a slightly lower energy level and be absorbed as light of longer wavelength by the primary pigment. Whether or not a single pigment is responsible for the entire photosynthetic process, there is little doubt that accessory pigments can transfer their energy to the primary one (see Fig. 9-3). Accessory pigments, even if they serve only to transfer absorbed energy to a primary pigment, afford physiological advantages for certain organisms. In weak light, of course, photosynthetic cells generally profit by being

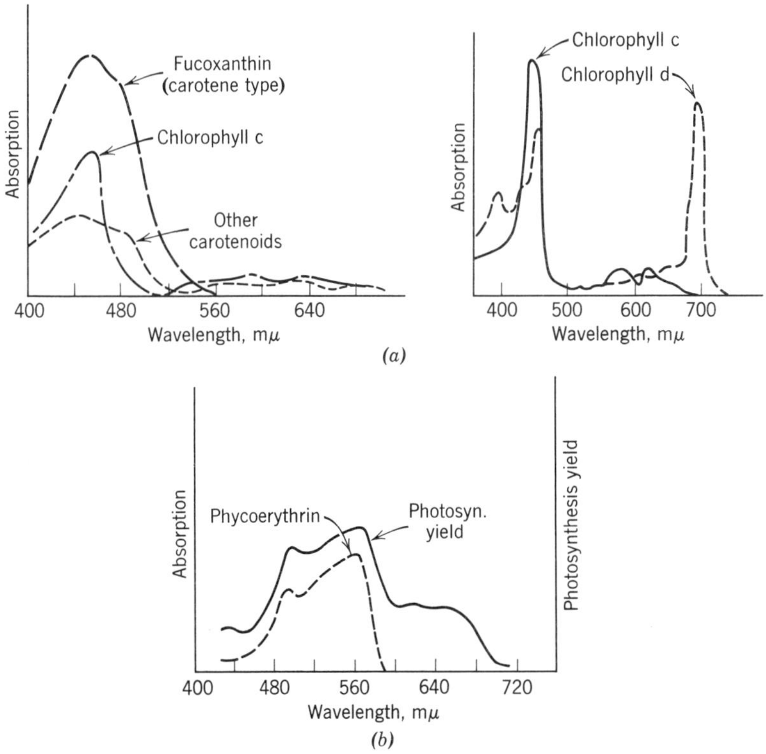

Fig. 9-3. Accessory pigments in photosynthesis. *(a)* Absorption spectra in methanol of pigments in the alga *Navicula minima*. *(b)* A comparison of the yield of photosynthesis for a red alga illuminated at different wavelengths with the absorption spectrum of its predominant pigment phycoerythrin (a phycobilin).

able to use light energy of a broad variety of wavelengths. Specific light problems arise for aquatic organisms because of the quality of light at different depths. Shorter wavelengths, because of their higher energy, penetrate more deeply, and blue light predominates at distances well below the water surface. Red algae grow more successfully in deeper waters because phycoerythrin makes possible an efficient utilization of blue light. A comparison of the spectra of chlorophyll and phycoerythrin will make this point clear (Fig. 9-3*b*).

THE CONVERSION OF LIGHT ENERGY

The heart of the photosynthetic problem lies in the mechanism whereby light energy is utilized to transfer hydrogen from a donor such as water to an acceptor such as NADP. The production of ATP, though equally dependent on light energy, may be assumed to occur in much the same way as described for electron transport in oxidative systems. Given an initially reduced substance, the progressive transfer of electrons to acceptors of higher oxidation-reduction potential releases energy which could be converted to ATP. The primary event must therefore be the production of some component with a very high reducing potential, the oxidation of which could lead to the two forms of energy essential to the synthesis of carbohydrate.

Using the photosynthetic process in green plants as example, the thermodynamic problem may be simply stated. The oxidation-reduction potential of the system $H_2 + O = H_2O$ is approximately $+0.820$ volts; that of $NADP^{2-} + 2H^+ = NADPH_2$ is $-.320$ volts. In the presence of suitable catalysts hydrogen would therefore be transferred from reduced NADP to oxygen, and this, of course is what occurs in respiration. To reverse this process two substances are required: one with a high enough oxidation potential to oxidize water, and the other with a low enough oxidation potential to reduce NADP. The two substances must be intimately related since, as pointed out in the discussion of oxidations, if a substance acquires an electron, another substance must lose one. Thus the essence of the light effect is an oxidation-reduction reaction in which one component acquires electrons and becomes highly reducing, while another loses electrons and becomes highly oxidizing. The main question is how the absorption of light by chlorophyll achieves this result.

A satisfactory answer is not yet available. The phenomenon of photosynthesis can therefore only be defined but not resolved in terms of the known components of photosynthetic systems. The activities of these components must be examined from two standpoints—the behavior of the individual molecules, and the interactions of these molecules in the organized photosynthetic system.

On absorbing a photon, one electron in the chlorophyll molecule becomes excited. One of two conditions may result. The energized electron may migrate out of the molecule and be absorbed by a neighboring substance, or the electron may acquire a new orbit such that a "hole" is created and an additional electron is absorbed from a neighboring substance. In the first case chlorophyll becomes oxidized; in the second, it becomes reduced. The new configurations are both unstable, so that the energized chlorophyll molecule becomes either a strong oxidant or a strong reductant. The substances accepting or donating electrons from or to the chlorophyll become correspondingly good reductants or oxidants. Directly or indirectly water becomes oxidized by the oxidant, and NADP reduced by the reductant. Both processes return the primary products of the light reaction to their normal configuration.

Of the two processes, the release of free oxygen from water is understood the least. Experimentally, oxygen release has been separated from the other photosynthetic reactions, simply by illuminating isolated chloroplasts in the presence of a suitable electron acceptor such as ferricyanide. The "Hill reaction," as it is called in honor of its discoverer A. V. Hill, is the production of oxygen and the reduction of ferri- to ferrocyanide by a suspension of illuminated chloroplasts. The significant result of this discovery was the demonstration that chloroplasts could use light energy to remove electrons from water, and could transfer these electrons to one of several suitable electron acceptors (Fig. 9-4) without the occurrence of any of the other characteristic photosynthetic reactions. Whether oxygen production occurs by a direct interaction between chlorophyll and water is unknown. Mn^{2+} appears to be essential to oxygen release, and this observation has suggested to some investigators that H_2O_2 may be the initial product which is decomposed by a manganese-requiring enzyme into water and oxygen.

The sum of the light reactions in photosynthesis could be simply represented as follows. An excited chlorophyll molecule emits an electron which is captured by a substance X; X^- reduces NADP, the hydrogens associated with NADPH being derived from a simultaneous reaction between the oxidized chlorophyll molecule and water. (Fig. 9-4*b*).

The transfer of electrons which occurs in the course of the return of the initially excited molecules to their normal, or "ground," state must be accompanied by a release of free energy. Part of this energy is used to produce oxygen and NADPH, but the remainder is utilized for the

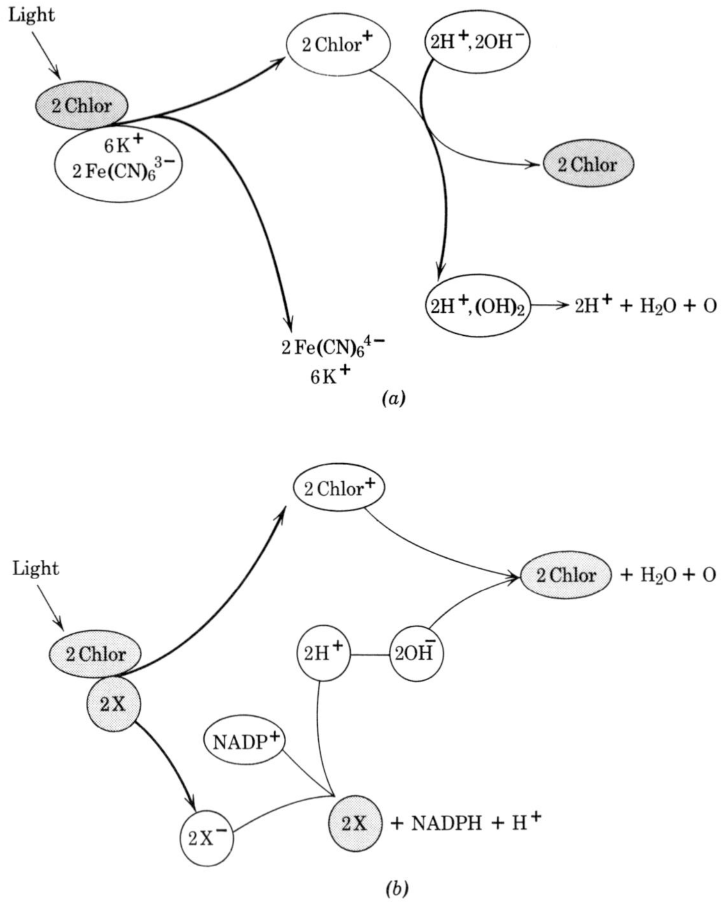

Fig. 9-4. *(a)* Simplified representation of Hill reaction; net reaction is: Light + $6K^+ + 2Fe(CN)_6^{3-} + 2H^+ + (OH)^- \rightarrow 6K^+ + 2H^+ + 2Fe(CN)_6^{4-} + O$. *(b)* Simplified representation of total light reaction.

phosphorylation of ADP to ATP. Isolated chloroplasts may be treated in several ways so that they do not produce oxygen or NADPH when illuminated (Fig. 9-5). Under such conditions the excited electron is returned to the chlorophyll molecule, the energy released being used only to form ATP. The phenomenon is called "cyclic phosphorylation," and in some organisms or under certain conditions in many organisms this process may actually occur. Conceivably, for example, if CO_2 is lacking and NADPH cannot be oxidized, the excited electron may return to chlorophyll by a route that allows only ATP formation. The

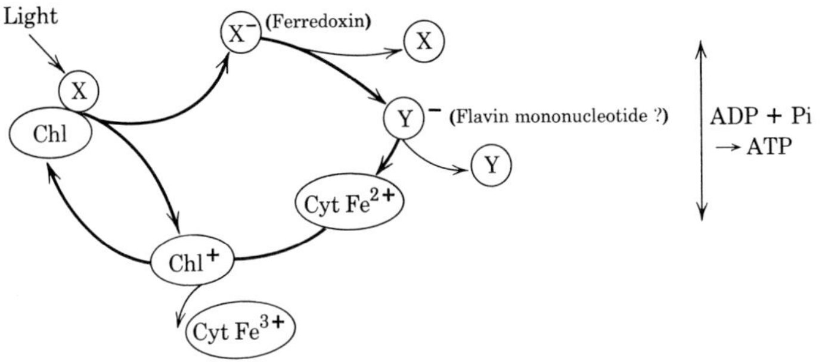

Fig. 9-5. "Cyclic" photosynthetic phosphorylation. This partly hypothetical scheme illustrates the fact that following chlorophyll excitation, the ejected electron may return to the molecule of its origin via an electron transport system which converts the energy released into ATP. Ferredoxin is one of the recently discovered components of chloroplasts. It contains iron, has an oxidation reduction potential of —0.43 volts, and readily reduces NADP (—0.320 volt). In the above scheme NADP is bypassed; neither reducing coenzyme nor oxygen is formed. Compare with electron pathway in Fig. 9-6 in which NADP is reduced by ferredoxin. The bypass is achieved by use of the metabolic poison *p*-chlorophenyldimethylurea.

reaction could be summarized as follows:

$$\text{Chl} + \text{X} + \text{photon} \rightarrow \text{Chl}^+ + \text{X}^-$$

$$\text{X}^- + \text{Chl}^+ + \text{ADP} + \text{P} \rightarrow \text{X} + \text{Chl} + \text{ATP}$$

Basic to all these reactions is the fact that the excited electron must directly or indirectly be returned to the chlorophyll molecule. In the course of such return the energy of excitation must be released. The nature of the release is determined by the pathway through which the electron travels. From the discussion of electron transport in respiration, it should be clear that the energy of excitation will not be released in a single step. The fact that ATP is produced suggests that at least some of the steps have free energy changes of the order of 15 kcal/mole, whereas the photon absorbed in photosynthesis has a value of about 45 kcal. These considerations have prompted a search for intermediates in the electron transport process, and a number of intermediates have been found.

The possible role of intermediates in photosynthesis can be determined to some extent from their oxidation-reduction potential. If, for example, intermediate A has a redox potential of -1 v and intermediate B has a potential of 0 v, then the reduced form of A will have a strong tendency to transfer electrons to the oxidized form of B, but the reduced form of B will not transfer electrons to the oxidized form of A. If two such components were found in the chloroplast, we would infer that the primary effect of light absorption was either a reduction of A which then led to a reduction of B, or an oxidation of B which then led to an oxidation of A. Whether all the components capable of undergoing oxidation-reduction which have been found in chloroplasts are actually essential to photosynthesis cannot be stated with certainty, if only because the complete mechanism is unknown. However, the presence of some components is suggestive of possible mechanisms, and these mechanisms have been put together in a single scheme which is illustrated in Fig. 9-6.

The most important feature of the scheme rests on the principle, for which there is experimental evidence, that the absorption of one photon by chlorophyll has insufficient energy to yield an oxidant capable of oxidizing water and a reductant capable of reducing NADP. The argument behind this principle may be briefly stated as follows. The standard redox potentials of NADP:NADPH and O:H_2O are -0.32 v and $+ 0.82$ v respectively. The primary reduced component must have a potential well below -0.32 v in order to allow for ATP production during electron transfer from the primary reductant to NADP. One mole of electrons transferred in the direction of free energy release over a voltage difference of about 0.7 would yield one mole of ATP. This may be calculated from the formula given in Fig. 8-3*c*. The consensus is that the primary reductant must have a redox potential of about -1.0 v or less. A similar calculation leads to the conclusion that the primary oxidant should have a potential of about $+ 1.5$ v. Even allowing for relatively broad margins of error, a difference of about 2.0–2.5 v between primary oxidant and reductant must be assumed. This difference can be no less than the energy content of the photon absorbed if a single photon effects the change. But a 2.5 v difference is the equivalent of 58 kcal/mole, and the photon primarily absorbed has a value of about 45. Moreover, 100% utilization of energy is most unlikely.

The solution to this problem has taken several forms, but essentially all of them require the use of two photons and two forms of chlorophyll. One photon is presumed to cause the transfer an electron from chlorophyll (Chl in Fig. 9-5) to a reductant (X). The resultant oxidized chlorophyll is estimated to have a potential of about $+0.4$ v, and the reductant of about -1.0 v. Simultaneous, or nearly so, with chlorophyll oxidation another photon is presumed to be absorbed by a different chlorophyll molecule which becomes reduced by withdrawing an electron from the oxidant (Y). Some believe that the two chlorophyll molecules are of one kind but differently associated; others believe them to be chemically distinct. Either interpretation would fit the experimental observation that wavelengths close to 7000 Å can effect the reduction of

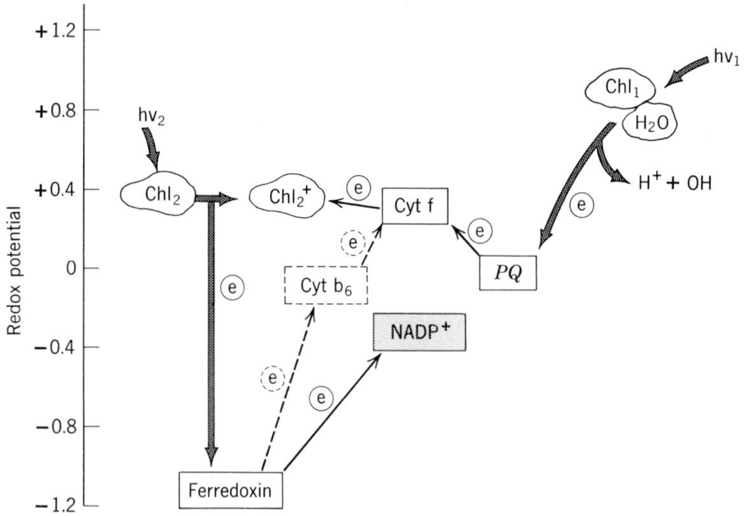

$$2\,(\mathrm{Chl}_2 + h\nu_2 + \mathrm{Ferredoxin}) \longrightarrow 2\,\mathrm{Chl}_2^{+} + 2\,\mathrm{Ferredoxin}^{-}$$

$$2\,(\mathrm{Chl}_1 + h\nu_1 + H_2O) \longrightarrow \textcircled{2e} + 2\,H^{+} + 2\,OH$$

$$2\,\mathrm{Chl}_2^{+} + \textcircled{2e} \longrightarrow 2\,\mathrm{Chl}_2$$

$$2\,\mathrm{Ferredoxin}^{-} + 2\,H^{+} + NADP^{+} \longrightarrow 2\,\mathrm{Ferredoxin} + NADPH + H^{+}$$

$$2\,OH \longrightarrow H_2O_2 \longrightarrow H_2O + O$$

Fig. 9-6. Schematic representation of the light reactions in photosynthesis. This scheme is based primarily on the recent studies of D. I. Arnon, although the fundamental ideas have been developed by a number of investigators. The original observation that photosynthesis requires two different wavelengths of light is due to the studies of the late Robert Emerson at the University of Illinois. Some believe that Chl_1 is chlorophyll b and that Chl_2 is chlorophyll a. The effective wavelength for Chl_1 activation is about 6800 Å, and that for Chl_2 is about 7050 Å. A photon absorbed by Chl_1 results in the pulling of an electron from water and the transfer of that electron through a series of intermediates—plastoquinone (PQ), cytochrome *f*—to Chl_2^{+}. The oxidation of Chl_2 is caused by the absorption of another photon, the electron being transferred to an iron-containing protein called ferredoxin. Chl_2^{2+} is highly unstable and the electron transfer, as discussed in the text, occurs very rapidly. ADP is phosphorylated to ATP at various stages in the electron transfer pathway; these stages are still not precisely known. Cytochrome b_6 is considered by some to be involved only in cyclic phosphorylation. In this scheme only Chl_2 is active in light absorption during cyclic phosphorylation. The electron expelled by Chl_2 returns via cyt b_6, and is not transferred to the stable electron acceptor, NADP.

NADPH, but not the production of oxygen. Shorter wavelengths catalyze both reactions. Since photons of shorter wavelengths can be absorbed and transformed into photons of longer wavelengths, but not *vice versa*, the observation has been interpreted as indicating that oxygen production requires a shorter wavelength of light than does NADP reduction. The principle of light activation implicit in this sophisticated scheme is identical with that implicit in the simple one previously described. The mechanisms differ in that two different types of chlorophyll are required for the interactions drawn in Fig. 9-6. One chlorophyll molecule becomes oxidized, the other becomes reduced, and both are returned to their ground state by a transfer of electrons between them via a cytochrome system. The requirement for highly reducing and highly oxidizing primary products is thus satisfied by using two photons simultaneously to generate opposite reactions in similar but not identical chlorophyll molecules.

CHLOROPLAST ORGANIZATION

The appearance of chloroplasts has already been described in Part I, and this discussion will be restricted to possible relationships between structure and function. The most general statement which can be made in this respect is that in all photosynthetic organisms, chlorophylls are associated with membranes. In blue-green algae such membranes extend through most of the cytoplasm; in green algae and all higher plants the membranes are restricted to chloroplasts. Since chloroplasts are specialized for photosynthesis, the relationship between membranes and ground substance may be better studied in these than in the unbounded membrane system of blue-green algae. Biochemically, this relationship may be simply defined. The membranes are the seat of light reactions and the ground substance (or the "stroma" of chloroplasts) is the seat of chemical synthesis. Isolated chloroplasts can mediate the full photosynthetic reaction resulting in starch formation; chloroplasts in which the peripheral membrane has been ruptured lose the capacity for carbohydrate synthesis but not for the primary light reactions. A parallel does exist between mitochondria and chloroplasts insofar as the location of the electron transport system is concerned; in both organelles this system is part of the membrane structure.

The membrane organization of chloroplasts also varies with evolutionary level. In green algae the membranes are more or less uniformly distributed through the chloroplast; in higher plants membranes are concentrated in regions called "grana" (hence the grain-like appearance of these chloroplasts under the light microscope). See Fig. 9-7. The functional significance of this difference is not clear. However, grana may be isolated by breaking chloroplasts, and if such grana are, in turn, disintegrated they yield small spherical-shaped granules (100 × 200 A) which can be seen under the electron microscope; these granules appear to be attached to the membranes of the grana. The granules contain the

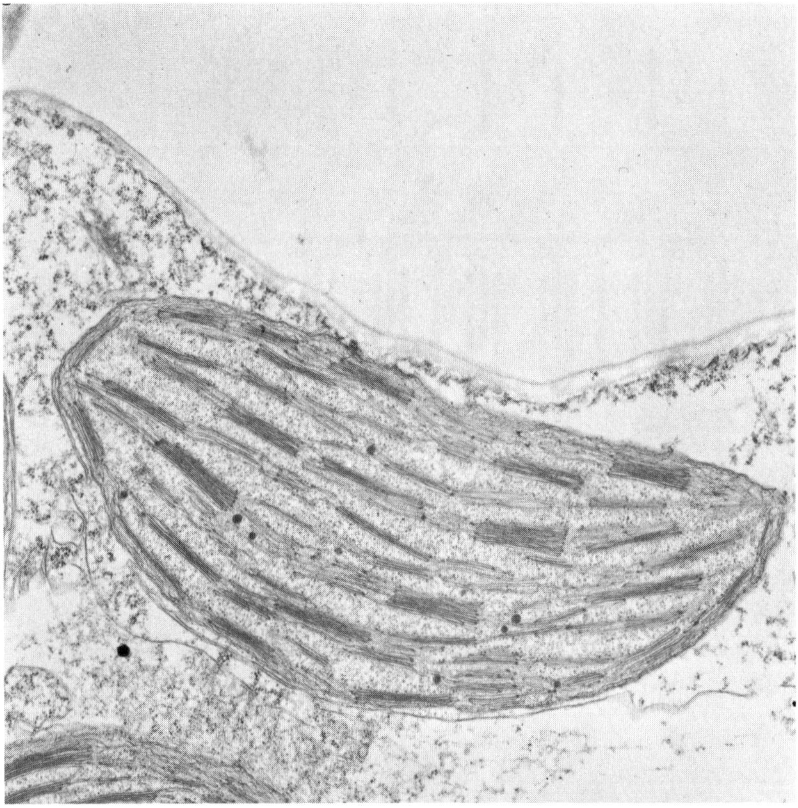

Fig. 9-7. A section through a chloroplast of corn. Note that the lamellae are stacked in certain regions. These regions are called "grana." Although their structure was unknown, grana were first recognized under the light microscope and were so named because of their granular appearance. Recent studies of lamellae indicate that they consist of spherical subunits called "quantosomes." Such subunits contain the various oxidation-reduction components referred to in Fig. 9-6 and are capable, when isolated, of performing the primary steps in photosynthesis. This particular photograph is of a relatively young chloroplast in which the membranous layers are not fully developed. At this stage numerous small granules may be seen in the regions between the membranes. These granules are considered to be similar to ribosomes in composition and function. (Courtesy of A. Jacobson and L. Bogarod.)

basic components of the light reaction system. For each molecule of lipid and protein (500,000 m.w.), there are about 100 chlorophyll molecules, 25 carotenoid molecules, 1 molecule each of cytochrome b_6 and f, and several other oxidation-reduction intermediates. These granules are capable of catalyzing the light reaction of photosynthesis.

One suggested arrangement for the molecules within the membranes is shown in Fig. 9-8. The important question is whether this or any other

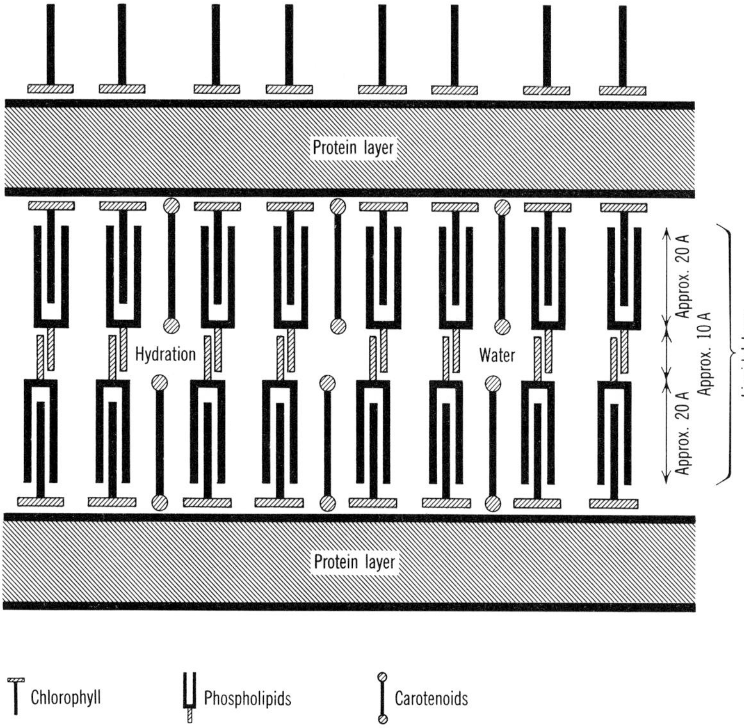

Fig. 9-8. Sketch of an individual lamella. The lipoidal layer is shown to contain chlorophyll, phospholipids, and carotenoids. Model of Hubert, as modified by Frey-Wyssling and by Brown (*Plant Physiology,* Vol. 1A, 3, 1960, ed. by F. C. Steward, Academic Press, New York). This model is basically the same as the one shown in the classical picture of membrane structure. Serious questions have been raised as to whether such a model is an adequate representation of molecular organization. Just as in the case of mitochondrial membranes, recent studies are revealing a more complex structural pattern than the above model would allow. One valid criticism of the model is that most of the lipids in the chloroplast are not phospholipids, but galactolipids. Another criticism is that the model has become too vague with respect to the recent studies of the chemical composition of "quantosomes" (see text).

similar arrangement has a bearing on the nature of the light reaction. A fundamental requirement of the light reaction would suggest that such a bearing does exist. The primary event in the light reaction is the simultaneous formation of a strong oxidant and a strong reductant. Since this event occurs by the transfer of an electron from one to the other, we must somehow account for the fact that the electron does not bounce back. Were the primary products stable, we might suppose that a return could only be effected in the presence of any enzyme. But these products are not stable, and some mechanism must therefore exist for their separation. In this respect the layering of the membranes could be effective, although a precise picture of how this could be so is difficult to draw. Conceivably, if the electron is moved from one layer into another, the recipient of the electron could interact with components present in that layer, but not in the original one, and thus prevent the electron from returning.

On issues of electron movements in living matter, whether they relate to photosynthesis or respiration, our understanding is limited. Cell biologists have been highly successful in studying the properties of individual molecules, and wherever possible of fitting them into the integrated activities of cells. The resolution of metabolic pathways is a model of such success. But, compared to an electron, even the smallest atom is a mammoth and sluggish piece of matter. The rapidity and energy with which electrons move, the displacements they can effect on entering the clouds of electrons which surround cellular macromolecules are properties which are alien to those surrounding the exchange of atoms or groups of atoms in biochemical reactions. If electrons are conducted along metals, can they not also be similarly conducted along the organized components of membranes? The answer to this cannot be found in the mechanisms of today, but it will undoubtedly be found in the studies of tomorrow.

CHEMICAL TRANSFORMATIONS

As pointed out in the beginning of this chapter, given ATP and NADPH, CO_2 and water can be used to generate carbohydrate. Most of the enzymes necessary to this transformation are found in many types of cells which are not photosynthetic. The only unique reaction is perhaps the enzyme system which "fixes" CO_2. Yet even that enzyme has a counterpart in other enzymes, discussed previously, which can attach carbon dioxide to the end of a carbohydrate molecule. The sequence of enzyme-catalyzed reactions is illustrated in Fig. 9-9. Elucidation of that

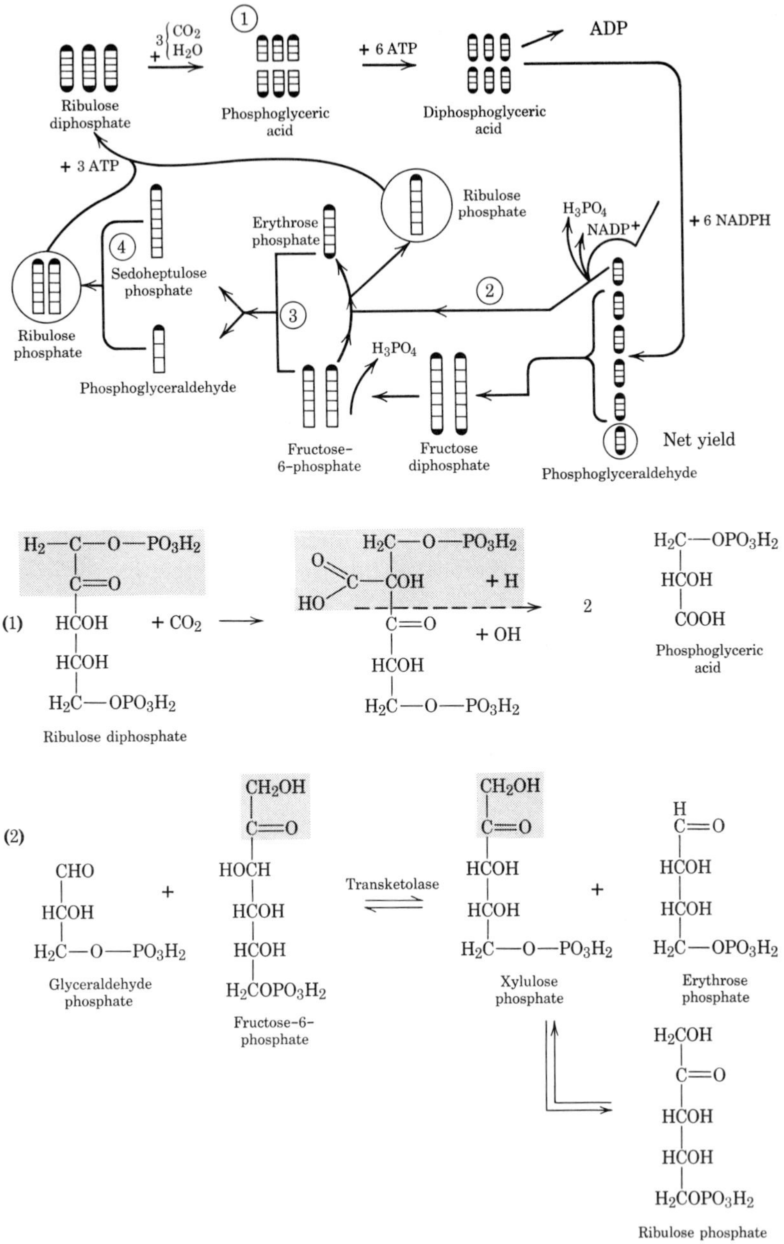

Fig. 9-9. Schematic biochemical reactions in photosynthesis. The solid "caps" represent phosphate. Principal reactions are given by equations (1) through (6).

(3)

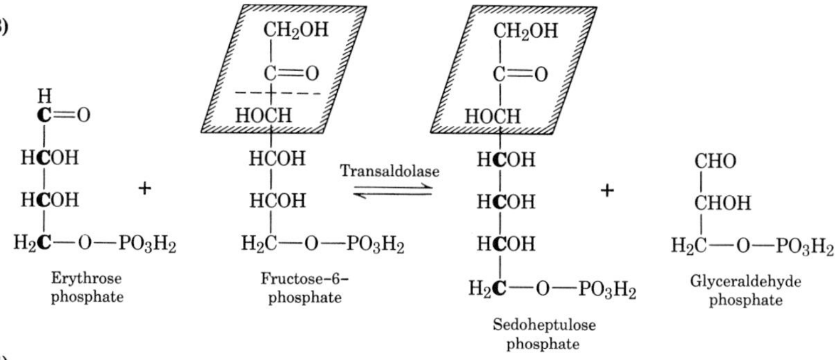

(4)

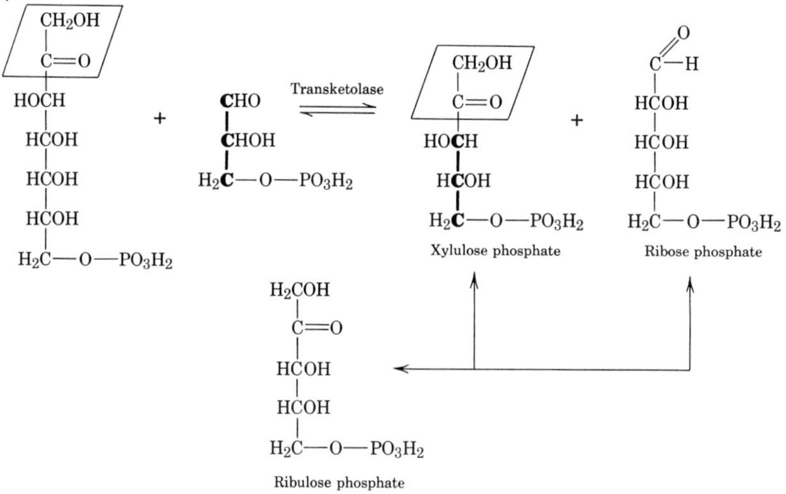

(5) Diphosphoglyceric acid $\xrightarrow[\text{glycolysis}]{\text{Reverse}}$ Fructose diphosphate

(6) Net reaction:

$$3\,CO_2 + 3\,H_2O + 9\,ATP + 6\,NADPH \longrightarrow \text{Phosphoglyceraldehyde} + 9\,ADP + 3\,NADP^+ + H^+$$

sequence is due primarily to the studies of a group of biochemists headed by Calvin. Using radioactive isotopes, they initially established that the photosynthetic product first accumulating was phosphoglyceric acid, a familiar intermediate in glycolysis. The source of phosphoglyceric acid was later established—the splitting of a six-carbon compound formed by the attachment of CO_2 to the compound ribulose diphosphate. Just as oxaloacetate serves to carry acetate groups through the Krebs' cycle, so does ribulose diphosphate serve to build CO_2 into carbohydrate. The reaction system is complex because ribulose diphosphate must be continuously regenerated. Phosphoglyceric acid is converted to carbohydrate by a reversal of the glycolytic pathway. All the phosphoglyceric acid cannot, however, be thus converted. Some of this primary chemical compound must be used to regenerate ribulose diphosphate. The net result of all these reactions is that 18 moles of ATP and 12 moles of NADPH must be expended to synthesize 1 molecule of fructose diphosphate. This is the equivalent of 54 moles of ATP. The complete oxidation of fructose diphosphate yields only 40 moles of ATP, and this serves to illustrate a thermodynamic characteristic that the chemical energy necessary to synthesize carbohydrate is greater than the energy yield from its degradation.

SUMMARY

This discussion of photosynthesis brings to a close the survey of energy acquisition by cells. Despite the complexity of enzyme systems essential to the acquisition, the fundamental principles are very few. In the formation of living substance cells must counter two thermodynamic trends—oxidation and hydration. The one is due to the abundance of oxygen, the other to the abundance of water. To overcome these trends living systems have evolved two chemical configurations—the reduced nicotinamide group and the pyrophosphate bond. With these, cells can build all the organic molecules of life; the exceptions to this are incidental to the general principle. All cells can form these organic compounds given a source of energy, and for most cells this source is carbohydrate. Carbohydrate, however, is not given to the living world, but is made by it. And the capacity to synthesize this fuel of life resides in a limited number of cells, those with photosynthetic pigments. Different though the processes of synthesis and degradation may appear to be, their ultimate energetic transactions rest on one and the same mechanism—the capacity of porphyrin molecules to catalyze the transfer of electrons. Out of the catalytic capacity of enzymes grow the diverse organic molecules of life; out of the special catalytic capacity of porphyrin enzymes grow the two forms of chemical energy which make the diversity possible.

10

The Control of Diffusion

Forces of diffusion promote disorder in aqueous systems. Local differences in solute concentration cannot be maintained without intervening barriers. If cells were not equipped to control diffusion, their distinctive internal environment could not exist and they would not function. Just as cells must overcome the forces of oxidation and hydration to effect molecular syntheses, so must they overcome the randomizing effects of the forces of diffusion to control molecular distribution. Cells absorb nutrients against concentration gradients, they distribute nutrients intracellularly to meet localized metabolic needs, they maintain ionic concentrations both above and below those in the external environment, and they secrete compounds which do not ordinarily penetrate their limiting membranes. In all such activities energy must somehow be expended, either to counteract the forces of free diffusion or to overcome the structural barriers which cells possess to prevent such diffusion.

The full span of the problem cannot be considered under one heading. For cells not only limit free diffusion, they also facilitate diffusion in certain directions. These contradictory requirements are not met by separate mechanisms; both operate through the agency of membranes. The broad purpose of this chapter is therefore to discuss the ways in which energy is used to control the passage of materials across cellular membranes.

MEMBRANES AS BARRIERS

Membranes are films of molecules, one or several molecular layers in thickness. The more closely the molecules are packed in a layer, the more effective a barrier they constitute to the passage of other molecules, unless these molecules can interact with the molecules of the film and penetrate by displacement. Although films may be liquid, for example, an oil interface between two aqueous solutions, in cells they are closer to being solids than any other state of matter. The insoluble films of cells, more properly called "membranes," are the structures primarily responsible for minimizing random diffusion (see Fig. 10-1). No living system is known which lacks membranes and yet functions in an inanimate environment. The only membraneless organisms are some viruses, and these grow only within the confines of a living cell.

The evidence that cellular membranes can, by virtue of their structure, prevent random diffusion is impressive. Examples on this point have already been given in Part I, especially in relation to osmotic behavior. Alongside these examples a second, occasionally overlooked aspect of membrane function should be emphasized—the capacity of membranes to segregate solutions within the cell. This point too has been discussed in connection with mitochondria and chloroplasts. Both subcellular organelles are bounded by limiting membranes, and injury or undue expansion of these membranes is accompanied by a loss of various soluble components from the interior of these structures. Vacuoles are still another example of membrane-bounded structures which segregate solutes. Especially in plants vacuoles may contain toxic substances which, if released, would instantly kill the cell. The vacuoles of succulent plants store high concentrations of organic acids, and if the cells of these plants are broken the acid released frequently denatures many of the cytoplasmic proteins. To obviate denaturation, investigators frequently expose succulent tissues to ammonia in order to neutralize the acids prior to breaking the cells.

Associated with the function of membranes in segregating solutes is their function in controlling the movement of water. Not only does the ionic composition of the cell interior differ from that of its environment, but so does its total concentration of solutes. In aquatic organisms such concentration differences constitute a major physiological problem. Fresh water has an osmotic pressure lower than that of most, if not all, cells; the reverse is true for sea water. If membranes

Fig. 10-1. Surface of a human red blood cell showing the "unit membrane." Compare this with diagram in Fig. 4-12. This is a beautiful illustration of a smooth limiting membrane. Its dimensions are consistent with the requirements for a bimolecular leaflet of phospholipids bordered by protein. Whether, in fact, proteins of the membrane interact with the polar ends of phospholipids still remains an open question. (Electron micrograph by J. David Robertson.)

were acting solely as semipermeable barriers, and were unaided by other mechanisms, freshwater organisms would become excessively hydrated and marine organisms would become dehydrated. In plant cells this difficulty is overcome by the combination of the rigid cell wall and the vacuole. The wall limits cell expansion and thus prevents excessive uptake of water; the vacuole traps both excess water and solutes so that the cytoplasmic components remain in a relatively constant aqueous environment. Many unicellular organisms, however, have neither a rigid wall nor a large vacuole. The most impressive example is that of amoebae, which, despite their near-fluid texture, neither swell nor appear to be physiologically disturbed when placed in distilled water. Assuming that all cellular membranes are more or less permeable to water, and evidence to the contrary has not been found, we must infer that organisms like amoebae have a mechanism to remove excess water. The "contractile" vacuole of protozoa in contrast to the "fixed" vacuole of plant cells, appears to be the mechanism. Just how such vacuoles pump water out of cells is still not fully understood.

Thus, superimposed on the property of cell membranes as physical barriers to diffusion, which affects the *rate* of solute and even water movement across cell boundaries, are two other well-established properties: the capacity to alter the degree of resistance to molecular diffusion, a phenomenon most apparent in a contractile vacuole which periodically releases its contents; and the capacity to control the *direction* of molecular movement, a property which is characterized by the selective secretion or absorption of various substances. The mechanisms underlying these properties fall into two categories: those which relate to membranes as modifiable structural units, and those which relate to the internal molecular organization of membranes.

MEMBRANES AS DYNAMIC STRUCTURES

Structure is usually taken to mean a fixed arrangement of parts. This meaning was certainly implied in discussing the role of membrane systems in respiration and photosynthesis. The dynamic qualities of metabolism were associated with the movements and transformations of relatively small substrate molecules; the catalytic machinery—the electron transport system, for example—was pictured as a fixed membranous arrangement of enzymes. An impression of fixity is also given by electron micrographs. The regularity of membrane patterns and the similarity of these patterns from cell to cell reinforce the view of membrane function, discussed in Part I, as a static diffusion barrier permitting a selective passage of molecules.

That certain membrane systems of cells do in fact function as more or less fixed structures is probable. Of the various types shown in Fig. 10-2 the mitochondrial and chloroplast membrane systems clearly fall into that category. But just as clearly other membrane systems do not. Until the electron microscope was used to study membrane dynamics, the evidence for membrane plasticity was indirect, for membranes cannot be seen with the light microscope. Yet this evidence was nevertheless convincing. One of the most prominent properties of membranes is their capacity for self-repair. If a microneedle is used

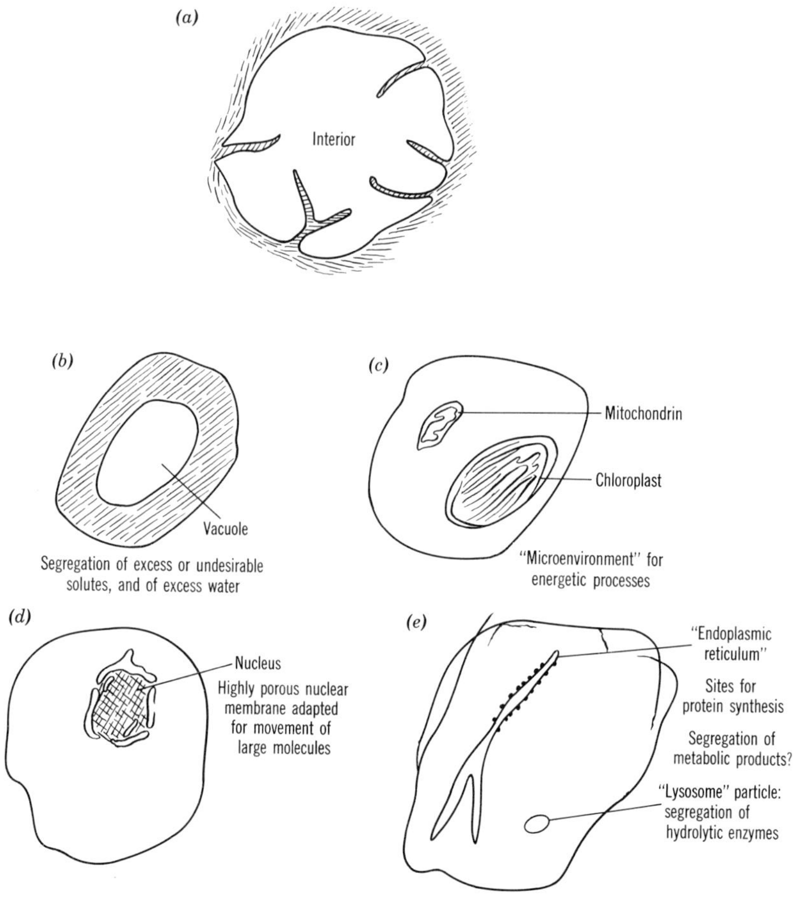

Fig. 10-2. Location of membranes in cells and their relations to physiological functions. *(a)* Outer limiting membrane; effects exchanges with external environment. *(b, c, d, e)* Internal membranes. The pores in nuclear membranes *(d)* have been commonly regarded as structures facilitating an exchange of substances between nucleus and cytoplasm. Nevertheless, the physical nature of the pores remains in doubt.

to puncture the surface of a living cell, the puncture is quickly sealed unless the damage is extensive enough to kill the cell. Many unicellular organisms may be cut in two or more parts, each of which becomes enclosed by a peripheral membrane possessing the same permeability properties as that of the original cell. A foreign nucleus may be inserted into a cell without damage to viability. The bursts of contractile vacuoles suggest a cycle of rupture and repair. The extensive membrane system of the cytoplasm—the endoplasmic reticulum—fragments when a cell is broken, but the ends of these fragments fuse giving rise to spheres or "vesicles" of submicroscopic size commonly called "microsomes."

Is such plasticity an incidental property of membranes, or does it serve some physiological function? The answer to this question is unequivocal. Cells utilize this property of membranes for bulk transport of solutions and insoluble particles either inwardly or outwardly. The principle of such utilization may be simply illustrated as in Figs. 10-3 and 10-4. The important feature of this mechanism is the irrele-

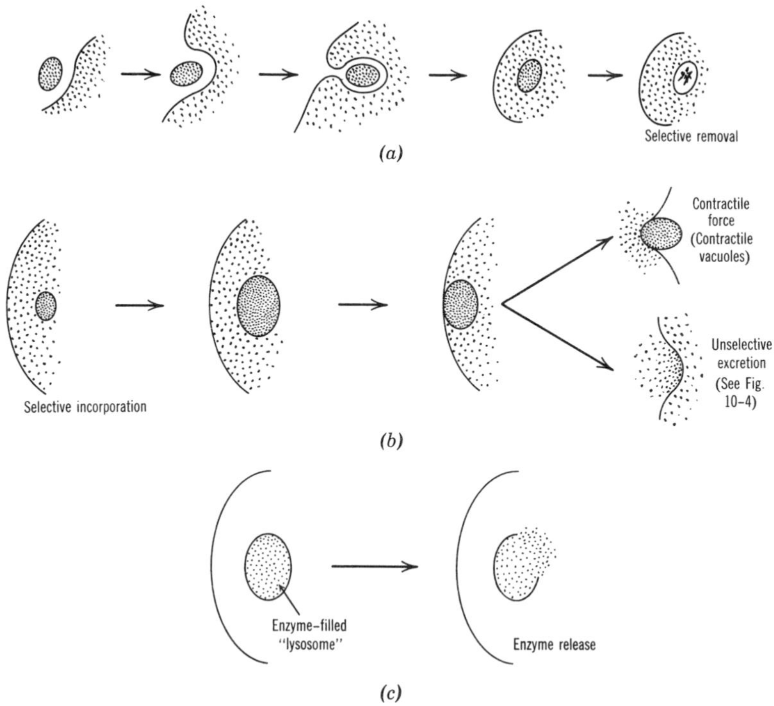

Fig. 10-3. Schematic representations of the adaptation of membrane plasticity to transport in cells.

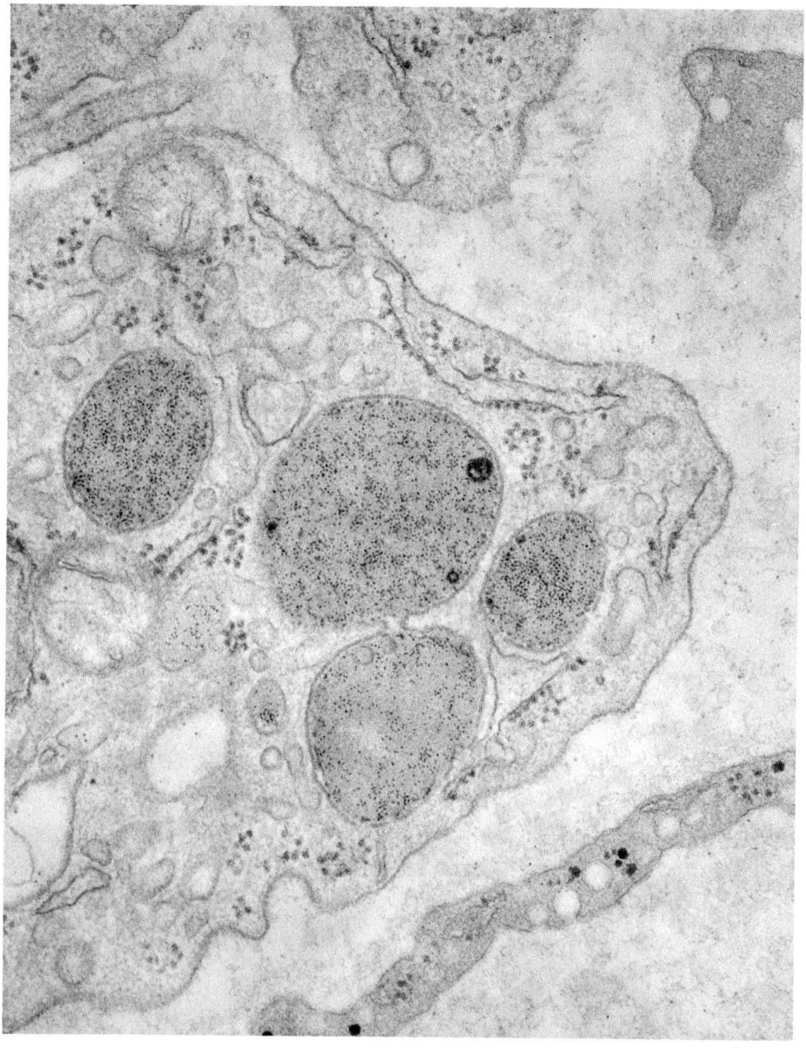

Fig. 10-4. Section through a macrophage removed from the diaphragm of a rat which had previously been injected with ferritin. The "macrophage" belongs to the amoeboid group of cells which are widely distributed in vertebrate bodies and function to remove foreign particles or debris formed as a result of tissue injury. Ferritin is a protein containing numerous atoms and iron and normally functions as a storage molecule for iron. The presence of iron makes the molecule easily recognizable under the electron microscope. On injecting ferritin prepared from another source into the blood stream of a rat, the ferritin molecules are ingested by phagocytosis as illustrated in Fig. 10.3*a*. In this photograph the membrane-bounded vacuoles loaded with the dark ferritin granules may be easily seen. (Micrograph courtesy of George Palade.)

vance of the permeability properties of the membrane to the kind of material being transported. The membrane either encloses the material by fusing or exposes the material by rupture. The advantages of such a mechanism are offset by important limitations. The bulk transport is unselective. Ingestion can be selective only to the extent that the stimulus to form a vesicle may be relatively specific, but the material absorbed represents all that is present in the vicinity of vesicle formation. The selective removal of the contents of the vesicle once inside the cell must be mediated by a different mechanism, just as the building up of vacuolar contents (10-3*b*) must be effected by a mechanism distinct from that of final excretion.

Leaving aside the phases of selective removal or accumulation indicated in Fig. 10-3, we may regard the break-repair mechanism as a device that enables a cell to segregate and transport substances by means of a membrane to which these substances are impermeable. Energy is thus not expended directly to move substances across an otherwise impenetrable barrier, but is expended for a break-repair system which obviates the problem of impermeability. How much energy is expended in the process is unknown, but the value of the mechanism cannot be fully judged in terms of energy costs. This is made evident by three phenomena which will be briefly described.

Contractile Vacuoles

These vacuoles are frequently visualized by exposing protozoa to colored solutions and observing the periodic excretion of the absorbed fluid. Although such experiments suggest a function of removing waste or superfluous solutes, other lines of evidence point more strongly to water removal as the principal role of the contractile vacuole. Strains of some protozoa which have lost the capacity to form contractile vacuoles cannot survive in a hypotonic environment. Such organisms accumulate water and either burst or become internally disorganized. Osmotic force drives the water into the cell because the membrane behaves as a typical semipermeable barrier. To remove this water an equivalent force must be applied. No mechanism is known, nor could one be easily conceived, by which the water is forced back along the same route through which it entered. What is observed is a steady enlargement of one or more small vacuoles within the cell which eventually expel their contents to the cell exterior. The more dilute the surrounding medium of the organism, the more pronounced is the vacuolar enlargement. How water is segregated internally is not clear; possibly this occurs because of a localization of ionic solutes in the vacuole which draws water by osmotic force from the neighboring regions of the cytoplasm. Whether or not this is so, the final removal of water is a mechanical process requiring energy to effect contraction. Poisons which inhibit respiration also inhibit vacuolar functions. The immediate effect of vacuolar contraction on the cell membrane is also uncertain. A distention or rupture of the membrane would be equally effective in permitting the discharge of vacuolar contents. In either mechanism the plasticity, rather than the sieve-like property of the membrane, would be the essential factor. In "pinocytosis" (See Fig. 10-3) estimates have been made of the rate of membrane replacement; one such estimate indicates a replacement of 25% of the plasma membrane every 5 minutes. For phagocytosis a value of 100% replacement every 5 minutes has been proposed.

Secretion of Enzymes

Transport of proteins across cell membranes is a phenomenon of limited occurence. All cells make their own characteristic complement of proteins, and no limiting cytoplasmic membrane is known which has pores large enough to permit the free passage of protein molecules. Such pores would have to be of the order of 100 Å diameter or greater and could be easily identified under the electron microscope. Moreover, if a porous membrane of this type did exist, the retention of soluble proteins, substrate molecules, and ions would be virtually impossible. The only large membrane pores which have been identified are those associated with nuclear membranes (Fig. 10-2) and their presence presumably assures a facile passage of macromolecules from nucleus to cytoplasm. The loss of some proteins from nuclei isolated in sucrose media is in fact due to the porous structure of the nuclear membrane.

Even though cells are so organized as to form all their proteins endogenously, situations do exist in which the transport of proteins across cell boundaries is essential to the survival of the organism. Many microorganisms secrete enzymes which hydrolyze organic macromolecules into products small enough to be absorbed as nutrients. Cellulose, lignin, and proteins are among the targets of such organisms. Animals also require mechanisms for protein transport. Liver cells synthesize serum albumin, which must be transferred into blood vessels. Pancreatic cells synthesize a variety of enzymes which hydrolyze proteins, nucleic acids, and polysaccharides, and these enzymes must be moved from their site of origin into the alimentary tract. Insectivorous plants also possess cells which secrete proteins to digest their prey.

And, on the negative side of the survival ledger for many organisms, is the ability of viruses to penetrate host membranes without immediately destroying the cell.

Whether other types of macromolecular transport occur between cells is unknown (except for the relatively small polypeptide hormones of animals), but the possibility cannot be excluded. However, even if the instances cited were the only ones, the mechanism for effecting these would be of widespread importance. There are, in fact, two fundamental issues of diffusion associated with secretory phenomena. The first of these is one of segregating the enzymes intracellularly so that they do not attack native cell components. Enzymes attacking proteins (proteases) or nucleic acids (nucleases), if introduced into a cell, quickly produce abnormalities. The second issue is one of transporting the segregated enzymes across the limiting cell membrane.

Although the mechanisms underlying secretion are not fully understood, a few generalizations can be made which at least point to the principal agents in the process. The most intensively studied phenomenon has been the synthesis and secretion of pancreatic enzymes. These studies lead to the conclusion that the membrane system of the cytoplasm is the structure which effects enzyme segregation. The sequence of events from enzyme synthesis to secretion is diagrammatically represented in Figs. 10-5 and 10-6. As in all other cell types, the enzymes are synthesized at the ribosomes which are attached to the endoplasmic reticulum. These enzymes do not diffuse into the matrix of the cell, but are somehow transferred into the space bounded by the membranes of the reticulum. The enzymes accumulate within this space, and shortly after initiation of synthesis small granules appear which are aggregates of the enzymes. The granules grow as synthesis continues. As growth proceeds, the granules move along the membrane-bounded cavities toward the end of the cell which faces one of the pancreatic ducts. Secretion itself is under a separate control mechanism. Starved animals retain the granules within the pancreatic cells, but feeding triggers their release. The mechanism of such triggering is also poorly understood, but the ultimate effect is what has previously been described as the break-repair process. The granules are ejected across the cell surface, but despite their relatively large size the surface is left intact.

Capillary Transport

A more striking example of the break-repair process is provided by the tiny blood capillaries of mammals which penetrate the tissues

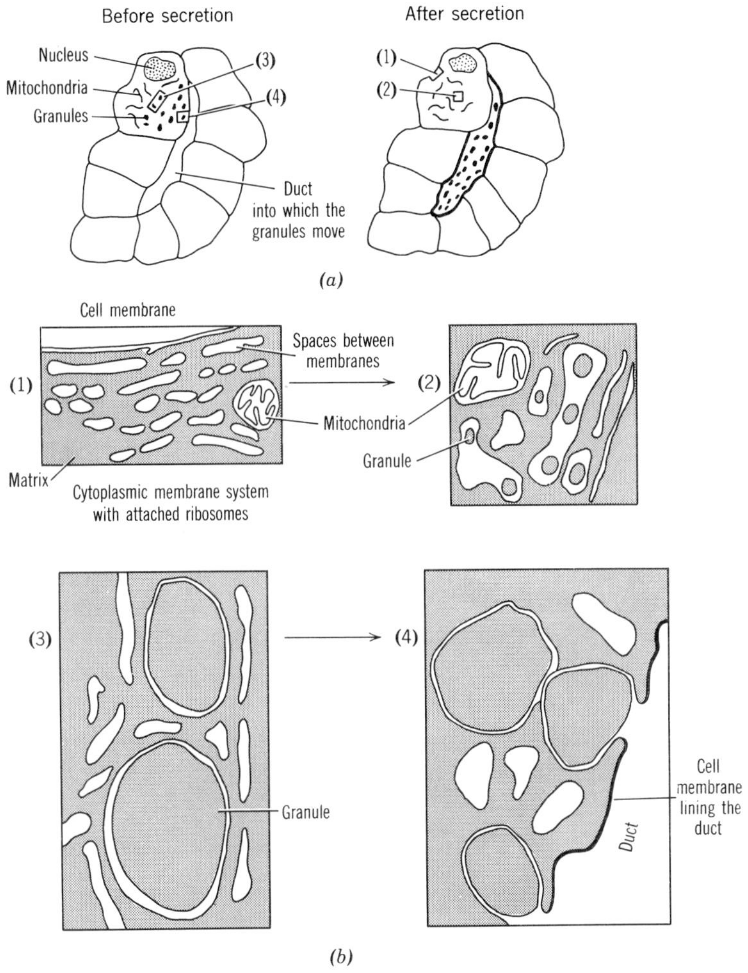

Fig. 10-5. Membrane behavior in a secretory process. *(a)* Two stages of cell group in pancreas as seen with the light microscope; details for one cell. *(b)* Segments of cells at different stages in granule formation as seen under the electron microscope: (1) Beginning of process. Enzyme is formed in ribosomes. (2) Enzymes presumably move into membrane spaces, there forming granules. (3) The granules grow and move toward end of cell adjacent to duct; note absence of ribosomes on membrane surrounding granule. (4) Fusion of membrane surrounding the granule with cell membrane; a pore is thus created through which the granule escapes.

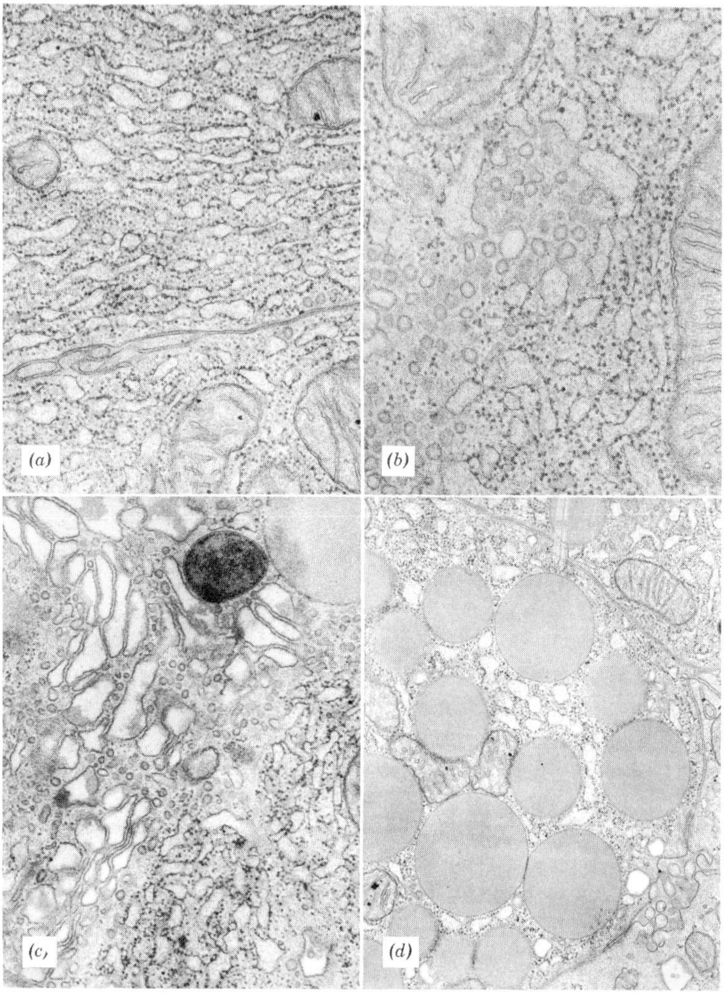

Fig. 10-6. Sections through different regions of a secretory ("exocrine") cell from the pancreas of a guinea pig. (Compare the photographs with diagrams in Fig. 10-5.) *(a)* The basal region. Numerous membranes associated with ribosomes are evident. This region of the cell is the most active with respect to synthesis of enzymes. *(b)* The transition region in which "smooth" and "rough" membranes are present. The "smooth" membranes are not associated with rough membranes and are more concentrated at the apical region of the cell where proteins are secreted. *(c)* The Golgi region. This type of region was long recognized by light microscopists who regarded it as a kind of network strongly stained by silver or osmium (see Chapter 1). Under the electron microscope, however, the "network" appears to be an association of smooth membranes and is found in most cells. It is in the vicinity of the Golgi region that the zymogen granules are channeled for secretion. *(d)* The apical region is filled with zymogen granules. Note the smooth membrane surrounding the granules. (Electron micrographs courtesy of George Palade.)

of every organ. Through the walls of these capillaries must pass not only the respiratory gases but also all those substances which serve as nutrients or as special physiological agents (e.g., hormones, serum albumin, globulins). Under the light microscope details of the wall structure cannot be discerned, for the walls are about 0.3 μ in diameter. Under the electron microscope, however, the walls are seen to consist of long flat cells, three of them being sufficient to form the circumference of a capillary. Despite their small size and unusual shape, these cells have the same subcellular components as other cells, but are distinguished by the profusion of membranes which fill the cytoplasm. The membranes are in the form of vesicles (Fig. 10-7), some

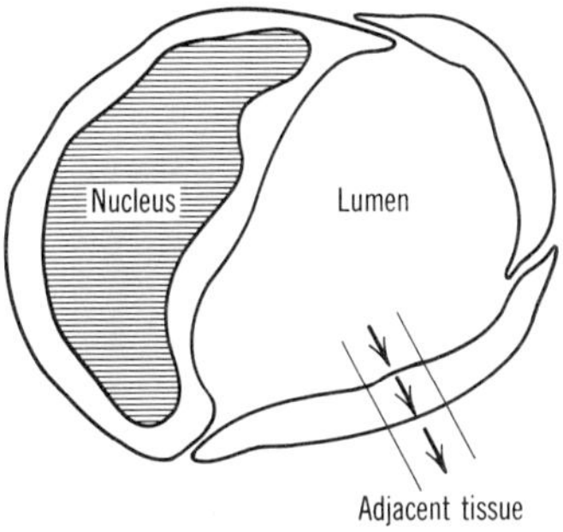

(a)

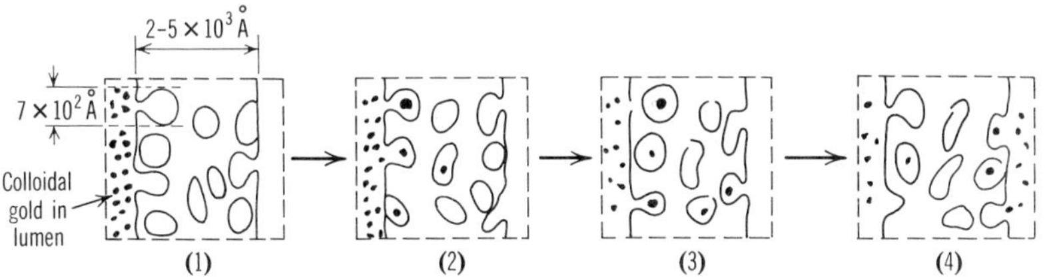

(b)

Fig. 10-7. Passage of foreign substance (colloidal gold) through cells of a blood capillary. *(a)* Cross section of a capillary. Three flat and elongated cells encircle the capillary. In the region of the nucleus, the cell is thicker and the "lumen" of capillary is reduced in size. Normally, substances circulating in blood vessels move in lumen and are transferred across the wall to the tissues. Cells must therefore be able to absorb and secrete materials. *(b)* Diagrammatic representation of events at different intervals following injection of collodial gold into blood. The sequence of events as shown by electron microscope is: (1) appearance of colloidal gold particles in lumen; (2) intake of particles at points where membrane appears cup-shaped; (3) presence of particles in cell by virtue of membrane being pinched off; (4) extrusion of particle to tissue side of capillary by fusion of internal membrane with external one. Note that the membranes which attach to the outer cell membrane are not associated with ribosomes, as were many membranes in the pancreatic cells.

of which are attached to the surface membrane, and the others lying free in the cytoplasm.

The behavior of the membranes has been ingeniously followed by injecting colloidal gold into the bloodstream. Particles of colloidal gold are small enough to be transported, but they are easily identified under the electron microscope because of the high electron-scattering power of gold atoms. The observations made by this procedure are illustrated in Fig. 10-5. The particles are ingested by vesicles located at the membrane on the inside of the capillary. The vesicles break loose from the surface, move across the cytoplasm, and release the particle through the membrane adjacent to the body tissues.

SUMMARY

At all levels of evolutionary organization—from microorganisms to mammals—a need clearly exists for bulk transport. Food ingestion, water or solution excretion by contractile vacuoles, enzyme secretion are examples of this need. The quality of membranes as molecular sieves appears to be inadequate for this function; the substances moved may be too high in molecular weight or they may have to be kept segregated from the cytoplasmic matrix and periodically discharged in bulk. To fulfill this need cells have utilized the plastic quality of membranes—the capacity to break and fuse or temporarily distend without otherwise disturbing the structure. The limitations of this mechanism have already been briefly discussed. However advantageous it may be to the function of many cell types, it cannot displace a universal and fundamental need for the selective control of solute transport. How such transport is effected is the subject of the next section.

MOLECULAR MECHANISMS IN MEMBRANE FUNCTION

The passage of molecules or ions across cell boundaries is governed by two different factors: the diffusion barrier and the force sustaining diffusion. The first is indifferent to direction of movement; the second is independent of resistance to passage. Cell membranes may be almost impermeable or entirely permeable to a particular molecule, but neither extreme determines the direction of molecular movement. Commonly, a directional movement of molecules is caused by a difference in concentrations; the greater the concentration difference, the greater the force causing the molecules to diffuse toward the more dilute phase. The resistance which a membrane offers to the passage of molecules determines the magnitude of the force necessary to move the molecules at a given rate, but this force, which can usually be expressed in terms of concentration differences, need not originate in the membrane. However closely related the overcoming of barriers to diffusion and the provision of a force to generate diffusion may appear to be, the basic distinction must be kept in mind.

Although we have yet to understand the mechanism by which molecular passage across cell membranes is effected, several factors relating to the structural characteristics of cell membranes and to the generation of forces for molecular movement are known. No single scheme can be proposed which would take all these factors into account and explain diffusion control. Certain basic problems can, however, be defined, and these should help to clarify the phenomenon as a whole.

Perspective from "The background of Ideas." Much of the early work on cell permeability, which was discussed in Part I, can be classified as studies of cells under extreme conditions. The techniques of plasmolysis and deplasmolysis usually required exposing cells to abnormally high concentrations of solutes. Cells were overwhelmed with solutes, and the responses they showed almost entirely reflected the physical rather than the physiological properties of their membranes. The normal uptake of sucrose by a plant cell, for example, could not be observed because such uptake would be negligible compared to the rapid loss of water following exposure of the cells to osmotic concentrations several times that of the cell interior. Whereas a 0.001*M* solution of sucrose or glucose would be ample for metabolic needs, cells were commonly exposed to concentrations ranging from 0.2*M* to 1.0*M* or higher.

These "unphysiological" studies were nevertheless most revealing, for they laid bare certain basic physical characteristics of cell membranes. The "leaky" nature of membranes was established, and with it the recognition that such membranes had pore-like openings which permitted the passage of small molecules but not of large ones. These studies also established the presence of charged components in the membrane and thereby pointed to the penetration of ions as a major problem in cell function. Finally, the presence of a lipid layer in the membrane was correctly inferred from the penetration characteristics of fat-soluble substances.

Electron microscopy has not made evident the postulated pores. Considering the degree of resolution which the electron microscope affords, we cannot consider this negative result to be significant. Pores with radii of the order of 5–10 A would not be detected, and these are the magnitudes which would account for the classical observations. That all cell membranes are to some degree leaky is commonly accepted as a fact. Membranes are not at all uniform in this respect, and some types, those surrounding plant vacuoles, for example, have extremely retentive properties. Although the discussion to follow will not consider this aspect of membrane behavior, we should bear in mind that perfect membranes do not exist in nature, and that all cells must expend a variable amount of energy to compensate for their imperfect boundaries. This imperfection may indeed be an unavoidable consequence of their organization. For membranes function not only to retain solutes, but also to absorb them, and a perfect barrier to the flow of nutrients would be physiologically disastrous.

Despite the apparent accuracy of early models of membrane structure based on proteins and phospholipids as the principal components, modern studies have not yet succeeded in clarifying how the two components interact functionally. This important question will have to be neglected—not for want of speculation, but for want of fact. Molecular diagrams of membranes (Fig. 4-12) generally show the phospholipids as unbroken films. Whether polar solutes move through these films or whether they move through pores in the film is unknown. And if the answer to this question is unknown, so must be that to the question relating to the general mechanism of solute penetration. Of all the structures in a cell, membranes have provided the most questions and the fewest answers, a stricture which is as true for electron transport as it is for solute movement.

Electron microscope studies have made their greatest contribution in revealing the morphology of membrane systems. Before the intracellular disposition of membranes was visualized, it was presumed that cells were bounded by a more or less smooth peripheral membrane. The perimeter of the cell was regarded as the boundary between the internal and external environment. The incorrectness of this picture is now recognized. The outer membranes of cells have finger-like invaginations which may extend deeply into the cell interior. With improvements in methods of fixation, evidence is accumulating for a continuity between many of the spaces bounded by the membranes of the endoplasmic reticulum with the cell exterior. Even the central portions of a cell may be in close contact with the outer environment. Although diagrams of solute movement into cells generally locate the movement across the cell periphery, such diagrams should not be interpreted literally.

Possible Mechanisms in Diffusion Control

The problem of cellular control of diffusion can be expressed in two questions. In what ways may molecules or ions pass through a barrier-like membrane? In what ways may force be generated to cause a movement of solutes in the desired direction? Four schemes of molecular transport are illustrated in Fig. 10-8, and together they cover most mechanisms which have been proposed. With respect to the first question each of the schemes in the figure is based on one of two alternatives. Either the membrane is analogous to a plate with mechanical pores (*a* and *c*), or the pores in the membrane are of secondary importance and transfer occurs by a combination between the entering solute and a component of the membrane called a "carrier." The alternatives are not mutually exclusive. Since no carrier molecule has been identified, we may picture it as large as a protein molecule or as small as a coenzyme. The main difference between the two alternatives is that the pore system would discriminate according to size, whereas the carrier system would discriminate according to chemical affinity. The one selects by purely physical parameters, the other by chemical ones. The carrier hypothesis nevertheless introduces a physical problem, for its presumed mobility within the membrane must be explained. This problem has not been extensively considered, perhaps to avoid adding speculation to speculation. One supposition is that the carrier-solute complex is lipid soluble and thus facilitates the passage of hydrophilic substances through the lipid layer.

The second question concerning the generation of directional diffusion forces has been answered in three ways, although the mechanisms illustrated in Fig. 10-8 would have to be modified in certain details to account for the movement of ionic substances. All three mechanisms are based on the operation of concentration gradients. The simplest of these, shown in (*a*), generates inward movement by a high extracellular concentration of solute. Such a force may operate with a pore or carrier mechanism. For the carrier mechanism one must postulate that formation of the carrier-solute complex is equally reversible on either side of the membrane. The limitation of a mechanism dependent upon relatively high external solute concentration is that it leads to an equilibrium condition. This limitation can however be overcome by metabolic removal of the solute (*c*). Polymerization (e.g., glucose to polysaccharide) or chemical modification (e.g., glucose to glucose phosphate) would remove the solute from the inner side of the mem-

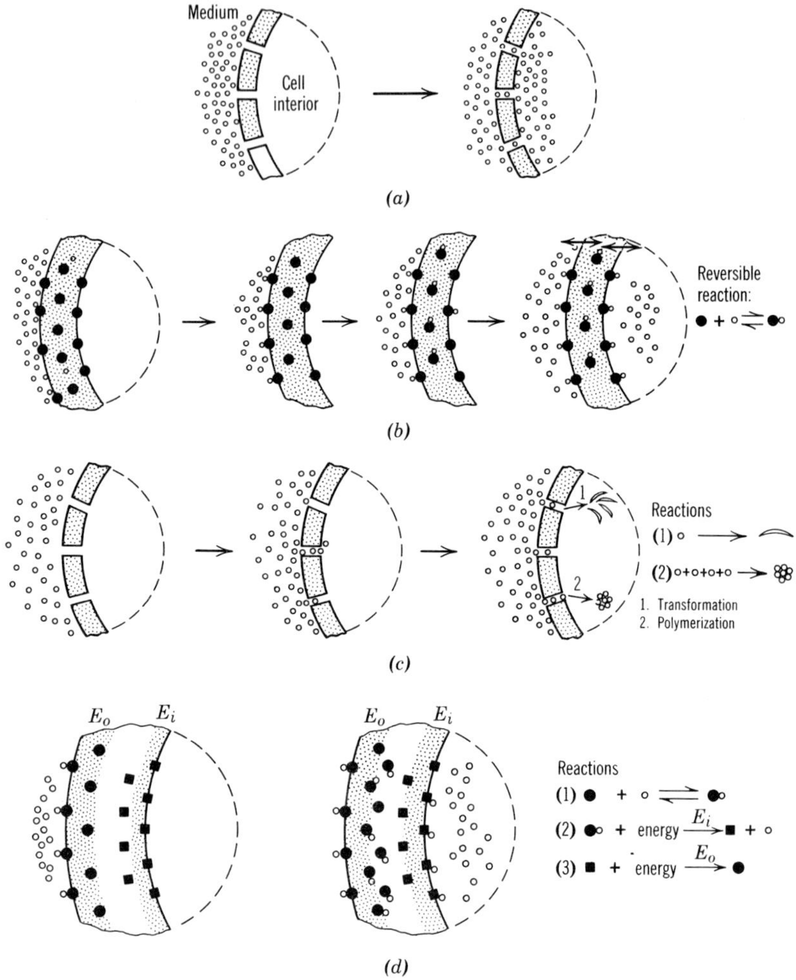

Fig. 10-8. Patterns of molecular passage through membranes. *(a)* Free diffusion through porous membrane. Process continues until external and internal concentrations are equalized. *(b)* Diffusion via carrier molecule. Force of diffusion as in *(a)*, but if carrier combines with specific compounds, entry through membrane is not random. *(c)* Free or carrier-mediated diffusion, but force generated by the action of cytoplasmic enzymes in transforming incoming molecule. Here cells may appear to be absorbing a substance against a concentration gradient. *(d)* Carrier-mediated diffusion with force generated by asymmetric positioning of enzymes within membrane. Here the molecule appears unchanged within the cell. The intracellular molecules do not recognize the concentration gradient because they cannot exchange directly with the molecules in the external solutions. The effectiveness of this type of mechanism depends on the ability of cells to utilize energy in order to effect a transformation of the carrier. The asymmetric distribution of enzymes localizes the changes and thus creates a favorable diffusion gradient for entry.

brane, and a concentration gradient would be maintained as long as metabolism continues. This mechanism may be regarded as an indirect expenditure of energy to generate a concentration gradient. Whether called direct or indirect, the fact is that this situation commonly prevails in cells. One of the greatest difficulties in studying the forces governing the penetration of metabolizable substrates is to separate experimentally the diffusion forces generated by the utilization of substrate from those generated by other presumed mechanisms. A very common, though not universal, characteristic of cells is their rapid transformation of otherwise unused small molecules into polymers. Cells do not maintain large reserves of low molecular weight precursors in their native form. Even phosphate ions are usually stored as polyphosphates. Thus although cells may appear to be concentrating solutes against concentration gradients, they are actually moving the solutes along a favorable gradient. Except for the special case of inorganic ions, we may well wonder whether rapidly metabolizing cells, such as those of bacteria, require any type of selective barrier to the movement of nutrient substances.

The most difficult mechanism to explain is one in which the solute is known to be present in the same form on either side of the membrane and yet moves in the direction of the higher concentration. This is the rule for many inorganic ions, but in some instances this is known to be true for sugars and amino acids. Since solute molecules will not diffuse against a concentration gradient, the membrane must functionally isolate the two solutions. The isolation is unlikely to be complete because of membrane leakiness, but it can be effective if the rate of entry of solutes against the concentration gradient exceeds the rate of leakage. Various schemes have been proposed to account for the role of the membrane in moving solutes "uphill," but they are basically similar in two respects. The first of these is the postulate that the solute must be transported by a carrier (the "leaky" process excluded); the second is the concept of what may be called the "asymmetric" membrane. If we assume, as shown in 10-8*d*, that two different enzymes are located at either edge of the membrane and that these enzymes catalyze different reactions involving either the carrier or the carrier-solute complex, then we may easily devise a scheme whereby the carrier-solute complex moves *with* a gradient in the membrane. By making these reactions energy-requiring, and hence spontaneously irreversible, solute may be released to the cytoplasm without any tendency to return except via the leaky outlets in the membrane. Specific evidence favoring this mechanism will be considered at the end of the chapter.

Pores vs. Carriers: Experimental Evidence

The introduction of isotopically labeled compounds into permeability studies made possible experiments which were beyond the reach of early investigators. Radioactive compounds can be detected in extremely small amounts, and their passage across cell membranes can therefore be followed over very brief intervals of time. This permits measurements of the penetration of a metabolizable compound before appreciable conversion of the compound is effected by intracellular metabolism. Furthermore, if such conversion does occur, it can be easily detected even if the cell already contains large amounts of the same compound. Whereas previously the movement of a particular substance could be measured only if it resulted in an appreciable change of intracellular or extracellular concentration, it can now be measured in the virtual absence of any concentration change. The use of isotopically labeled compounds in permeability studies consequently revealed a new parameter of cell behavior—exchange reactions. Substances can move back and forth across cell membranes unaccompanied by changes in concentration. Such movement is independent of the concentration gradient because it is not directional. How molecules of a particular chemical substance can interchange across a cell membrane with apparent indifference to the concentration gradient cannot be explained in terms of a pore mechanism. The concept of some form of carrier becomes indispensable to the explanation, but even so, the process is not self-evident despite the model shown in Fig. 10-8*b*.

Three characteristics of cell permeability point to the operation of a carrier mechanism: specificity of solute movement, kinetics of solute uptake, and nondirectional solute exchange. In certain types of cells and under certain conditions, the absorption of a particular substance may be measured whether or not that substance is being utilized for metabolic purposes. By eliminating the factor of metabolic utilization, we eliminate the factor of concentration gradients which would favor the uptake of the substance preferentially utilized. The clearest situation of this kind occurs in some bacterial mutants. A variety of mutations have been found which affect the uptake but not the metabolism of a particular substance. Certain mutants, for example, lack the capacity to absorb specific sugars (lactose, arabinose) and yet have all the enzymes essential to their metabolic utilization. Each mutant type lacks the ability to absorb one specific sugar; the mutation does not affect transport in general. Thus a mutant bacterium which grows very poorly on lactose because of a failure to absorb the sugar grows normally on glucose. Some lactose does penetrate due to membrane leakiness, and the amount of penetration can be increased by exposing the cells to relatively high external concentrations of the substrate. However, even under such conditions the rate of penetration is far below that achieved in the presence of the specific absorbing mechanism.

Specific absorption may be demonstrated more directly without using growth as a criterion of penetration. The bacterial strains commonly studied do not normally utilize galactose, and the various mechanisms essential to lactose utilization are formed only after the bacteria have been exposed to this substance. Mechanisms thus formed are said to be "induced" (see Chapter 18). Bacteria may be stripped of their walls by one of several techniques to yield "protoplasts" which are sensitive to the osmotic pressure of the medium because of the absence of mechanical protection against swelling. If protoplasts of noninduced and induced bacteria are compared in a solution of galactose, we observe the type of behavior illustrated in Fig. 10-9*a*. The induced protoplasts gradually swell because their absorption of galactose increases their internal osmotic pressure and causes water to be drawn in; the noninduced protoplasts maintain a constant volume over the time period tested; internal and external osmotic pressures are equilibrated on initial exposure, and no change occurs thereafter.

The operation of mechanisms which facilitate the passage of specific substances across cell membranes are also demonstrable in other organisms, although not as unambiguously as in bacteria. Two types of behavior are especially suggestive of carrier-mediated transport: substrate competition and responses to hormonal stimulation. Various substances exhibit a mutual antagonism to transport across cell membranes. 2-Deoxyglucose antagonizes glucose entry (Fig. 10-9*b*); uric acid competes with hypoxanthine; the basic amino acids compete with one another, as well as do the neutral and acidic amino acids. None of these, or other instances of competition, can be satisfactorily explained in terms of pores. If solute penetration depended on a mechanical passage through pores, antagonism between two substances should not lead to any reduction in the total number of molecules transported but only in the relative amounts of each substance transported. Theoretically, a similar result should be obtained if both molecules used the same carrier and the number of carrier molecules was limiting. In many cases, however, (as in Fig. 10-9*b*) antagonism leads to a reduction in the total number of molecules absorbed, suggesting that one carrier-substrate complex is transported more rapidly than

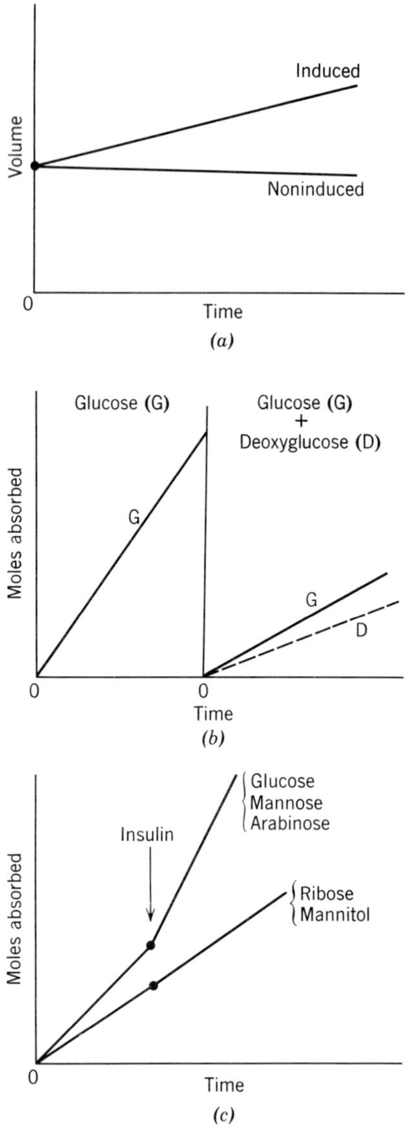

Fig. 10-9. Selective absorption of organic molecules. The graphs in this figure are approximate representations of experimental data obtained by different investigators. Each graph illustrates a particular aspect of membrane behavior, and all the graphs are consistent with the concept of specific molecular carriers. (a) *E. coli* protoplasts exposed to a galactose medium. Protoplasts prepared from induced forms have the capacity to absorb galactose. As the intracellular galactose concentration increases, so does the internal osmotic pressure, and this causes a swelling as a result of water absorption. Such a difference would not be observed if glucose replaced

the other. This type of observation combined with the evidence that a hormone like insulin specifically accelerates the transport of certain sugars (Fig. 10-9*c*) strongly supports the view that cells have an elaborate mechanism for specifically controlling the diffusion of substances. The specificity of the mechanism is not as high as that which prevails in enzymatic reactions. Glucose, mannose, and arabinose, for example, all respond to insulin even though glucose is the respiratory substrate; ribose penetration is unaffected by insulin, and yet this sugar is a major contituent of RNA.

The apparent lack of a highly specific chemical affinity in carrier molecules has been used to explain other anomalies of solute transport. The "exchange reactions" referred to earlier are characterized by the fact that appreciable quantities of a substance may penetrate a cell by exchanging with molecules of the same substance which are present in the interior. Such exchange is probably universal. Cells which do not show appreciable uptake of solutes (e.g., sugars, phosphate) by chemical measurements of net uptake nevertheless show an extremely rapid penetration of isotopically labeled compounds. Such penetration is especially evident in cells which already have a high internal concentration of the substance tested. If the internal concentration is low the rate of exchange is lower than if it is high. This relationship is contrary to expectations based on movement generated by concentration gradients and is, in fact, independent of gradients. The relationship may however be explained if we assume that substrate-carrier complexes move more rapidly across the membrane than free carriers. This interpretation is supported by other observations of cell behavior. Exchange is a more rapid process than accumula-

---

galactose. The capacity to absorb galactose is inherited as a specific trait unrelated to other absorption mechanisms and independent of the enzymes metabolizing galactose. *(b)* Nerve cells exposed to a fixed concentration of glucose in the presence or absence of deoxyglucose absorb much more glucose if deoxyglucose is absent. The physical properties of the two substances are nearly identical, and if pores were their principal mode of entry through the membrane they should penetrate at equal rates. The suppression of glucose entry by deoxyglucose cannot be explained in terms of pores because the total number of both molecules which have penetrated is less than that of glucose alone. A simple explanation is that both substances compete for the same membrane carrier, but that the carrier-deoxyglucose complex is less rapidly transported. *(c)* The stimulated transport of specific sugars into muscle cells by the hormone insulin. The selective stimulation is not directly related to metabolic utilization. The experiment points to a specific transporting mechanism in the membrane which is influenced by insulin. It also indicates, as do other experiments, that the transporting component does not have the degree of specificity characteristic of enzymes.

tion, and in accumulation we would have to suppose that free carrier must be transported from the internal to the external face of the membrane. Moreover, the exchange rate may be increased not only by increasing the internal concentration of the substance being exchanged but also by loading a cell with a substance which is distinct from the one being exchanged but which appears to have an affinity for the same carrier. Various groups of substances have been shown to behave in this way, and one example is provided in Fig. 10-10.

The characteristics of solute passage through cell membranes may be summarized by comparing the two graphs in Fig. 10-11. One graph indicates the type of relationship expected between concentration gradient and solute movement if the membrane acted as a purely mechanical barrier. The other graph indicates the relationship to be expected between solute movement and external solute concentration if a carrier molecule were necessary for transport across the membrane. Experimental measurements of solute movement follow the relationship predicted from a carrier mechanism. In virtually all cases studied the rate of transport across a membrane is proportional to external concentration over a very limited range; beyond that range, the availability of carrier molecules becomes increasingly limiting until a maximum rate is reached, and further increases in external concentration have no detectable effect on rate of transport even though they have a marked effect on net accumulation. In summary, the classical picture of the porous membrane is correct but inadequate. Cells counter the randomizing effects of diffusion, not only by forming mechanical barriers to diffusion but also by investing membranes with a chemical configuration which facilitates the passage of specific types of molecules. How accurately the carrier mechanism accounts for the configuration is still undetermined, for if such carrier molecules are present they should be identifiable. That they have not yet been identified may be due either to technical difficulties or to a misjudgment of the chemical capacities of membranes. Whichever is true, an explanation must ultimately be found for a basic property of cells—the facilitated passage of specific solutes across membrane barriers.

The Generation of Diffusion Forces

Neither of the two mechanisms for solute passage through cell membranes can alone account for the directional flow of solutes. Yet the supplementary mechanisms which might be envisaged for effecting directional flow are not indiscriminantly applicable to either mode of passage. A membrane which behaves purely as a porous barrier can-

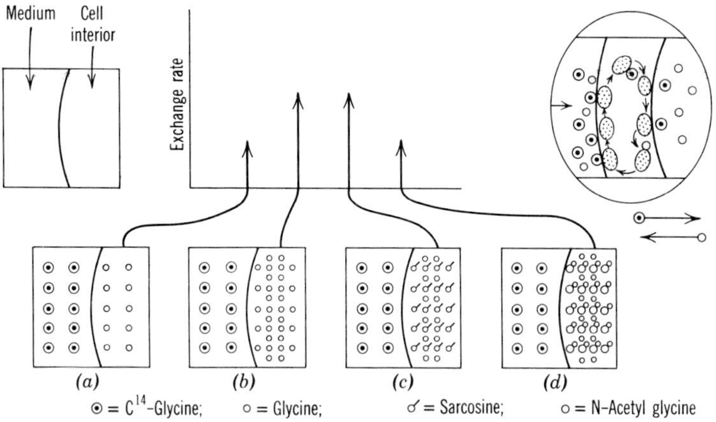

Fig. 10-10. Facilitated diffusion of an amino acid in tumor cells. In this type of experiment the movement of a substance is detected by exposing the cells to a radioactive form of the substance and determining the level of radioactivity in the cells after a given period of time. Changes in the intracellular concentration of the substance are irrelevant to the measurement. In many cases no detectible concentration change occurs despite the fact that a large proportion of the externally supplied radioactive form is found inside the cell. Such a result can only be explained by assuming that for each radioactive molecule which has penetrated, a nonradioactive form of the molecule has been released into the medium. Certain types of tumor cells (Ascites tumors) are particularly suited for studying amino acid exchange because they normally store amino acids. For purposes of comparison, an arbitrary exchange rate is shown for tumor cells with normal intracellular concentrations of glycine (*a*). If tumor cells are first exposed to a high concentration of nonradioactive glycine so as to increase their intracellular pool, and then exposed to C^{14}-glycine, exchange is more rapid. The apparent anomaly of molecules moving in faster against a higher concentration gradient is easiest to explain in terms of a carrier which moves rapidly when complexed. If the intracellular concentration of glycine is low, the probability of a carrier complexing with a glycine molecule is correspondingly low; as the internal concentration is increased so are the numbers of carrier-glycine complexes which shuttle back and forth across the membrane. A more striking result is shown in (*c*) where the cell has been loaded with sarcosine (methyl glycine). The exchange rate is as high as in (*b*) although C^{14}-glycine molecules are exchanging for sarcosine. The presumed carrier does not therefore discriminate between the two substances. If, on the other hand, the cell is loaded with N-acetyl glycine (*d*), the exchange rate remains the same as in the control cells. The carrier has no affinity for this compound and neither accelerates nor depresses the rate of exchange.

not play an active role in transporting substances against a concentration gradient. On the other hand, a membrane which complexes with penetrating substances has the potential for such an active role. For, to the extent that a particular solute can penetrate the membrane only by complexing, a direct interaction does not exist between in-

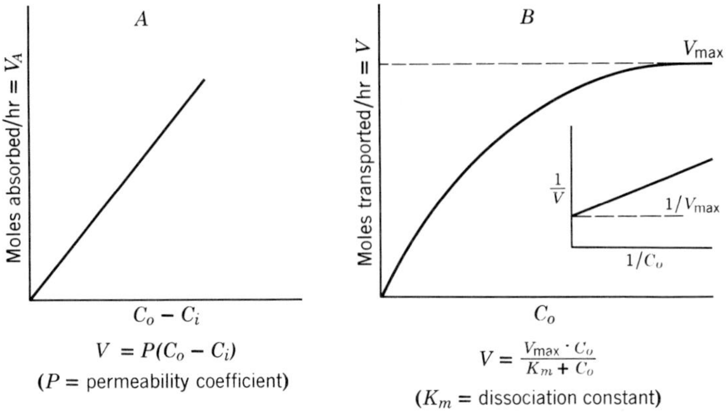

Fig. 10-11. Kinetic characteristics of solute transport. The relationship shown in *A* is that expected if cell membranes behaved like mechanical barriers to solute movement. The permeability coefficient is a constant and combines the diffusion characteristics of the solute and the diffusion resistance of the membrane. The rate of movement in the direction of the gradient V_A is directly proportional to the difference in concentrations between outside and inside of the cell $C_o—C_i$. The relationship in *B* is that expected if molecules were transported across membranes by forming a complex with a specific carrier. If the internal concentration is kept constant and the outside concentration varied, the rate of transport V yields a curve when plotted against C_o. A maximum rate is reached when the carrier molecules become saturated. A straight line is obtained if the reciprocals are plotted against one another. Such relationships are characteristic of enzyme kinetics. The dissociation constant K_m, analogous to the dissociation constant of the enzyme substrate complex, may be calculated from the equation shown or derived from the graph. Actual measurements of solute absorption by cells follow the relationships shown in *B*. Kinetic data therefore favor a carrier mechanism, but mathematical similarities cannot be regarded as proofs of similarities in mechanisms.

ternal and external solutions. A control of the complexing mechanism could therefore lead to an arrest of solute movement which would otherwise lead to equilibrium. No similar mechanism is possible in a purely porous membrane; such a membrane could retard diffusion, but it could not eliminate a direct interaction between internal and external solutions.

Although a carrier system would appear to be the more attractive mechanism for control of solute passage, the evidence is clearly against its being the exclusive mechanism. The classical evidence for the passive penetration of solutes according to molecular size, charge, and lipid solubility is unequivocal. Moreover, during the period when interest in this aspect of permeability was strong, studies of developing cells showed that porosity of the membrane could undergo change. Cells giving rise to pollen, for example, are almost freely permeable to glucose at one stage of development and virtually impermeable to it at another. Since cellular environments differ considerably, especially for unicellular vs. multicellular organism, one should allow for a variability in the relative functional values of pore and carrier mechanisms.

In all cells studied there is a consistent relationship between solute accumulation and energy utilization. Interruption of energy transformations, whether they arise from fermentation or respiration, causes an arrest of solute accumulation. The basic relationship between energy and solute accumulation is shown graphically in Fig. 10-12. On thermodynamic grounds the relationship could not be otherwise, although physiologists commonly refer to it as "active transport" to distinguish it from transport which appears to be effected by a concentration gradient from exterior to interior. As mentioned earlier, however, the challenging aspect of the problem is not to prove an energy requirement, but rather to demonstrate the mechanisms by which energy is utilized.

Active transport across a porous membrane is possible only if energy is utilized to alter the penetrating molecule. The alteration may either lead to an insoluble product, or may result in a transformed molecule. If, for example, the enzyme hexokinase transfers the termi-

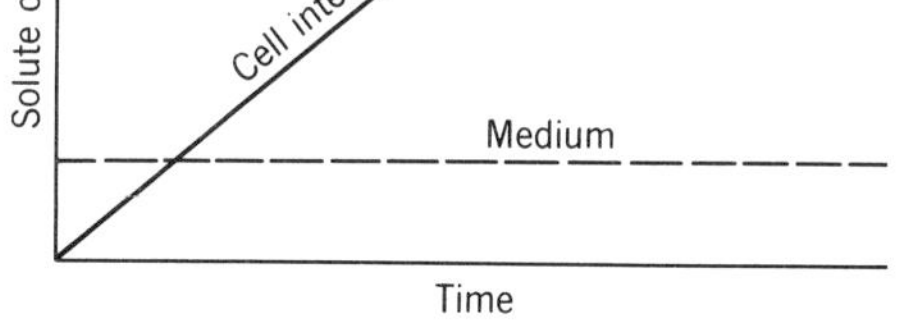

Fig. 10-12. Energy-dependent transport. The graph is an arbitrary plot summarizing results obtained with a variety of cellular systems. The main features to be noted are: The persistent uptake of solute despite the unfavorable concentration gradient; and the arrest of uptake on interrupting the flow of energy. In some cells such interruption does not lead to an outflow of solute, whereas in others it does (dotted line). The rate of outflow may be taken to reflect the leakiness of the membrane system. Although straight lines have been used to accentuate the nature of the processes, actual experiments would yield curved lines. A fall in rate of uptake would become manifest as the unfavorable concentration gradient increased. Complete retention of absorbed solutes in poisoned cells is also unlikely, although most plant vacuoles have a very high retentive capacity.

nal phosphate of ATP to glucose, the product glucose phosphate would remain within the cell because the charged nature of phosphorylated compounds makes them impermeable to the membrane. Cells metabolize phosphorylated derivatives present in the external medium. Detailed analyses of such utilization indicate, however, that the phosphate groups are hydrolyzed by enzymes present at the cell surface, and that the organic moiety penetrates in the dephosphorylated form. The presence of enzymes at cell surfaces is in itself a subject of considerable interest, and the role of surface enzymes in solute transport is subject to a great deal of speculation. Nevertheless, with respect to the generation of favorable diffusion gradients across porous membranes, enzymes can act effectively only by transforming the penetrating molecule in the cell interior.

The value of porous channels could be judged on a rate basis. If, as appears to the case, passage by complexing is much more rapid than free passage through pores, then cells are likely to depend on the complexing mechanism for provision of solute. The major objection to a pore mechanism, however, does not apply to rate characteristics, but rather to the associated mechanism responsible for directional diffusion. Pores and carriers are equally responsive to intracellular transformations of penetrating solutes, but the pore mechanism is difficult to reconcile with the movement of solutes against an actual concentration gradient. The most outstanding examples of such movement pertain to the monovalent cations of sodium and potassium. These are not the sole examples. Many cells have been found to accumulate sugars or amino acids and to retain them as such in the cell interior. A mutant of the bacterium, *E. coli,* for example, which lacks the enzyme to phosphorylate galactose, can accumulate that sugar until it is 2700 times as concentrated internally as externally. No mechanism has yet been envisaged whereby molecules can move from a dilute to a concentration solution without an intervening barrier, which not only isolates the two solutions but which also contains some mechanism to modify the molecule in the course of passage. The universal capacity of cells to move solutes against an actual concentration gradient thus requires a universal property of membranes to effect such movement.

Transport of Inorganic Ions

The movements of inorganic ions or of organic molecules against a concentration gradient pose the same thermodynamic problem. The reason that inorganic ions have received much more attention in studies of uphill transport arises from the distinctive role which ion concentration *per se* plays in cell function. Organic molecules are substrates for metabolic transformations; inorganic ions are not. Cells which must regulate their internal osmotic pressure generally do so by controlling their electrolyte concentrations, for the major contributors to such osmotic pressure are the inorganic ions and not the organic molecules. Apart from this nonspecific but critical role, the qualitative as well as quantitative composition of intracellular electrolytes has a direct effect on membrane potentials. These potentials are the seat of nerve impulses, a property which will be considered in connection with cell behavior. But not only must cells regulate total and individual ion concentrations, they must also provide for an exchange of H ions. As carbohydrates are transformed into various organic constituents, H ions are released. If the formation of organic acids in general and amino acids in particular were unaccompanied by an exchange of H ions for the monovalent K ion, the pH of a growing cell would drop to lethal levels. Control of osmotic pressure, membrane potential, and pH are thus three critical functions in which the directional flow of inorganic ions plays a key role.

Associated with and indispensable to the uphill movement of ions is the preservation of electroneutrality. Generally, the critical ions are the cations. The levels of K^+, Na^+, Mg^{2+} appear to have a much more sensitive and immediate effect on cell function than the levels, say, of sulfate, nitrate, chloride, or even phosphate. Except for chloride, the anions can seriously impair cell function if their concentrations become limiting to biosyntheses, but there is little evidence for an exacting requirement in anion composition such as exists in the case of cations. The total ionic composition of a cell is not always a true indicator of the ionic requirements of cytoplasm and nucleus. In plants the large vacuoles act as traps for surplus ions, and the ion content of such vacuoles may far exceed that of the cytoplasm. Some aspects of vacuolar storage reveal a great deal about the process for maintaining electroneutrality. Metabolizing plant cells accumulate K^+, and such accumulation is generally accompanied by an intake of one or several of the four anions listed. If, however, bromide is substituted for these physiological anions, K^+ accumulation is unimpeded (provided that intracellular phosphate is adequate) and bromide is drawn in to preserve electroneutrality. Various marine algae frequently accumulate iodide in their vacuoles in the course of maintaining electroneutrality.

The apparent indifference of certain cell types to the nature of the balancing anion can be misleading. The requirements for anions like sulfate and phosphate are absolute, and cells generally accumulate these anions along with the cations. However, cell membranes generally have a positive charge, due to the predominant ionization of the basic amino acids in the proteins, which has the effect of repelling cations. Red blood cells, for example, require about 31 hours for half their K^+ to exchange with extracellular K^+, but only about 0.24 seconds for a similar exchange of chloride ions. If red blood cells are suspended in a sulfate-rich medium, intracellular chloride is replaced by sulfate; by contrast, Na^+-K^+ exchanges are highly specific and energy dependent. Observations such as these and those discussed in the preceding paragraph lead to the conclusion that energy expenditure for uphill transport is mainly directed at the cations.

Although anion accompaniment of cations is a common mechanism for maintaining electroneutrality, it is not the only one. An important mechanism, already briefly referred to, is ion exchange. Excess H ions are frequently exchanged for the monovalent cations, electroneutrality thereby being maintained without anion absorption. The exchange mechanism is particularly prominent in nerve cells, where stimulation leads to an outflow of K^+ and an inflow of Na^+. Stimulation has the effect of temporarily destroying the membrane barrier, and the ensuing flow of ions is in the direction of the concentration gradient. Reconstitution of the original ionic composition is effected by a pumping out of Na^+ ions and a replacement of these by K^+. The mechanisms for maintaining electroneutrality are distinct from even though associated with those effecting transport against concentration gradients. The forces of electrostatic attraction are so strong that unbalanced ion diffusion except over extremely short distances is inconceivable. In considering uphill transport, we may therefore set aside the problem of electroneutrality, which is a condition and not an agent of the process.

The primary question asked concerning the utilization of energy for ion transport is what chemical changes could be effected by an input of energy which would result in the complexing and release of cations. Two types of changes have been considered, one involving electron transport, the other involving ATP. The first is an integral part of the energy-yielding system in all respiring cells; the second is the major energetic product in all cells. Highly simplified schemes illustrating how these two systems could lead to a complexing and release of K ion have been assembled in Fig. 10-13.

The potential of electron transport systems for ion complexing lies in the fact that each of the electron carriers in the chain becomes capable of binding an additional monovalent cation on reduction. The complex falls apart if the carrier is reoxidized. Since the progressive transfer of electrons toward higher oxidation potentials releases free energy, we could suppose that part of that energy is used in complex formation. Whether the porphyrins are themselves ion carriers or whether they transfer electrons to other specific intermediates is unknown. The fact that this potential for complexing exists as a consequence of electron transport, and the fact that some type of cytochrome is present in all cellular membranes, have made the scheme attractive. This attractiveness is increased by the possibility which the scheme has for H ion exchange (Fig. 10-11.1*b*). Since the coenzymes which furnish electrons to the cytochrome system do so by releasing H ions, an exchange could occur between the ions thus released and K ions in the external medium. It should be noted, however, that this process is of limited value unless $(OH)^-$ is exchanged for another anion. Ultimately, H ions are used to form water with reduced oxygen (see Chapter 8.), thereby leaving free hydroxyl ions; if these ions are counterbalanced by K ions, an increase in pH results. A pure exchange of hydrogen for potassium ions is advantageous only to the extent that it disposes of the surplus H ions associated with newly formed organic acid intermediates.

Electron transport cannot be a direct source of energy for uphill ion movement in all cells, if only because many fermenting organisms do not have a true electron transport mechanism. Moreover, some cells, like those of muscle, can accumulate ions by utilizing the energy of fermentation if their respiratory channels are blocked. The generality of ATP as a source of cellular energy has therefore lent itself to speculation with respect to ion transport. Support for the schemes involving ATP function in ion transport comes from the fact that cell membranes contain enzymes which hydrolyze ATP. The interpretation commonly given to this finding is that the enzymes do not normally function solely to hydrolyze ATP, but that the energy of hydrolysis is normally tied to the formation or dissociation of cation complexes. Consistent with this interpretation are the findings that certain inhibitors of these hydrolytic enzymes also inhibit cation accumulation and that rates of ATP hydrolysis by membrane enzymes are increased by the addition of potassium or sodium ions.

Two possible mechanisms for ATP utilization in ion transport are illustrated in Fig. 10-13.2. In the one mechanism, (*a*), phosphate addition to an organic molecule makes possible cation binding; in the other mechanism, (*b*), phosphate removal makes the binding possible. Some specific suggestions have been made as to the nature of the carrier molecules, but as yet no convincing evidence on their identity has been presented. The essential point is that circumstantial evidence favors ATP as an immediate agent in transport, and that various chemical configurations involving ATP which would serve to complex ions are conceivable.

None of the molecular mechanisms discussed is by itself adequate to account for uphill transport. Without an appropriate spatial arrangement of the enzymes involved, formation and dissociation of complexes would lead to a random distribution of ions. In this respect the mechanisms which cells must use to counter diffusion tendencies are distinct from those which cells use to effect metabolic transformations. Even in electron transport, which appears to be the most coherently organized structural system, a random mixture of the component enzymes can be made to function in the desired direction, however inefficient such function might be. In ion transport, however, substance must be absorbed on one side of a membrane and released on the other. The need for a structural arrangement is indispensable

Fig. 10-13. Sample mechanisms for complexing and releasing cations.

1. Electron transport system

(*a*) Generation of anionic carrier substance

Reducing energy

(1) $Cyt_1Fe^{++} + X \rightarrow Cyt_1Fe^{+++} + X^-$

(2) $X^- + K^+ \rightarrow KX$

(3) $KX + Cyt_2Fe^{+++} \rightarrow K^+ + X + Cyt_2Fe^{++}$

Oxidation

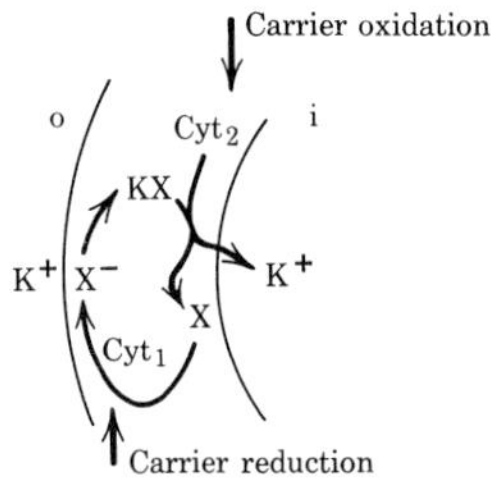

(*b*) H-ion exchange

(1) $FADH_2 + 2Cyt \cdot Fe^{+++} \rightarrow$
$2K^+ \rightleftarrows$
$2H^+ + FAD + 2CytFe^{++}$

(2) $2CytFe^{++} + \underbrace{2H^+ + 2(OH)^-}_{} + O \rightarrow$
$\rightleftarrows 2H_2O$
$2CytFe^{+++} + H_2O + 2(OH)^-$

Note that unless cell has a mechanism for simultaneously exchanging $(OH)^-$ for another anion, it will become alkaline.

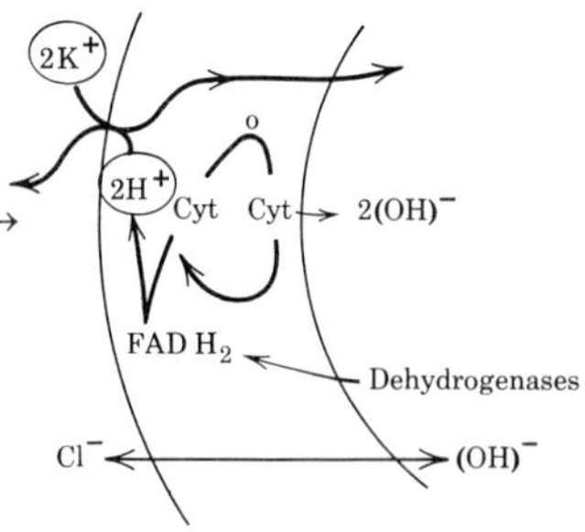

2. ATP utilization

(*a*) Phosphorylation of ROH

energy + H_3PO_4

(1) $R{-}CH_2OH + ATP \rightarrow R{\cdot}CH_2{-}O{-}P(=O)(O^-){-}O^- + ADP$

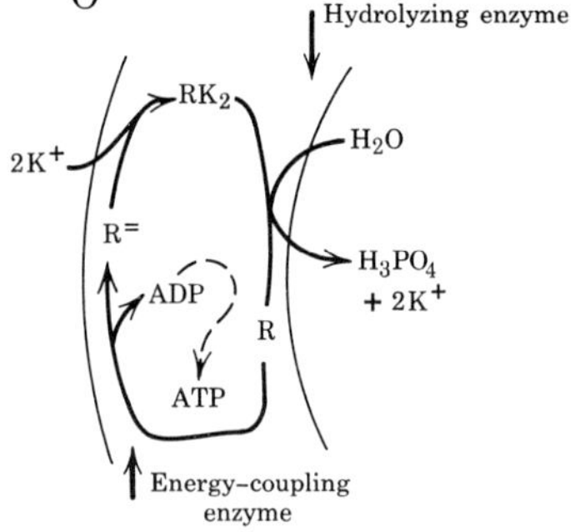

(2) $R{-}CH_2{\cdot}O{-}P(=O)(O){-}O^- + 2K^+ \rightarrow$

$RCH_2OP(=O)(OK){-}OK$

(3) $R{\cdot}CH_2{-}O{-}P(=O)(OK){-}OK + H_2O \rightarrow$

$RCH_2OH + H_3PO_4 + 2K^+$

(*b*) Release of cation by phosphorylation

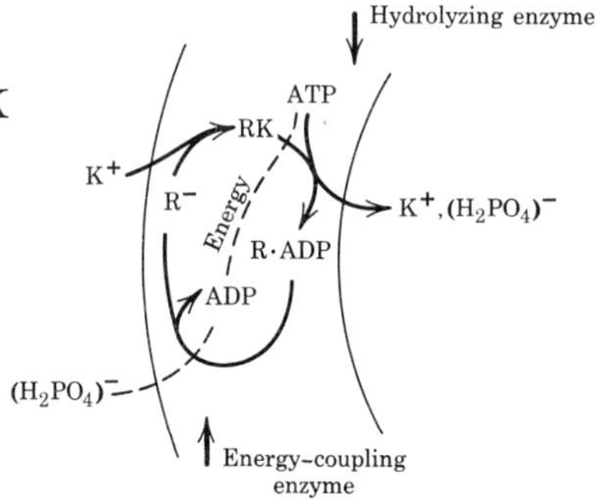

(1) $R{-}C(=O){-}O^- + K^+ \rightarrow RCOOK$

(2) $R{-}C(=O){-}OK + ATP \rightarrow$

$RC(=O){-}O{-}P(=O)(O){-}P(=O)(O){-}\text{adenine} +$

$H_3PO_4 + K^+$

(3) $RC(=O){-}O{-}P{-}O{-}P\ \text{adenine} \rightarrow RC(=O){-}O^- + ADP$

(4) $ADP + H_3PO_4 + \text{energy} \rightarrow ATP$

regardless of the molecular mechanism. The failure of modern cell biology to provide an explanation of this arrangement marks the fork which separates the avenues to clear success and those to misty progress. Insofar as cellular phenomena can be reduced to molecular terms, their resolution appears to be unimpeded; insofar as cellular phenomena implicate structural aggregates of molecules, their resolution must await a clarification of fundamentals.

To accompany these remarks with a reference to the simplified diagrams of uphill transport (Fig. 10-13) seems incongruous. Yet the diagrams do serve to clarify the basic concepts currently applied to the phenomenon. And among these concepts the asymmetric pattern of membranes is the most significant. How soon we will arrive at an understanding of the detailed molecular pattern of membranes cannot be predicted, but the challenge clearly lies in this area. When this understanding is achieved, we will be able to discuss the capacity of cells to overcome the randomizing effects of diffusion in terms as rigorous as those now applied to the overcoming of oxidative and hydrolytic tendencies. The terms will not by-pass the established observations of membrane leakiness, porous structure, and facilitated diffusion, but hopefully the observations will be set into a single logical framework.

11

Energy and Order: Information Theory

Ordering may be defined as the reciprocal of randomization. Because thermodynamics deals only with energy states, the process of randomization, symbolized by ΔS, applies specifically to energy distribution. Familiar examples of randomization are the diffusion of solutes from high to low concentrations, or the flow of heat from higher to lower temperatures. The total energy remains unchanged, but the distribution becomes equalized so that the energy within either system is unavailable for work. Thermodynamic law categorically maintains that no reaction is possible unless accompanied by a net randomization of energy. In a complete account of the energy transactions between a cell and its environment, the sum of the energies randomized must exceed that of the energies localized in ordered systems. This cosmic relationship, however, is only of incidental or philosophical interest to the biologist. In terms of cell function the significant question is how the energy trapped by a cell is directed toward the achievement of order.

Although the theme of this section is "maintenance of order," "order" as manifested in cells has been left undefined. The thermodynamic definition of order as the reciprocal of energy randomization is the sole rigorous and quantitative expression available in natural science. A molecule containing 1000 carbon atoms may seem to be much more ordered than a molecule containing 6 atoms; a branched molecule may seem to be more ordered than an unbranched one; but, appearances notwithstanding, the difference in degree of ordering is equal to difference in energy randomized when the forms being compared are transformed to the same state.

If thermodynamic equations were drawn up for each of the reactions that have been discussed in this section, the ΔS component would generally be found to be relatively small but positive (indicating randomization). Reactions with a $-\Delta S$ can and do occur, but are not a characteristic of the energy transactions within the cell. The primary and secondary processes involved in the formation of cellular substances and the acquisition of energy through respiration or photosynthesis are obviously highly ordered processes, but by thermodynamic criteria they are not *ordering* ones. If we adhere to the rigorous thermodynamic definition of order (and fundamentally there is no choice in the matter), all the categories of metabolism hitherto discussed essentially represent transfers of energy from one type of chemical bond to another. Enzyme catalysis appears to be a mechanism for the transfer but not for the ordering of energy. The orderliness of the enzymes themselves is incidental to this consideration and will be discussed later in the chapter.

If these basic metabolic reactions are not ordering ones, which ones are? That such ordering reactions must exist is obvious from the nature of cell organization, but are they identifiable by thermodynamic criteria? Theoretically, it should be possible to determine the amount of ordering involved in the formation of a particular protein molecule or a multimolecular structure by taking all the component processes into account and inserting values for ΔF and ΔH into the equation $\Delta F = \Delta H - T\Delta S$. But this procedure would have little value in practice, if only because of the very many accessory reactions to be included in the equation. In its final form the equation would be closer to a summary of the whole cell than to that of a single ordering reaction. The relationship between thermodynamics and order becomes increasingly obscure at increasingly higher levels of biological organization. Speech, literature, music, design are all clear instances of order, but few would be willing or interested to measure such order by a thermodynamic yardstick.

Yet if order is a real property of matter, it should be susceptible to measurement whether it is manifested as a sequence of amino acids within a protein molecule or as words in a poem. And if order can be measured, ordering reactions should be identifiable. The significance of such measurements to biological studies is an open question. To some biologists, the measurement of order is an intellectual exercise of marginal interest; to others, it is an approach which is pregnant with possibilities for the future. The reason for this divergence of opinion will become apparent as the nature of cell orderliness is discussed.

THE MEASUREMENT OF ORDER

The simplest measure of order is a generalization of the thermodynamic definition equating orderliness with the reciprocal of randomness. Commonly, order is recognized as a spatial or temporal arrangement of objects (atoms, dots, words, sounds). If the arrangement is a product of chance, as from tossing cans of paint on a canvas, it may be considered interesting, but not ordered. If the arrangement is a deliberate product of human effort or of a machine, it might be found unattractive, but it is nevertheless ordered. In general, the more ordered an arrangement, the less likely is it to have arisen by chance alone. If a pattern consists of subunits which could be randomly arranged in n ways, its orderliness could be expressed as the reciprocal of $1/n$ (the probability of its random occurrence) or n. If a second pattern consists of subunits which could be arranged in $2n$ ways, its orderliness could be expressed as $2n$. By such calculation, the second pattern is twice as ordered as the first. No basic objection can be raised against the validity of this simple comparative method. Nevertheless, to those interested in the measurement of order, this method appears to be inadequate. To understand this inadequacy, we must examine the concepts underlying "information theory."

Information theory was developed in response to a need for theoretical treatment of modern communication systems. Even the simplest machine may be regarded as a communications system in which a "message" (e.g., the pressing of a lever) is communicated to a receptor which responds in a predetermined way (e.g., the opening of a valve). In a complex machine the message may contain numerous instructions, may be transmitted by several different physical mechanisms, and may be translated by a receptor which is responsive not only to the original message but to other modifying factors independently sensed by the machine. In such a machine an efficient use of symbols (or impulses) to transmit unambiguous messages is of prime importance. Information theory concerns the relationship between the *process* of ordering and the orderliness achieved in a message (its "informational content"). Instructional messages are a special instance of order; in principle the concepts applied to the ordering of messages should be applicable to any type of ordering.

The basic difference between measuring order as a pure function of improbability, and measuring it by the rules of information theory lies in the term "process." In the first type of measurement the intervention of a selective process is ignored; the frequency of occurrence of a particular pattern of symbols is compared with the probability of its occurrence by random associations. Clearly, if the frequency of actual occurrence is highly improbable on the basis of random association, a selective or ordering mechanism must be presumed. Indeed, without an ordering mechanism order is impossible. The question raised by information theory is whether a comparison of the orderliness of different systems as a pure function of improbability is quantitatively meaningful. The answer provided by information theory is that such measurements cannot be meaningful unless they take an ordering or selective mechanism into account. If taken into account, orderliness becomes a measure of the number of steps required to select a particular arrangement from a pool of random arrangements.

Information theory is based on the simplest of conceivable selective mechanisms, "binary choice." This is defined as a mechanism which can in any one step discriminate only between two equal possibilities. In familiar terms this may be described as a "yes-no" or "sense-nonsense" mechanism. From the standpoint of the theory the choice is arbitrary even though some believe that this mechanism operates in natural systems and is not entirely hypothetical. Whether or not the selective mechanism is purely hypothetical, the important point is that the relative values for orderliness obtained by including a selective mechanism are different from those obtained by a direct comparison of improbabilities. And if binary choice is replaced by a more complex selective mechanism, the differences become even greater.

The basic equation of information theory states that if a structure consists of subunits which can be arranged in n possible configurations, then a binary choice mechanism would require $\log_2 n$ steps to select a particular configuration; if the subunits could be arranged in $4n$ ways, $\log_2 2n$ selective steps would be required. Thus a structure which is four times as improbable as another structure is only twice as ordered, given the simplest selective mechanism. The process of choice can be illustrated schematically (see Fig. 11-1). The illustrations become complicated if the total number of possibilities is not an integral power of 2, since the values obtained are mixed numbers, and fractional steps cannot be illustrated schematically or practiced operationally. Nevertheless the principle of selecting between any two equal possibilities at each selective step is consistently applied.

The equation governing the informational content of a particular arrangement is generally expressed as $H = \log_2 n$, where n designates the number of possible configurations, and H is defined as the infor-

Fig. 11-1. The principle of binary choice.

(*a*) Random arrangements of A, B, C, and D in pairs: the selection of AB.

Step 1: Accept those beginning with A or B
Reject those beginning with C or D

AA	BA	:	CA	DA	
AB	BB	:	CB	DB	
AC	BC	:	CC	DC	$\mathrm{H} = \log_2 16 = 4$
AD	BD	:	CD	DD	

Step 2: Accept those beginning with A
Reject those beginning with B

AA	:	BA
AB	:	BB
AC	:	BC
AD	:	BD

Step 3. Accept those ending in A or B
Reject those ending in C or D

AA	:	AC
AB	:	AD

Step 4: Reject A endings
Accept B endings

$AA : AB$

(*b*) Pool of colored balls: selection of black (⊗). $H = \log_2 8 = 3$

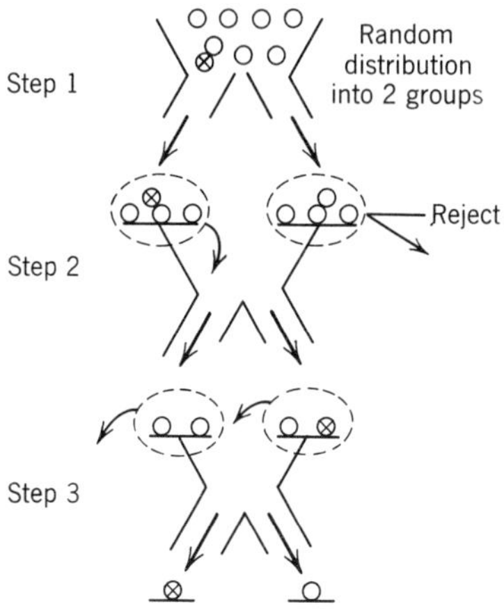

mational content. The units of H are called BITS, a contraction of BInary digiTS. One bit of information implies a degree of ordering requiring one selective step. In general, the number of informational bits in any particular arrangement increases as the square root of the number of possibilities.

The application of information theory to parlor tricks has both didactic and entertainment value. If, for example, the "parlor magician" were asked by an admirer to select a particular but unnamed card from a deck, he would not attempt to do so directly because the probability of picking the correct one would be 1/52. If, however, he made use of binary choice, he would require $\log_2 52$ or 5.7 choices to select the right card. (Since $\log_{10}$ tables are readily available, it is useful to remember that $\log_2 n = 3.322 \log_{10} n$.) By blending sleight of hand with feigned casualness, the magician could alternately flip cards into two piles and innocently ask his admirer to indicate the pile containing the card. The first step in the binary choice sequence will have been performed for him, and he need only be deceivingly casual about keeping the "yes" half on top as he reassembles the pack and repeats the operation. In this second step, as he flips the cards into two groups, he will be concerned only with the first thirteen cards in each of the groups, although he might confuse his admirer by complex manipulations while sorting the remainder of the deck. The admirer will however oblige with a second binary choice, and the magician will innocently locate the "yes" group of thirteen cards on top. At the sixth operation the choice will be unambiguous to the expectation of the magician and the puzzlement of his admirer. A more enterprising magician might use a system of ternary choice, in which case he would require $\log_3 52$ steps, although he could make more spectacular use of the procedure if he had a deck of 81 cards since both would require the same number of steps for selection of the one card. With only 52 cards a fourth step would be unnecessary for some of the "guesses."

APPLICATION OF INFORMATION THEORY TO MEASUREMENT OF CELL ORDER

The method of information theory for the quantitative evaluation of orderliness cannot be applied with equal facility to all aspects of cell organization. The information required to build a protein molecule can be easily calculated, as is frequently the case, if the number of amino acids in the molecule is known. If, for example that number is 100, the number of possible patterns is 20^{100} (unlimited numbers of each of the 20 amino acids taken 100 at a time), and the number of bits required is $100 \log_2 20$. To calculate the information necessary for the building of a mitochondrial membrane would at present be impossible because we do not know how to evaluate n. If the properties of the individual molecular components of the membrane are such that they spontaneously aggregate to form only one type of membrane then $n = 1$ (no other configuration being possible) and $H = 0$; the conclusion would therefore be drawn that the information for building mitochondrial membranes is inherent in the molecules, and that a cell which has the information to build the molecules requires no additional information to build membranes. On the other hand, if it were known that the component molecules do not aggregate spontaneously into the natural configuration, we would have to implicate an ordering mechanism which places the molecules in the desired configuration and hence excluded the other $n - 1$ possible arrangements. The number n could be arrived at from a knowledge of the physicochemical properties of the individual molecules. But this calculation would not provide a complete answer because we would also have to know the amount of information required to form the ordering mechanism. Difficulties of this or a similar type would arise in analyzing most aspects of cell organization. Calculation of the information required to build macromolecules such as proteins is possible because the physicochemical properties of the subunits are sufficiently known, and the absence of any tendencies among the amino acids for preferential association is assumed. A similar calculation for multimolecular structures is at present impossible because we cannot make that assumption about most macromolecules, nor do we know the actual arrangement of these molecules in most structures of interest.

Even if the orderliness of the various organized elements in a cell could be determined, the usefulness of the values obtained in calculating the energy required by a cell for each ordering process is doubtful. One obvious reason for this conclusion is that the value H which is expressed in BITS, must be equated with standard energy terms (calories). Theoretically, such an equation could be set up on the assumption that the entropy term S is a function of probability. Use of the equation $S = k \ln n$ (k being a constant of undetermined value) has been suggested, and by combining it with $H = \log_2 n$, a third equation $H = S/k \ln 2$ is deduced. Some attempts have been made to apply this equation, but the results can hardly be considered as significant revelations about cell behavior. Undoubtedly, the limited value of such measurement is due to the nature of the mechanisms which cells employ in ordering reactions. These mechanisms, which will be fully explored in the next section, distribute the energetic cost of ordering in many reactions. Some of these reactions are utilized in the formation of structures which serve as reaction sites. The nature of the mechanisms appears to be a far more significant aspect of the problem than the means by which energy is provided. The means are no different from those utilized in the metabolic reactions already described. Membrane structure is a good example of an energy investment which is not recorded in the energetic reactions postulated for the uphill transport of ions.

Although the main purpose of information theory—to provide a measure of the informational content of organized systems—appears to have a limited value in elucidating cell behavior, certain concepts associated with this theory have had an appreciable influence on approaches to the study of cell behavior. The most obvious influence has been on terminology. "Information" is now a commonplace term in cell biology; whereas formerly we spoke of genetic influence or genetic determination, we now speak of genetic information. Related to the use of information are terms such as messenger, message, transcription, translation, and recognition. All these terms have a precise meaning when used in discussing communications systems, and they are acquiring correspondingly precise meanings when applied to cell behavior.

One of the benefits of information theory has been to promote a more objective evaluation of complexity. The adjective "complex" is now used less generously when referring to substructures of cells. Even though, for reasons already given, the measurement of complexity or orderliness is frequently difficult, the possibility of such a measure has, at least to some extent, discouraged subjective evaluations. Visual complexity is especially deceiving. A mitochondrion or chloroplast looks extremely complex under the electron microscope, whereas a fiber of DNA or protein looks extremely simple. Yet the information required to form a specific protein molecule from amino acids is probably very much higher than that required to form subcellular organelles from their constituent molecules. The phospholipid lecithin forms "complex" myelin films when dispersed in water (see p. 154), yet the information required to form such films is zero because lecithin molecules spontaneously orient themselves in the presence of water molecules. The question of how much information is built into an organic molecule is receiving increasing attention. Several examples have already been found in which natural organic molecules spontaneously aggregate into structures similar to the natural ones. A solution of tobacco mosaic virus protein can form virus-like particles; a solution of the protein components of collagen can be induced to form collagen-like fibers (Fig. 11-2). We should not assume that structures such as these are spontaneously formed under natural conditions. A catalytic factor, for example, may be present in cells to accelerate the process. Whether or not this is so, we may safely assume that the amount of information necessary to form these structures is small compared to the information necessary to form the molecules. And if so, the energy necessary to form the apparently complex is minor compared to what is required in forming the apparently simple.

"Informational capacity" of a molecule has thus acquired a fairly specific meaning in cell biology. It is a succinct way of stating that different types of molecules have correspondingly different limitations and possibilities with respect to the orderly properties of cells. In considering the informational capacity of molecules, another new term has been introduced, "molecular recognition." The meaning is literal—what molecules recognize. The term may at first seem superfluous because molecules can only "recognize" by virtue of chemical affinities. Yet, even though this is true, chemical affinity and recognition are not used interchangeably. Chemical affinity identifies specific types of chemical interactions; recognition is a more general term covering all categories of interactions without implying a specific type. Thus if proteins are said to recognize small molecules, no specific type of bonding between protein and small molecule is implied, but a physiologically significant interaction is suggested. The use of familiar lay expressions is a refreshing departure from the coining of esoteric terms.

MOLECULAR RECOGNITION—THE POTENTIAL OF MOLECULES FOR MAINTAINING CELL ORDER

A selective agent, be it a molecule or a large machine, can have information only about items it recognizes. A sieve assembly which grades peas according to size has no information about apples or any other objects which are beyond the capacity of the sieve mechanism. To say that the sieve assembly has an informational capacity of $\log_2 n$ where n is the number of all objects that might be tossed into its hopper is to miss an important point. So far as the operation of the sieve is concerned n is equal to two—peas vs. all the other objects which cannot enter the sieve assembly. In practice, this particular property of the sieve is of no interest; the significant informational content of the sieve is measured by the number of size classes of peas it discriminates. The building of the sieve obviously required information, but such information is contained in the building mechanism and not in the operation of the sieve. Extending the analogy to cells, a distinction must always be made between the informational *capacity* of a molecule and the information *content*, which is a measure of the ordering necessary for its synthesis.

Throughout the discussion of metabolic processes, only one basic mechanism was implicated, that of enzyme catalysts. The whole of the orderliness of metabolic transactions was based on enzyme specificity whether the reactions involved group exchanges, electron transfer, anhydride formation, or hydrolysis. In certain cases, such as solute and electron transport, the significance of structural arrangements was discussed, but these are in addition to and not in place of the fundamental role which enzymes play. Questions concerning the informational capacity of enzymes and the information necessary for their synthesis may now be raised.

Enzymes as a class can recognize small molecules and a variety of chemical bonds. These two properties are actually aspects of a single one because small molecules are recognized by virtue of their bond patterns. Enzyme substrate complexes are formed because of a correspondence in spatial arrangements of their respective reactive groups. Since, basically, individual enzymes consist of single polypeptide chains folded into sphere-like configurations, the site of recognition must lie somewhere along the surface where interactions are sterically possible. The geometry of the enzyme molecule thus limits the area over which recognition mechanisms can operate. Molecules spanning distances greater than 30 Å are probably not recognized in their entirety by enzymes. Enzymes do in fact recognize other macromolecules, but the recognition is restricted to specific regions. This is the basis for the specificity of various enzymes in hydrolyzing only those peptide bonds which are between certain amino acids of a polypeptide chain. The inability of an enzyme to recognize long stretches of a macromolecule is a point of cardinal importance which will be elaborated upon shortly.

If all the enzymes in a cell are considered as a single mechanism, then this mechanism may be characterized as having a very high informational capacity. The kinds of small molecules which are metabolized by a cell are very large in number; they are all recognized and selectively transformed by the enzyme system. The enzyme system is not a unit, however, except in a classificatory sense. Enzymes are produced individually and independently of one another; they may also be considered as acting independently. From the standpoint of the mechanisms by which cells maintain order, the significant question is the informational capacity of each type of enzyme. In this respect the enzyme is not impressive. A particular enzyme does not recognize many kinds of substrate molecules and hence discriminates between few. Indeed, an enzyme which has absolute specificity for a substrate contains no information about the selection of a substrate molecule (see Figs. 11-2, 11-3). An enzyme thus endowed cannot recognize similar molecules and cannot be influenced by them. Few enzymes are absolutely specific; most enzymes respond with varying degrees of efficiency to other small molecules. The extent to which the enzyme discriminates between such molecules is a measure of its informational capacity. The total informational capacity of an enzyme must, however, also include the ability of an enzyme to discriminate between different types of chemical bonds within a molecule. If an enzyme lacked this ability, it could not catalyze reactions of specific groups within the substrate molecule.

The capacity to recognize a substrate molecule and to catalyze a chemical change do not always go together. Natural or synthetic "competitive inhibitors" are molecules which are recognized by the enzyme but are not metabolized by it. Broadened recognition is not necessarily a defect of enzyme structure. Many cases are known in which natural inhibitors serve to control the catalytic activity of an enzyme. The lower specificity of the enzyme is thus balanced by higher informational capacity. The hypothetical carrier in solute transport may be similar to an enzyme in its recognition of substrate molecules but different from it in that complexing effects no catalytic transformation. In the final section on the regulation of cell behavior, examples will be considered in which the activities of enzymes are affected by small molecules which complex with the enzyme at a locus different from the catalytic one. Although individual enzymes do not have a large amount of information about metabolizable substrates, they carry much more information than one might suppose from their relatively high degree of catalytic specificity. Figure 11-3 is closer to a caricature than to a diagrammatic representation of the informational capacity of enzymes.

Two sweeping generalizations may be made about the role of enzymes in the maintenance of cell order. (1) The formation of all chemical bonds leading to the production of specific cellular molecules is catalyzed by enzymes. Without enzymes no structures could be formed. (2) Although enzymes can recognize and discriminate between closely spaced chemical configurations, they are incapable of recognizing molecular patterns over extended distances. Since the specificity of proteins is due to the sequential arrangement of amino acids in a relatively long chain, enzymes can, and must, catalyze the linkage formation between adjacent amino acids, but cannot direct the ordering of an amino acid sequence. Enzyme catalysis, however specific, has limited value as an ordering mechanism. To the principle of catalysis must therefore be added some other principle to account for molecular ordering in cells. The principle is the subject of Section B.

Looking at the cell as a whole, we may confidently assert that cells utilize energy to maintain order. Yet the more closely we examine the individual reactions within a cell, the more difficult does it become to

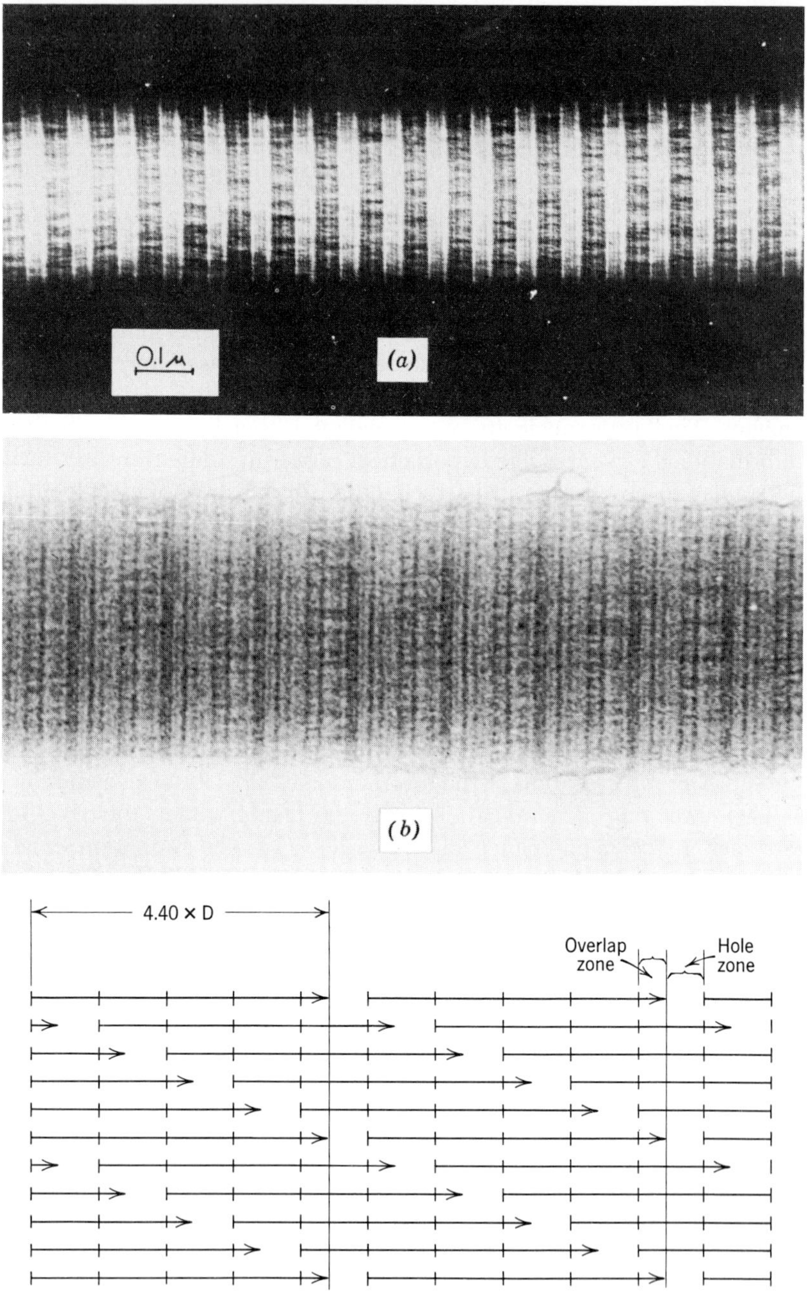

(c)

Fig. 11-2. The orderly aggregation of "tropocollagen" molecules under natural and artificial conditions. The substance "collagen" is present in many kinds of "connective tissues" of animals. The tissues, as their name implies, connect different organs and tissues within the organism. There are many kinds of connective tissues and correspondingly many functional types. Although they vary in structural organization they all contain collagen fibers. When such fibers are examined under the electron microscope they have a regular pattern of striations. Chemical studies of collagen have shown it to be composed of large protein molecules called "tropocollagen." These molecules may be dissolved and, under appropriate conditions, may be caused to reaggregate into small fibers or fibrils which closely resemble those of the natural form. The point of interest in connection with the problem of order as discussed in this chapter is that the "conditions" are very simple (e.g., gradually altering the acidic solution in which tropocollagen molecules are dispersed to a solution which contains 1% NaCl). Thus, in the case of collagen, the constituent molecules appear to have sufficient information in their own structure to form the highly patterned aggregate.

(a) A collagen fibril from a rat-tail tendon negatively stained with phosphotungstic acid. "Negative staining" is a technique used in electron microscopy to stain the background rather than the specimen. This technique accentuates the relatively dense regions of the specimen since these regions contain less of the electron-scattering stain. The micrograph thus indicates that the collagen fibril consists of repeating regions which are alternately dense and diffuse. (Courtesy of A. J. Hodge.)

(b) A fibril reconstituted from a solution of tropocollagen molecules obtained from calf-skin collagen. This fibril was stained with phosphotungstic acid which binds preferentially to the positively charged R groups of the protein molecules. The dark areas in the specimen represent clusters of such groups. Fibrils stained with uranyl acetate would show a similar pattern. Since the uranyl ion binds preferentially to negatively charged groups, we may infer that the polar amino acids are clustered in the tropocollagen molecule. A native collagen fibril stained in the same way would appear very similar to the reconstituted one shown here. (Courtesy A. J. Hodge.)

(c) A hypothetical scheme to explain the appearance of the collagen fibril when negatively stained. By studying the staining patterns of aggregated and disaggregated tropocollagen under the electron microscope, it became possible to establish that the molecule was asymmetric in that one of the ends had a denser localization of polar residues. The arrowhead in the diagram indicates the asymmetry. The short crossbars indicate regions of interacting polar residues. To explain the fact that the densest regions in the native fibril recur at a distance one-tenth less than the length of the individual molecule, the assumption is made that the molecules overlap in the native structure. To explain the image obtained by negative staining, the assumption is made that each length of the molecule bounded by the interacting groups (crossbars) is associated with two regions, the shorter one representing an "overlap zone." In the fibril the overlap zone would contain more molecular segments than the "hole zone." The bright areas in (a) correspond to the overlap zones in the diagram. (Courtesy A. J. Hodge.)

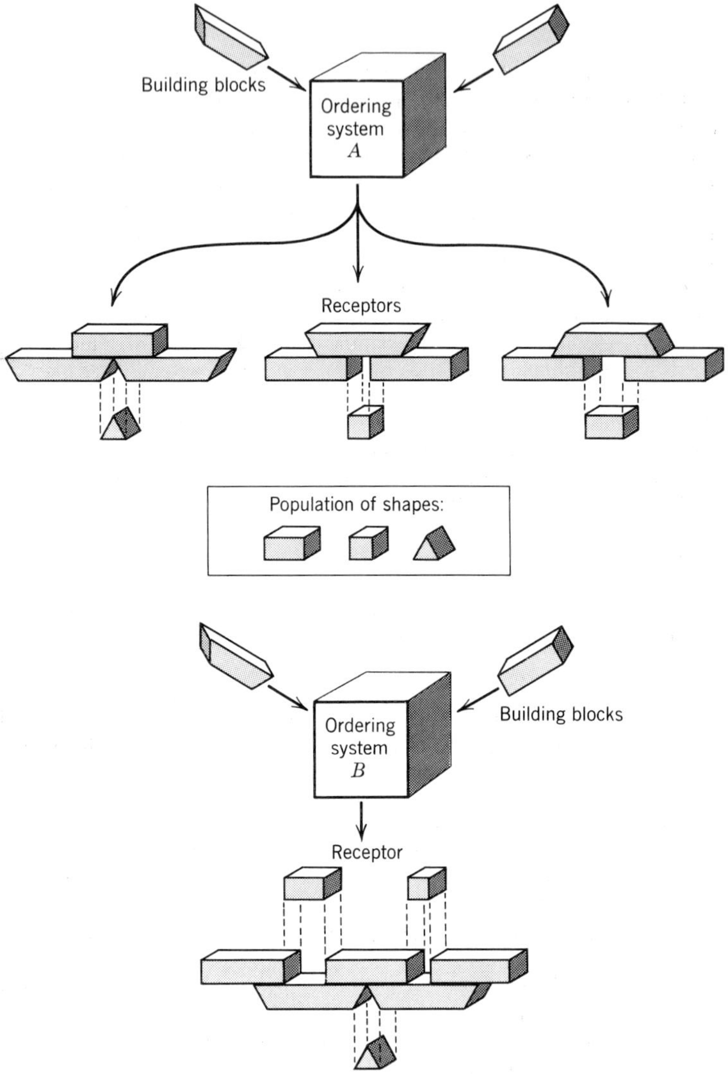

Fig. 11-3. Ordering systems and informational capacity of products. Each of the total systems has the same informational capacity—to discriminate between the three overall shapes shown. In system *A*, the capacity has been organized into three specific receptors; in *B*, one receptor is used. The receptors in *A* are highly specific but carry information about one shape only, and hence no information about the population of shapes. Ordering system *A* carries more information about the shapes because it can selectively produce specific receptors. Ordering system *B* carries less because selectivity is lodged in the receptor mechanism. If the shapes are considered analogous to the population of small molecules within a cell and the receptors to the enzymes, system *A* more closely approximates the pattern of cell function. The *A* receptors are "ordered" but not "ordering."

find any reaction in which the specific utilization of energy for increasing order can be identified. Even for protein synthesis the identifiable energy transactions are of the same kind as those discussed earlier in the section. If we ask how so much order can be created in a protein molecule without the investment of correspondingly large amounts of energy, the answer (as will be made clear in Section B) is that the order is already there, the cell expends energy only to translate it. We should marvel not at the cell but at evolution as the source of orderliness. For evolution creates order; cells maintain it.

To be sure, even in the maintenance of order, relatively disordered molecules must be brought into organized relationships. And such ordering must be paid for by an expenditure of energy. The expenditure is a very diffuse process manifesting itself in the elaborate enzyme systems which are required to effect chemical changes; in the structures which have to be formed to accommodate the enzymes and to regulate their activities; in the membranes which have to be formed and kept functional to counter the randomizing tendencies of diffusion; and, as will be seen, in the templates which are formed and discarded in the course of translating the information for the building of protein molecules. In the course of all these activities energy is transferred from one type of bond to another, but never with complete efficiency. Inevitably, as predicted by thermodynamic law, part of that energy is dissipated as heat; and once dissipated, it is randomized and lost for all time. Cells copy order and through their enzymes flows the energy necessary to maintain it.

EXERCISES*

The term "variety" is defined as the number of distinguishable elements in the group under study. Thus, if sexes are being considered in a population of mice, two elements are distinguishable—male and female. If, on the other hand, coat color is being considered, the number of distinguishable elements will equal the kinds of coat colors that can be identified (e.g., white, gray, black, tan, brown, etc.). As discussed in the text, variety may be measured in several ways. For these problems, two measurements are used: (1) The arithmetic sum of the number of distinguishable elements; (2) the $\log_2$ of the number of distinguishable elements. This measurement is referred to as "variety in bits."

1. With 26 letters to choose from, how many 3-letter combinations are available for motor registration numbers?
2. If a farmer can distinguish 8 breeds of chicks, but cannot sex them,

* Adapted from W. Ross Ashby, *An Introduction to Cybernetics*, Wiley, 1956, pp. 125–127.

while his wife can sex them but knows nothing of breeds, how many distinct classes of chicks can they distinguish when working together?

3. A spy in a house with four windows arranged rectangularly is to signal out to sea at night by each window, showing or not showing a light. How many forms can be shown if, in the darkness, the position of the lights relative to the house cannot be perceived?

4. Bacteria of different species differ in their ability to metabolize various substances: thus lactose is destroyed by *E. coli* but not by *E. typhyi*. If a bacteriologist has available ten substances, each of which may be destroyed or not by a given species, what is the maximal number that he can distinguish?

5. If each personality test can distinguish five grades of its own characteristic, what is the least number of such tests necessary to distinguish the 2,000,000,000 individuals of the world's population?

6. In a well-known card trick, the conjurer identifies a card thus: he shows 21 cards to a bystander, who mentally selects one of them without revealing his choice. The conjurer then deals the 21 cards face upwards into three equal heaps, with the bystander seeing the faces, and asks him to say which heap contains the selected card. He then takes up the cards, again deals them into three equal heaps, and again asks which heap contains the selected card; and similarly, for a third deal. The conjurer then names the selected card. What variety is there in (1) the bystander's indications, (2) in the conjurer's final selection?

7. In Exercise 6, 21 cards is not, in fact, the maximal number that could be used. What is the maximum, if the other conditions are unaltered?

8. In exercise 6, how many times would the bystander have to indicate which of the three heaps held the selected card if the conjurer were finally to be able to identify the correct card out of the full pack of 52?

9. If a child's blood group is O and its mother's group is O, how much variety is there in the groups of its possible father?

10. In example 4, how much variety, in bits, does each substance distinguish?

11. In example 5, how much variety in bits does each test distinguish? What is the variety in bits of 2,000,000,000 distinguishable individuals? From these two varieties check your previous answer.

12. What is the variety in bits of the 26 letters of the alphabet? What is the variety, in bits, of a block of five letters (not restricted to forming a word)? Check the answer by finding the number of such blocks, and then the variety.

13. A question can be answered only by Yes or No. What variety is in the answer? In twenty such answers made independently?

14. How many objects can be distinguished by twenty questions, each of which can be answered only by Yes or No?

15. A closed and single-valued transformation is to be considered on six states:

a	b	c	d	e	f
?	?	?	?	?	?

in which each question mark has to be replaced by a letter. If the replacements are otherwise unrestricted, what variety (logarithmic) is there in the set of all possible such transformations?

16. If the closed transformation had n states, what variety is there?

17. If the English vocabulary has variety of 10 bits per word, what is the storage capacity of 10 minutes' speech on a phonograph record, assuming the speech is at 120 words per minute?

18. How does this compare with the capacity of a printed page of newspaper (approximately)?

ANSWERS

1. $26 \times 26 \times 26 = 15{,}576$.
2. 16.
3. 11.
4. $2^{10} = 1024$.
5. 5^X must be as great or greater than 2×10^9. The value of X may be determined in the following way (using logs to base 10).

$$\log 5^X \geqq \log 2 \times 10^9$$
$$X \log 5^X \geqq \log 2 + 9 \log 10$$
$$x \geqq \frac{(\log 2 + 9 \log 10)}{\log 5}$$
$$x \geqq 13.3$$

14 tests would be necessary.

6. (1) 27; (2) 21.
7. 27.
8. $3^3 =$ and $3^4 = 81$. $\therefore$ 4 indications necessary.
9. 3: A, B, or O.
10. 1 bit.
11. 2·32 bits; 30·9 bits.
12. 4·7 bits; $5 \times 4{\cdot}7 = 23{\cdot}5$ bits.
13. 1 bit; 20 bits.
14. $2^{20} = 1{,}048{,}576$.
15. The replacement of each question mark has a variety of $\log_2 6$ bits. Total variety $= \log_2 6$ bits $= 15{\cdot}5$ bits.
16. $n \log_2 n$ bits.
17. 12,000 bits.
18. A page of 5000 words would carry about 50,000 bits.

SECTION B

The Maintenance of Specificity

Cells and all other living systems possess two fundamental but opposing properties which must be maintained and kept in balance if the system is to persist and prosper. First, they must be able to make more systems like themselves. If cells could not make more cells, the accidents to which they are exposed would eventually exterminate them. If they did not make more *like themselves* (i.e., if the progeny were not like the parents), natural selection would have no lever for manipulating evolution, and life would have very different characteristics than those with which we are familiar.

We have already discussed at some length the molecular basis of biological specificity; many of the special properties of cells are referable to differences in the kinds of molecules which make them up. Differences in proteins are of especial significance since the catalysts essential for biological activities are chiefly proteins. We might think, therefore, that the problem of heredity could be reduced to that of understanding the mechanisms whereby identical proteins are manufactured generation after generation in cell lineages. This is a part of the problem, and perhaps the major part, but it is not all. It is certainly the part we understand most fully. But a cell is not simply a bag of enzymes. It represents a higher level of organization in which the relations of the molecular constituents to each other are also of great importance. The study of heredity must therefore be concerned eventually not only with the details of molecular syntheses, but also with the integration of the molecules in significant and persistant patterns. In this section we shall focus attention on those mechanisms responsible for molecular specificity. We should be aware, however, that additional mechanisms are involved in the persistence of cellular types and we shall return to these in Chapters 18 and 19.

Heredity is the conservative principle in biology, the result of all the mechanisms operating to maintain the status quo through reproductive activities. But conservation is not sufficient for long-range persistence, much less for advancement. Cells must also be able to change their properties. The qualities required for existence at one time and in one place are not the same qualities required in other times and in different places. Living forms required both heredity and the capacity to change to invade the multitudinous biological niches which they inhabit and to produce the great variety of adjustments to those environments which are so apparent. The long-range evolutionary modifications in cells and organisms are only a part of the story, however. Cells must also change in a restricted time scale to accommodate short-term shifts in the environment or to achieve particular biological objectives. The capacity for change is fully as characteristic of living systems as their ability to conserve their specificities. Heredity and variation are two aspects of the regulation of specificity and cannot be considered in isolation. Again, in this section, we shall be concerned primarily with genetic change in the sense of changes that result in altered molecular specifications, and we shall return later to a discussion of epigenetic alterations in cellular properties.

12

Molecular Replication

THE MEANING OF HEREDITY

One problem which constantly arises in considering biological variation is in determining whether observed differences are hereditary. An exploration of this problem as it applies to cellular traits may clarify somewhat the fundamental nature of heredity. We may begin by considering two cells which are different. The fact that they are different does not necessarily imply that the differences are hereditary. The cells might have had very different histories and could be distinctive, for example, because of variations in the materials they have taken up from or produced in response to their environments. An artificial but instructive example is provided by two cells supplied with different isotopes of nitrogen in their food supply over many cell generations. One of the cells might then contain only N^{14}, and the other only N^{15}. The cells would be different, and the differences would extend into many of their molecular constituents; but the differences would not need to be hereditary. Clearly, one of the requirements for a demonstration of hereditary differences is that the differences persist for a period of time in the same or identical environments.

How long must the differences persist—4 hours? 10 days? 3 years? Time measured in a strictly chronological fashion does not provide an adequate basis for a determination. In the absence of cell division some of the materials accumulated in prior periods of growth might persist indefinitely in the cells where they are localized. The ability to retain a part of the past environment does not apply in the same way to cells which are dividing. Before a cell divides, new materials have to be taken up, and approximately half of a cell's original substance passes to each daughter. As this process continues, the original materials of the cell are distributed among more and more daughters until cells are eventually produced which carry no material relics of their previous history. Even though the daughter cells "inherit" materials from the parental cells, this is a passive transmission, much like the inheritance of a family clock or a silver teapot. The transmitted material is not reproduced but only passed on and is associated with a progressively smaller fraction of the descendents.

We must therefore inquire through how many cell generations materials might be passively transmitted, though progressively diluted, and still be present in sufficient quantity to be detected. If observed differences are found to persist past this point, we may be certain that the determinants of the differences are themselves being multiplied, that the differences are being actively maintained, that the differences are "hereditary."

A limiting condition may be approached with the example already raised: the cells grown in different isotopes of nitrogen. The two kinds of cells might be raised for a number of generations in N^{15}, and the generations required for the originally pure N^{14} cells to become pure N^{15} cells could be calculated. The amount of N^{14} in each descendent of the N^{14} cell is reduced by 1/2 at each division: to 1/2 after one division, to 1/4 after two divisions, to 1/8 after three divisions, etc. To determine the number of atoms this would represent, we need to know the number of N atoms present in the original cell. This certainly varies greatly with different kinds of cells, but even large cells usually contain fewer than 10^{13} atoms of N. Thus if the cell gave rise to 10^{13} cells, only one atom of the original N^{14} would on the average be found in each; and the number of cell divisions required to produce this number of cells is between 40 and 50 ($2^{47} \cong 10^{13}$). Hence if all the progeny of two cells maintained in the same environment continue to manifest a difference after fifty or so cell divisions, we may be reasonably certain the differences are not due to differences between the nitrogen atoms contained in the original cells. We could extend this approach to all the atoms of the original cells without greatly affecting the required dilution period. Of course, biological specificity does not normally reside in atomic specificity, but in molecular specificity and more particularly in the specificity of large molecules. Since the number of macromolecules in a cell is much lower than the number of atoms of the common kinds, the dilution of macromolecules is accomplished much more quickly than that of atoms (assuming the atoms to be randomly assorted) and a dilution test based on fifty cell generations is a conservative test. If differences persist past this point one may be reasonably sure that the basis for the differences is itself being replicated and that the differences are hereditary in a meaningful biological sense.

This may seem like a strange approach to a definition of heredity, but it focuses attention on the fundamentals; and it is in fact the most widely applicable of definitions at the cellular level. In particular, it involves no breeding study and requires no sexual manipulations. In tissue cultures, where breeding analysis is not yet possible, in somatic tissues of higher organisms, and in certain microorganisms whose sexual activities have not been identified or controlled, this definition of heredity is the only applicable one. Other current definitions of heredity are less general, being applicable to particular kinds of organisms. They refer specifically to techniques of breeding analysis and are limited to situations where these can be applied. They are also more restrictive in that they are more likely to identify hereditary differences with chromosonal bases.

A definition of hereditary differences by the operation of dilution separates heredity conceptually from its sometimes confusing handmaiden, sex. Sex has important biological functions, but the chief of these is not reproduction. Indeed, the terms "sexual" and "asexual" used in connection with reproduction are only confusing. At the cellular level, all reproduction is asexual. A sexual process is a process of uniting cellular elements from different sources; it is not a process of producing more cells or cellular elements and hence is not a process of reproduction. We may argue even that sex is antagonistic to reproduction since it introduces complexity and reduces reproductive potential. We will consider later the biological significance of sexuality, which allows it to be such a pervasive biological activity in spite of its cost.

Although sexual processes are only incidental to reproductive activities and to the essential hereditary phenomena, they are of fundamental significance in the analysis of those activities and phenomena. They permit, in effect, the dissection of the hereditary apparatus with utmost delicacy and precision.

EXERCISES

1. If cells of the same strain have become different as a consequence of growth in different environments, the differences may persist for a limited number of cell divisions in the same environment. On the basis of certain assumptions we may estimate the original number of molecules responsible for the differences. For example, subpopulation A may possess a number of enzyme molecules while subpopulation B contains none. Under certain growth conditions this enzyme may not be produced but is simply diluted out. If after eight cell divisions half of the cells in subpopulation A have no enzymatic activity, how many molecules did each cell have at the beginning of the experiment? Determine the average

number of enzyme molecules per cell at each generation. Plot these numbers on (a) standard graph paper, and (b) semilog paper. What assumptions must be made to make such calculations?

2. Bacterial viruses occasionally introduce into a bacterial cell a fragment of the chromosome from another bacterial cell; this process is known as "transduction." In some cases the chromosome fragment is incorporated into the chromosome of the bacterium, but in other cases it simply remains in the cell, influencing the phenotype of the cell but not dividing (abortive transduction). When the bacterium divides, the fragment goes to one of the daughter cells but not the other. If two cells differ from each other in the presence or absence of such a fragment, are the differences between them hereditary? Are they genetic?

3. Some strains of Paramecium (killers) produce a substance (paramecin) which kills other strains (sensitives). Grown at 26°C under precisely the same conditions the differences between the strains persist indefinitely. When the strains are grown at 32°C, however, the killer cells all become sensitive after about 10 cell divisions and remain sensitive after they are returned to 26°C. Are the differences between killers and sensitives hereditary? Killer cells are found to contain in their cytoplasm bacterium-like particles (kappa particles). How might you determine the number of such particles without actually counting them?

THE DELIMITATION OF THE GENETIC MATERIAL

We have already seen how breeding analysis led to the discovery of hereditary determinants and permitted genes to be arranged in linear order on the chromosomes. We must now inquire about the chemical basis of heredity. In spite of the brilliant achievements of classical cytogenetics, certain fundamental questions could not be answered by breeding analysis, nor through a combination of breeding studies and cytological observations. In particular, cytogenetics explained heredity in terms of the chromosomes, but the chromosomes themselves were not "explained," either in their essential chemical constitution or in their unique ability to make more of themselves. Much of the recent work in cellular biology has been focused on these questions with remarkable and dramatic results. But these newer developments must not be construed as marking a sharp discontinuity in biological science, rather they must be seen as the continuation and culmination of efforts to delimit the genetic material which began long ago.

The first step in the identification of the hereditary material was the recognition that the material bridge between generations is cellular.

Although both the egg and the sperm are highly atypical cells and differ from each other greatly in size and structure, the essential features of cellular organization are discernible in both. Whatever is passed from a parent to its progeny must be incorporated within cellular structures.

Before the development of modern micoscopic techniques and differential staining procedures, even the identification of sperms and eggs was a sufficiently remarkable feat, and an adequate understanding of the structure and behavior of subcellular components was impossible. In the absence of knowledge concerning cellular functions, several interpretations of the roles of sperms and eggs were possible. One group of biologists (ovists) considered that the sperm functioned simply as a stimulant to development, and that it contributed nothing of significance to the progeny. Another group (spermists) believed that the entire determination of the progeny's traits was accomplished by the sperm, and that the egg only supplied nourishment to the developing organism. A third group, relying perhaps more completely on the evidence of their eyes, believed that both the father and the mother contributed significantly to the characteristics of the offspring. The matter was settled for many biologists less than a hundred years ago by Sir Francis Galton, who showed by statistical analysis that the metric characteristics of children were correlated equally with the mother and the father. Experiments such as those of Mendel were, of course, decisive.

The demonstration of coordinate roles of males and females in inheritance was of importance in further identifying the hereditary material. When good biological stains were developed in the 1880's, a more precise comparison of sperms and eggs was possible. Although the egg might be thousands of times as large as the sperm of the same species, the sizes of their respective cellular components do not all differ correspondingly. Specifically, the major differences are due to inequalities in the cytoplasm and not in the nucleus. The large size of the egg is due to an enormous increase in the extranuclear materials. The cytoplasm of the sperm is, on the contrary, reduced to a thin covering except for a portion which is modified into a propulsive organ. The nucleus of the sperm, though condensed during transit, on entering the egg cytoplasm quickly swells to the same size as the egg nucleus. In view of the equality of the hereditary contributions of the egg and the sperm, it appeared likely that the nucleus—the chief cellular component contributed equally by sperm and egg—was the primary hereditary organelle of the cell.

Differential staining permitted not only discrimination of nucleus and cytoplasm, but also the identification of other structures within both. In particular, we have seen that at certain times in the life of the cell the ordinarily homogeneous nuclear material takes the form of discrete bodies (chromosomes) which stain intensely with basic stains. Evidence was soon developed to show that the chromosomes were not temporary structures but permanent cellular organelles, coiling and condensing at certain times in the life of the cell and uncoiling into structures below the limits of optical microscopy at other times. At this time the question could be asked as to whether all nuclear components or only the chromosomes were significant in hereditary phenomena.

Initially the answer to this question had to be indirect, but it could be suggested by an argument based on economic considerations. Even before their behavior was fully worked out, cytologists discovered that the chromosomes were involved in remarkably complex maneuvers. And a little later they discovered that the number of chromosomes within a particular species was maintained with great precision in nearly all the cells of nearly all individuals of that species. Since complexity and precision are expensive, whether in mechanical or biological systems, it is apparent that organisms expend important resources in maintaining their chromosomal constitutions. If so, the chromosomes must be important to the organism. And what could be more important than the hereditary organelles? Such arguments are not, of course, decisive, but even before the turn of the century many biologists believed in a general sort of "chromosome theory of inheritance."

With the rediscovery of Mendelism in 1900 and the conjunction of breeding studies with cytological observations, the role of the chromosomes was decisively established. The delimitation of the genetic material did not stop at this point, however, for now it could be asked whether the entire chromosomes or only part of the chromosomal constituents were the critical materials. To answer this question new techniques had to be devised and applied. Methods for determining the composition of microscopic structures and for discovering their biological roles had to be developed. The major advance in the identification of chromosomal components came through cytochemistry, particularly quantitative cytochemistry. Dyes were developed which stained specifically certain classes of chemical compounds, and optical methods were perfected which permitted an estimation of the amount of dye bound in a particular cellular structure. By the 1930's the two major classes of chromosomal components were known to be proteins and nucleic acids. Since the chromosomes contain several kinds of proteins and two major classes of nucleic acids, many candidates for the genetic material were available, even if it were assumed that only one of them was of primary significance. Moreover, only intact chromosomes could be shown to have biological activity, and a discrimination among the chromosomal constituents appeared very difficult.

Biologists were at first inclined to believe that the hereditary factors were proteins, but gradually indirect evidence began to shift attention to the nucleic acids, particularly to DNA. For one thing, DNA was found in the chromosomes and usually only in the chromosomes, while RNA and proteins were found not only in the chromosomes but in the cytoplasm. This evidence was not strong, however, because some special kind of protein or some particular class of RNA might also occur invariably and exclusively in the nucleus.

Another kind of evidence suggested that nucleic acid, though not necessarily DNA, was of special significance in mutagenesis. Mutations, changes in the genetic material, were found to be induced by ultraviolet light. When the absorption spectra of different cellular components were examined, nucleic acids were found to absorb ultraviolet light most effectively at the wavelengths which gave the most mutations. This evidence again was not critical, since the possibility existed that the energy absorbed by nucleic acids was secondarily transmitted to some other material which was "genetic." Moreover, different organisms sometimes showed unaccountably different mutagenic spectra.

Finally, quantitative studies were highly suggestive of a special role for DNA. We might expect, in view of the constancy of hereditary properties generation after generation, that the genetic material itself would be unusually well regulated not only in quality but in quantity. When the amounts of proteins and nucleic acids are measured in various kinds of cells under various conditions, great variations are observed in the amounts of the proteins and also in the amounts of RNA. But the amount of DNA in a liver cell is, within experimental error, identical to that in a kidney cell or a pancreas cell. Two major exceptions to DNA constancy are known, but these are important in indicating its significance. Some cells in many organisms occasionally undergo chromosome replication without cell division and come to have twice or more the usual number of chromosomes. Such cells (polyploid cells) are also found to have an amount of DNA proportional to the number of chromosome sets. Second, the sex cells (the eggs and the sperm) have half as many chromosomes as normal somatic cells and also half the amount of DNA. DNA is the only cellular component which shows this degree of constancy and a direct quantitative correlation with the chromosomes. One protein component, the histones, shows a similar constancy in somatic cells, but it may be entirely missing from sperm cells. Actually, the absolute constancy of DNA is not firmly established, and we will discuss later some indications of slight DNA variations and explore their significance.

All this evidence was, however, only inferential, and few biologists were convinced of the critical significance of DNA in maintaining specificity until a more direct proof was established. This came first in genetic studies of a bacterium, pneumococcus. This organism exists in two forms: a form which is enclosed in a polysaccharide coat and produces smooth colonies on an agar plate; and a form which lacks the coat and produces rough colonies. Different smooth strains produce antigenically different coats, that is, coats which can be distinguished through the use of antiserum prepared against the strains. By the early 1930's F. Griffith and his colleagues had demonstrated that rough strains of pneumococcus could be changed into smooth strains by growing them in the presence of heat-killed smooth strains. And, most significantly, the antigenic characters of the modified cells were identical to those of the dead cells with which they had been mixed. This transfer of hereditary characters from a dead strain to a live strain was called *transformation.*

In the early 1940's O. T. Avery and his co-workers undertook an investigation to determine what the dead cells contributed to the live cells to induce them to change their hereditary traits. They found that transformation could be brought about by DNA extracts and only by DNA extracts. This was the first successful extraction of genetic material from a living cell which still maintained its significant biological activity. And at least for these traits, in this organism, and employing this technique, DNA was the only material which carried hereditary information from strain to strain.

These studies have subsequently been greatly extended. Essentially any trait in pneumococcus, whether this involves the formation of an enzyme, resistance to an antibiotic, or a requirement for a particular nutrient, may be transferred by DNA extracts from one strain to another. Moreover, the process of transformation has been demonstrated in several other bacterial species. Hence for bacteria, and apparently for nearly any trait in bacteria, DNA can serve as genetic material.

Unfortunately, cell extracts from other organisms have not ordinarily been found to possess transforming ability, perhaps because they are inactivated by the preparatory techniques or because they are incapable of entering into the critical sites in the cells to be treated. Quite possibly the significant difference between bacteria and other organisms lies in the type of cellular organization; bacteria lack a discrete membrane surrounding the nuclear apparatus, and their "chromosomes" have no proteins regularly associated with the nucleic acid.

The only other organisms whose genetic material has been chemically identified are viruses. Two classes of viruses may be distinguished on the basis of their chemical constitution: those which contain DNA and protein, and those which contain RNA and protein. In both the protein can be shown to be dispensable in the reproduction of the virus; under special circumstances viral nucleic acid alone is infective, but viral protein alone has never been shown to be infective. Even when both nucleic acid and protein are present in virus particles, it is the characteristics of the nucleic acid which are significant in determining the hereditary characteristics of the progeny virus. This is demonstrated by artificially combining nucleic acid and protein from different viral sources to create synthetic viruses; these yield progeny viruses of the type from which the nucleic acid was derived.

Although such studies confirm a special genetic role for nucleic acid, they also raise difficulties. There is no reasonable doubt that under different circumstances both DNA and RNA can carry genetic information. But in the bacteria, where both DNA and RNA are found, only DNA has transforming ability. A likely explanation holds that two major classes of RNA exist: those with and those without the ability to synthesize more of themselves. Perhaps only the RNA's in RNA viruses are capable of autosynthesis. We do not yet understand the chemical basis for the differences in these kinds of RNA. It appears likely that DNA is the usual basis for persistent biological specificity and that RNA has significant but different roles in the cellular economy.

THE MOLECULAR BASIS OF AUTOSYNTHESIS

The identification of DNA as the primary genetic material immediately made possible an investigation of what is possibly the most fundamental problem in genetics—the manner in which genes make more of themselves. Before this time several speculations concerning this process were discussed, but they were incapable of experimental analysis. If, however, genes can be equated with DNA, genetic replication can be approached through a study of DNA synthesis.

Several possibilities of DNA synthesis might be entertained. One of the most fruitful ways of classifying these possibilities (Fig. 12-1) is

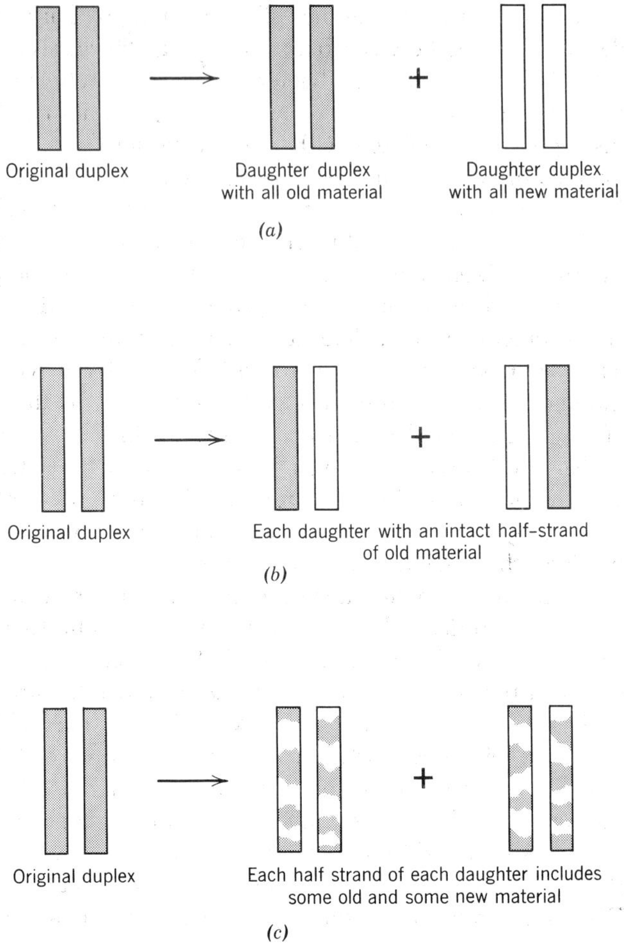

Fig. 12-1. Possible schemes of DNA replication: (*a*) conservative; (*b*) semi-conservative; (*c*) dispersive.

based on the distribution of the atoms in the original molecule among the progeny molecules synthesized under its influence. For example, the original molecule might be maintained intact and new molecules synthesized out of entirely new material. The original molecule would be conserved, and such a mechanism would be called a *conservative* mechanism. Taking into consideration the fact that the DNA molecule is a duplex molecule made up of two complementary polynucleotide chains, another possibility is that the original molecule comes apart and that each daughter molecule receives one of these. Here a part of the original atoms, those in one half of the duplex, are conserved in the same structure; and this mechanism is termed *semiconservative*. This was the mechanism first suggested by J. D. Watson and F. H. C. Crick when they proposed the now accepted model for the structure of DNA. Finally, the original template molecule might be broken down completely in the process of synthesis, and the original atoms distributed more or less at random among the daughter molecules. Such a mechanism would be termed *dispersive*.

A distinction among these mechanisms requires techniques for following the distribution of atoms during DNA replication. Biological tests, of course, are not adequate for making such distinctions because the essence of the replicating system is the biological equivalence of the products. The atoms can be followed, however, by employing isotopes to mark "old" and "new" materials. M. Meselson and F. W. Stahl, for example, found that they could distinguish DNA made by cells supplied with N^{14} from DNA made from cells supplied with N^{15}. The two DNA's would be expected to differ from each other only very slightly in density but this slight difference was sufficient to separate them during equilibrium centrifugation in a cesium chloride gradient. Each kind of molecule was located at a characteristic level in the gradient. The ability to separate DNA molecules made with different isotopes, and even molecules made with different proportions of the isotopes, made possible a study of the distribution of old and new materials during DNA replication.

This was accomplished as follows. Bacterial cells were grown in the heavy isotope N^{15} until the DNA molecules were uniformly labeled. These heavy molecules are then the "old" materials in the experiment. Such cells are transferred to ordinary N^{14} for further growth. Any newly synthesized DNA must incorporate N^{14} into its structure. After new DNA has been synthesized, a conservative replication would lead to the detection of two kinds of molecules—the original molecules containing only N^{15}, and the new molecules containing only N^{14}. This result was not obtained. Instead, (Fig. 12-2) after one replication a single band appeared in the gradient system at a level approximately halfway between the levels for pure N^{14} and pure N^{15} DNA. Hence the original molecules are not preserved intact; some of the parental atoms are distributed to each of the daughter molecules. The mechanism of replication is not conservative.

These facts did not, however, distinguish between the other two mechanisms of replication. Either a semiconservative or a dispersive mechanism could lead to a distribution of parental atoms to both daughters, but the manner of distribution would be different. In the first each daughter molecule would consist of one N^{14} strand and one

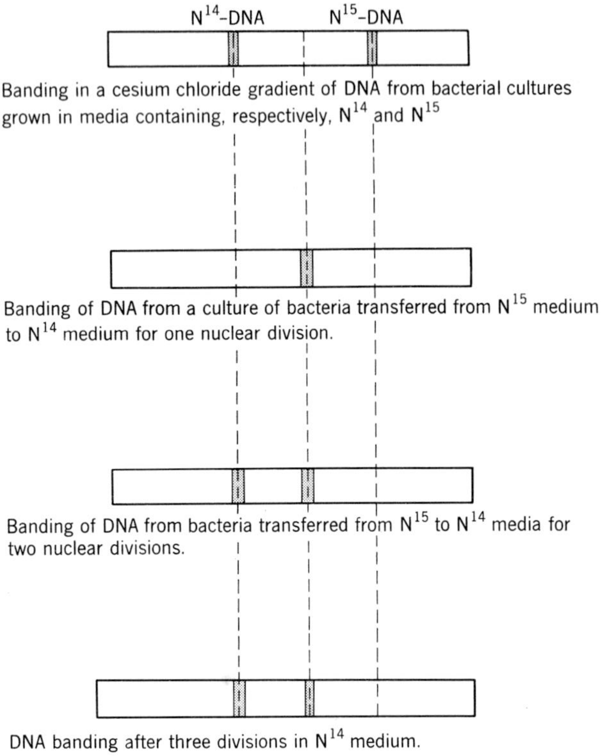

N14-DNA
N15-DNA
Banding in a cesium chloride gradient of DNA from bacterial cultures grown in media containing, respectively, N14 and N15
Banding of DNA from a culture of bacteria transferred from N15 medium to N14 medium for one nuclear division.
Banding of DNA from bacteria transferred from N15 to N14 media for two nuclear divisions.
DNA banding after three divisions in N14 medium.

Fig. 12-2. The distribution of heavy nitrogen atoms (N^{15}) in DNA during growth in media containing only N^{14}.

N^{15} strand; in the second both strands would have both isotopes, perhaps in roughly equal amount. A distinction between these distributions is made by allowing the cells to continue synthesizing DNA through another cycle. If half-molecules retain their integrity, even after an additional replication some of the molecules should consist of an N^{14} strand and an N^{15} strand, but others should now consist entirely of N^{14} strands. Hence a semiconservative mechanism would predict the formation of two classes of DNA after a second replication, N^{14}/N^{15} molecules and N^{14}/N^{14} molecules. The dispersive mechanism on the other hand would predict a uniform dilution of the N^{15} label and the production of a single class of molecules containing about $\frac{1}{4}$ of their atoms as N^{15}. This was not observed. Two clearly distinguished bands were formed, one corresponding to the first generation DNA (N^{14}/N^{15}) and one corresponding to pure N^{14} DNA. The results are entirely concordant with a semiconservative mechanism. In replicating, the two strands of the DNA molecule are separated from each other and the complementary strands to complete the double helices are newly synthesized (Fig. 12-3).

This ability of DNA to produce more of its kind is not known for any other subcellular compounds (except RNA in certain viruses), and is perhaps the most important device for preserving specificity during reproduction. Interestingly enough, Arthur Kornberg and his associates have demonstrated the same capacity in cell-free systems. Preparations including DNA precursors and a special enzyme, "DNA polymerase", are found to form no DNA in the absence of DNA, but when the mixture is "primed" with a small amount, more DNA is formed. This new DNA appears to be similar to the primer, in that the base ratios in the new DNA correspond to those in the priming material. These studies were carried out with bacteria, but similar studies have been undertaken with a variety of organisms and similar results have been obtained. Replication by a semiconservative mechanism appears to be a common property of DNA molecules in all kinds of cells.

Studies on the replication of DNA molecules do not, however, solve the problem of chromosome replication, and that of cellular replication is even more remote. Understanding the relationship of DNA synthesis to chromosome synthesis requires first an exploration of the relationship of the DNA molecule to the larger chromosome structure (Fig. 12-4). One question which has not been satisfactorily answered so far

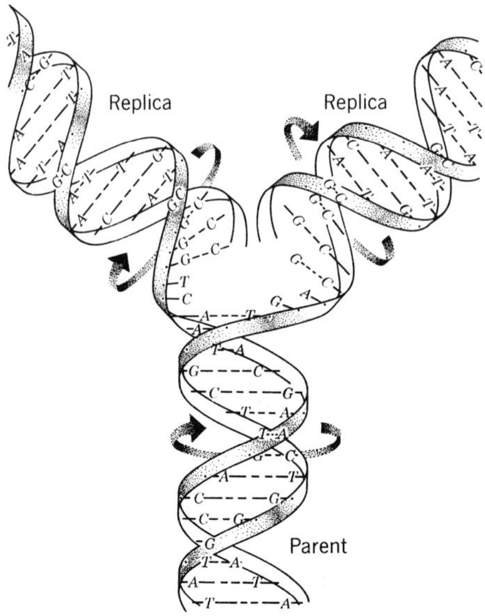

Fig. 12-3. The Watson-Crick hypothesis for DNA replication.

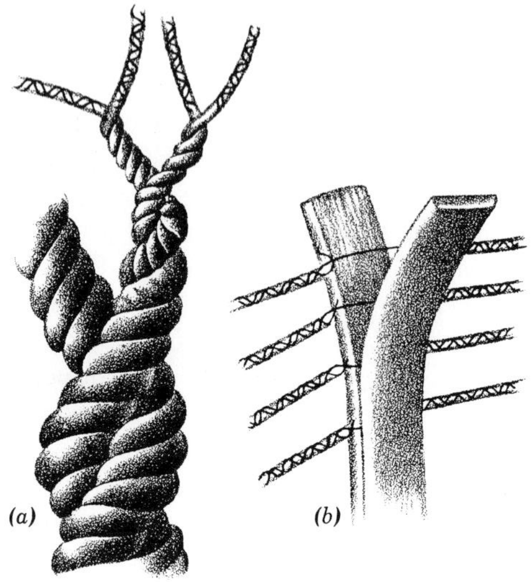

Fig. 12-4. Hypothetical models of chromosome structure. (*a*) Multistrand or "rope" with eight helices of DNA-histone. (*b*) Protein backbone with lateral DNA-histone fibers. (*c*) Alternating DNA-histone fibers and proteins. (*d*) Differential coiling of single DNA-histone fibers. (From H. Swift, in *Molecular Control of Cellular Activity* (ed. by J. M. Allen) McGraw-Hill, 1962.)

is the number of DNA molecules incorporated into a chromosome. Although extracted DNA molecules are usually several orders of magnitude smaller than would be expected of a single DNA strand per chromosome, DNA is notoriously fragile and is almost certainly greatly fragmented by most preparative procedures. Nevertheless, the possibility remains open that each chromosome, at least in forms more complex than bacteria and viruses, includes several or even many DNA molecules. If this is so, the manner of organization of the molecules is critical in evaluating chromosome replication. The individual molecules might, for example, be joined in linear arrays by "connecting links" made up of other materials, perhaps proteins. Or the DNA molecules might project as side chains from a continuous protein backbone. In either case the individual molecules might replicate by a semiconservative mechanism, but the chromosome as a whole might not.

Another critical question is that of strand multiplicity. Does a chromosome consist of a single strand, or is each type or section of DNA represented by several copies in as many separate strands? This question cannot be answered confidently on the basis of direct cytological examination. In their condensed state, when they can be studied by light microscopy, the chromosomes are hundreds of times wider than a single nucleoprotein strand, and some cytologists believe they can see a multiplicity of strands even with a light microscope. Electron micrographs of sections of chromosomes reveal a bewildering array of smaller fibrils apparently organized in several hierarchies of associations (Figs. 12-6, 12-7). Such observations could indicate that the

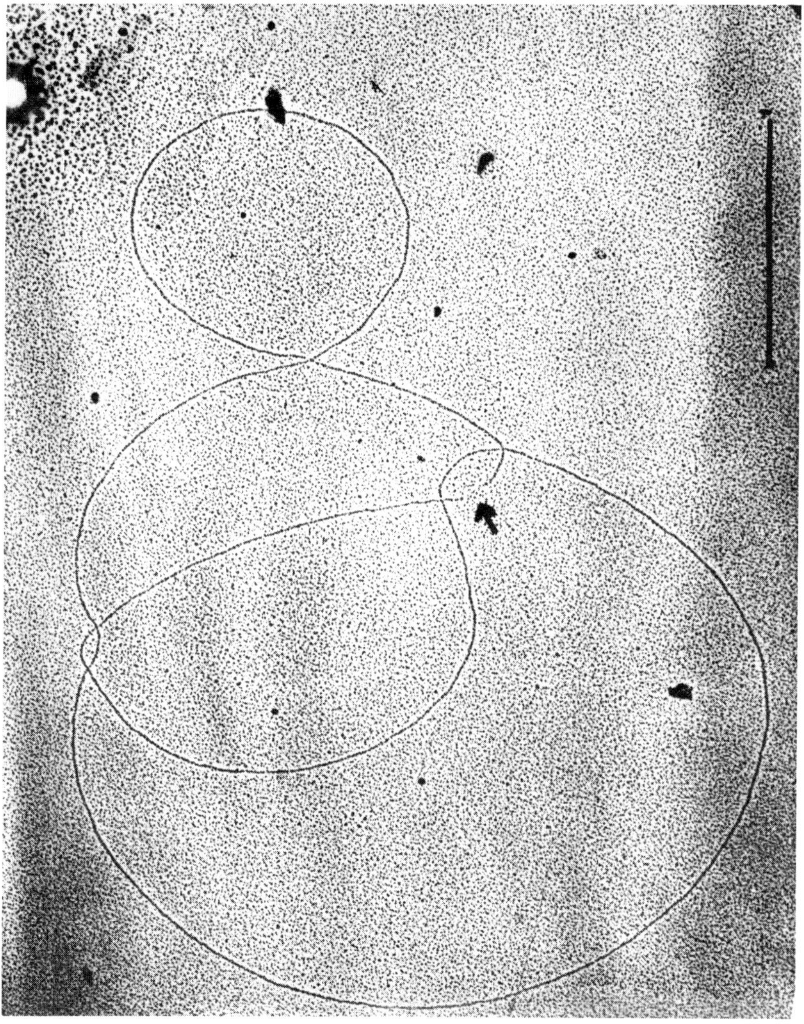

Fig. 12-5. The DNA molecule of lambda phage. To provide sufficient contrast, the molecule has been "shadowed" (that is, coated with a metal). The circularity of the molecule which consists of two complementary strands is evident. A probable break in the ring is indicated by the arrow. (Courtesy of Hans Ris.)

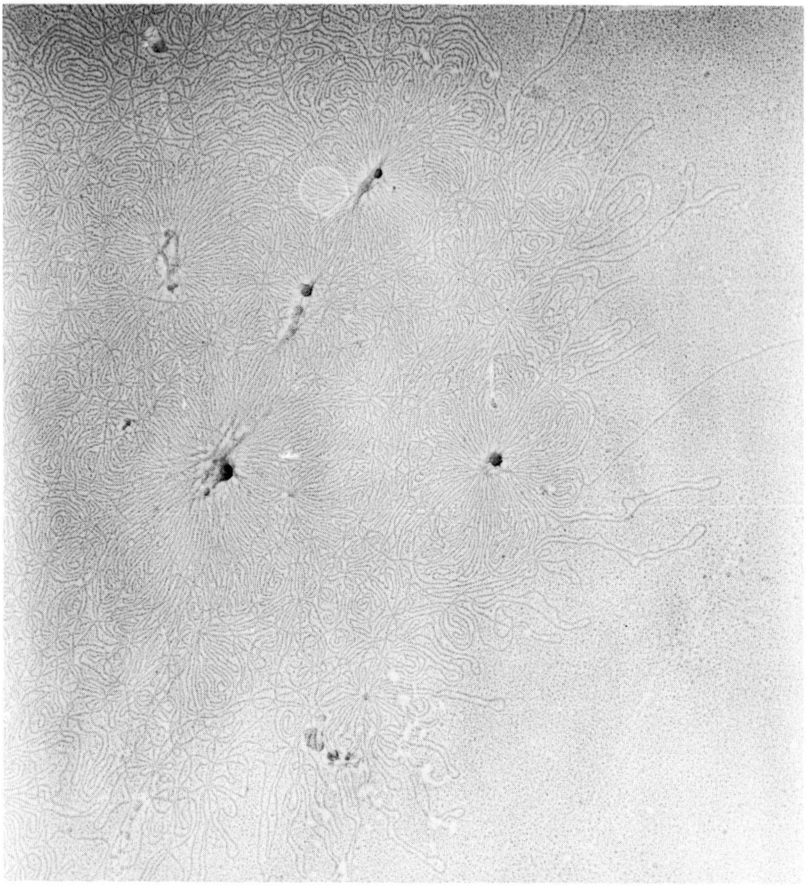

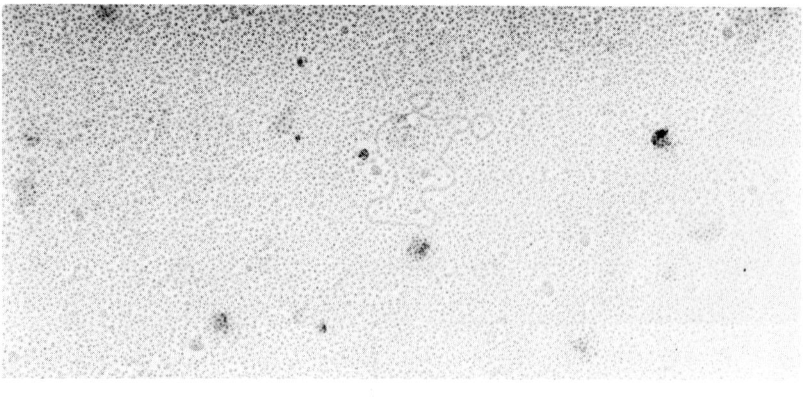

chromosomes are compound, perhaps highly compound. Nevertheless the possibility cannot be excluded that the apparent multiplicity of the chromosome is a consequence of the coiling and packing of a single continuous strand of nucleoprotein. Indeed, some students of "lampbrush chromosomes," the large unraveled chromosomes found in the oocytes of many organisms, believe this to be the best explanation of their observations.

Recent studies also suggest another interpretation of chromosome structure. The DNA of bacteriophage has been shown by electron microscopy (Fig. 12-5), as well as by genetic evidence to consist of a single closed circle. The bacterial genome has been shown to have a similar organization. Occasional rings are even found in DNA preparations from higher organisms (Fig. 12-6). The possibility exists that these rings reflect a fundamental pattern in the organization of DNA, but the precise relationship between the rings and the larger chromosomal organization is still completely obscure.

In view of the lack of agreement among experienced cytologists concerning the structure of chromosomes, an attempt to determine the mode of replication of chromosomes might seem premature. Indeed, had certain results been obtained they would have been most difficult to interpret. Surprisingly, the studies on chromosome replication have not only provided information on how chromosomes makes more chromosomes but also strongly support one interpretation of chromosome structure: The chromosomes behave as if they contain single uninterrupted DNA molecules. The evidence was obtained in the following manner. Cells were provided with tritium-labeled thymidine, which had been shown to be incorporated only into DNA. The tritium label was then a marker for new material. The label was detected by allowing the cells to duplicate in the new medium, placing them on microscope slides, and covering them with a photographic emulsion. When the tritium decayed, it gave off short beta rays which were capable of blackening silver grains in the emulsion. Since the beta rays are weak, they do not penetrate far, but leave blackened grains very near their origin (Figs. 12-8, 12-9). This technique is sensitive

---

Fig. 12-6. DNA filaments prepared from pig sperm. The tangle of threads is difficult to loosen without breaking the filaments. How long an individual filament might be is yet to be determined. The lower photograph shows a circle of DNA from the sperm of a boar. DNA rings of various sizes have recently been isolated from chromosomes of higher organisms. Whether chromosomes actually consist of a series of DNA rings is still an open question, but the presence of similar molecules in a bacteriophage and in a pig is an excellent illustration of the universality of certain basic molecular configurations. (Courtesy of Alix Bassel.)

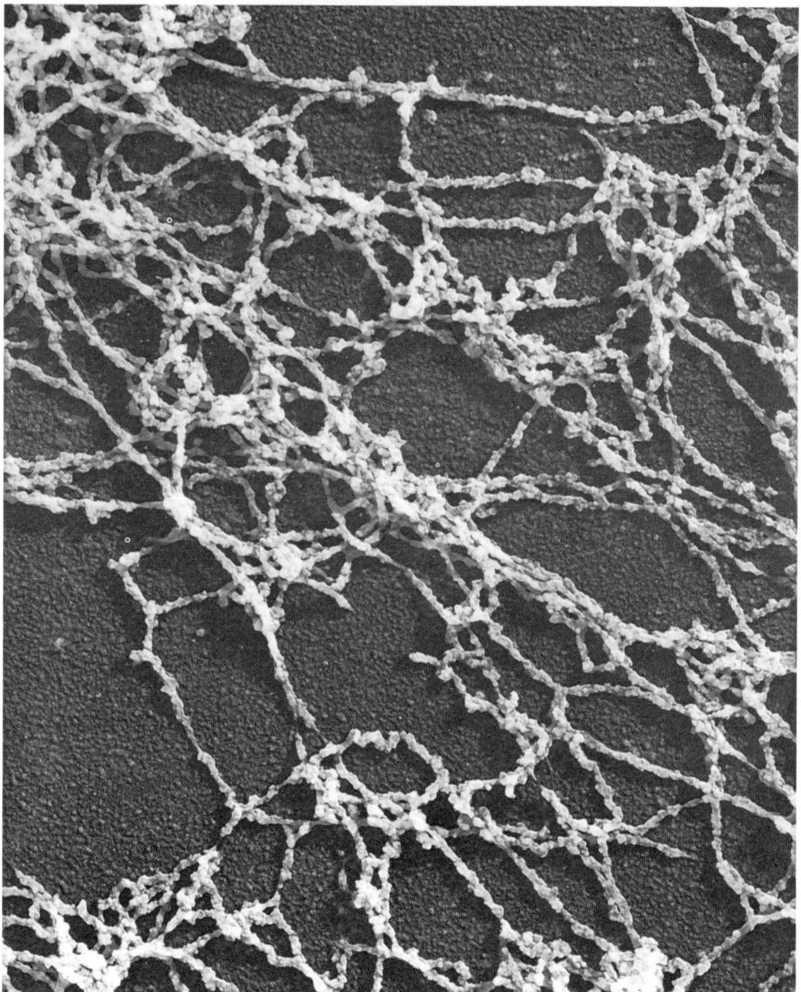

Fig. 12-7. Chromosomal filaments from the nucleus of *Triturus viridescens*. In this preparation proteins have not been removed. This photograph should be contrasted with Fig. 12-5. DNA of bacteriophage, like that of bacteria, has very little protein associated with it. DNA from "eucaryotes" (cells with a defined nucleus), on the other hand, are always associated with protein. Not only are the filaments thicker than a double-stranded thread of DNA, but if the filaments are freed of protein by means of a proteolytic enzyme, the residue has a diameter large enough to account for 4 double-stranded threads. The original filaments must remain undisturbed during proteolysis. If DNA is prepared by bulk extraction methods, the picture obtained is that shown in Fig. 12-6. (Micrograph by Hans Ris.)

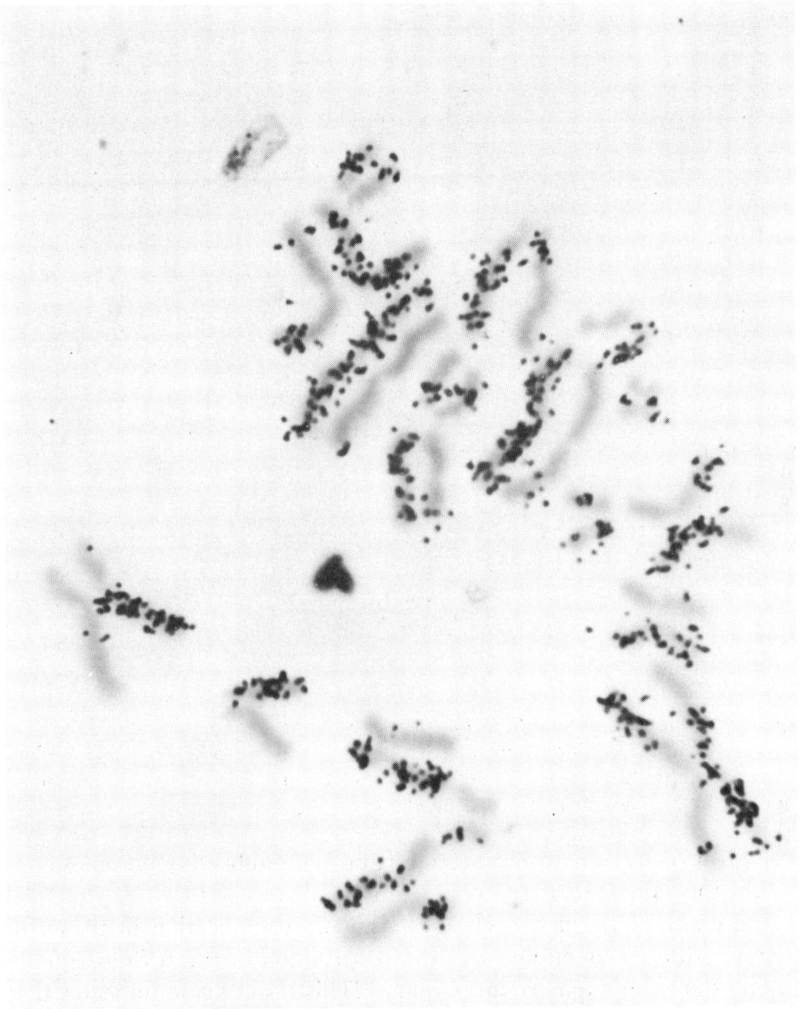

Fig. 12-8. The distribution of radioactive labels incorporated into chromosomes after two subsequent replications in the absence of label. Each set of sister chromosomes (still in close association) consists of one labeled chromosome, indicated by the silver grains deposited in the photographic emulsion, and an unlabeled one. Occasional exchanges result in partial labeling of both sisters. (Photograph supplied by D. Prescott and M. A. Bender.)

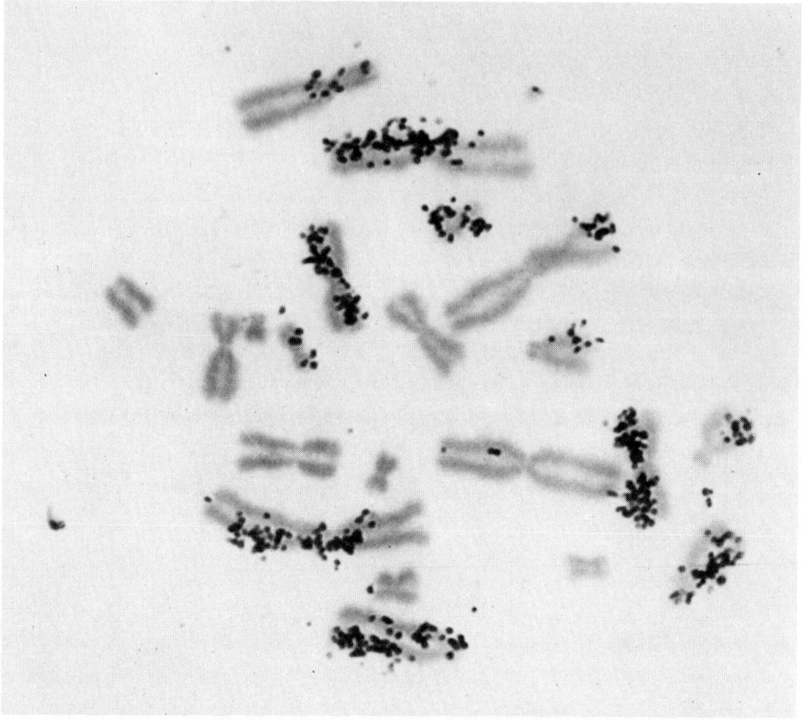

Fig. 12-9. The distribution of radioactive labels after a third replication. Some of sister sets are now entirely unlabeled. Others consist of a labeled and an unlabeled sister. The effects of exchanges are again evident here. (Photograph supplied by D. Prescott and M. A. Bender.)

enough to discriminate radioactive from nonradioactive chromosomes within the same cell if the cell is properly flattened.

The result of the first cycle of chromosome replication in the presence of tritiated thymidine was the distribution of the tritium equally among the two daughter chromosomes. This result again indicates that chromosomes, like DNA molecules, do not replicate conservatively. Each daughter chromosome receives some of the newly synthesized material and, presumably, some of the original material. A test of the semiconservative and dispersive alternatives is again provided by examining an additional generation. If replication of chromosomes is semiconservative, the first cycle chromosomes should consist of a "hot" half-strand and a "cold" half-strand. If such a chromosome were allowed to replicate in the absence of a thymidine label, one daughter chromosome would receive the labeled half-strand and the other the unlabeled. Since the new half-strands would be unlabeled, the results expected would be a "hot" daughter chromosome and a "cold" one. A dispersive mechanism at the chromosome level would yield two labeled daughter chromosomes, with approximately half of the radioactivity going to each. The results obtained were those anticipated on a semiconservative mode of replication.

The simplest interpretation of these results is, as already suggested, that a chromosome may contain a single uninterrupted DNA molecule. Since some cytological evidence supports this interpretation, and since most genetic results are most readily interpretable on this basis, such a structure may be accepted as the simplest working hypothesis available. Certainly any chromosome model which requires a multiplicity of strands encounters great difficulty in the face of the genetic evidence. Recombination in meiosis almost invariably is interpretable in terms of unitary chromatids, and crossing-over between sets of multiple strands in precisely the same point is difficult to visualize. Moreover, when mutations are induced in sperm cells by ionizing radiation, most of the mutations are "complete" mutations, or "half" mutations, either of which is consistent with a duplex molecule. A multiple-stranded chromosome might be expected to receive mutations in only one or a few of its copies, and these would be expected to yield offspring in which the different strands are sorted out among the tissues to yield a genetic mosaic. The requirement that precisely the same molecular rearrangement will occur in each member of a bundle of DNA molecules is difficult to accept.

In spite of such arguments, the simple view of chromosome structure and replication is by no means universally accepted, and the simple view sometimes is proved to be a simple-minded view.

CELLULAR REPLICATION

An understanding of the replication of DNA and of chromosomes does not explain cellular replication, but it is an important first step. Although all cellular constituents must be replicated in the course of the cell cycle, only the DNA is known to govern its own reproduction in a direct way. And, as we shall see, the genetic materials not only are capable of autosynthesis, but also through their heterocatalytic functions regulate the qualities and quantities of other cellular compounds. Before we can adequately discuss cellular replication, we must first consider gene action and its regulation.

13

Molecular Transcription

GENES AND BIOCHEMICAL PATHWAYS

The studies so far discussed are concerned with the nature and the transmission of hereditary determinants, but they do not elucidate the mode of action of these determinants. The problem of gene action may be approached in a variety of ways. The objective is to uncover the chemical basis for persistant biological specificities, but the procedures may be very different. One approach is to start with phenotypic end-points which show hereditary variation, flower colors or developmental anomalies, for example, and to trace back step by step the biochemistry of these variations until the role of the gene is elucidated. This, however, proves to be difficult, since many interrelated chemical functions are involved in most phenotypic characteristics. No simple one-to-one correspondence is usually found between complex phenotypes and particular genes. Many different genes on several chromosomes influence the eye colors in Drosophila, for example. And most genes have effects on a variety of an organism's characteristics.

More substantial progress has been made through a different approach. Instead of starting with known genetic differences and trying to solve their chemistry, studies were begun with known chemical events in an attempt to understand their inheritance. This approach was first employed successfully by G. W. Beadle and E. L. Tatum with Neurospora and is now commonly utilized with many microorganisms. It may be useful to outline the methods employed in locating genetic variants controlling chemical events in Neurospora.

With Neurospora, as with any other organism, genetic variants (mutants) may be induced by mutagenic agents—any of a variety of physical or chemical agents including ionizing radiation, ultraviolet radiation, nitrogen mustard compounds, analogs of natural purines and pyrimidines, etc. However, even under the best conditions a large fraction of the treated cells are killed and the frequency of particular mutants among the survivors is very low. Methods for screening the treated cells are necessary; automatic selection devices are desirable. Short cuts are possible only for certain classes of mutants and for certain kinds of organisms. Often the search for the necessary mutants is a long slow process. In the early work with Neurospora this was the case.

Neurospora (see Chapter 2) requires little in the way of preformed constituents from its environment—certain salts, a carbon and a nitrogen source, and the vitamin biotin; from these it can synthesize its constituent organic compounds. The normal wild-type strains can thus be grown on a relatively simple "minimal medium." Mutants with biochemical defects might, however, be expected to lose the ability to make certain compounds and to require additional nutritional supplements. To obtain such mutants the following procedures were followed. Suspensions of conidia were treated with a mutagen with a dosage which killed 90–99% of the spores. Since the conidia are usually multinucleate cells, another step had to be taken to assure that the cultures to be tested were derived from a single nucleus. The simplest procedure for this was to cross the treated spores to another strain of the other mating type. Only one nucleus enters each fruiting body, and within the fruiting bodies the ascospores are initially uninucleate. Thus if we select single ascospores from many separate fruiting bodies, a group of independently derived single nuclei are obtained for further screening.

The screening consists first of growing the spore cultures on a so-called "complete medium," a medium containing a complex mixture of organic compounds which might be required by the strains—amino acids, nucleic acid constituents, and vitamins. This procedure does not discriminate mutants from normals, since nonmutated strains grow on this medium; and the procedure does not permit the growth of all mutants, but only those which can be "repaired" by the added supplements. The cultures which develop on the supplemented medium are then transferred to a minimal medium which discriminates the nutritionally deficient strains from the normals; the normal strains grow and the defectives either grow poorly or not at all. For ease of analysis mutant strains are usually chosen for further study which grow at approximately normal rates in complete medium and which do not grow at all in minimal medium.

The mutants so selected are then analyzed to determine which components of the complete medium are necessary for their growth. This analysis can be carried out in a series of stages. The cultures might, for example, be supplemented with mixtures of amino acids in one test, with mixtures of purines and pyrimidines in another test, with mixtures of vitamins in another test. Usually they would be found to respond in one such situation; some of the mutants would grow when given amino acids, but not when supplied with vitamins or nucleic acid precursors; other mutants would respond to the vitamin mixture.

These tests identify the classes of defects in the various mutants. Their specific defects can be further identified by supplementation with one or a few of the compounds in the effective mixture. Thus a mutant which grows on the amino acid mixture might be exposed singly to an array of the amino acids tryptophan, arginine, isoleucine, etc. Each mutant is usually found to respond to one of the supplements. Eventually large numbers of mutants of various kinds have been assembled: tryptophanless (requiring tryptophan), pantothenicless (requiring pantothenic acid), etc. After the mutants are characterized, they are analyzed further through breeding studies. They are crossed with normal strains, asci are removed from the perithecia, and the ascospores are isolated in order. In nearly all cases each ascus yields four spore cultures which grow on minimal medium and four which require the supplement identified as a necessity for the mutant. Hence mutants so developed and so characterized usually behave as though their abnormal behavior is due to a change in a single gene.

These results are usually interpreted as follows. Each of the mutants has lost the ability to synthesize a particular kind of organic molecule. The formation of such a molecule invariably involves a series of enzymatic reactions; the loss of the ability to form the required molecule is therefore probably due to a defect in or absence of an enzyme. Each of the defects is correlated with a change in a gene, as demonstrated by breeding analysis. Hence it appears that genes act through their control over the formation of enzymes.

This interpretation was formalized as the "one gene-one enzyme hypothesis." It states that each gene specifies one and only one enzyme and that each enzyme is specified by one and only one gene. A number of further studies raised critical tests for this hypothesis, and we will review some of them briefly.

For example, several independently derived mutants might respond in the same way to preliminary nutritional supplementation, but would be found in breeding analysis to be changed as a result of mutations of different normal genes. Did this indicate that some enzymes were controlled by two or more genes, so that mutations of any one of the normal genes could result in a defective enzyme? Or did the different normal genes actually control different enzymes, both of which were required for the fabrication of the critical compound? Studies in cellular metabolism demonstrate that biochemical reactions often occur in sequence, with each step involving a minor modification of a molecule and each step catalyzed by a different enzyme. However, a decision cannot be based entirely on general considerations and further study is required. When more extensive studies of growth requirements of apparently identical mutants are carried out, it is often possible to discriminate between them. For example, five independent mutants (m_1 to m_5) might grow when supplied with substance a, but respond differently when supplied with compounds b, c, d, and e, which are structurally similar to a.

Mutant	Growth on Supplement				
	a	b	c	d	e
m_1	+	−	−	−	−
m_2	+	+	−	−	−
m_3	+	+	+	−	−
m_4	+	+	+	+	−
m_5	+	+	+	+	+

Such results indicate that all the genes are involved in the same biochemical pathway, but at different steps. Mutant m_1 represents a defect in the conversion of substance b into substance a; it will not grow when supplemented with any compound in the sequence prior to a. Mutant m_2 involves a block in the transformation of substance c into b, but if it is given b it can also synthesize substance a. Similarly, mutants m_3, m_4, and m_5 represent blocks at progressively earlier steps in the sequence. In agreement with this interpretation, some of the mutants accumulate in their culture medium compounds whose utilization is blocked. Mutant m_3, for example, might be found to accumulate substance d.

Normal Sequence

Gene	m_5^+		m_4^+		m_3^+		m_2^+		m_1^+	
	↓		↓		↓		↓		↓	
Enzyme	E_5		E_4		E_3		E_2		E_1	
	↓		↓		↓		↓		↓	
Substrate	⟶	e	⟶	d	⟶	c	⟶	b	⟶	a

Mutant m_3

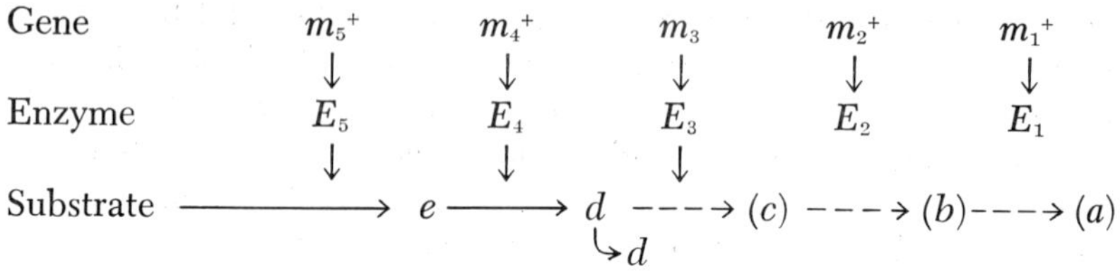

These studies showed that the occurrence of mutations at a number of genetic loci which could be remedied by an identical supplement did not indicate that all these genes were concerned with a single enzyme. Indeed, further investigation of these cases provided impressive evidence for the one gene–one enzyme hypothesis.

Other situations also needed to be explained. Occasionally mutants were obtained which would not grow on minimal medium, but which would grow on complete medium just like the mutants thus far discussed; however, they were found not to respond to any one supplement and required not one but two different compounds for normal growth. Such results would be expected if some genes were responsible for not one but two enzymes, and the existence of such genes would contradict the universality of the one gene–one enzyme hypothesis. Other interpretations were possible, however, and these had to be explored. One possible explanation is that these were not single mutants, but represented strains which had by chance received mutations in two different genetic elements. This possibility could be tested by breeding analysis, and indeed some of these mutants were found to segregate, producing some progeny requiring one of the supplements and some requiring the other. Some such mutants, however, did not segregate for the two requirements and appeared to have sustained an alteration in only one gene.

Satisfactory explanations for these cases emerged eventually when the biochemical bases for the deficiencies were explored. Biochemical pathways not only proceed in sequences, but occasionally they converge or bifurcate and eventually yield interlocking networks of biochemical reactions. A bifurcation is produced when a compound can be modified in several ways by different enzymes, or when a larger molecule is broken by enzymatic action into two smaller molecules; if both of the smaller molecules are required for growth, a double nutritional supplement is required, even though only a single enzymatic lesion has been incurred.

This is not the only explanation for double requirements. We have mentioned that a genetic block may not only lead to a requirement for some compound past the block in a biochemical sequence, but it may also result in the accumulation of a precursor. In some cases the accumulated precursors have secondary effects and may effectively block a reaction in some other pathway, by interfering with another enzyme, for example. To repair the damage of a single defect, an organism may need to be supplied with two different compounds.

Still another mechanism is occasionally encountered. Although enzymes are notoriously specific, occasionally a single enzyme may participate in more than one reaction; it may catalyze similar reactions of related molecules. If such a double-functioned enzyme is defective, two different supplements might be required to restore normal growth.

Thus, many of the apparent exceptions to the one gene–one enzyme hypothesis are readily explained when the chemical basis of the traits is subjected to study. Other exceptions, however, are not so readily reconciled and may lead to further refinements in the interpretation.

One kind of exception is illustrated by human hemoglobin, the pigment found in the red blood cells. One of the first hemoglobin modifications was found through the study of individuals afflicted with sickle cell anemia, a blood disease so designated because of the peculiar form taken by the cells under lowered oxygen tension. The hemoglobin from such individuals (*S* hemoglobin) was first differentiated from the normal hemoglobin (*A* hemoglobin) by its behavior in an electrophoretic apparatus. Eventually a detailed analysis of this molecule showed that it differed from normal hemoglobin by having a single amino acid substituted in its structure. Genetic studies, meanwhile, demonstrated that sickle cell anemia was inherited as if due to a single mendelian gene. Individuals of type Hb^A/Hb^A have normal hemoglobin; those of genotype Hb^A/Hb^S have some hemoglobin of each type and a moderate anemia; those with the genotype Hb^S/Hb^S have only S hemoglobin and are severely anemic. These facts are, of course, readily encompassed in the one gene–one enzyme hypothesis. A single gene appears to be responsible for the formation of hemoglobin, and a single change in a gene is correlated with a single change in the amino acid sequence of the protein molecule.

However, when other hereditary blood disorders were examined, they yielded facts which complicated the interpretation. Several of these were found to be correlated again with modifications of the hemoglobin, but modifications different from those encountered with sickling. In family studies the hemoglobin variations were found to be transmitted as if due to single gene differences. If a single gene controls the structure of the hemoglobin molecule, these different mutants should be allelic to each other, that is, they should be different modifications in the same gene. However, in some families with two different blood disorders the traits were found to be segregating independently; the genes responsible for the different hemoglobin modifications were not allelic; at least two genes seemed to be involved in the specification of the hemoglobin molecule.

In this case an explanation and reconciliation emerged from a more thorough knowledge of the chemistry of the molecule. Hemoglobin is not a single uninterrupted protein molecule, but a complex of four polypeptide chains. These chains are of two kinds, designated as the alpha chain and the beta chain; the hemoglobin molecule consists of two of each. The available evidence suggests that different genes are responsible for these two polypeptide chains. Although a single gene does not specify the entire hemoglobin complex, it does seem to specify a single polypeptide chain. This relationship between genes and polypeptide chains is more fundamental than that between genes and units of enzymatic activity, for many enzymes may consist of two or more kinds of polypeptides.

Other apparent exceptions to simple gene-enzyme relations also occur, and some of these have not been completely explained. Certain mutants, for example, appear to be blocked in a certain reaction step and apparently lack the relevant enzyme. However, the enzyme may be recovered in extracts of the cells; the enzyme is present, but enzyme *activity* is absent; the enzyme, indistinguishable from the enzyme in normal cells, is incapable of carrying out its characteristic reactions in intact mutant cells. This observation does not deny that *some* gene is responsible for the specification of this enzyme, but the gene under study is apparently not that gene, and the action of the mutant is not known. It might indirectly alter the intracellular environment, perhaps through enzymatic action of an unknown sort, so as to render ineffective the first enzyme. Or it might interfere directly or indirectly with the production of some essential cofactor. In any case, the mode of action of the gene is not established, and it may not be involved in specifying an enzyme or other protein at all.

Several situations similar in principle to this are known. Mutations at any of several different genetic loci may markedly modify the amount or activity of a particular enzyme or even reduce it below the limits of detection. Perhaps one of these genes is *the* gene determining the specificity of the enzyme, but no simple means of determining which is the "structural" gene have been developed. Moreover, the mode of action of the "modifying genes" remains unspecified. For these cases the one gene–one enzyme hypothesis remains a possible but thus far incompletely established interpretation.

Another set of studies may also be mentioned as suggesting that several genes may be involved either directly or indirectly in controlling the structure and/or activity of a single enzyme and, conversely, that single genes may be involved in the regulation of several enzymes. These studies are concerned with "suppressor mutations." If many cells from a nutritionally deficient strain are plated on a minimal medium, a certain small fraction are often found to grow. On analysis some of these "reversions" are shown to be due to changes at the original locus of the mutation, they are true "back mutations." Others, however, prove to be due to changes at completely different loci; the reverted strains contain the original mutation, but also contain another mutation which cancels the effect of the first. When crossed, the original mutant may be separated from its suppressor and reappear in unchanged form. Several possibilities exist for explaining the mode of action of suppressors.

One possibility is that some other gene undergoes mutation and as a result begins to produce the missing enzyme. Aside from the inherent improbability of this *de novo* production of an enzyme indistinguishable from the original, this interpretation is rendered unlikely on other grounds. First, a single suppressor mutation is occasionally found to suppress mutants involving several different enzymes. We would have to postulate that in a single mutational step a gene acquires the ability to synthesize several different enzymes. Second, a suppressor gene is not able to accomplish its action autonomously; it requires a gene at the original locus. This is shown by the fact that a particular suppressor is able to suppress (i.e., repair) only certain mutants at a locus. And these suppressible mutants are generally those which continue to produce a protein similar in structure to the normal enzyme, as demonstrated by immunological techniques, but without the activity. This suggests that some mutations result in the formation of modified products and that the action of other genes may restore the activity of these products to normal. But if, in these special cases, modifying genes can make significant modifications in gene products, it would appear not improbable that similar secondary modifications might be a common occurrence in the production of enzymes. If so, the one gene–one enzyme hypothesis would have to undergo a modification to account for the facts.

Thus at the present time the one gene–one enzyme hypothesis appears to summarize adequately many of the relationships between genes and biochemical functions. Indeed, by a not uncommon shift in definitions, a gene is sometimes defined as that element which prescribes an enzyme. However, we may not be perfectly certain that all genes operate by prescribing enzyme (or protein) structures. In particular, we must be aware that the very technique which enabled a probing of gene-enzyme relations were designed for locating enzymatic differences in cells. The possibility that other genes operate on other principles is yet open.

Finally, studies such as those just discussed only partially bridge the gap between phenotype and genotype; they establish a bridgehead at an intermediate point, but do not solve all the problems relating to the interval between genes and enzymes or that between enzymes and cellular or organismic phenotypes.

EXERCISES ON GENE ACTION

Studies of inheritance in higher forms frequently lead to demonstrations of various kinds of genic interaction and various phenotypic ratios, even when genic transmission is regular. Many of these patterns and ratios can be rationalized by models of metabolic pathways. Consider the following hypothetical example:

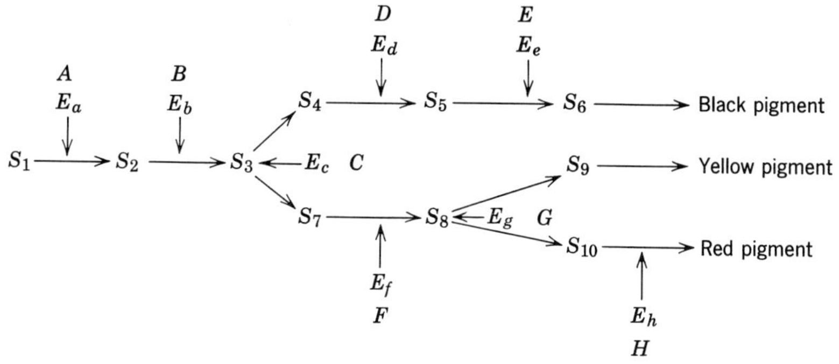

Assume that this is a sequence of molecular transformations leading to the production of pigment in a mammal. Assume further that in the wild-type mammal all the genes are the dominant genes, that all the enzymes and all the pigments are produced. However, if black is produced, the other pigments are masked. When black is not produced, the animal may be either red, yellow or brown (red + yellow).

Now consider the consequences of substituting various homozygous recessive genotypes into an organism which block the formation of particular enzymes. If a genotype including a large A is replaced by the homozygous recessive condition *aa*, substance S_1 is not converted to S_2; none of the subsequent reactions can occur, and no pigments are formed; an albino results. If gene *D* is absent, S_4 is not transformed into S_5; and no black pigment is produced; the organism can still produce red and yellow pigment, however, and has brown fur.

Now determine the phenotypic ratios expected from the following crosses; assuming that unless otherwise specified all the relevant genes are present in the homozygous dominant form.

1. *AaBb* × *AaBb*
2. *AaDd* × *AaDd*
3. *DdFf* × *DdFf*
4. *EdHh* × *EeHh*
5. *CcGg* × *CcGg*

THE TRANSLATION OF GENETIC INFORMATION

The primary activity of genes was a problem which could scarcely be phrased in molecular terms until the chemical nature of genes was understood. Now that DNA is established as the usual reservoir for genetic information, the possible modes of action can be probed. First we must recapitulate the chemical nature of the genetic material for the insight it provides into the language of genetics.

DNA is what we might designate as a "differentiated" polymer, a long chain made up of (usually) four kinds of nucleotides in many different sequences and ratios. It is essentially a one-dimensional structure; although it is made up of two strands, one is a complement of the other and could be specified if the sequence of the first were known. It might appear that such a structure is incapable of a large amount of variation and hence cannot be a reservoir of much "information." That this is not the case can be demonstrated very readily. Considering only one chain of the duplex molecule, we can see that at any point in the chain only one of the four possible nucleotides may be placed. But this is all that is necessary. The Morse code contains only two elements (or three if spacing is considered), but *Gone With the Wind* could be transmitted by Morse code. Consider how many different chains can be built even if the chain is only ten nucleotides long. For nucleotide 1 we have four choices. Each of these might be combined with any of the four types in position 2, thus yielding sixteen kinds of dinucleotides. Again each of these could be combined with any of the four nucleotides in position 3 to give a total of 64 kinds of trinucleotides. This progression may be symbolized by the expression 4^N, where N is the number of positions and 4 the number of possibilities at each position. Hence 4^{10} or 1,048,576 different kinds of ten-unit polynucleotides could be constructed if we start at a particular end. Yet most estimates of gene size would involve nucleotide sequences not of 10 units, but of at least a thousand and possibly a million or more. And even 4^{1000} is an extremely large number, larger than the number of seconds in the ten billion or so years of the earth's existence and larger by far than the number of electrons in the visible universe. The amount of information which could be encoded in DNA, even though it is limited to four kinds of units at a given position, is essentially limitless.

Granting that information can be stored in DNA, we have yet to consider the precise manner in which the information is translated from DNA structure to cellular specificities. We have seen that there is a close correspondence between genes and enzymes, or more generally between genes and polypeptides. And changes in gene structure are known to be reflected in amino acid changes in proteins. Proteins are, of course, built from different building blocks than nucleic acids; amino acids are used in one case and nucleotides in the other. And instead of four elementary units for DNA, twenty or so amino acids may be used in the same protein.

Since changes in genes result in changes in amino acid sequences, it appears likely that the arrangement of amino acids is in some way specified by the arrangement of nucleotides. Both proteins and nucleic acids are linear structures, and we must investigate the method of translating linear codes built up of different numbers of elements. Some information comes from the application of formal analytical techniques, such as those involved in cryptanalysis, but formal considerations are primarily useful in clarifying possibilities and in framing questions which can be approached experimentally. It may be useful to discuss briefly a few of the theoretically possible means of coding and translating chemical information.

We have mentioned that DNA contains four kinds of nitrogenous bases (adenine, A; thymine, T; guanine, G; cytosine, C), and have suggested that the differences among these are all meaningful. It is conceivable, however, that the significant differences are fewer than four—that the two purine bases (A and G) are equivalent and that the two pyrimidine bases (C and T) are equivalent. In this case the DNA code would be a two-letter code, much like a Morse code, rather than a four-letter code.

A second problem concerns the "coding ratio," the number of units in one system which is required to specify one unit in the other system. Certainly there is no one-to-one correspondence between nucleotides and amino acids. If each kind of nucleotide specified a single amino acid, only proteins consisting of four amino acids could be constructed. If the unit in DNA corresponding to an amino acid consisted of two nucleotides (AA, AC, AG, AT, etc.), a larger number of possibilities would be generated, but still not enough—only 16. If a three digit code is employed, however, a total of 64 kinds of units is established (AAA, ATA, CGA, etc.), more than enough to encode twenty amino acids. The surplus triplets create a minor problem, but several possibilities concerning their functions are open. Some of them might be used as "punctuation" in a chemical code, designating the beginning or the end of a chemical message. Some of them might be simply meaningless, in the same way that many combinations of letters in the English language are meaningless. Quite possibly the same amino acid could be designated by two or more DNA triplets. These are all problems for experimental analysis. The formal cryptanalytic approach merely indicates that at least three DNA units are required to encode an amino acid; a coding ratio of four or more is also conceivable, or even a coding ratio which varies from amino acid to amino acid. Thus phenyl alanine (or some other amino acid) might be specified by three nucleotides, while cysteine is specified by four.

Another question raised by formal analysis is whether the code is "overlapping." A sequence of bases of the following sort, A T A C C G T A G, could theoretically be read as ATA CCG TAG, thus yielding three amino acids; this reading would be nonoverlapping. Alternatively the same sequence could be read as ATA TAC ACC CCG CGT GTA TAG; each base is used in specifying more than a single amino acid and the code is overlapping. The two kinds of codes have different properties. The overlapping code is restrictive in this special sense; the amino acid specified by ATA must be followed by an amino acid specified by TA (×), where only (×) is free to vary among the four possible bases. Thus certain sequences of amino acids (as, for example, an amino acid specified by ATA followed by one specified by CCG) are not possible. Some freedom from these restrictions is possible if an amino acid can be specified by two or more different triplets (in which case the code is said to be "degenerate") or if the overlap is less extensive; the sequence above for example might be read as ATA ACC CGT TAG, providing a partially overlapping code.

One may also question whether the code is identical in all kinds of organisms. Related organisms would certainly have similar codes, but codes might be expected to change in the course of evolution so that identical triplets would be translated differently in different forms. We may speculate on the evolutionary pressures which might enhance either change or stability of the code, but this is a question which requires direct experimental approaches. A universal code would be a great convenience for biologists, of course, but its convenience for other organisms is not entirely obvious.

Formal analysis also poses other problems. So far we have treated DNA as if it consisted of a single strand of ordered bases. But the DNA molecule is duplex, and the strands are not identical but complementary, and they also have different polarities. Is the genetic information read from both strands? Is the reading from a single direction?

Many of these problems raised by a general consideration of biological codes have been solved in recent years through direct experimental attack. Before discussing these studies, however, we must first consider in greater detail the chemical mechanisms involved in the transfer of information. These mechanisms are not as direct as might be suggested by our previous discussions. The genetic material is located in the nucleus, but the bulk of protein synthesis occurs not in the nucleus but in the cytoplasm. Amino acids labeled with radioactive tracers may be incorporated into protein without passing into the nucleus. Cells may even continue to fabricate proteins for a period of time after the nucleus has been removed. In cell-free systems protein synthesis can often be demonstrated, even though it occurs at a much slower rate than intact cells, and this protein synthesis may continue even when DNA is absent. Such evidence demonstrates that DNA does not regularly participate directly and personally in the fabrication of proteins. Rather, the information encoded in the genetic material must first be transferred to an intermediary which is less localized geographically.

The nature of this intermediary has been the subject of much study and the evidence indicates that the intermediary is a special kind of RNA. Although RNA is found throughout the cell, little if any of it is synthesized in the cytoplasm; RNA precursors labeled with isotopes appear first in nuclear RNA and appear only later in cytoplasmic RNA. Moreover, and more importantly, all cell-free systems for protein synthesis contain RNA and are inactivated when they are hydrolyzed with RNAase. RNA certainly plays an essential role in protein synthesis and is a likely candidate for the role of "messenger." If so, the transfer of information from DNA to protein involves two steps, DNA to RNA, and RNA to protein, and two translations have to be considered. First, how is the DNA information transferred to RNA and in what form?

The translation of DNA information to RNA appears not to constitute a major problem. The simplest translation conceivable would be little more than a transcription, an RNA strand could be manufactured by a DNA strand in much the same way that a complementary DNA strand is formed. A cytosine base in DNA would be opposed by guanine, and a thymine base in DNA by adenine. The major difference between autosynthesis (forming more DNA) and heterocatalysis (forming RNA) would be that the complementary strand would be fabricated of ribonucleotides rather than deoxyribonucleotides, and that uridine rather than thymidine would be apposed to adenine. The immediate factor determining whether DNA is to be copied or transcribed would be the enzyme involved.

The validity of this interpretation has been strongly supported by recent enzymatic studies. The biochemical techniques which had earlier proved successful in elucidating pathways of cellular metabolism have proven equally effective in elucidating nucleic acid synthesis. Two types of enzymes have been isolated and purified, DNA polymerase and RNA polymerase; both are dependent on the presence of DNA for activity. The DNA polymerase synthesizes DNA in the presence of preexisting DNA; the RNA polymerase synthesizes RNA only in the presence of DNA. Most importantly, the characteristics of the newly synthesized nucleic acids are dependent on those of the "primer."

From the point of view of energetics, these enzymes catalyze reactions similar to those discussed in Section A. The substrates are the deoxyribonucleotide and ribonucleotide triphosphates, and the enzymes utilize the energy released by hydrolysis of pyrophosphate bonds to form phosphate ester linkages (see Fig. 6-9). From the standpoint of the chemical characteristics of the products, the reactions are unique among cellular biosyntheses. The enzymes are normally inactive if only the triphosphates are present; they will not catalyze the formation of ester linkages unless DNA is present in the solution. Furthermore, unlike most enzymes which have a specific affinity for certain substrate molecules and complex with them to yield reaction products, the specificity of the polymerase with respect to a particular nucleotide triphosphate is determined by the DNA. The enzyme thus determines whether ribonucleotides or deoxyribonucleotides are to be incorporated, but the DNA determines the sequence in which they are to be linked to one another.

These polymerases are not simply test-tube curiosities, for their activities may be demonstrated in living systems. Immediately after infection by a virulent bacterial virus, a bacterial cell is found to produce a new kind of RNA which has a specific relationship to the DNA of the infecting virus; the ratios of the nitrogenous bases in the RNA mimic the ratios of the bases in one strand of the DNA of the virus, with the exception that thymine is replaced by uracil. (This observation suggests the answer to a question raised earlier; in this case at least only one strand of the DNA is transcribed.) Moreover, this specific RNA is capable of forming a duplex structure with one of the viral DNA strands. Presumably this "hybrid" strand of DNA-RNA is possible because of the detailed complementarity of the structures of the strands. Such hybrid molecules can also be detected under special circumstances in normal cells and presumably represent newly synthesized RNA strands still in association with the master templates.

Considerable illumination on the problem of protein synthesis has been developed through disrupting cells and putting the components together again in reaction mixtures. Such studies now demonstrate that RNA is not only an essential element in protein synthesizing systems, but that at least three different kinds of RNA are involved. A fraction of low molecular weight RNA, designated as soluble or transfer RNA (tRNA), is essential. So also are the ribosomes which contain a different kind of RNA, called ribosomal RNA (rRNA), and protein. In addition, the complete system requires an energy source (ATP and GTP), amino acids, and a variety of enzymes. Even with all these components the cell-free system is incapable of producing protein unless it is supplied with "instructions" in the form of message RNA (mRNA), an unusually labile RNA fraction.

To be effective in protein synthesis mRNA must become attached to the ribosomes. This attachment is spontaneous but requires energy, at least in the form of GTP. Enzymes may be involved in the attachment, but the mechanism is still being investigated. The ribosome itself is a double structure consisting of two separate parts distinguishable by their sedimentation coefficients of 30 S and 50 S. The mRNA appears to attach to the 30 S particle.

The amino acids are bound to the other ribosomal particle, but only through the mediation of tRNA. The attachment of amino acid to tRNA represents the only direct interaction between nucleic acid and protein. Each amino acid forms an "acyl" linkage (see Chapter 6) with a specific type of tRNA. ATP provides the source of energy. However, the attachment must be catalyzed by a specific enzyme so that the number of specific attachment enzymes is at least as large as the number of amino acid-tRNA complexes.

Despite the fact that amino acid sequences are derived from nucleic acid sequences, the cell achieves the ordering in a remarkable way. Earlier the question was posed of how cells reduce entropy. The mechanism of protein synthesis provides a good example. A single molecule of ATP provides sufficient energy to form a peptide bond, but the restrictions provided by the ribosomal mechanism assure that the order of the amino acids preserves that specified by the nucleic acids.

All the details of protein synthesis have not yet been clarified, though the broad outlines are apparent. Each tRNA molecule seems to have two specific sites, one appropriate to a particular amino acid and another which "recognizes" (perhaps through hydrogen bonding) a code sequence in the mRNA. The placement of tRNA molecules at adjacent sites on the ribosome coincidentally places their attached amino acids in positions appropriate for the formation of peptide bonds. The mRNA appears to travel across the surface of the ribosome, having its "message" translated into sequences of amino acids. Eventually the completed polypeptide is released from the template. Further changes occur in many proteins; they coil into their characteristic three-dimensional structures and cross-linkages are established.

The cell-free systems for protein synthesis offer excellent opportunities for exploring the problem of information transfer. The removal of message RNA (by depletion) and the inactivation of the DNA (by DNAase) which could supply more message RNA leaves the system available for examining the consequences of introducing messages from various sources. Intact cells are ordinarily carrying out so many different syntheses that the effect of a particular addition is difficult to assess. The artificial systems, in contrast, can be used to determine the effects of messages isolated from diverse species, or even of messages which are constructed artificially.

These artificial messages have proved of greatest significance in solving the genetic code. S. Ochoa and his collaborators isolated an RNA polymerase with properties very different from the DNA-dependent RNA polymerase previously discussed; it had the capacity to catalyze the formation of polyribonucleotides without a "primer." By means of this enzyme he was able to synthesize a variety of polyribonucleotides; by controlling the nucleotides available to the enzyme, he was able to construct polynucleotides (homopolymers) containing only uridine (poly-U) or cytidine (poly-C), or mixtures (copolymers) of uridine and cytidine (poly-UC), of uracil and guanine (poly-UG), etc. Copolymers consisting of the same bases but with different ratios of those bases could be constructed by regulating the ratios of the bases in the reaction mixtures.

These synthetic polynucleotides were first shown to be utilizable as messages in cell-free systems by M. W. Nirenberg and his co-workers. A cell-free system depleted of message RNA was found capable of incorporating amino acids into proteins when it was supplied with synthetic polynucleotides. Most significantly, the kinds of amino acids which were incorporated depended on the kind of polynucleotide supplied. When poly-U was employed, polypeptide chains composed entirely of phenylalanine were fabricated. The other three homopolymers appeared to be less effective in promoting polypeptide formation, but a variety of kinds of copolymers were used and many of these were found to promote polypeptide formation and the kinds of amino acids included in the polypeptides were specific for the kind of copolymer used.

Nirenberg, Ochoa, and their collaborators have subsequently established a series of correlations between nucleotide compositions and amino acid incorporations. The manner in which these correlations were established can be set forth simply. Poly-U was found to promote the incorporation of phenylalanine; if the coding ration is three, the triplet specifying phenylalanine appears to be UUU. If a copolymer of U and A is used, polypeptides containing not only phenylalanine but also tyrosine and isoleucine are constructed. This qualitative result yields some information. The possible triplets in poly UA are UUU, AUU, UAU, UUA, AAU, AUA, UAA, and AAA. Of this total of eight triplets the UUU is assigned; and the low activity of poly-A indicates that AAA can be neglected.

The six remaining triplets consist of two groups which can be distinguished quantitatively. Three contain two U's and one A, and the other three contain two A's and one U. The simple probability considerations discussed earlier permit us to assign tyrosine and isoleucine to one of these groups. The procedure is the following. The ratio of the nucleotides in the copolymer is determined; it might, for example, be composed of $\frac{3}{4}$ U and $\frac{1}{4}$ A. If the U and A units are distributed at random in the polynucleotide, the relative frequencies of the triplets can be calculated. The UUU triplets should have a frequency of $\frac{3}{4} \times \frac{3}{4} \times \frac{3}{4} = \frac{27}{64}$; the AUU triplets should have a frequency of $\frac{1}{4} \times \frac{3}{4} \times \frac{3}{4} = \frac{9}{64}$, as should also the UAU and UUA triplets. The triplets with two A units (AAU, AUA, UAA) should each have a frequency of $\frac{1}{4} \times \frac{1}{4} \times \frac{3}{4} = \frac{3}{64}$. The AAA triplets should make up $\frac{1}{64}$ of the total. Such calculations lead to the expectation that an amino acid coded by one of the triplets with two U and one A unit should be incorporated with this polymer $\frac{1}{3}$ as often as phenylalanine; those coded by a triplet with one U and two A units, in contrast, should be incorporated only $\frac{1}{9}$ as often as phenylalanine. On such grounds the amino acids tyrosine and isoleucine were assigned triplets with two U and one A unit.

Copolymers differing both quantitatively and qualitatively have been used, and all the amino acids have been assigned unordered triplets. However, not all the 64 possible triplets have been assigned yet. Some may be meaningless, or may be significant in providing punctuation rather than in specifying amino acids. Enough triplets have been assigned (see Table 13-1), however, to demonstrate that many and perhaps all of the amino acids are represented by two or more different triplets.

The demonstration that a single amino acid can be coded by two or more combinations of bases establishes that the genetic code is degenerate. Parallel studies suggest a plausible basis for degeneracy. Several techniques have been exploited to separate the tRNA molecules into different classes and separable groups with the same amino acid specificities have been isolated. This observation indicates that transfer RNA's with equivalent amino acid sites have distinguishable sites of affinity for portions of the mRNA.

The use of synthetic polynucleotides has provided the clearest indications of the nature of the genetic code yet available. It has, moreover, provided strong evidence for the universality of the code; synthetic polynucleotides appear to perform in essentially the same manner in cell-free systems prepared from a variety of organisms. Somewhat anomalous results are sometimes obtained, however, when the components of the cell-free systems (enzymes, tRNA, ribosomes) are combined from several sources; these results may suggest that, although the language of genetics is universal, the dictionaries involved in its translation have undergone subtle changes in the course of evolution.

*Table 13-1 Tentative Assignments of Some RNA Triplets to Amino Acids**

Amino Acid	Triplets	Amino Acid	Triplets
Alanine	CUG, CAG, CCG	Leucine	UAU, UUC, UGU
Arginine	GUC, GAA, GCC	Lysine	AUA, AAA
Asparagine	UAA, CUA, CAA	Methionine	UGA
Aspartic Acid	GUA, GCA	Phenylalanine	UUU
Cysteine	GUU	Proline	CUC, CCC, CAC
Glutamic Acid	AUG, AAG	Serine	CUU, ACG
Glutamine	AGG, AAC	Threonine	UCA, ACA, CGC
Glycine	GUG, GAG, GCG	Tryptophane	UGG
Histidine	AUC, ACC	Tyrosine	AUU
Isoleucine	UUA, AAU	Valine	UUG

* From A. J. Wahba et al., *Proc. Natl. Acad. Sci.,* **49**, 116-122 (1963).

These assignments are tentative in two respects. Some triplet assignments are still subject to verification in independent studies and under varied conditions. The possibility also exists that some triplets may not perform precisely the same way in living cells as they do in cell-free systems. Recent studies, for example, indicate that streptomycin may cause a misreading of the code so that an "illegitimate" amino acid is incorporated into a polypeptide chain. Since a variety of other conditions might modify the protein synthesizing equipment under these artificial conditions, some caution is required in extrapolation.

Other problems also remain, and other approaches are being exploited. Methods must yet be perfected for determining the order of the nucleotides in assigned triplets, for example. Improved procedures for constructing synthetic polyribonucleotides with defined rather than statistical compositions should be a major factor in this effort. Another approach is based on the increasing knowledge of the action of certain chemical mutagens; specific changes in DNA may be correlated with known alterations in amino acid sequences in proteins. Such studies may eventually allow a reasonably complete sequence determination for an entire gene. Such a completely defined gene (or its message) with its completely defined product (polypeptide) could prove to be a Rosetta Stone, making clear not only the alphabet but also the vocabulary and the grammar of information transfer in the cell.

ENZYMES AND CELLULAR PHENOTYPES

We remarked earlier that the first attempts to bridge the gap between the phenotype and the genotype were directed centripetally, from the phenotype toward the gene. We also noted that these efforts were not remarkably successful. The identification of gene-correlated metabolic events established a beachhead at an intermediate position. The chemical characterization of the genetic material, followed by studies on protein synthesis, provided a solid bridge between genes and enzymes. For the sake of symmetry we should give some attention to the remaining gap, that between protein structure or enzyme activity and cellular and organic phenotypes.

To be sure, enzyme activity is in itself a legitimate phenotype, but a special sort of phenotype. A cell is so much more than a membrane-surrounded sack of enzymes that to stop our discussion with enzymes is to neglect vast areas of biological structures and function. Enzyme molecules are not randomly dispersed in the cell but are organized into associations with other enzymes, with structural proteins, with the lipids and carbohydrates which compose the fabrics of cells, and finally into the distinctive organelles required for proper cellular function. A critical question about this organization is the extent to which it is determined by the structures of the component molecules and by their relative proportions.

Little doubt exists that some aspects of the organization are implicit in the molecules. Membranes, for example, can be fabricated artificially by combining certain kinds of lipids and proteins. And some structural proteins, such as collagen, can be dissolved and then reprecipitated into fibers essentially identical to those observed in cells. Without question the component molecules have properties, perfected by natural selection, which make them suitable constituents of cellular structures.

To so assert, however, is not the same as to say that all cellular organization is derived by the "self-organization" of the molecules. As we shall see later, a large variety of structural and functional patterns may be developed on a single genetic basis. The genotype, particularly the genotype acting on primary structure of proteins, is only one of the factors governing biological specificity and organization. The immediate environment is also important, as is the "memory" of environments of the past. With our present rudimentary understanding of biological organization we are not certain about how significant these other factors may be or how they work. And even if we were certain that their contributions were trivial, we could scarcely be satisfied with a general dogma that all biological specificity is the consequence of the unfolding of the molecular patterns stored in nucleic acid. We would need to know the mechanisms involved in that unfolding.

An understanding of the manner in which DNA specifies proteins is a first step. A detailed understanding of the relationship between the primary structure of a protein and its tertiary structure is only beginning to be worked out. The significance of tertiary structure in the function of macromolecules is a problem about which we have some hints. Still other questions arise when we consider the associations of molecules into fibers, membranes, and granules, and beyond this into the complex organelles of an integrated cell. We may have a faith that the mechanisms involved are the basic physicochemical mechanisms governing the behavior of matter throughout the universe, but on many of these questions our faith is not much better informed than that of the mechanists when they were clashing with the vitalists a century ago. The exploration of biological mysteries is not approaching completion, but is only well begun.

14

Molecular Diversification

Heredity and variation are two sides of a single coin. We have already discussed obliquely the major sources of biological variation in our consideration of hereditary mechanisms. Indeed, variations are essential materials for probing the hereditary apparatus. Nevertheless, a more explicit treatment of diversification may be useful.

In a general sense any of a large variety of hereditary changes may be designated as a mutation. Several systems of classification of mutations are employed for a variety of purposes. Mutations may be categorized on the basis of their effects on the phenotypes: some are lethals, killing the cells or organisms in which they occur; some are subvitals and reduce the probability of survival of the organism in which they are found; some small fraction are supervitals and improve the biological fitness of the organism under some specified circumstances. In addition to having effects on general vitality, some mutations have clearly defined phenotypic effects and are classified as "visibles."

Mutations may also be classified in terms of the extent of change involved in the chromosomal materials. In some cases cytological examination reveals no obvious alteration in the chromosomes; these are called "point mutations." Others may be detected as microscopic deficiencies or duplications, or even as exchanges of pieces between chromosomes (translocations). Finally some mutations involve the loss or addition of whole chromosomes (aneuploid variations), or even of whole sets of chromosomes (euploid variations). Although the distinctions are sometimes arbitrary, and a continuous gradation of sizes of mutations may be demonstrated, such a classification is useful for some purposes.

Another way of classifying mutations is by their modes of origin. Those which occur without deliberate treatment are designated as "spontaneous" mutations, while those arising from exposure to ionizing radiation, ultraviolet radiation, mutagenic chemicals, etc., are called "induced" mutations. The induced mutations are sometimes separately designated according to the treatment under which they have arisen. Again, this classification is somewhat artificial, but significant differences among the several categories are sometimes demonstrable.

SPONTANEOUS MUTATION

Before considering induced mutations, we should first try to characterize spontaneous mutations and particularly spontaneous point mutations. These are usually said to be rare, but rarity is relative and we should try to be more precise. Unfortunately, this is not easy. Although mutation rates have been empirically determined for a number of genes in organisms as diverse as viruses, fruit flies, corn, and men, we cannot be perfectly certain that they provide a basis for extrapolation of other genes in these organisms. For example, most of the mutation rates determined for genes in corn are of the order of 1 in 10^4 to 1 in 10^6/gene/generation. This is a considerable range in rates and raises the possibility that an even greater range might exist if even lower rates were explored. But determining rates of mutations which occur, say, only once in 10^8 or 10^9, sex cells would require an enormous investment in time, personnel, and equipment. The examination of so many different individuals makes such an undertaking out of the question. Therefore, the average rate of mutation for genes in corn might well be only 1 in 10^7, even though certain genes—those which have been studied—certainly mutate much more frequently than this.

On the other hand, we might plausibly argue that most methods for determining mutation rates seriously underestimate their frequency. The mutations detected are usually those which give a sharp change in phenotype. Yet we know that a single gene may mutate to a variety of forms whose effects grade imperceptibly into the normal. And other mutations are cell lethals which leave no progeny and are easily missed. If a gene is capable of an extremely large number of changes, only a few of which modify phenotype in a manner suitable for them to be detected in the rough screening required for large scale work, the frequency of mutations for a particular gene might be underestimated by a factor of 10, 100, or even 1000.

Some but not all these difficulties are overcome by special selective systems which have been applied in microorganisms. One mutant in

10^9 bacterial cells may be detected in a single petri dish if the mutation is from sensitivity to resistance to an antibiotic. Such studies again demonstrate a wide variety of apparent mutation rates for different genes and make difficult a meaningful calculation of an "average" mutation rate. And such studies do not circumvent the possibilities that many "slight" mutations and cell lethals are missed.

Regardless of the exact value for an average mutation, mutations are in another sense not rare at all. For purpose of illustration let us assume an average rate of mutation per gene per generation of 10^{-6}. This would mean that only 1 to 10^6 gametes produced by an individual would have this particular mutation. However, most estimates of numbers of genes for higher organisms like corn or men range from 10,000 to 100,000. If 10^5 genes had mutation rates of 10^{-6}, 10% of the gametes would have a mutation for some gene. This figure would also apply to the somatic cells so that many of the cells in a multicellular organism would contain mutant genes, most represented in only a small fraction of its cells, and most of little biological importance, except perhaps in cases of malignancy.

Another statement often made about mutations is that they are "random," but they are random only in a special sense of this word. Certainly mutations are not equally likely to occur for any gene in the genome; the rates of *detected* mutation in any case vary greatly. Nor is it true that mutations are equally likely to occur at any time or in any place. The frequencies of mutation have been shown to vary in dependence on a variety of external circumstances and even in different tissues of the same organism. Finally, mutations are not equally likely to be harmful or beneficial to the organism; most detected mutations impair viability to a greater or lesser degree.

Perhaps a better word than "random" to describe mutations would be "blind." The molecular alterations occurring in the genetic material are consequences of chemical and physical events in their immediate vicinity. No reason exists for believing that they are intelligently directed so as to accomplish specific alterations in gene products and subsequent specific effects on cellular and organismic properties. Indeed, the "intelligence" which might be capable of extrapolating the consequences of a particular nucleotide alteration in DNA would have to be much better informed about the intricacies of genetic translation than any current biologist. And locating a particular nucleotide in a certain nucleotide sequence among the tangled nuclear threads would appear to require some special kind of magic.

In spite of such considerations, the belief that mutations are in some sense "directed" according to the requirements of the organism has been current in biology at least since the time of J. B. P. Lamarck early in the last century and has not yet entirely disappeared. The reasons for its persistence, if we neglect the early naiveté about hereditary mechanisms, are twofold. The first of these is concerned with some observations which appear to require a direction of mutations. For example, the insecticide DDT is an effective control device for mosquitoes, at least for a few seasons, but eventually the areas treated with DDT are found to be populated with mosquitoes which are resistant to it. When breeding studies are carried out, the resistant mosquitoes are found to be genetically different from the original sensitive mosquitoes. The mosquitoes appear to have changed their genetic constitution in response to a particular environmental hazard, and in a manner which permits them to circumvent that hazard.

At another level of biological organization, similar phenomena have been observed with bacteria. When a population of bacteria is first exposed to an antibiotic, it is decimated, but eventually a population reappears which is now resistant to the antibiotic. Again breeding analyses show that resistance is due to a change in the genetic material. No doubt exists that the genetic composition of the populations has changed, nor can any doubt be entertained that the change is a consequence of a specific environmental modification. However, a demonstration that a *population* changes its genetic character as a direct response to an environmental variable is not the same as a demonstration that a particular *gene* changes its character in such a direct response. The possibility is open that mutants constantly produced by blind chemical events are always present in the population, and that the environmental modifications serve only to permit their differential multiplication. In this case the insecticides and antibiotics would serve only as selective agents, not as specific mutagens.

A test of these alternative interpretations requires an exploration of the hypothesis of preexisting mutants. Do the variants occur in an untreated population in sufficient quantity to account for its subsequent resurrection after exposure to the toxic agent? With the mosquitoes the question is easy to answer, for the resistant variants do indeed occur in the populations, often to the extent of making up several per cent of a virgin population. Answering the question for the bacteria requires more elaborate and less direct approaches. The population of bacteria is reestablished from a very small fraction of the original cells. When a culture of bacteria is plated on an agar plate containing antibiotic, a colony develops from not 1 in 100 original cells, or even 1 in 1000, but usually from only 1 or fewer cells in 10^7.

Do these cells which produce the population of resistant cells exist as resistant mutants prior to treatment with antibiotic? Or does the antibiotic induce a small fraction of the cells to mutate? If the mutants preexisted, they could be detected directly only by extremely laborious means. A large number of individual cells might be isolated, say, 10^7 or 10^8, and each allowed to develop into a large population. The resulting populations could then be sampled and tested for antibiotic resistance. If none of the populations was completely resistant, it would appear that antibiotic treatment is required to produce the resistant mutants; if resistant populations are observed in the frequencies required to explain the population resurgence, the existence of "preadapted" variants—produced by blind molecular alterations—is established and specifically directed mutations do not need to be invoked. This experiment is not technically feasible, however, and less direct approaches have to be used.

Several such approaches have been developed, but perhaps the simplest and most dramatic is that of "indirect selection," which uses the "replica plating" technique. If a culture of bacteria is spread in proper dilution on an agar plate, each cell begins to multiply and produces a small colony of a few hundred cells separated from other microcolonies on the same plate. If at this time a piece of velvet cloth is carefully placed over the surface of the plate, a fraction of the cells in each colony will adhere to the fibers of the cloth. When the cloth is removed from the first plate, it may be placed on the surface of another, and a part of the cells picked up will be transferred to the second plate. In this way replicas of the original colonies will be established and they will occupy the same relative positions on the second plates. On further incubation the colonies will develop and the two plates will appear essentially identical; each colony on one plate will have an identical "twin" on the other.

How does this technique permit the exploration of the origins of the resistant variants? If the second agar plate is impregnated with the antibiotic, the expected small number of resistant colonies will develop, while all the colonies of the original unimpregnated plate will grow. If antibiotic resistance is a consequence of exposure to antibiotic, none of the colonies on the first plate should be resistant. If, on the other hand, antibiotic resistance is due to a mutation which had occurred prior to exposure, an equal number of resistant colonies should arise on the unexposed plate. And, more significantly, many of those resistant colonies should be identifiable by reference to their twins which had been exposed to antibiotic. To test this expectation, the unexposed twins of resistant colonies may be "picked" with a needle and transferred to separate culture dishes. Samples of the new populations may then be tested for antibiotic resistance.

Such studies reveal a very high proportion of cells resistant to antibiotics in the indirectly selected colonies. Not all the cells in any particular colony need be resistant, because in many cases the mutations have occurred during the development of the initial microcolony and mixed populations are produced. These studies and others employing different principles, different organisms and different traits characteristically yield results expected of a blind origin for genetic variants and lead to the conclusion that the specific role of the environment lies in the selection of the variants and not in the induction of the changes.

We mentioned earlier that two factors are responsible for the persistent belief in environmentally directed hereditary changes. The first is a confusion between changes in the genetic constitution of populations brought about by selective forces, and changes in individual genetic elements brought about by blind molecular accidents. The second factor also involves a source of confusion, but of a somewhat subtler nature. We attempted to develop a generalized operational definition of "heredity" as the sum of the mechanisms which are responsible for the persistence of specific differences through a dilution process. Meanwhile we have gone to some pains to document the evidence that hereditary differences usually involve changes within the chromosomes and particularly within the nucleic acid molecules of the chromosomes. But not all hereditary differences involve changes in the "primary genetic materials." Some "hereditary" alterations may indeed be directly induced by a particular environmental agent, and these alterations are of critical significance in microbial adaptation and developmental differentiation. We will return later to a more thorough treatment of these modifications, but for the present we may simply indicate a resolution of an apparent conflict. Most such modifications are the result not of changes in the information content of the central genetic library, but of alterations in the use of such information. The hereditary differences are established in the face of identical nucleic information.

MUTAGENESIS BY IONIZING RADIATION

Most of the things we might say about spontaneous mutations are equally applicable to induced mutations. The mutants span the same range of phenotypic effects and are all likely to reduce biological fitness except under exceptional circumstances (as when a population of bacteria is challenged with a toxic agent). The mutants in both cases range from invisible point mutations to gross rearrangements of chromosome structure. A variety of physical and chemical agents may induce mutations. The first demonstration of artificial mutagenesis by H. J. Muller in 1926 involved x-rays. These radiations impart energy to the cells by removing electrons from the outer shell of atoms and thus ionizing cellular components; the ions so formed result in secondary chemical transformations. Most local disturbances of molecular structure are inconsequential, because most cellular components are represented many times, and any single molecule is expendable. However, if the molecular structure of the genetic material is modified, the effects are multiplied many times over and become hereditary.

The general characteristics of radiation mutagenesis should be discussed briefly, for cultural purposes if for no other in a civilization relying more and more on atomic energy as a source of power. All ionizing radiations—x-rays, alpha rays, gamma rays, neutrons—dissipate their energy in matter by forming ion pairs. The quantity of radiation is measured in roentgen or r units; an r unit is the amount of radiation which would form 2.082×10^9 ion pairs in one cubic centimeter of air under standard conditions. The various kinds of ionizing radiations have quantitatively different biological effects (dependent largely on the distribution of the ionizations in the material irradiated), but we need not be greatly concerned with such differences here. Rather, we should try to understand some of the generally accepted quantitative and qualitative relations between the amount of radiation delivered and the number of mutations induced.

We will start with the induction of point mutations; gross chromosome rearrangements are produced in a somewhat different manner and will be discussed separately. Perhaps the first principle established in radiation genetics was the linearity of the dose response curve for x-ray induced point mutations. The frequency of induced mutations is directly proportional to the dose measured in r (Fig. 14-1). A treatment of adult male fruit flies with 400 r will yield approximately twice as many mutants among their progeny as a treatment with 200 r; and 200 r will give twice as many mutants as 100 r, etc. Some question has been raised about the applicability of this relation to very low doses, doses from 0.1 r to 10 r in multicellular organisms. This is an important practical question in view of the greater anticipated exposures of the human population to low level radioactivity. However, studies of mutation rates at these low levels of exposure are complicated by the large numbers of individuals which have to be screened to obtain significant numbers of mutants, and by the "background" mutations of spontaneous origin which make up the bulk of the mutants detected at these dosage

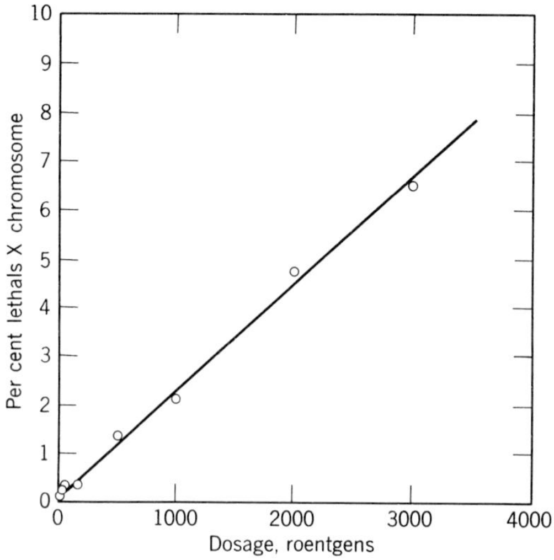

Fig. 14-1. Dose response curve for point mutations. (From Wagner and Mitchell, *Genetics and Metabolism,* 2nd ed., Wiley, 1964.)

levels. Final answers to the question of low level linearity will have to await further studies.

The linearity of the dose response is usually interpreted as indicative of a "single hit" process of mutation induction. What is meant by this term is that a single elementary effect of the radiation, perhaps the formation of a single ion pair, is sufficient to induce a mutation if it occurs in the right place. If two or more elementary events were required, and particularly if the events are distributed more or less at random in the cell, cells treated with twice as much radiation would incur more than twice as many mutations. This can be appreciated on simple probability grounds. Assume that two "hits" of some kind are required to give a mutation, and that on the average cells exposed to 100 r receive a hits, cells exposed to 200 r receive $2a$ hits. Then the probability that any cell will receive two hits is a^2 in the first case and $(2a)^2$ in the second case. If, for example, $a = 0.1$, the frequency of mutants in the sample treated with 100 r would be 0.01, while the frequency in the second case would be 0.04. The frequency of mutants would not be directly proportional to the dose, but rather proportional to the square of the dose.

Two other observations regarding radiation mutagenesis support this one-hit interpretation. The frequency of mutations induced by a given dose of radiation is not dependent on the intensity of the radiation. That is, 400 r may be given in 5 minutes, 5 hours or 5 days with identical results. Although the energy absorbed by a cell is rapidly dissipated, so that ion pairs formed at one time are not available even a few minutes later to reinforce the activity of other ion pairs, the same total amount of genetic damage is incurred. If the induction of a mutation required the coincidence in the same cell of two or more elementary radiation related events, the intensity of the exposure should be a critical factor, but it is not. Similarly, fractionation of the dosage by giving it in a series of short treatments rather than giving it in a single uninterrupted exposure yields the same frequency of induced mutations.

Dose responses for gross chromosomal rearrangements do not follow these same rules. In particular, doubling the dosage more than doubles the frequencies for such mutations; the same dosage given at a higher intensity yields more mutations; and fractionating the dosage reduces its mutagenic effectiveness. These differences between point mutations and gross chromosomal mutations are readily rationalized.

If pieces of chromosomes are to be rearranged in new patterns, the chromosomes must first be broken. While radiation in some cases modifies the chemical constituents of chromosomes, in other cases it disrupts the integrity of the backbone without otherwise modifying the chromosomal chemistry. The consequences of such breaks depend on their involvements; they may or may not result in gross chromosomal rearrangements. Broken chromosomes are known to have "sticky" ends at the points of fracture which may unite indiscriminately with other broken ends. However, if only one break occurs in a cell, only two ends are available for reunion and the only possible union of broken ends reconstitutes the original chromosomal pattern. Such "restitutions" do not produce detectable hereditary alterations. If, on the other hand, two or more breaks occur in the same cell at about the same time, several different combinations of ends are made possible. If, as appears likely, chromosome breaks are directly proportional to dosage, the number of breaks produced by a given dose would be the same regardless of the intensity of the radiation and regardless of the manner of fractionation. But the patterns of reunion of broken ends would be very different. If, for example, a dose of 50 r results in an average of one break per ten cells, exposure to a single acute dose of 50 r would yield one cell in a hundred ($\frac{1}{10} \times \frac{1}{10}$) with two breaks and a possible rearrangement of pieces. If the 50 r were given instead in two doses interrupted by an interval long enough to allow restitution, during the first period only $\frac{1}{20}$ of the cells would have one break and only $\frac{1}{400}$ ($\frac{1}{20} \times \frac{1}{20}$) would have two. Repeating the 25 r exposure later would again yield only 1/400 cells with two simultaneous breaks and a total of only one in 200. From such considerations it should be apparent that if the number of chromosome breaks is directly proportional to the dose, the number of detected chromosomal rearrangements should increase as the square of the dose. And this relationship approximates the dose-effect curves actually observed (Fig. 14-2).

CHEMICAL MUTAGENESIS

Chemical mutagenesis was demonstrated later than radiation mutagenesis, but it has developed into an active field of investigation. The realization that DNA is the genetic material has aided greatly in locating mutagens and in exploring their mode of action. Although many mutagens are known, their activities are generally poorly understood and may be very remote from the actual site of genetic change. The activities of a few chemical mutagens are, however, capable of explanation in molecular terms.

Before discussing particular mutagens, however, we have to consider the kinds of alterations possible in DNA molecules. A classifica-

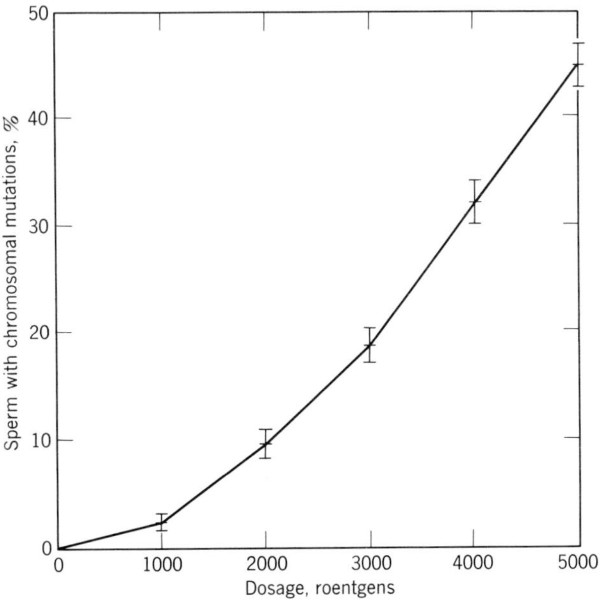

Fig. 14-2. Dose response curve for chromosome aberrations. (From Wagner and Mitchell, *Genetics and Metabolism*, 2nd ed., Wiley, 1964.)

tion of "micromutations" turns out to be very similar to a classification of mutations at the chromosomal level. Assuming a sequence of base pairs 1–8, for example (see Fig. 14-3*a*), we could imagine that the bases at position 4 have been altered so that a new sequence is established (14-3*b*). This would be a point mutation even in molecular terms: the substitution of an AT, a GC, or a TA pair for an original CG pair, for example. Another possibility is that a nucleotide pair is lost rather than replaced (14-3*c*); this would be the smallest possible deletion, but deletions involving any number of nucleotide pairs can be imagined. And just as nucleotide pairs might be lost, so might they be gained by insertion to lengthen the total segment by one or more links (14-3*d*). Finally, we might imagine that, as in whole chromosomes, segments of DNA might be removed and reinserted in a reverse direction (so long as the polarities of the DNA strands are retained) to give a molecular equivalent of an inversion (14-3*e*).

In terms of current thinking about translation mechanisms, each of these changes has predictable consequences. A substitution of a base pair would ordinarily lead to the substitution of an amino acid in a protein, or if the new base pair contributed to a "nonsense" triplet, to an interruption of the protein backbone at the point of the alteration. A deletion or insertion of a new base pair might result in a very different kind of protein beyond the point of the change; if the reading of the genetic message is in triplet units, the gain or loss of one or two base pairs would throw all subsequent coding units out of register and an entirely new message would be produced. How serious this might be would depend on whether the change occurred near the beginning or close to the end of a gene. Only for a gain or loss of three nucleotide pairs (or multiples of three) could a message which is approximately normal be produced; a protein with an additional amino acid or deficient in a single amino acid might be expected. An inverted region would result in an atypical amino acid sequence of a size proportional to the inverted section of DNA.

Although the theoretical consequences of specific alterations are readily perceived, the demonstration that a particular kind of change occurs preferentially with a given mutagen is more difficult to achieve. Most mutations are detected through changes in phenotypes rather than through detailed modifications of a protein's structure. And some mutagens cause drastic modifications in many cellular components so that their action may be very nonspecific; they may interfere with the pathways supplying DNA precursors, they may convert other molecules into mutagens, they may interfere with enzymes essential for orderly replication. Some mutagens, however, are suspected on reasonable

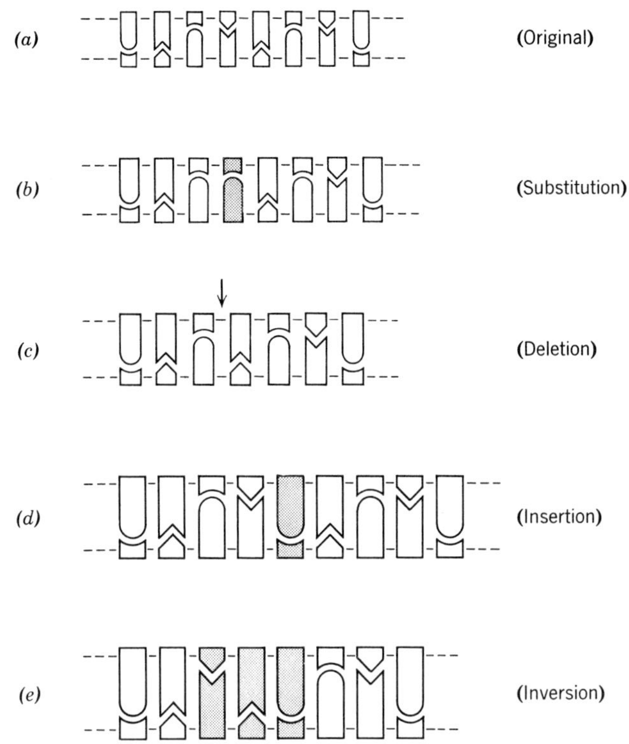

Fig. 14-3. Schematic representation of molecular events in micromutation.

grounds of highly specific action directly on the genetic material. One such compound is 5-bromouracil (BU), a pyrimidine analog sterically very similar to thymine. It is so similar to thymine that a cell discriminates poorly between them and will incorporate BU quantitatively into DNA if the supply of thymine is restricted (as in a thymine-requiring mutant). If BU were equivalent in all respects to thymine, this incorporation would be of little consequence, but BU is highly mutagenic. Its activity as a mutagen is attributed to its ability to shift frequently (relative to thymine) from the normal keto state to an enol state; this shift is accompanied by a change in its hydrogen bonding capacity so that it can now pair effectively with guanine rather than adenine (Fig. 14-4). Thus as a segment of DNA containing BU replicates, it occasionally makes a mistake and complementary strands are produced which contain guanine at sites originally occupied by adenine. At a following replication the BU might pair normally (with adenine), but the strand with the substituted guanine would pair with cytosine

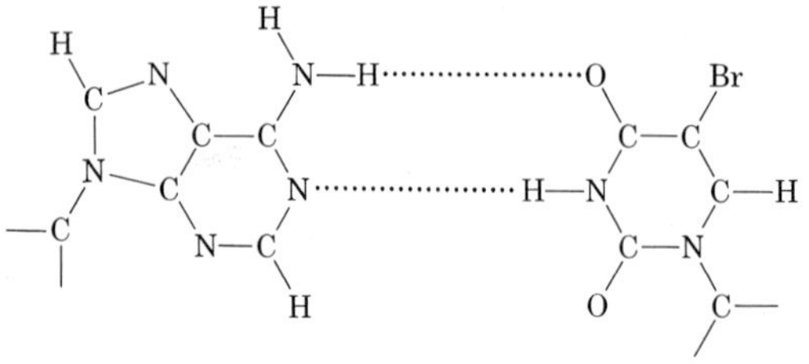

Adenine (normal amino state) 5-Bromouracil (normal keto state)

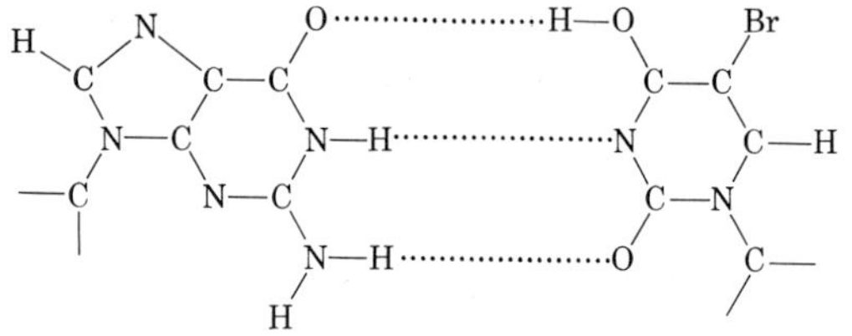

Guanine (normal amino state) 5-Bromouracil (rare enol state)

Fig. 14-4. Pairing properties of 5-bromouracil. (From S. Benzer, *Proc. Natl. Acad. Sci.* **47,** 410 (1961).) See also discussion of keto-enol states, p. 144.

and the original AT pair would be replaced by a GC pair. Alternatively, a BU base in the enol state might be incorporated into a new strand opposite guanine and later shift to the keto state and pair with adenine at the next replication. In this way an original GC pair would be replaced by an AT pair. Other base analogs, such as 2-amino purine, are believed to act in a similar fashion, but we might expect some differences in specificity; some analogs are incorporated more readily than others, and some undergo keto-enol shifts which might permit specific kinds of changes.

In contrast with the base analogs, certain dyes such as proflavine and acridine orange produce mutations with somewhat different properties. Generally, mutations induced by base analogs can be stimulated to revert to their original state by treatment with base analogs again, but not with proflavine. Conversely, mutants induced by proflavine commonly revert in the presence of proflavine, but not when treated with base analogs. Apparently the changes induced by the two treatments are of different character and can be undone only by a similar kind of treatment. Since base analog changes appear to be substitutions, the dye-induced changes would appear to fall into another category. The best supported interpretation is that they represent changes by addition or substitution of single base pairs. The dyes are known to combine with DNA in vitro and to distort their structure; comparable distortions in living cells could well lead to errors in duplication.

The analysis of acridine mutants has provided some of the strongest evidence for a triplet genetic code. Such mutants often back-mutate to a form similar to but not identical with the wild type; fine genetic analysis (see p. 425) indicates that the revertants have undergone a second change at a "site" near the original mutation and that the second alteration has canceled the effect of the first. In view of the suspicion that acridine mutagenesis occurs via gains or losses of base pairs, the first mutation might be arbitrarily designated as a "plus" (or "minus") mutation and the second as a "minus" (or "plus") mutation. Sets of such "sign" mutants may be assembled and their interactions examined. Generally, a plus and a minus mutation in the same DNA segment produce an effect similar to that of the wild type; single plus or minus mutations or combinations of two plus or two minus mutations are defective. When, however, three plus mutations or three minus mutations are placed in a single segment, again an effect similar to that of the wild type is achieved.

An explanation of these results suggests that sign mutants are "frame shift" mutants. A sequence of bases, such as ATAGCTCCG would ordinarily be read (if the code is triplet) as ATA GCT CCG, and some mechanism must exist for bringing the code words into register. If one of the bases (say, the first T) were removed as the result of a mutation, the consecutive triplets from the left would be AAG CTC CG . . . ; every code word beyond the point of the change would be different from that in the normal protein. If now a second mutant of the plus sort occurs nearby, say, by insertion of a T between the two C's, the reading of this segment would become AAG CTC TCG. The second mutation does not "correct" the code words in this limited segment, but beyond the third triplet, the code words are back in their original register and a protein with relatively normal amino acid sequences might be possible. If the amino acids in the disturbed region played relatively minor roles in the protein's activities, the double mutants might be functionally normal.

It should be apparent that the addition of several bases (by combining plus mutants) or the subtraction of several bases (by combining minus mutants) could also bring the reading frames back into register; in addition to having scrambled amino acids in one segment, the protein would also differ from the normal in containing one more or one less amino acid. The significant fact that three plus or three minus mutants often restore normal function finds its simplest explanation in a triplet code.

Base analogs and acridine dyes are effective mutagens only when applied to replicating systems, but other mutagens can act on resting DNA, removing side groups from bases or interrupting the sugar-phosphate backbone. The actions of some of these compounds are beginning to be rationalized, and the time is perhaps not too distant when specific kinds of changes can be induced at will. This does not imply, of course, that a specific gene can be changed in a particular cell. The specificity of mutagens is apparently directed to small molecular groupings (such as a base, or at most a small number of bases), and these groupings are probably found in nearly all the genes of an organism. Nevertheless, these studies hold great promise as experimental tools, and certain practical applications can even now be envisioned.

RECOMBINATION

We have seen that the genetic material is remarkably stable in its composition and that spontaneous changes in genetic elements are rare and haphazard. Although, in a sense, mutations are the ultimate source of all genetic variability, they are not efficient means of providing the variability which all organisms require. The reason for this is that the effect of a gene depends only in part on its primary action. Each gene's activities must be integrated with the activities of other genes in the cell, and its biological significance is in large part determined by these secondary interactions. In the extreme case the same gene may be lethal in one genetic background and improve viability in another. Consequently, a single mutation may be exploited to generate a large number of meaningful variants if devices can be developed to place it in a large number of genetic backgrounds. The recombinations produced in sexual processes must not be viewed simply as the results of idle card-shuffling, but as biological variants of profound evolutionary significance. Even though a particular mutation is rare, once it has occurred and been established in an individual, sexual processes permit the exploration of a multitude of combinations in which it might be useful.

To achieve recombination two events are required: diverse genetic material must be brought together and then it must be sorted out again in new combinations. In higher forms the first stage is accomplished by the union of differentiated cells, the eggs and the sperms. In some lower forms—in many algae, fungi, and bacteria, for example—the cells which unite are not morphologically distinct, and the organisms are said to manifest isogamy rather than anisogamy. Even here, however, cell fusions are usually carefully regulated. Although sexes—(i.e., individuals producing eggs and sperms) are entirely missing, only certain classes of cells are capable of sexual union. In most cases the union of identical haploid cells is discouraged, probably because such unions are incapable of generating recombinants. And in all cases the union of very distantly related cells is prevented because so large a fraction of the recombinants represent inharmonious combinations. The particular patterns of sexual activity vary from organism to organism, but they presumably represent means established by natural selection for providing the amounts and kinds of genetic variants required for particular organisms living in particular kinds of environments. A detailed exploration of sexual systems lies, however, beyond the province of this book—in the rationalization of life cycles and in the explication of genetic economies, problems of primary significance at levels of organization beyond the cell. The importance of sex is underscored, however, by the fact that essentially all living organisms engage in some form of sexual activity. And the maintenance of the sexual machinery is not a trivial economic consideration; any organism could produce more progeny at smaller risk if it circumvented the requirement for fusing and recombining genetic material from different sources. It is true that a few forms appear to have abandoned sexuality, but some of these have been found to engage in cryptic fusion and recombination processes (parasexuality) which are detected only with considerable effort. And others may be forms which have sacrificed long-term evolutionary plasticity for a short-term exploitation of a limited biological niche.

The sexual processes with which we are most familiar are those in which whole cells, or at least whole nuclei, fuse to form a zygote. In recent years, however, several mechanisms for introducing smaller portions of a genome into another cell have been described. We have mentioned the phenomenon of transformation already—the incorporation by one bacterial cell of fragments of DNA released upon the rupture of other cells. Still other methods are known for bringing small samples of genetic material from one cell to another; we will return to these methods in a moment.

Regardless of the precise mechanisms for uniting diverse genetic elements in the same cell, the next step is that of assorting them in new combinations. In higher organisms this assortment is achieved by standardized maneuvers in connection with meiosis, and these have been discussed earlier. We may reiterate here that two kinds of assortment occur during meiosis. In organisms with multiple chromosomes, the assortment of whole chromosomes is accomplished by the random alignment of centromeres at the metaphase plate in the first meiotic division. Although incompletely efficient, this device yields considerable variety. To illustrate the power of chromosomal assortment in generating diversity, consider the number of genetic combinations possible among the progeny of two human beings. Let us assume for purposes of calculation that the two parents are each heterozygous for one gene on each of their 23 pairs of chromosomes (ignoring for present purposes the special situation in regard to the sex chromosomes). We may readily calculate the number of different genotypes these genetically identical parents could generate—theoretically, of course, since the number of genotypes far exceeds the reproductive potential of any two humans. For one pair of genes (*A* and *a*) heterozygous in both parents, three genotypes are produced—*AA*, *Aa*, and *aa*. For a second pair of genes (*B* and *b*) three additional genotypes are generated—*BB*, *Bb*, and *bb*. But in combination with the first three genotypes, these yield nine combinations—*AABB*, *AABb*, *AaBB*, *AaBb*, etc. For each additional pair of genes included the number of combinations is increased three times, giving a series of 3, 9, 27, 81, 243, 729, etc., or in general terms, 3^n, where n is the number of chromosome pairs involved. For 23 pairs of genes, one on each pair of chromosomes, the figure is 3^{23}, or more than 8×10^{10} different genotypes. This number is larger than the total human population. Obviously, even with the limitation of one pair of heterozygous genes per chromosome (and these identical in the two parents) essentially no possibility exists for the production of two identical offspring, except under the special circumstances involved with identical twins. Chromosomal assortment is a remarkably effective device for generating hereditary diversity.

Even so, the assortment of whole chromosomes accounts for only a part of the genetic shuffling which occurs in meiosis. We have discussed at some length the gross features of crossing-over, the meiotic process responsible for recombining genes on the same chromosome. The mechanisms of crossing-over are still not established in detail, but its significance is clear enough. Moreover, processes similar to crossing-over occur in organisms which do not undergo regular meiotic reduction. Because of their central position in modern biological investigations these organisms, chiefly bacteria and viruses, deserve special attention even though their life histories depart in many details from those in higher forms. We should perhaps begin with the viruses.

Earlier we noted that viruses are small; at first this was their most distinguishing feature. They were identified as agents capable of passing through filters which would keep back bacteria and hence were called "filterable agents." With further study this became of less significance, and some overlap was found in the size ranges of bacteria and viruses. Of greater importance is their inability to multiply and metabolize autonomously. Thus far no artificial culture medium has been devised in which viruses can multiply; they all require a specific kind of living cell in which to reproduce. Many different kinds of viruses have been studied: those causing polio and influenza in man, mosaic disease in tobacco, and lysis in bacteria are notable examples. Although they show great variation in size and morphology, they all contain nucleic acid, surrounded by a protective protein coat, and contain only a limited supply of enzymes—chiefly those necessary to obtain entry into a cell and to initiate their reproduction. We will consider in a little more detail one group of viruses, those which attack bacterial cells and are called bacteriophages or simply "phages."

Bacteriophages are classified into two major groups: the temperate and the virulent. These differences are not fundamental since the same virus may be virulent with one bacterial host and temperate with another, and a single mutation may convert a temperate into a virulent form. Nevertheless, the classification reflects an important difference in the life histories. The virulent forms attach to a cell, inject their nucleic acids into its interior, block the normal synthetic machinery of the cell, initiate the production of viral proteins and nucleic acids, and after a short period of time, usually 20–40 minutes, cause the cell to lyse. At lysis dozens or even hundreds of new virus particles complete with their nucleic acid, protein coats, and essential enzymes are released, ready to start the cycle anew.

The temperate phages have a different history; they enter a bacterial call and disappear. Even if an infected cell is artificially lysed, the virus cannot be recovered in an active state. Nevertheless, the virus is present and is capable of multiplying as its host multiplies and at a later time of completing its cycle. Occasional descendants of the infected cell will spontaneously lyse and release large numbers of viruses, or an entire population of infected cells may be induced to lyse by special treatment; a mild exposure to ultraviolet light, for example, is sufficient to stimulate the completion of the cycle in some strains. In many cases the presence of the virus can be demonstrated indirectly also; cells carrying one kind of virus may be resistant to subsequent infection by another virus, or they may manifest other physiological differences from uninfected cells.

Since bacterial strains carrying a virus in this manner are susceptible to lysis, either naturally or artificially, they are designated as lysogenic strains. The virus whose complete life cycle has been postponed is said to be in the prophage state. A natural question to ask is where the prophage is located and what it is doing. Before we answer this question, we have to consider the life cycle of the host bacterium.

Until recently both bacteria and viruses were considered by many to be life forms distinctly different from organisms with conventional cellular structures. They appeared to be without nuclei, without sex, and perhaps even without a genetic apparatus comparable to that in higher forms. This opinion was based in part on their small sizes, which made direct observations difficult and which made necessary a different kind of experimental analysis. Another significant factor, however, was the fact that most investigators of bacteria and viruses, for obvious reasons, were more concerned with killing the objects of their study than with understanding them. This situation has changed. Even with optical methods nucleus-like bodies containing DNA can be identified in bacteria, and the genetic systems of both viruses and bacteria have been thoroughly exploited. In many ways bacteria and viruses are the most thoroughly understood of biological systems, and the fundamental similarities between them and higher forms are ever more compelling.

The modern era of bacterial biology began with the demonstration of genetic recombination in *Escherichia coli* by Joshua Lederberg and E. L. Tatum in 1946. These workers developed techniques for detecting recombinational events which might occur only very infrequently. They established a series of strains which differed from each other in a number of metabolic capacities. One strain required biotin and methionine; another required threonine, leucine and thiamine (vitamin B_1). These two strains may be represented as $B^- M^- T^+ L^+ B_1^+$ and $B^+ M^+ T^- L^- B_1^-$. When the two strains are mixed and plated on a culture medium lacking all of the five compounds required by one or the other of the parents, a small fraction of the cells, about 1 in 10^7, gave rise to colonies; they had the genetic formula of $B^+ M^+ T^+ L^+ B_1^+$.

Initially several explanations for the appearance of these "wild type" or "prototrophic" colonies could be advanced. Mutation, for example, of B^- to B^+ and of M^- to M^+ in the same cell of one of the strains would yield a prototroph. And the simultaneous mutation of three genes in the other strain would also give the results observed. However, probability considerations were against this interpretation. Although single mutations for these markers occurred with a frequency of the order of 1 in 10^7, mutations are independent events. Hence, if single mutations occur once in 10^7 cells, two mutations should occur only once in 10^{14} cells (1 in $10^7 \times$ 1 in 10^7), and three mutations only once in 10^{21}. Perhaps more impressive was the fact that neither of the original strains when plated on the selective medium individually produced detectable prototroph colonies. Hence mutation did not provide a satisfactory explanation.

Before concluding that mating occurred between cells in the two population, other forms of interaction had to be eliminated. For example, the lysis of some of the cells and the release of DNA might lead to transformation. But if the two cultures were kept separated by a sintered glass filter which prevented cell contact but permitted the macromolecular constituents of the medium to flow freely, no prototrophs were formed. Direct contact between cells of the two cultures was required, and transformation was eliminated.

Still other interpretations were considered and subjected to test, but all were systematically excluded except the interpretation that cell to cell contact, followed by genetic recombination, occurred in this bacterial species. This was the beginning of formal bacterial genetics. Subsequent studies demonstrate that three different classes of *E. coli* strains can be distinguished, and that genetic recombination occurs only with certain combinations. More specifically, some strains, designated as F^- strains, behave as recipient or "female" cells. When in contact with appropriate donors they receive genetic material, but are incapable of contributing genes to their mates. In contrast, other strains designated as F^+ strains when mixed with F^- strains donate but do not receive genes. A more thorough study of the F^+ strains indicates that they contain a small proportion of distinctly different cells which mate avidly and are in fact responsible for the donor capacity of the strains. Pure lines of these distinctive cells have been established and they are designated as *Hfr* lines (for high frequency recombination); when they are mixed with F^- strains the frequency of mating increases from the one in 10^7 observed with $F^+ \times F^-$ crosses to nearly 100%, and mating pairs can be readily observed.

Once bacterial mating could be controlled, formal genetic analysis could be conducted in much the same way as that in higher forms except that special techniques are required to isolate and score the recombinants. In addition, the bacteria have provided another means of genetic mapping which has demonstrated an unusual organization of the genetic material. The technique involves interrupting mating at various times after the strains are mixed. This is accomplished by mechanical agitation. The results of interrupted matings are different from those observed in the normal course of events. Ordinarily essentially all of the "marker" genes of the *Hfr* cells can be shown to be transmitted to the F^- cells, though in very different frequencies; some genes are often transferred, others with intermediate frequency, and still others only occasionally. If matings are interrupted soon after the cells are mixed, only a few of the genes are transmitted at all and these are the same ones ordinarily transmitted in high frequency. With progressively later interruptions, progressively more genes are found to be transferred and always in a definite order that corresponds to the frequency of transfer in uninterrupted matings. These observations suggest that the genes of the *Hfr* cell are injected in a linear sequence and that the time of the injection can be determined and used as an indication of the location of the gene in a genetic map. In crosses with one *Hfr* strain, for example, the gene *T* (for threonine independence) appears in the F^- cells first after about 8 minutes, the *Lac* gene (for lactose utilization) only after 18 minutes, the *Tr* gene (for tryptophane independence) after 33 minutes, etc. Such data indicate that *T* is near the end of the chromosome which is transferred first, the *Lac* gene further away and between *T* and *Tr*. And the time difference in injection can be used as a measure of the distance between genes.

This technique provides an unambiguous ordering of the genes from a single injection point designated as O for the origin. The genome of the *Hfr* bacterium is slowly pulled into the F^- cell through an area of contact between the cells and can be broken easily by mechanical agitation. The different frequencies of transmission in ordinary crosses are believed to be due to spontaneous disruptions of the process which make less likely the injection of genes far from the origin. Commonly, therefore, the *Hfr* cell after mating contains a greater or smaller fragment of the genome, as well as one or more complete genomes, since *E. coli* cells are commonly multinucleate. The fate of the fragment in the *Hfr* cell has not been followed; presumably it is destroyed or diluted out at subsequent cell divisions. The fragment which enters the F^- cell, however, can be studied further since it may contain genes not present in the original F^- genome. By some process not yet fully elucidated, but perhaps similar to crossing-over in higher forms, the *Hfr* genes may be incorporated into a genome of the F^- cell, replacing the genes originally present. Since the F^- cell is also multinucleate, genes may be incorporated into one F^- genome and not into another and an assortment of genes occurs for several cell divisions after mating.

The observations just summarized indicate that the bacterial genome consists of a single uninterrupted strand with a definite orientation with reference to the point of insertion. Further observations, however, modify this picture somewhat. Although different *Hfr* strains derived from the same F^+ strain behave in a similar way, the sequences of gene injection are specifically different for different *Hfr*'s. The

Table 14-1 *Linkage and Order of Genes in Several Different Hfr Types of E. coli**

Types of *Hfr*		Order of Transfer of Genetic Characters																			
Hfr	H	O	*t*	*l*	*az*	T_1	*pro*	*lac*	*ad*	*gal*	*try*	*h*	*S-G*	*Sm*	*mal*	*xyl*	*mtl*	*isol*	*m*	B_1
	1	O	*l*	*t*	B_1	*m*	*isol*	*mtl*	*xyl*	*mal*	*Sm*	*S-G*	*h*	*try*	*gal*	*ad*	*lac*	*pro*	T_1	*az*
	2	O	*pro*	T_1	*az*	*l*	*t*	B_1	*m*	*isol*	*mtl*	*xyl*	*mal*	*Sm*	*S-G*	*h*	*try*	*gal*	*ad*	*lac*
	3	O	*ad*	*lac*	*pro*	T_1	*az*	*l*	*t*	B_1	*m*	*isol*	*mtl*	*xyl*	*mal*	*Sm*	*S-G*	*h*	*try*	*gal*
	4	O	B_1	*m*	*isol*	*mtl*	*xyl*	*mal*	*Sm*	*S-G*	*h*	*try*	*gal*	*ad*	*lac*	*pro*	T_1	*az*	*l*	*t*
	5	O	*m*	B_1	*t*	*l*	*az*	T_1	*pro*	*lac*	*ad*	*gal*	*try*	*h*	*S-G*	*sm*	*mal*	*xyl*	*mtl*	*isol*
	6	O	*isol*	*m*	B_1	*t*	*l*	*az*	T_1	*pro*	*lac*	*ad*	*gal*	*try*	*h*	*S-G*	*Sm*	*mal*	*xyl*	*mtl*
	7	O	T_1	*az*	*l*	*t*	B_1	*m*	*isol*	*mtl*	*xyl*	*mal*	*Sm*	*S-G*	*h*	*try*	*gal*	*ad*	*lac*	*pro*
AB	311	O	*h*	*try*	*gal*	*ad*	*lac*	*pro*	T_1	*az*	*l*	*t*	B_1	*m*	*isol*	*mtl*	*xyl*	*mal*	*Sm*	*S-G*
AB	312	O	*Sm*	*mal*	*xyl*	*mtl*	*isol*	*m*	B_1	*t*	*l*	*az*	T_1	*pro*	*lac*	*ad*	*gal*	*try*	*h*	*S-G*
AB	313	O	*mtl*	*xyl*	*mal*	*Sm*	*S-G*	*h*	*try*	*gal*	*ad*	*lac*	*pro*	T_1	*az*	*l*	*t*	B_1	*m*	*isol*

*From F. Jacob and E. L. Wollman, *Sexuality and the Genetics of Bacteria,* Academic Press, 1961.

orders of transfer for 18 genes in several *Hfr* strains are shown in Fig. 14-5. The arrangements of the genes in the different *Hfr* strains are the same, except that the insertion points (O) are different. If the first and last genes in each sequence are joined—to create a circle, the only differences among the *Hfr's* are differences in the placement of O and in the direction of transfer. Strains H and 4 have their origins at approximately the same point in the circle, but the first gene injected in H is the last in strain 4. This suggests an origin of *Hfr* strains by breaking a circle in different places in different cases, and providing one of the broken ends with an insertion point (Fig. 14-5).

To evaluate this interpretation more adequately we should return to a fuller consideration of the mating classes of *E. coli* strains. An early and puzzling observation concerned events which occurred when F^+ and F^- strains were mixed. Although relatively few matings were consummated (and these presumably due to the presence of a small number of *Hfr* cells in the F^+ strains), all the F^- cells in the mixture are fairly quickly converted to F^+ cells, simply by coming in contact with F^+ cells. This observation suggested that F^+ cells might contain an infective agent capable of being transmitted by cell contact and then multiplying in the new cells it invaded. This agent was called the *F* factor and, since it could be transferred without other genetic markers, it was considered to be separate from the rest of the bacterial genome.

This behavior is to be contrasted with that observed in mixtures of Hfr and F^- strains. The F^- cells are not converted into F^+ cells; indeed most of the recombinants—those F^- cells which have received gene markers from the *Hfr* cells—are still F^-. Only the relatively rare recombinants which receive an entire genome are *Hfr*. The *Hfr* does not then have the *F* factor, at least not in the same condition in which it exists in the F^+ cells, and the genetic determinant for the *Hfr* condition is located at the very end of the chromosome at the farthest distance from O.

This entire set of observations is rationalized as follows. The bacterial genome is considered to consist of a single closed circle in both F^- and F^+ cells (14-6). F^+ differs from F^- however, in containing a separate genetic element capable of being transmitted readily to F^- cells on cell contact. The F^+ cells are converted to *Hfr* cells by specific interaction between the *F* factor and the circular genome; the *F* factor attaches to some point in the genome and a break is made in the circular structure. Simultaneously the now broken genome is given a polarity; the end at which the *F* factor is not attached becomes the origin. This is a consistent interpretation of the observations and is supported by other facts which we will not take time

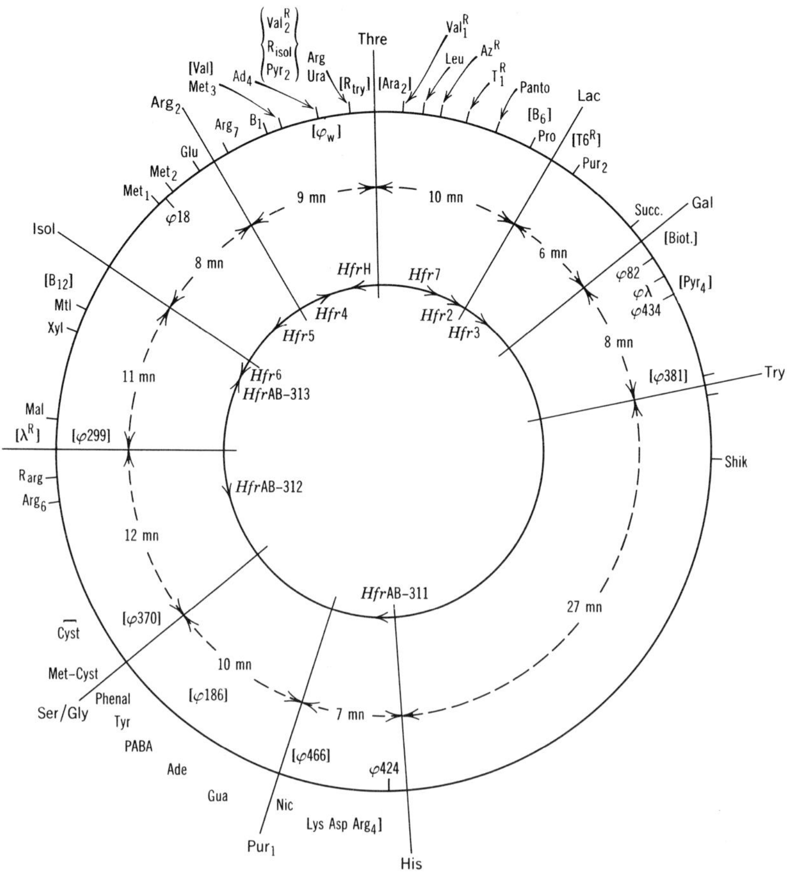

Fig. 14-5. Schematic representation of the linkage group of *E. coli* K-12. The outer line represents the order of the characters (not their absolute distances). The dotted lines represent the time intervals of penetration between pairs of markers corresponding to the radial lines. The inner line represents the order of transfer of different *Hfr* types described in Table 14-1. Each arrow corresponds to the origin of the corresponding *Hfr* strain.

Symbols correspond to synthesis of threonine (thre), leucine (leu), pantothenate (panto), proline (pro), purines (pur), biotin (biot), pyrimidines (pry), tryptophan (try), shikimic acid (shik), histidine (his), arginine (arg), lysine (lys), nicotinamide (nic), guanine (gua), adenine (ade), para-aminobenzoic acid (paba), tryosine (try), phenylalanine (phenal), glycine (gly), serine (ser), cysteine (cyst), methionine (met), vitamin B_{12} (B_{12}), isoleucine (isol), thiamine (B_1), valine (val); to fermentation of arabinose (ara), lactose (lac), galactose (gal), maltose (mal), xylose (xyl), mannitol (mtl); requirement for succinate (succ), aspartate (asp), glutamate (glu); resistance to valine (val^r), to sodium azide (az^r), to phages T1 ($T1^r$), T6 ($T6^r$), λ (λ^r); repression for arginine (R_{arg}). (From F. Jacob and E. L. Wollman, *Sexuality and the Genetics of Bacteria,* Academic Press, 1961.)

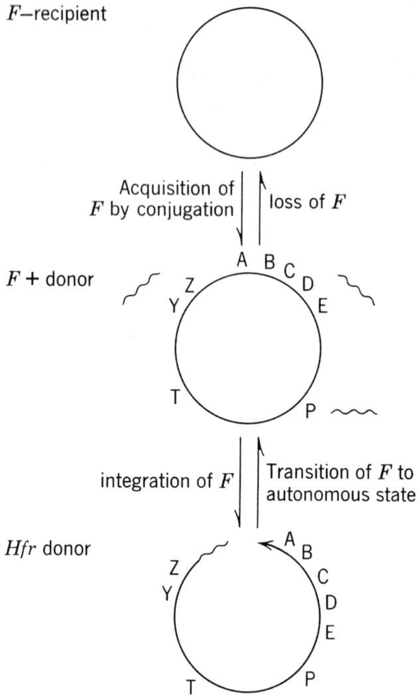

Fig. 14-6. Diagrammatic representation of how the sexual types of *E. coli* are determined by the state of the sex factor, *F*. The sex factor is represented by the short, wavy line. The letters indicate hypothetical chromosomal markers, and the arrow the leading chromosomal extremity of the *Hfr* type. (From W. Hayes, F. Jacob, and E. L. Wollman in W. J. Burdette (Ed.), *Methodology in Basic Genetics,* Holden-Day, 1963.)

to present. It suggests the existence of a class of genetic elements which have two possible cellular locations: they may be attached to the chromosome and replicate in synchrony with it, or they may be detached and replicate autonomously. Such elements as a class are designated as episomes. A few other examples of episomes, besides the *F* factor, have been demonstrated in bacteria and are concerned with a variety of cellular characteristics. Similar elements have been suspected in higher forms, but their status is more equivocal. In any case episomes provide a supplementary means of transferring genetic material from cell to cell which can be very specific. We should note that episomic transfer is not restricted necessarily to the genetic materials of the element itself. The *F* factor, for example, occasionally detaches from the chromosome in *Hfr* cells and carries with it a fragment of the chromosome which was near it. Since in different *Hfr* cells the *F* factor is attached at different points, different detached *F* factors carry different gene markers. These encumbered episomes are transmitted to F^- cells in the same way as other F factors and provide an efficient means (F-duction) of spreading a gene in a population as well as a useful experimental tool.

The stated purpose of this excursion into bacterial genetics was to provide a basis for discussing the location of prophages in lysogenic strains of bacteria. By now it should be apparent that a chromosomal location is a possibility and that the possibility is capable of being tested. And indeed, when appropriate crosses are carried out, prophages can often be unambiguously assigned to chromosomal positions. Some viral strains are highly specific in their points of attachment and these attachment sites have been useful markers for sections of the bacterial chromosome. Other viruses are less fastidious and may, like the F factors, attach more or less at random. These attachment sites are important in evaluating the role of bacteriophages in bacterial recombination. Like the F factors, when prophages are separated from the bacterial chromosome and complete their lytic cycle, they occasionally retain portions of the bacterial genome—at least in some cases portions which were near their chromosomal attachments. The mature virus particles may then include bacterial genes, packaged for distant delivery. When the virus particle enters another bacterium the bacterial genes may be incorporated into the host cell's genome and thus modify its characteristics. Perhaps because the bacterial genes occupy essential space in the virus particle and exclude viral nucleic acid, most such transducing viruses are in fact defective and cannot complete their cycle in the new host. The transfer of bacterial genetic material by means of virus particles is called transduction and is an important supplementary method of genetic recombination in these forms.

We may now return to a consideration of recombination in the life cycle of the virus itself. In a sense a bacterial virus behaves much like a bacterial gene, although it has some very special properties, and its size is comparable to that estimated for genes in many organisms. For some time, in fact, viruses were considered to be "naked genes" and for this reason ideal objects of genetic study. Although the simplicity of viruses is only relative, their study has greatly advanced our understanding of genetic organization and has forced a redefinition of many terms. Much of the genetic work on viruses has been focussed on the virulent bacteriophages T_2 and T_4, using techniques first developed by S. E. Luria and Max Delbrück.

The first requirement for genetic studies is genetic differences which can be quantitatively assayed. These were provided by searching for distinctive colony characteristics. When phages are plated on a solid sheet of bacteria in a petri dish, each virus enters a bacterial cell, and after a characteristic latent period lyses the cell releasing the progeny. The results of this first cycle of replication are not even microscopically visible, but the progeny viruses now attack other bacterial cells in the vicinity and cause them to lyse also. Eventually a sufficiently large number of bacteria are destroyed to leave a clear area (or plaque) in the otherwise opaque sheet of bacteria. The number of initial particles can be determined by counting the plaques produced in appropriate dilutions. The use of plaques extends beyond their value in quantitation, however, for occasional mutant viruses produce plaques of distinctive appearance. In some cases the plaques are smaller or larger than usual, or they may be turbid rather than clear, or have halos surrounding them. The bases for many of the plaque differences are unknown, but their distinctiveness and constancy make them useful genetic markers. Another set of markers is concerned with host specificity. Phages of a particular strain may attack some bacterial strains and not others, but mutants arise with modified host ranges. Other variants are also known, but these are sufficient for our purposes.

The first unequivocal evidence for virus recombination came from studies in which viruses with different mutations were allowed to enter the same bacterial cells. For example, a mutant with a distinctive plaque morphology (r as opposed to r^{+}) might be used with a mutant with a modified host range (h vs. h^{+}); one virus might be designated as $\underline{r \quad h^{+}}$ and the other as $\underline{r^{+} \quad h}$. A mixed infection is achieved when suspensions of the two viruses are added to a bacterial culture simultaneously in concentrations high enough so that most of the bacteria receive at least one of each of the virus particles. When the lysate of the bacterial culture is assayed, four types of progeny viruses are characteristically recovered: the parental types $\underline{r \quad h^{+}}$ and $\underline{r^{+} \quad h}$, and also the recombinant types $\underline{r \quad h}$ and $\underline{r^{+} \quad h^{+}}$. The recombinants appear in frequencies characteristic for the particular mutants employed and provide a means for mapping the virus "chromosome."

The frequencies of recombinants do not, however, provide so direct a means of mapping as do those in a cross of higher organisms. The reason for this is that the individual recombination events are not so well controlled as in a mating of other forms. In a mixture of bacteria and viruses the viruses attach more or less at random; if the average multiplicity of infection for one kind of virus is 4, some bacteria receive none, or 2, or 6, etc., in frequencies determined by chance. Similarly, even if the same numbers of the two kinds of phages are used, some bacteria will be infected with more of one kind and some with more of the other kind. These distributions of multiplicities of infection and input ratios for individual cells have to be taken into account in evaluating recombinant frequencies.

Another factor which has to be considered is the events which occur after infection. Both multiplication and recombination occur between infection and lysis, and several facts demonstrate that recombination occurs not once but several times in this interval. If, for example, three different kinds of virus are introduced into a cell, some progeny are produced with markers from all three parents; such products would not be expected if recombination occurred only once in associations of two. Moreover, if infected cells are artificially lysed prematurely during the growth period, the frequency of recombinants is lower than at the normal lysis time; and if lysis is artificially delayed, recombinants may be more frequent than usual. These observations suggest that mating occurs repeatedly during growth and that it tends to dissociate closely linked markers more effectively as time goes on. The events occurring within a bacterium are analogous to those which might occur in a large population cage into which two fruit flies are introduced. If the population is sampled several months, and several generations later, the recombinant frequencies will provide evidence of linkage relations, but will not be directly comparable to recombinant frequencies in single crosses. Means have been developed, however, to convert gross frequencies to meaningful map distances. A large number of mutants affecting a variety of traits have been analyzed and the phage genome has been extensively mapped. Indeed, the phage genome represents the most thoroughly mapped section of genetic material available. A single linkage group is represented and it appears to be, like the bacterial chromosome, a closed circle.

This brief excursion into viral recombination was not undertaken entirely because of its intrinsic interest. It provides the basis for a discussion of the organization of the genetic material. The gene, as classically conceived, was the irreducible unit of transmission. It was detected through observed phenotypic differences, but these were not sufficient to define it. Identical phenotypic effects might be achieved by changes in various parts of the genome. Hence the gene was defined as the smallest element of the genetic material which could not be further subdivided by recombination mechanisms. When the functions of the genetic material were probed further, the gene was inferred to be also the basic unit of biochemical activity—a portion of the DNA which specifies the amino acid sequences in a single polypeptide chain. And, finally, the gene was considered to be the unit of mutation, the smallest portion of the genetic material in which a change affecting genetic expression could occur. This unified concept of the gene—the unit of transmission, of function, and of mutation—was intellectually satisfying because of its simplicity and coherence. Nevertheless, certain observations on higher organisms raised questions about its validity and studies on virus recombination have demonstrated its inadequacy. The issues can perhaps be illustrated best by a case history.

We need to consider in a little more detail the mutants which give the phenotype designated as *r* (for "rapid lysis") in the bacteriophage T_4. When these are plated on the bacterial strain *B* of *Escherichia coli*, they yield plaques larger than those produced by wild-type phages and have certain other characteristics. If many independent r mutants are isolated, they may be separated into three large groups by their behavior on other bacterial strains. We are concerned primarily with the group designated as *r*II which produces no plaques on strain *K* and wild-type plaques on strain S. These *r*II mutants represent a phenotypically homogeneous set which may be expected to have arisen as the result of changes in the same genetic region. And indeed, when crosses are performed to locate the sites of mutation, all the *r*II mutations are localized in a small portion of the genetic map.

Further study of the *r*II mutants shows, however, that they are a diverse collection. We may first note that two major classes may be distinguished on the basis of their ability to produce wild-type mutants. Some fail to produce any revertants and in addition show mapping peculiarities which indicate that they have undergone extensive modifications; most of them are interpretable in terms of deletions of sections of the DNA. These deletion mutants have been employed in an ingenious fashion to facilitate the genetic mapping of the *r*II variants, but we will not consider their use in detail. The other mutants show appreciable rates of back mutation and are considered to be point mutations, relatively small alterations which can be "repaired" by accidents in replication. We are first concerned with the mapping of these point mutations. In most cases when two independent *r*II point mutants are used in a mixed infection, a small fraction of the progeny is found to be wild type. Although the fraction of wild-type progeny is small, it is higher than that observed as the result of mutation and is characteristic for the particular mutants used in the cross. Presumably the mutations have occurred at different positions in the *r*II region and the mutants may be described by the symbols $\underline{r_a \quad r_b^+}$ *and* $\underline{r_a^+ \quad r_b}$. By an act of recombination the wild type, $\underline{r_a^+ \quad r_b^+}$, as well as the usually undetectable double mutant, is produced. In those cases where two mutants do not produce wild-type recombinants, they are assumed to have undergone alteration at precisely the same "site." By procedures comparable in principle to those employed in higher organisms a total of over 2400 independent mutants have been distributed among over 300 separate sites and arrayed in linear sequence. Mutations at some sites are relatively common, but many sites are identified by only a single mutation and many more might be found with further search. See Fig. 14-7.

What is the meaning of this multiplicity of genetic changes at separable sites? If a gene is defined as a unit incapable of subdivision by recombination, the *r*II region would have a minimum of 300 genes within it. Since the *r*II region occupies less than 1% of the genetic map of T_4, a total of at least 30,000 such "genes" would be expected if the entire genetic map were similarly explored. Since we know something about the total size of T_4, we can readily compute the maximum size of the units which are being discriminated. The total DNA of a T_4 particle consists of about 4×10^5 nucleotides, or 2×10^5 nucleotide pairs in a Watson-Crick double helix. If 200,000 nucleotide pairs are distributed among 30,000 units of recombination, this permits less than 7 nucleotide pairs to be assigned to each unit. And this must be emphasized as a maximum size estimate. If we consider the high probability of unmapped sites within the *r*II region or compute the total number of recombination units from the closest map distances between sites, a larger number of units is provided and a correspondingly smaller size. Quite possibly, the recombination units are no larger than a single nucleotide pair.

The significance of this analysis should be obvious. If a gene is defined as a unit of recombination, it cannot also be a unit which specifies the amino acid sequences in a long polypeptide chain; recombination can apparently occur between any two nucleotide pairs in DNA. The unit of recombination, at least in T_4, is not a large and complex entity. Seymour Benzer, who with his associates is responsible for most of the foregoing study, suggests that a new name should be given for the unit of recombination and has proposed the term *recon*. We may properly question whether a recon, as determined in a phage, is equivalent to a unit of recombination in a higher organism. It is at least conceivable that recombination in an insect or in a mammal is restricted to a limited number of positions along the chromosome, perhaps between functional units or at points where DNA is interrupted by protein linkages, and that the virus mechanism is special in permitting unrestricted crossing-over at any point in the DNA chain. The recon in Drosophila, for example, might correspond closely to the classical gene, even though it does not in lower forms. A com-

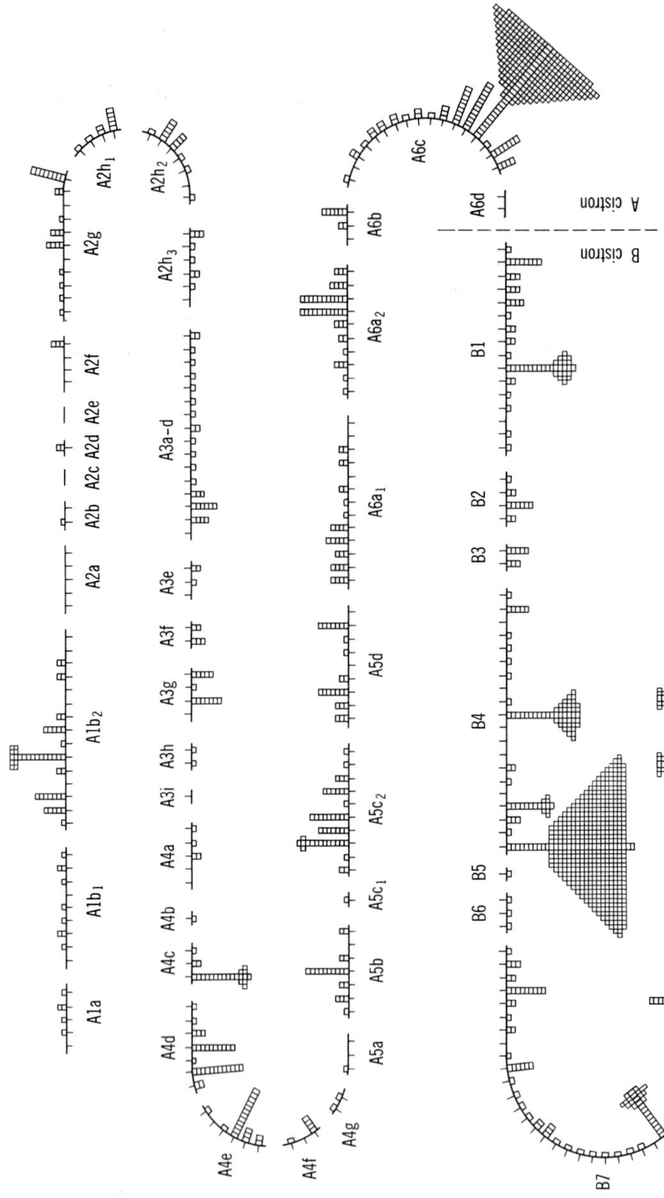

Fig. 14-7. Map of *r*II region of phage T_4. (From S. Benzer, *Proc. Natl. Acad. Sci.*, **47,** 410 (1961).)

pletely satisfactory answer to this question is difficult for technical reasons; the required numbers of mutants and the required numbers of tested progeny are difficult to assemble in higher forms. Nevertheless, situations comparable to that just discussed for T_4 have been discovered in bacteria and in several fungi. Moreover, "pseudoalleles," mutant genes with similar phenotypes located at corresponding positions on homologous chromosomes which can recombine to produce normal progeny, have long been known in higher forms; some of these may well find explanation as alterations at separable sites within a single functional unit. A final resolution awaits a fuller understanding of the mechanism (or mechanisms) of intrachromosomal recombination.

Since the unit of recombination, at least in some forms, does not correspond to a unit of function in the usually recognized sense, the question arises whether a unit of function can be adequately defined and whether it might correspond more closely to our idea of the gene. As remarked earlier, this shift in definitions has already taken place to a certain extent; to many biologists the word gene suggests a segment of genetic material which prescribes an amino acid sequence in a polypeptide chain. Neglecting the possibility that some genes have functions other than that of coding proteins, this definition is useful conceptually but is of limited value in practical situations. It can be applied rigorously only in those cases in which the protein product has been identified and shown to be modified as a consequence of an alteration in a particular segment of DNA. Mendel's gene for tallness in peas, the gene for forked bristles in Drosophila, the gene for taste sensitivity (for phenylthiocarbamide) in man, or the gene for *r* plaques in phage T_4 cannot be defined on this basis. And a definition which is applicable to so few of the items in a class is admittedly an impractical definition. For this reason several attempts have been made to develop definitions of functional units of wider application. No one of these has been entirely satisfactory, but the one with the greatest currency at the present time emerged from the study of the *r*II region.

The student will recall that *r*II mutants are characterized as showing the *r* phenotype on strain *B*, the wild phenotype on strain S, and a failure to grow on strain *K*. They can gain entry into *K*, but they are unable to reproduce once inside; presumably they have lost a function possessed by wild type phages required for multiplication in this particular environment. We may note, however, that when both *r*II mutants and wild-type phages simultaneously infect strain *K*, the mutants do reproduce and their genetic markers appear in the progeny;

the normal phage performs an essential function of the *r*II region and permits both phages to multiply. The reparability of defects permits a test which determines whether two mutants have defects in the same function. Although each *r*II mutant has lost a function, different *r*II mutants may have lost different functions. It is conceivable, for example, that the *r*II region includes segments of DNA responsible for the specification of several different polypeptide chains, all of which are required for multiplication. The test consists of allowing two independent *r*II mutants to infect the same *K* cells. If the two mutants are defective in the same function, they should not be able to cooperate to complete the lytic cycle. But if they are concerned with different functions, each could supply the function missing from the other and multiplication would be possible in *K*.

In practice, some combinations of *r*II mutants grow and some do not grow in *K*. The mutants can be classified into two groups, *A* and *B*, on the basis of such tests. Any combination of a group *A* mutant and a group *B* mutant results in growth, but no combination of an *A* with an *A* or a *B* with a *B* will grow. The *A* and *B* mutants are apparently defective in different functions and can cross-repair each other by supplying products which the other requires. When placed on the genetic map, the two classes of mutants are cleanly separated; the *A* mutants are located on one side of the region and the *B* mutants on the other, even though *A* and *B* mutants (close to the dividing line) may be closer together than two *A* mutants or two *B* mutants (at different ends of their subregions). The functional test reveals a discontinuity in the rII region which is not apparent simply by genetic mapping. The *r*II region consists of two sections with different functions.

Functional units defined by tests of this sort have been called cistrons, but the reason for this terminology requires explanation. We may represent the *r*II region as (___*A*___) (___*B*___) and an *r*II mutant defective in the *A* section as (___*x*___) (_______); similarly, a mutant defective in the *B* section will be (_______) (___*x*___). A doubly infected K cell will be similar to a heterozygote, and if it is infected with mutants of the two kinds it can be represented as

(___*x*___) (_______)
(_______) (___*x*___)

If, however, it is infected with two mutants, both defective in the *A* section, it will appear as

(___*x*___) (_______)
(___*x*___) (_______).

In each of these "heterozygotes" the defects are placed in different chromosomes, across from each other in what may be called the *trans* configuration. By recombination, however, both defects can be placed in the same strand and the double heterozygotes will appear as (—x—) (—x—) and (—x—x—) (———). The defects are now

(———) (———)   (———) (———)

in the same strands, in the *cis* position. The change from a *trans* to a *cis* arrangement does not change the composition of the elements but only their positions relative to each other. The change does, however, have consequences for the functioning of the material. In both *cis* arrangements a normal functional *A* region and a normal *B* region are available; but one of the *trans* arrangements is normal and the other defective. When two mutants are defective in the same function, a difference is detected between the *cis* and the *trans* configurations, but if the two mutants are defective in different functions, no such differences are seen in the *cis-trans* test. This *cis-trans* position effect was first recognized in Drosophila and is capable of being detected wherever two strands of genetic material can be placed in a common environment. Although the functional units, or cistrons, defined in this way may not be entirely equivalent in different cases, the term is applicable in many situations where fine structure analysis of genetic regions is involved and it appears to have acquired a firm position in the genetic lexicon.

We thus see that whenever detailed analysis is undertaken, when new relationships are exposed and new concepts are developed, a new terminology is required and produced. The concept of the gene is one of the most powerful and useful that biology has fostered, and it is still a satisfactory tool in most contexts, but it is gradually being replaced by a set of terms which designate more precisely certain aspects of genetic organization and function. The *recon,* the smallest unit of recombination, is in some organisms at least no larger than a single nucleotide pair in DNA double helix. The *codon,* has been proposed as the unit of DNA which is responsible for specifying a single amino acid in a protein; as we have seen, the codon probably consists of a sequence of three bases. The *cistron* is a unit of function defined by the *cis-trans* position effect and may correspond fairly well with the unit which specifies an entire polypeptide chain. We shall introduce a little later the concept of the *operon,* a sequence of cistrons which serves as a unit in transcription of the DNA messages. The coining of these terms, and their ready acceptance by biologists, reflects the need to characterize the organization of the genetic material. The older idea of a series of beads on a string is no longer adequate; genetic material is arrayed in a series of hierarchies ranging from the nucleotide to the entire chromosome.

15

The Origins of Molecular Order

Origins have always been intriguing to man. Even the earliest written records bear witness to his preoccupation with his own origins, the origins of other living things and of the cosmos which he inhabits. Historically the problem has been compounded because two different questions were entangled: the question of "why" and the question of "how." The early explanations were directed more to a rationalization of the meaning of life than of the mechanisms of its origin, and they were usually framed in philosophical or religious terms. We shall not consider these early interpretations, interesting and significant though they are in other contexts, but shall restrict ourselves to recent analyses which offer hope for explaining origins in mechanistic terms. This restriction does not imply a denial of a relationship between the two kinds of questions, only a doubt concerning the propriety of metaphysical discussions in an elementary biology text.

The scientific study of the origin of life belongs to two separate eras. The first began with the studies of the Italian biologist Francesco Redi in the 1660's, and ended with Pasteur's decisive experiments two hundred years later. Prior to Redi's studies many people believed in spontaneous generation, in the notion that life was constantly produced by the processes of decay; rats and mice appeared *de novo* in filth; maggots were produced by the putrefaction of meat. Redi's contribution consisted of an experiment whereby he demonstrated that meat, placed in containers covered with gauze, attracted flies but did not produce maggots. Only samples of meat directly exposed to flies produced maggots. He concluded that maggots were not produced directly from decaying meat, but arose from preexisting flies.

Such studies demonstrated that life, at least life of the larger variety, came from preexisting life, and that it did not generally appear spontaneously. They did not, however, answer the question categorically. At about the same time that Redi was examining the problem, Leeuwenhoek was discovering through his primitive microscopes the teaming universe of living things invisible to the naked eye. Although lice, mice, and flies were not spontaneously generated, the possibility was still open that simpler creatures were, and many observations seemed to indicate just this. Broths or infusions of various sorts could be prepared and boiled until no life could be detected in them; yet these infusions, if allowed to stand, produced a large array of microscopic organisms.

A hundred years after Redi another Italian, Lazzaro Spallanzani, performed an experiment analogous to that of Redi. He heated a broth and sealed the mouth of the vessel containing it to prevent contamination by microscopic organisms which might be present in the air. The sealed vessels produced no microbes until the seal was broken. Theodor Schwann, in the 1830's, went one step further and showed that boiled broth produced no living things even if the air over the broth was replaced, provided that the air had been heated. These studies appear in retrospect, to be decisive, but they were far from convincing in their own time. They were interpreted by some as demonstrating not a requirement for preexisting life, but rather the heat lability of a "vital principle" present in the air. Not until the 1860's did Pasteur settle the question. By simple but effective means he sterilized air without heating it (using curved glass air passsages that served as traps for microbial spores) and showed that broth exposed to such air did not generate living forms.

These studies were convincing in demonstrating the unbroken continuity of life, but they also exposed a difficult problem for evolutionists. Darwin, a contemporary of Pasteur, was concerned with the derivation of the multitudinous life forms which inhabit the earth. The concept of the continuity of life, combined with the ideas of heritable variation and natural selection, permitted a vision of evolution extending into the distant past. Existing organisms could be traced backward conceptually through the twigs and branches of evolutionary trees to ever simpler progenitors, until all life was embodied in a single primordial form—the ancestral cell. The problem posed was that of the origin of the initial entity; if life did not appear spontaneously in laboratory experiments, how could one account for the spontaneous appearance of life at the beginning of biological time?

At first the explication of life seemed to reside in an understanding of the nature of "protoplasm"; if this "stuff of life" could be properly analyzed, it could probably also be synthesized and the mystery of its origin could be removed. So long as cells seemed to be simple, the conceptual leap between an inorganic environment and a living thing appeared not to be great. It could readily be imagined that under some conditions, not yet achieved in the laboratory, life could appear by a fortuitous combination of circumstances. Those who held this position, however, based it on a faith in the inalterability of the laws of nature and not on a detailed understanding of which laws, applied in what ways, might have accounted for the origin of life. Indeed, until the properties of living systems were much better understood, it would not be possible to prescribe what events would be necessary for their establishment. However, as cell studies continued through the nineteenth and early twentieth centuries, the initial impression of cellular simplicity evaporated and was not immediately replaced by an understanding of the simplicities of organic principles. Cells were far more complex in both structure and function than the early cytologists had imagined. The gap between the living and the non-living seemed to increase with every advance, and the goal of rationalizing spontaneous generation appeared remote.

By the 1930's sufficient information had been acquired and assembled from various sources particularly by A. I. Oparin to permit a new assault on this question, and on a more experimental basis, but even today we are only beginning to approach an understanding of living systems which permits us to reconstruct tentatively some of those early critical steps. Before discussing these attempts to account for the origin of life we should pause briefly to consider an auxilliary question: that of the "Where" of life's beginnings. The implicit assumption up to this point has been that life originated on our planet Earth. Alternatively, life could have developed elsewhere in the universe and could have been transported secondarily to the earth either by intelligent beings or by natural cosmic events. Neither of these latter possibilities can be completely rejected. We know that in the near future we shall have the capacity to contaminate other planets in our own solar system, and if life has originated elsewhere in the universe, intelligent beings could have perhaps contaminated a barren earth. The current efforts to devise means of communication with intelligent beings across interstellar space are tokens of a strong belief in the distribution of life foci elsewhere in the universe. Nevertheless, even if this belief is substantiated, the problem of transportation across the distances between planetary systems is one which we have not solved, except by fanciful devices in science fiction, and one which may not be capable of solution either in principle or in practice.

The transportation of the seeds of life from another source by purely cosmic forces is considered by some to be more feasible. If life developed elsewhere in the universe, and if the planet of origin were fragmented in a cosmic accident, the spores of living organisms might on occasion have been imbedded within rocky masses and insulated from the radiation, the cold, and the emptiness of space. They might even have survived the fiery plunge of a meteor through another planet's atmosphere to be released on the surface in a viable state. These circumstances may seem fantastic, but some investigators are sufficiently intrigued by the possibilities to make serious studies of meteoric fragments in search of traces of life. Some of these studies have produced interesting results; complex organic compounds have been recovered from meteors, but thus far they can be explained at least as well through processes of cosmic chemistry as through the activities of living organisms.

An extraterrestrial origin for earth's life is usually discounted, however, on another basis. To assume that life originated elsewhere does not in fact answer the question of how life originated; it simply removes the problem to another time and to an unspecified place about which we know even less than we do about primitive earth. Most biologists would prefer to exhaust the possibilities of a terrestrial origin before considering the unlimited possibilities of conditions and circumstances throughout the universe. We shall not take the time to discuss the larger issues of cosmic evolution—the origins of the chemical elements, the establishment of local condensations of matter, the differentiation of solar systems—but shall move immediately to the time at which the earth was established as a planet moving in orbit about our sun. We might be reminded, nevertheless, that the events occurring on the earth's crust represent only a minor episode in the cosmic panorama, and that life itself may be considered, in Muller's words, as "merely a fancy kind of rust, afflicting the surfaces of certain lukewarm minor planets."

An evaluation of the prospects for the generation of life on earth requires information regarding the conditions which prevailed throughout the early history of our planet. We need to know the temperature, the energy sources, and the time span available; we must know whether water was present and, in general, the chemical composition of the earth's crust with regard to the critical elements of C, H, O, and N. Obviously this information is not going to be obtained directly, but reasonable inferences have been provided through the efforts of cosmologists, geochemists, and geologists.

What may we reasonably infer about the conditions prevailing at the time life first appeared? The earth is estimated by various means to be approximately five billion years old, but it was a very different kind of place at its origin than it is today. At first it was much warmer and it cooled only gradually. Initially it had a smaller mass but it grew bigger through its scavenging of interplanetary debris. Its chemical composition also changed in the course of time. The most common element in the universe is hydrogen, and the dust clouds from which the earth condensed contained high proportions of this element. The sun is even now about 85% hydrogen. A large portion of the earth's atmosphere was hydrogen and most of the molecules on its surface were compounds of hydrogen. Water (H_2O) was present, initially as steam; carbon was mainly represented in its most reduced form as methane (CH_4); and nitrogen occurred primarily as ammonia (NH_3). Chemical bonds were also established less frequently among the less common elements—between C and N and between C and O—so that a variety of simple organic molecules was established. This much is reasonably inferred, both from theoretical considerations and from spectroscopic examination of the larger planets in our solar system. Jupiter's atmosphere, for example, is even today largely composed of hydrogen and helium and its surface contains detectable amounts of ammonia and methane. But, although hydrogen continued to dominate the atmosphere of the larger planets, the earth possessed neither the mass nor the gravitational field required to hold the smaller atoms and these gradually escaped into space, altering the chemical composition of the earth's atmosphere and crust and thereby playing a significant role in the chemical evolution of the earth.

The major events in chemical evolution may be traced, even though many of the details are obscure. First we may mention that a hydrogen-dominated atmosphere is much more transparent to ultraviolet radiation than our present atmosphere, and initially the earth's surface was bombarded with light rays of short wave length and high energy. After oxygen evolved into a major atmospheric component the situation was greatly changed; the ozonosphere, populated with energetic oxygen atoms, creates an efficient screen for ultraviolet light some fifteen miles above the surface. Only the less energetic light waves now reach the surface of the earth in quantity. We may also note that far more radioactivity would have been detected on the primitive earth than now. And finally we should recognize that the establishment and settling of the earth's crust produced a variety of local differences in temperature and chemical constitutions. As the earth cooled the water in the clouds fell as rain on the hot crust and rose again in clouds of steam; torrential rains and violent storms must have continued over periods of geological time as mountains were lifted and eroded and as ocean basins were filled with the salt solutions leached out of the mountains. Eventually the atmosphere became quieter, though it continued to be disturbed by local volcanic activity.

The reason for this survey is to emphasize the variety of conditions which existed in the earth's early history; not only was an array of circumstances available at any one time—from mountain tops to shallow seas, from polar to arctic latitudes, and from desert plains to volcanic flows—but the chemical and physical properties of the atmosphere were undergoing slow progressive alteration and contributing to an almost limitless sequence of conditions for nearly three billions of years before life appeared. If life was generated on the earth, locating that special set of conditions which made it possible might seem to require an endless quest.

Fortunately the task is not as hopeless as it first appeared. Although very few conditions have been examined in the laboratory, and though life has not yet been generated under any of these conditions, sufficient results have been obtained to encourage further studies. In 1955 Stanley Miller attempted to establish conditions which would correspond generally to those assumed to exist during the time the earth had a reducing atmosphere. He prepared a mixture of ammonia, methane, carbon dioxide, and water and supplied energy by means of an electric discharge. From the reaction mixture he recovered in quantity a wide assortment of amino acids and other organic compounds. Other workers using different reaction mixtures and different energy sources, including ultraviolet light and radiation from atomic disintegrations, have recovered sugars, purines, and pyrimidines in addition to amino acids.

The many extensions of Miller's original studies have thus led to one general conclusion. Given that the early earth's atmosphere consisted mainly of CH_4, NH_3, N_2, H_2, and H_20, all the essential subunits of biological macromolecules could have been formed spontaneously.

The origin of macromolecules raises two problems, neither of which can be explained in terms of the conditions essential to the production of subunits. The first of these problems concerns the chemistry of the polymerization process—how a linking of the subunits into a molecular chain is effected. The essentials of this problem have already been discussed in Chapter 4. Regardless of the type of macromolecule formed, the equivalent of one molecule of water must be removed for each covalent link generated between two submits. Since in the presence of water thermodynamic conditions do not favor such a reaction, some special conditions must be postulated in which water would not affect the course of the reaction. Various suggestions have been made, but no single one has been generally accepted. One scheme is based on the probability that certain ponds would have lost water by evaporation, such loss leading to dehydrating conditions. Another scheme is based on the fact that phosphate esters are a universal mechanism for biological polymerizations and on the probability that cyanamide (H_2NCN) was present on the primitive earth. Cyanamide hydrolyzes very slowly, and if such hydrolysis were to occur by removing water from inorganic phosphate and simple organic molecules, phosphate esters could be generated. Cyanamide on the primitive earth would thus play the role which oxidation reduction plays today in the generation of phosphate esters. Several experiments have been performed by Melvin Calvin and his collaborators to test the scheme, and they have found substances such as glucose phosphate and adenylic acid formed even though the reagents were present in a dilute aqueous solution.

Hence the conditions necessary for the production of sugars, amino acids, and nitrogenous bases are not special and precise; and their polymerization can be achieved without living organisms or enzyme extracts. Before experiments such as these had been carried out, cosmologists had argued cogently that prior to the advent of life the processes of chemical evolution would have produced vast seas of organic compounds which would be protected from dissolution in the absence of oxygen and organisms. The experiments provide a striking confirmation of this deduction and lessen the apparent gap between the living and the nonliving. The organic solutions could serve both as a reaction mixture for the synthesis of living things and a nutrient medium for their development. Nevertheless, a gap still exists; an organic soup may well be a precursor of a cell, but it is far from a living system. The transition from the nonliving to the living raises the second problem concerning the origin of biological macromolecules.

One of the most obvious differences between a cell and a soup lies in the degree of organization of the components. A cell consists of membranes, fibers, and granules in which the molecular constituents have specially assigned positions and interrelations. We may, however, demonstrate that some compound structures, like their component parts, are capable of forming under relatively imprecise circumstances. Fibrous proteins like collagen may be dissolved and then reprecipitated into fibers similar to those observed in cells; membrane systems may be constructed of various components and may behave in major respects like biological membranes. Much of the organization of cells is thus implicit in the molecules which make it up, and we may readily imagine the spontaneous appearance of complicated structures from dissociated molecules. But complication and organization are not really enough to characterize life. What is critical is the establishment of an organization with certain essential properties which endow it with the capacity to participate in the long range processes of evolution. More specifically, no system of organization has the essential qualities of life unless it can replicate itself and undergo heritable variation.

As we presently understand self-replication, it is primarily a property of a certain class of molecules, the nucleic acids. Although some biologists have doubts concerning the time at which nucleic acids achieved their functions in the history of life, the simplest conceptual approach places them at the very beginning. Although one could scarcely expect the first life forms to have all the specialized equipment which modern forms have accumulated through thousands of millions of years of evolution, they must have had the capacity for reproducing in their own images before they could evolve at all. Our present understanding of nucleic autosynthesis actually makes the initiation of life a more easily visualized process than many of those which come after. We know that the precursors for nucleic acids can be generated in reaction flasks from simpler organic compounds and we know that they can be polymerized by appropriate manipulations. We have reason to believe that base pairing is the underlying principle for nucleic autosynthesis and that it might have functioned primitively even without enzymes. It is true that not all these reactions have yet been demonstrated in the absence of enzymes, and certainly they occur more reliably and more readily with the assistance of enzymes. But catalysts function by modifying the rates of reactions which are thermodynamically possible even in their absence.

Thus we can imagine a nucleic acid molecule serving as a template for ordering a series of nucleotides complementary to itself; and we can imagine the separation of the strands and the repetition of the process. To assume that nucleic autosynthesis came before directed protein synthesis is not, of course, equivalent to assuming that no catalysts facilitated the replication, only that molecules with the degree of specificity and rates of activity of modern enzymes were not available. Protein particles and surface films may have played a critical role in these events, but their production was not necessarily related to the nucleic acids; the autoreplication of nucleic acids had no specific consequences in the production of particular proteins.

At this stage life would have appeared to an intelligent observer as an erratic and unimpressive phenomenon with limited potentialities. Certainly so long as autosynthesis was separated from heterocatalysis its prospects were slim indeed. The nucleic acid molecules could replicate only so long as the immediately available supply of precursors was adequate, and at rates determined by the accidental availability of catalytic surfaces. About the only meaningful selective force would be that exerted by the supply of precursors; polymers with certain nucleotide ratios might reproduce more than others. Exactly how autosynthesis and heterocatalysis became linked we do not know, but that linkage represented the first step toward a control of the environment which continues to the present time and was absolutely essential for further biological advance. The kind of association between autosynthesis and heterocatalysis which is observed in modern organisms appears highly improbable; proteins are not regularly produced directly by nucleic acids but only indirectly through a fairly complicated set of machinery. DNA produces message RNA; message RNA is bound to ribosomes; transfer RNA carries amino acids to the message RNA template; a variety of specific enzymes facilitate the various steps. The important point here is that modern heterocatalysis is achieved by translation; no direct structural relationship is established between the master DNA template and the amino acid sequence. The current procedures may, however, represent an evolutionary elegance far removed from the primitive stages of heterocatalysis, when transcription rather than translation was the basic mode. By this is implied a direct structural relationship between the nucleic template and the protein configuration. Such structural relationships, at least among completed macromolecules, are implied by the close relationships between chromosomal DNA and the basic nuclear proteins and between ribosomal RNA and ribosomal proteins. These relationships might have provided a basis for linking directly nucleic structure and protein synthesis. This is certainly only a speculation at the moment, but it provides a rationalization of a critical step in evolution. In any case autosynthesis and heterocatalysis were linked so that the structures of proteins and nucleic acids were meaningfully related; with this step nucleic acids acquired the capacity to regulate the conditions under which they were replicated. The degree to which the controlled proteins facilitated DNA replication provided a powerful selective force driving life to further accomplishments.

The initiation of self reproduction and its linkages with heterocatalysis were perhaps the most important events in the history of life; in a sense, all the other properties and manifestations of life are direct or indirect consequences of these innovations. This is not to say that the course of life beyond this point was smooth, nor that we understand in detail the manner in which subsequent crises were met, but a reconstruction of all the successive steps is beyond our present capacity and constitutes a continuing problem for contemplation and experimentation. We recognize, for example, the crucial significance of tapping the available energy sources and channeling them to biological objectives, and we are aware that the techniques evolved for one period and one place were not adequate for all times and all places. The original anaerobic environment changed gradually into an aerobic environment; the stockpiles of organic compounds elaborated during earlier eons eventually were exhausted; the sources of high energy radiation were cut off and adjustments had to be made to capture and use the gentler light waves filtered through the modified atmosphere. New environments appeared and the mechanisms of Darwinian selection allowed the continued diversification of organisms and exploitation of niches. At some time in this sequence cells of reasonably modern design appeared, sex was discovered, the differences between plants and animals became established, and the possibilities of the multicellular state were explored. Many of the tactics employed in these developments remain to be explicated, but the major strategy of life on earth is becoming comprehensible.

SECTION C

The Regulation of Cell Behavior

This topic is extremely broad,—too broad to be subjected to a tight systematic discussion. Any activity of a cell is part of its behavior, and all behavior is an expression of regulation. Our present knowledge is certainly inadequate to explain such regulation, even though a great deal of information has been collected about factors and conditions which affect regulation. The discussion in this section is therefore an open-ended one. In some respects this section will serve to provide sundry pieces of information about cellular properties which have not been considered in preceding chapters. Topics such as cell movements and membrane conduction, which could not be conveniently fitted into earlier chapters, will here be discussed as aspects of regulation.

The section is organized under four headings: spatial organization; temporal organization, environmental regulation, and developmental regulation. The objective which underlies the treatment of all these topics is to expose some of the very many problems which are yet to be solved. By now, the reader should be familiar with most of the basic mechanisms and structures which are universal attributes of cell function. In this section he will be confronted with the many loose ends. Unfortunately, wherever there is looseness of understanding one also finds a surfeit of information. To do justice to ignorance, we should provide a maximum of information. In a general text this is patently impossible. The interested reader will have to dip into more specialized books.

16

Functional Organization of Cells: Spatial

In preceding chapters cellular function was reduced to three basic mechanisms: catalysis, sequence coding, and diffusion control. This is undoubtedly an oversimplification of cellular life, but it has the advantage of sharply defining those mechanisms which appear to be adequate for the operation of living systems. No attempt was made, however, to oversimplify the mechanisms themselves, and attention was frequently drawn to the uncertain boundaries between knowledge and ignorance. One general question was nevertheless left aside. What bearing does the spatial arrangement of subcellular components have on the performance of a cell? The answer which immediately suggests itself is that each of the subcellular components has a set of distinctive functions, and that the activities of a cell represent the interactions among these components.

For the sake of perspective the conventional picture of how the three basic mechanisms are spatially distributed within a cell may be briefly reviewed. Catalysis as such is not a localized activity. Wherever chemical linkages are formed or disrupted within a cell, enzymes are present. There may be some exceptions, but they are few enough and special enough to prove the rule. Thus whether it be the synthesis of macromolecules, the transformation of small molecules, or the transient changes in chemical bonding during membrane passage, enzymes are involved. With the exception of certain vacuoles virtually all other regions of the cell are seats of changes in chemical linkages. Enzymes are therefore ubiquitous in distribution. Localization is only evident with respect to certain types of enzymes or enzyme systems. The organization of energy-transforming systems in chloroplasts and mitochondria has been adequately documented. The presence of a wide variety of enzymes in the ground substance of cells has also been alluded to on several occasions. In general, the most prominent feature of enzyme localization is the organization of structurally coherent assemblies of enzymes. Such assemblies are particularly well recognized in membranous structures, especially those of mitochondria and chloroplasts. Not so readily recognized (and this point will be returned to later) are instances of localization in which the enzymes are not tightly bound to a coherent structure. Many enzymes are lost from nuclei when these are isolated in sucrose media, and unless special precautions are taken, enzymes are also lost from chloroplasts in the course of isolation. Uncertainties about enzyme localization are as much due to technical problems in subcellular fractionation as to the difficulty in defining how tight a chemical bonding between an enzyme and a structure is essential before the association can be considered as a localization. However, one generalization may be advanced. Enzymes which are localized appear to be functionally related to the biochemical activities of the structures present in the region of localization.

The principal site of sequence coding is within the nucleus because the nucleus houses the chromosomes. However, as soon as we touch on the issue of coding, the problem ramifies. The phenotype of a cell is essentially an expression of the kinds of proteins coded. With the exception of the nucleic acids, all other molecules which contribute to the structure or function of cells are entirely formed by the catalytic activities of protein molecules. Sequence coding and protein synthesis are too closely connected to be treated separately. If gene transcription is localized in the nucleus, then protein synthesis is at least largely localized in the cytoplasmic ribosomes. In a typical cell, however, this description of localization is incomplete. Both by autoradiographic and biochemical techniques, the nucleolus has been found to be the most intense site of ribonucleic acid synthesis. This observation should be tied in with the fact that two types of RNA, transfer and ribosomal RNA (Chapter 13), are parts of the general machinery of protein synthesis rather than representatives of specific coding assemblies typified by message RNA. One of the interpretations offered to join these separate observations is that the nucleolus functions primarily in providing the cytoplasm with ribosomal RNA and possibly also transfer RNA.

Diffusion is primarily controlled by membranes. All cells have a limiting surface membrane, and the relationship of this membrane to the control of solute exchanges between cell environment and cell interior has been discussed at length. The fact that various subcellular organelles—mitochondria, chloroplasts, nuclei, lysosomes—are membrane-bounded structures has also been mentioned but not the implications of this for cell function. A generalization which can be made is that except for the interior of the nucleus, membranous structures are prominent and universal components of the cell.

This brief sketch of the organization of a typical cell raises an important question. How mandatory are the spatial arrangements for the operation of the basic cellular mechanisms? The question must be raised because bacterial cells do not possess the elaborate organization found in cells of higher forms. If bacterial cells can regulate their behavior with less elaborate structures, what functions do the elaborate structures play in cells of higher forms? The answer to this question is not that the elaborate structures are more efficient, unless we are able to specify the nature of the efficiency. And to specify this, some analysis of the difference between these two classes of cells is necessary.

If bacterial and animal cells are compared under the electron microscope, differences in structural organization are not nearly as striking as under the light microscope. The basic types of molecular aggregates —membranes, particles, fibrils—are found in all cells. So too are the physiological activities of each of these primary structural systems. Diffusion is regulated by a limiting membrane; respiration and photosynthesis (if present) occur at the site of a membranous or lamellar system; genetic transcription is mediated by molecular filaments of DNA and RNA; genetic translation is effected at the site of a ribosomal particle; and cell movements (if the species is so endowed) are made possible by an organization of fibrillar protein elements. The primary molecular aggregates which form the architectural skeleton of bacterial cells also form the cellular skeleton of higher forms. Differences in structural organization cannot be attributed to new molecular units, but must be considered as elaborations or modifications of primary and universal molecular building blocks.

An unproved but reasonable generalization is that every subcellular unit found in higher forms has its homologous counterpart in bacteria. Many biologists have long held the view that even the most specialized of mammalian cell functions, as nervous transmission, muscular contraction, can be found to operate, although primitively, in unicellular organisms. The evidence in support of this view is formidable. We are therefore led to the conclusion that, appearances notwithstanding, the structural complexity of higher cell forms is a modification of a universal and basic structural design, and that such complexity facilitates or accentuates the expression of fundamentally universal processes. The discussion of complexity in Chapter 11 should be recalled at this point, for appearances may be highly deceptive in estimating the complexity of different structural systems.

SPATIAL ORGANIZATION IN "PRIMITIVE" CELL SYSTEMS

Whether bacteria are primitive in an evolutionary sense is a subject that is still debated. The remarkable adaptability of the group of organisms to diverse environmental conditions has been used as an argument against such a classification. This aspect of the question is, however, largely irrelevant to the present discussion. Bacteria may be considered "primitive" in the sense that their structural organization is less complex than that of other cell forms. The relative simplicity of their internal structure permits a few conclusions to be drawn about the relationship between spatial distribution of subcellular systems and total cell function.

The bacterial cell consists of a set of distinct and physically separated metabolic systems embedded in a cytoplasmic matrix which contains a variety of enzymes. This statement is based on the fact that several structural aggregates, each of which mediates a distinctive set of metabolic reactions, can be visualized as localized structures under the electron microscope. The complex of DNA fibrils is located at or near the central region of the cell and is not continuous with any other structural element. Claims have been made that the DNA filament within a bacterial cell is in contact with the peripheral membrane. Such continuity, if true, would nevertheless be limited in extent, so that DNA replication and transcription into RNA must therefore be regarded as a localized activity. Ribosomal particles are scattered through most of the cytoplasmic matrix, but inasmuch as protein synthesis is known to occur only on these particles, this process too is structurally localized. Respiratory activity is associated with membranous elements which are usually found only at the cell periphery. Evidence for different membrane systems which would separately govern solute diffusion and electron transport has been sought, but the evidence is inadequate to permit broad generalization. In many species of bacteria, however, the peripheral or "plasma membrane" shows small invaginations which frequently develop into elaborate patterns of tubules confined to a characteristic region of the cell. Their function is undetermined. In photosynthetic bacteria we find either small membranous vesicles which fill a large part of the cell interior, or a stack of lamellae similar to those in chloroplasts. In both cases the photosynthetic system is physically separate from other structural systems.

The conclusion from this very brief survey of bacterial organization is that even the most primitive cellular form is spatially differentiated into regions of specialized metabolic activity. The reference to bacteria as "bags of enzymes" is factually incorrect even though the structurally undefined ground substance contains a large variety of enzymes. Subcellular localization, which is so readily demonstrated in cells of plants and animals, is not an exclusive characteristic of advanced organization. What needs to be clarified, however, is the extent to which the spatial segregation of structural units promotes a corresponding segregation of metabolic activities. Enzymes which are a coherent part of a structural unit are easy to localize if the structural unit itself can be isolated. Enzymes which are associated with but not tightly bound to a structure are difficult to localize unless the structure is enclosed in a membrane, as in cells of higher forms. Despite this difficulty, certain loosely bound enzymes have been shown to be associated with discrete structures. Instances of this were discussed in connection with enzymes of the Krebs' cycle (Chapter 7). Although bacterial cells do not contain typical membrane-bounded mitochondria, the Krebs' cycle oxidative complex remains associated with membranous elements after disruption of the bacterial cell. A more striking example of localized molecular associations is provided by the enzyme complex essential to DNA replication. If bacterial cells are disrupted in a sucrose medium, molecules and/or molecular aggregates may be separated according to density and size by means of an ultracentrifuge. The DNA aggregate, although largely DNA, is found to be associated with various enzymes essential to DNA replication. These enzymes are clearly not an integral part of the DNA molecule, and the bonding between enzymes and DNA is weak enough to be ruptured under relatively mild conditions. How many similar types of association between a defined structural element and otherwise soluble enzymes exist in a cell is an open question; adequate techniques for demonstrating such associations are difficult to achieve.

At least three types of enzyme localization are thus observed in living cells: enzymes which are tightly bound to membranes; enzymes which are enclosed by membranes; and enzymes which are bonded to nonmembranous but discrete structures. In some cases the physiological significance of the association is clear. The electron transport systems in chloroplasts and mitochondria, and the DNA synthetase system in association with DNA are obvious examples. In other cases, however, information is too fragmentary for clear interpretation. Nevertheless, we may draw a significant conclusion about the spatial organization of cell functions from the very limited comparative study of respiratory and genetic systems. In cells of higher forms nuclei and mitochondria are membrane-bounded structures; in bacteria the homologs of these structures are not enclosed by membranes. On the other hand, some of the principal enzymes which are associated with but not tightly bonded to these subcellular structures in higher forms are similarly associated with the homologous structures in bacteria. The limiting membranes of subcellular organelles cannot therefore be considered as essential to the localization of metabolic machinery.

Two major roles have been assigned to membranous elements: electron transport and regulation of diffusion. With respect to the first of these roles, no clear difference has been found between bacterial and other types of cell except for location and enclosure by an additional limiting membrane. With respect to the regulation of diffusion, we may assume that the plasma membranes of both cell types function in more or less the same way. What must be explained is the profusion of membranes within cells of higher forms which not only enclose subcellular organelles but also fill many regions of the cytoplasm. Since these membranes are not exclusively associated with electron transport, we may infer that they play some role in the regulation of diffusion. Diffusion is probably a minor factor in the life of a bacterial cell. A glycine molecule travels about 100 μ per second in water at 20°C. Such a molecule would traverse the interior of a bacterial cell measuring 0.5μ in diameter about 200 times per second. And since the relationship between diffusion constant and molecular weight is such that a molecule of 500 times the weight of glycine would move at a little less than $\frac{1}{10}$ the speed, we may assume that all substrate molecules would move through the cell at a rate adequate to meet the needs of metabolic utilization.

MEMBRANE SYSTEMS IN CELLS OF HIGHER FORMS

As pointed out, the striking difference between cells of bacteria and those of higher forms is in the concentration of intracellular membranes. In bacteria, the principal subcellular units (the nuclear apparatus, the respiratory or photosynthetic systems) are not enclosed by membranes, whereas in cells of higher forms they are. The limited invaginations of the plasma membrane noted in some bacterial species are represented in the cells of higher forms by an extensive complex of membrane-bounded canals (the "endoplasmic reticulum") which link external with internal regions. Other types of membrane-bounded systems—lysosomes, vacuoles, contractile vacuoles, food vacuoles, plastids other than chloroplasts (chromoplasts, leucoplasts, amyloplasts)—are present in the cells of various groups of organisms but not in bacteria. With respect to spatial organization, the distinctive feature of more complex cells is not a localization of metabolic activities, but an insulation of these localized metabolic activities by means of limiting membranes. Unless the assumption that intracellular membranes have the capacity to regulate diffusion is in error—and experimental evidence makes this error most unlikely—the distribution of metabolites in complex cells cannot be governed by random diffusion alone.

Inasmuch as membranes regulate diffusion either by acting as barriers to solute passage or by facilitating passage of specific molecular types, they may be considered to function within cells in three different ways: to provide microenvironments for subcellular metabolic units, to isolate substances (including macromolecules) from interaction with other cytoplasmic elements, and to channel the flow of solutes. The first function has been amply discussed in connection with mitochondria and chloroplasts (Chapters 8, 9). Although we might suppose that the nuclear membrane falls under this heading, the evidence is contrary. The nuclear membrane differs from all other cellular membranes in that about 10% of its surface appears to be occupied by pores with diameters ranging from 400–800 Å. Such pores are large enough to permit the passage of protein molecules. Chemical evidence supports the observations made through the electron microscope. Nuclei exposed to sucrose media lose an appreciable fraction of their proteins. If electrolytes are added to the medium, the proportion of proteins lost increases with electrolyte concentration, indicating that many proteins are attached to the nuclear framework by electrostatic bonds. Such losses would not occur if the nuclear membrane were nonporous. The cytoplasmic ground substance and the nuclear apparatus are therefore not separated by the kind of limiting membrane present in mitochondria. The early impressions of cell physiologists that nuclei were insulated from cytoplasmic activities by an intervening membrane are incorrect. The open texture of the nuclear membrane is more in keeping with the physiological and genetic evidence that a continuous traffic of substances (including macromolecules like RNA) between nucleus and cytoplasm occurs during most of the life of a cell. If chromosomes do function in a microenvironment different from that of the cytoplasmic ground substance, the environment is not maintained by a limiting membrane.

The term "microenvironment" as used here designates a region of microscopic dimensions in which a differential composition of solutes is maintained by the activities of the localized metabolic apparatus and the selectivity of the enveloping membrane. The functional result is to restrict the random diffusion of metabolic intermediates to the volume enclosed by the membrane, and hence to increase the probability of interaction between those intermediates and component enzymes. The second function of membranes, solute isolation, is no more than a variant of the first. The most prominent example of solute isolation is the vacuolar system, which has already been discussed in Chapter 10. To designate a single function for vacuoles is impossible. In plants they serve as a storage organ for water and salts, and as a sink for unused metabolic by-products. In protozoa they function in water elimination, food retention, and decomposition. In animals they may serve as vehicles for bulk transport across cells (p. 313). All these examples illustrate the fact that a single mechanism may become modified in various ways and provide a corresponding variety of functions, many of which are accessory to the basic activities of the cell. Such accessory activities may, however, assume functions of primary importance in the multicellular system. The water economy and mechanical structure of all higher plants are inconceivable without highly vacuolated cells.

Lysosomes furnish another example of membrane-bounded regions which probably have general importance in cell behavior but have a special importance in cells of higher organisms. Lysosomes are defined as globules, usually about 0.6 μ in diameter (though ten times the size in kidney cells and even larger in the phagocytes of the blood stream), which are bounded by a limiting membrane. The definition is rather loose except for the requirement that all lysosomes contain hydrolytic enzymes. The interior of lysosomes varies according to cell type. In some cells it is uniformly dense; in others it is differentiated into a less dense inner zone and a dense outer zone; in still others the lysosome may enclose a small vacuole. Since vacuoles are also membrane-bounded globules, the distinction between a food vacuole and a lysosome is difficult to make on the basis of the definition. The distinction is perhaps unnecessary. A most important feature of the lysosome is that its limiting membrane presumably insulates the cytoplasmic ground substance from the activities of the enclosed hydrolytic enzymes. Lysosomes spill their contents in liver cells which have been injured, and hence accelerate the decomposition and removal of dead cells. In many developmental situations, metamorphosis, for example, the organism selectively stimulates the destruction of certain cell groups and the proliferation of others. One of the mechanisms which has been postulated for selective destruction, and to some extent experimentally demonstrated, is the controlled rupture of lysosome membranes.

The capacity of intracellular membranes to isolate different regions of a cell so as to prevent the randomization of metabolic precursors and products makes possible intracellular differentiation. The possibilities are most prominently realized among the protozoa, which, though unicellular, have developed many of the attributes found in multicellular organisms for exploiting their inanimate environment. But regardless of the degree of intracellular differentiation, all cells of higher forms must confront a challenge related to diffusion which is a function of their relatively large size.

The nature of the challenge may be made evident by comparing a bacterial with an animal cell having average diameters of 0.5 μ and 20 μ respectively. The ratio of their volumes is 1/64,000. If all metabolic reactions within a cell were dependent on random collisions between substrate and enzyme, the probability of a single macromolecule colliding with one substrate molecule would be 1/64,000 times as great in an animal as in a bacterial cell. The comparison would be meaningless if the animal cell had 64,000 times as many macromolecules of each type as did the bacterial cell. This is not the case. Increase in cell size is accompanied by an increase in diversification, and the target volume of a specific subcellular metabolic unit does not increase in direct proportion to the total cell volume. To take an extreme example, a specific protein is coded by one or two DNA segments in both bacterial and animal cells. If transcription were dependent upon the random diffusion of ribonucleotides to the DNA site, animal cells would be at an extreme disadvantage compared with bacteria in synthesizing specific proteins at appropriate times. No evidence for such a disparity in efficiency has been found. The one conclusion to be drawn is that control of intracellular diffusion becomes a more critical problem as cells increase in size and diversify in function.

This problem is intensified by surface to volume ratios. The surface increases as the square of the radius, whereas the volume increases as the cube. A 40-fold increase in volume is accompanied by a 40-fold decrease in surface/volume ratio. Assuming that the inherent capacity of the plasma membrane to facilitate solute entry is more or less the same in all cells, a decrease in transport capacity accompanies an increase in size.

The effective adaptation of cells to the challenge of diffusion appears to reside in the organization of the endoplasmic reticulum. The deep and extensive invaginations of the plasma membrane increase the surface area available for exchange of solutes between cell interior and environment. The spaces enclosed between the membranes of the reticulum limit the diffusion of solutes and channel them to different regions of the cell (Chapter 10). The nuclear membrane may be far more important as a mechanism for facilitating an exchange of solutes between the contents of the membrane-bounded canals and the nuclear apparatus than for separating that apparatus from the cytoplasmic ground substance. On the whole, however, relatively little is known about the regulation of diffusion by the system of intracellular membranes. That such regulation could exercise a strong influence on the regional metabolic activities of cells is patent, but this possibility must be regarded as speculative until more is known about the properties of the membrane network.

MEMBRANES AS MESSAGE TRANSMITTERS

Increase in cell complexity results not only from the formation of more elaborate associations between basic structural units, but also from their modification. More often than not, the two kinds of changes are inseparable, and singly or in combination they lead to the acquisition of new functional properties. The manifestations of these properties, if not the properties themselves, are evident to any observer, for they represent all the specialized activities found in the living world.

To man, the most impressive achievement of evolution is the operation of the nervous system. His very capacity to think, and to sense the peaks and valleys of his emotional experience, is an attribute of this system. Speech, vision, memory, muscular activities, and glandular secretions are among the traits controlled by the transmission of messages through the nervous system. No item of physiology has evoked and continues to evoke more admiring interest than the function of brain and nerve fiber. And no item is more poorly understood. We can trace the rudiments of nerve function to simple cells, but even the rudimentary processes elude adequate interpretation. The concepts and techniques so successfully applied to the formation of organic molecules seem to have no central target in studies of nervous transmission. The one solid fact about nerve behavior is that it is basically a membrane process.

The capacity of cell membranes to act as agents of nervous transmission has its origin in their selective permeability. A fact of primary importance to this whole subject is that in most cells the concentrations of specific ions are different on either side of the membrane. Since the thickness of a cell membrane is about 100 Å, the concentration differences represent steep concentration gradients. If an electrode which is sensitive to the presence of monovalent ions (see Chapter 8, Fig. 8-3) is placed on the inner side of the cell membrane and another on the outer, a voltage is recorded if the electrodes are connected by a metallic conductor through a voltmeter. The system is almost identical with that of a concentration cell (Fig. 8-3*E*). The difference is that the salt bridge is replaced by the membrane itself. Most cells will record a voltage of 50–100 mv if thus tested, the internal surface being negative to the external one. The magnitude of the voltage is determined by the difference in ion concentrations across the membrane. Only those ions which are mobile in the membrane under the conditions of testing affect the measured voltage. If the membrane were a perfect ion barrier no voltage would be detected, just as no voltage is measured if the salt bridge between the two electrodes of a concentration cell is removed. Commonly, K^+ is the mobile ion which manifests a concentration difference, and it is this ion which ordinarily determines the magnitude of the voltage. Under normal conditions the Na^+ is essentially impermeable. The expected voltage E across a cell membrane may be calculated if the ionic concentrations are known, by applying the basic formula for concentrations cells give in Chapter 8. If differences in chloride concentrations are negligible, the formula may be expressed in the following form after converting the different constants to numerical values:

$$E = -58 \log_{10} \frac{K^+_{\text{in}}}{K^+_{\text{out}}} \text{ mv.}$$

That the difference in K^+ concentration is the source of the voltage difference may be readily demonstrated by altering the external concentration of KCl; as the concentration is increased, the voltage difference decreases. The magnitude of the voltage drop across a cell membrane may be stated in an impressive way. Using 100 Å as the thickness of the membrane and 50 mv as the potential difference across it, a figure of 50,000 volt/cm is obtained.

What physiological significance does this measurement have? Cells do not have metallic conductors to form electrical circuits. Neither do they have the ready source of electrons which metals provide in artificial electrode systems. That cells can generate a high enough electrical potential to short-circuit electrons through an aqueous conductor is evident from the behavior of electric fish, but the mechanism for achieving this is no better understood than the rest of nerve physiology. In the context of our present knowledge, the basic point of interest is not the voltage measured, but the fact that voltage measurement reflects a unique set of reactions within the membrane. In all cells any factor which affects either the ionic concentration or the organization of the membrane affects the measured potential.

The dependence of electrical potentials on membrane organization and ion concentration is a fact of considerable interest, but by itself reveals little about the message-carrying capacity of membranes. This capacity only becomes evident when the pattern of response to a membrane stimulus is examined. If a cell is stimulated locally, for example, by electrical or mechanical shock, and as closely as possible to the position of one of the electrodes, the measured changes in voltage show a characteristic pattern of behavior. With very weak stimuli the potential change is small, and in common with induced changes of all magnitudes short of membrane injury the normal "resting" potential is quickly restored. The time interval is of the order of a few milliseconds (Fig. 16-1). With increasing intensity of stimulus an increasing change in potential is recorded until the potential reaches a maximum value. This value is referred to as the "spike potential," and is unaffected by additional stimulation. (If electric stimulation is used, the voltage required for a spike is much below that of the generated spike.) The force of stimulation is very small compared with the force of response. Once a spike potential is induced, a basic and primary property of cell membranes is revealed: successively adjacent regions of the membrane are stimulated to form virtually identical spikes, and these proceed along the entire length of the membrane away from the region of stimulation without any decrease in intensity. In effect, the initial signal which is operationally measured as a voltage change is propagated

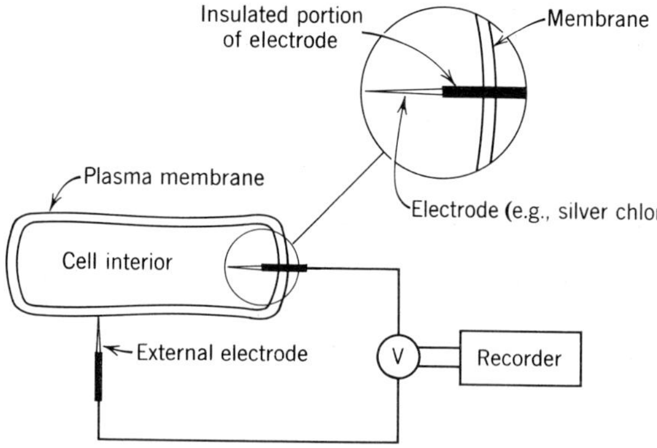

(a)

Fig. 16-1. Membrane potentials and their adaptation to message conduction. (*a*) Measurement of a membrane potential (highly simplified representation). (*b*) Pattern of voltage change recorded at a single point on the cell surface following stimulation. (*c*) Propagation of a spike as manifested in voltage records of three external electrodes placed at different positions on cell membrane. (*d*) A "train" of spikes measured at one point along the conducting portion of a neuron after stimulation of its receptor end.

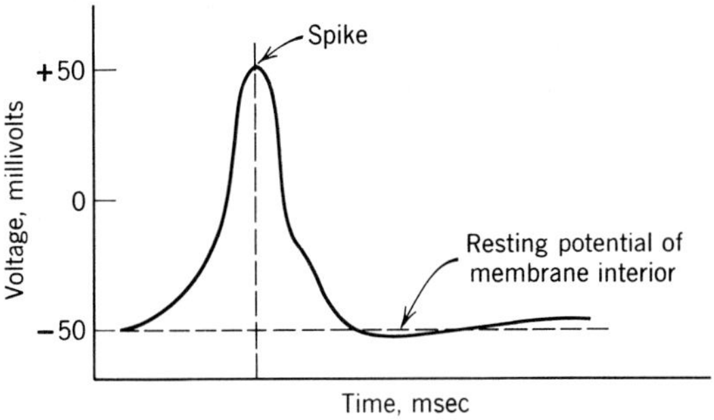

(b)

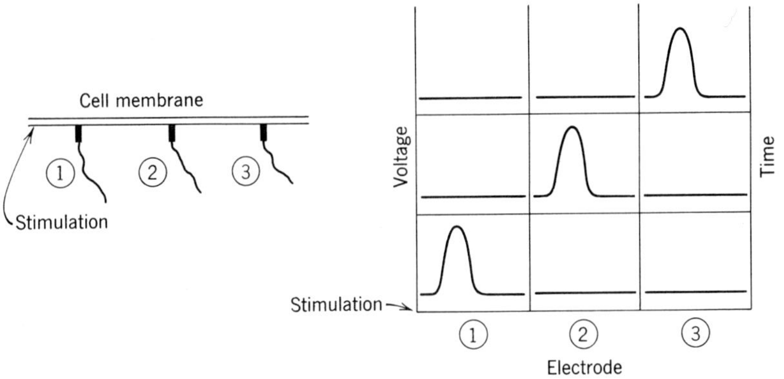

(c)

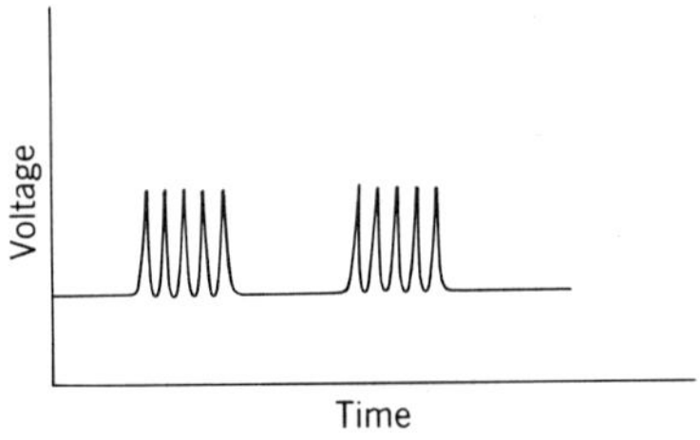

(d)

along the surface of the cell. The propagation is an "all-or-nothing" phenomenon; in this respect, the cell operates by the binary choice principle. The rate at which the signal is propagated varies with the cross-sectional area of the conducting layer. In some giant nerve fibers the value may be as high as 150 m/sec, but rates as low as 0.1 m/sec have been recorded in some of the simple nerve nets of invertebrates. Even these low values, however, are exceedingly high when compared with the diffusion rates of small molecules (e.g., glycine, 1×10^{-4}

m/sec). From the standpoint of speed of communication, molecules are very sluggish messengers. They are probably entirely adequate for effective communication within the confines of a cell boundary, but they cannot be effective in multicellular systems where rapid coordination is required.

The question which must now be raised is what kind of messages can membranes conduct. Although we have some knowledge of the changes which accompany the propagation of a spike, we do not know the actual configuration of the impulse as it exists in the membrane. To answer the question at all, we must make what appears to be a reasonable assumption, that the *pattern* of measured voltage changes has a close correspondence to the pattern of actual changes along the membrane. If this is so, we may examine such patterns to determine whether they have the capacity to carry information.

An effective messenger system must be equipped with at least three mechanisms: one that translates the original stimulus (itself a message) into a communicable form, a second that provides for transport of the message, and a third that effects a translation of the message into a chemical vocabulary by which the affected cell reacts. Since the most specialized cell for conducting nerve messages is the neuron, studies of the configuration of such messages have been most intensively pursued with this type of cell. Basically, only two variables are found in the character of a message traveling along a particular nerve fiber—the number of spikes and their frequency. Thus for any single pulse passing along a neuron, we may, by means of appropriate sensing instruments, classify the pulse with respect to spike number and the frequency with which the spikes are sensed at a single point along the membrane (Fig. 16-1). Messages cannot be superimposed and hence distorted, because following a stimulus a "refractory period" ensues during which time (0.4–2.0 msec) no additional stimulus can be provoked.

In a giant algal cell a spike is propagated along the cell membrane in direct response to a local disturbance. The formation of the spike is not a discriminate one because it can be produced by various stimuli which have the common property of disturbing the membrane equilibrium. Nor is there any evidence for the existence of receptors in the cell which would induce special activities in response to the stimulus. That large unicellular organisms may transmit membrane messages is conceivable but very few examples are known. One of these is the naked dinoflagellate, *Noctiluca,* in which a change in membrane potential is believed to induce luminescence. On the other hand, the specificity of nerve messages in animals is unquestionable. The mechanisms for transmitting such messages are complex and beyond the scope of this text. The principle by which these mechanisms operate, however, may be simply described. The specific stimulus is recognized either by a specialized end of the neuron or by a specific receptor cell (e.g., a photoreceptor). The stimulus induces a potential change in the membrane of this specialized region of the cell. This "generator potential" is more variable in character and of much lower intensity than the spike; it is not propagated but remains confined to the region of origin. This potential, however, acts in the same way as the artificial stimulus described earlier to induce a train of spikes in the conducting region of the neuron. The configuration of the generator potential, which is determined by the intensity and duration of the stimulus, determines the number and frequency of spikes. The acceptor mechanism at the other end of the nerve line translates the spike train into a specific response. Compared to the information contained in a DNA sequence, the information contained in the train of spikes is small. By itself, the train of spikes carries no information about the specific nature of the stimulus, but only about its intensity and duration. The specificity of the nervous system resides in the vast network of nerve fibers which carry messages from specific receptors to specific acceptors. The most intriguing question is how a specific receptor is structured so as to respond to only one type of stimulus.

The apparent gap between the primary function of membranes as regulators of solute diffusion and their specialized function as message conductors is very large. Yet if we examine the chemical events which accompany stimulation, a common underlying mechanism is evident. When stimulated, a membrane temporarily loses its selective property (it becomes "depolarized"). K^+ flows out and Na^+ flows in; the potentials recorded reflect this extremely rapid transition in ionic composition. That membrane structure is simultaneously affected may be demonstrated by measuring the electrical resistance of the membrane during stimulation. This resistance drops sharply, and its return to the resting state is accompanied by a return to the original ionic composition. To achieve this return the cell must expend energy, part of which at least must be directed toward the pumping out of Na^+ against a concentration gradient, and possibly also toward the absorption of K^+. The maximum spike voltage is predetermined by the energy originally invested in creating the asymmetric ion distribution. As each local chemical disturbance stimulates the adjacent region to a similar transient loss of membrane selectivity, much the same spike is produced because the same amount of energy has been invested at different points along the membrane. The flow of a nerve message is not a current flow such as occurs in a metallic conductor because that would have the speed of light. Instead, it is a set of discrete chemical events tied together by the fact that the event at one locus triggers the same event at its adjacent one. And these chemical events appear to be no different in nature from those which every cell experiences in its regulation of ion content. If we examine a neuron under the electron microscope, the elaborate nature of its membranous organization is apparent. And, since membranes are basically identical in organization, we might be tempted to say that the complex is really simple. We should rather say that complexity is a compounding of the simple, and the compounding may at times present a greater scientific challenge then elucidation of the simple fundamentals.

FIBRILLAR ELEMENTS AND CELL MOVEMENTS

The ability of cells to convert chemical energy into mechanical work extends from the flagellated bacteria to man. Cells without a capacity for some kind of movement are unknown. Some cells, such as those of the nonflagellated bacteria, show little or no evidence of deliberate movement during most of their life cycle, but during cell division movements are involved in separation of the nuclear network and in fission of the cells. For cells in general movements may be either internal and localized, or they may encompass the cell as a whole. The most common forms of intracellular movements are protoplasmic streaming and those associated with cell division. In many plant cells protoplasmic streaming is an essential physiological function, and its value is most prominently displayed in large vacuolated forms. In Spirogyra, for example, the nucleus is suspended by means of cytoplasmic strands in the center of the vacuole. Streaming may be seen within these strands, often in two directions within a single strand, the streams carrying cytoplasmic granules as well as ground substance in the course of their flow. The total picture of a circulation of cytoplasm from cell periphery to the nucleus and back to the periphery suggests another mechanism by which large cells meet the problem of diffusion. The rate of movement in some cells has been measured as 250–300 μ/minute, a rate considerably lower than that of random diffusion (100 μ/sec for glycine), but its important feature is that it is directional. Cytoplasmic streaming occurs not only in green algae, but in very many types of cells in higher plants, in slime molds, in mycelia of fungi, and in a number of protozoa. Less striking forms of intracellular movements probably occur during the vegetative life of most cells, but such movements have only occasionally been documented. There is no reason to doubt that some type of basic mechanism to effect movement is found in all cells.

Examples of cell locomotion abound. Movements encompassing the whole cell have been classified as amoeboid, ciliary or flagellar, and contractile. Amoeboid movement derives its name from the characteristic locomotion of amoeba, and such movement is possible only in cells which have a more or less fluid structure. The process is essentially one of a forward flow of the fluid cytoplasm at the anterior end, and a contraction of the cytoplasm at the posterior end of the cell. Such locomotion is not confined to unicellular organisms. Blood phagocytes move in this way. An example of widespread interest is found in the development of animal embryos; at a characteristic stage of development some cells detach from their neighbors and migrate to specific regions of the embryo. Of all forms of locomotion, the amoeboid one is distinguished by the apparent absence of any structural element to effect movement. The underlying mechanism—and there is no agreement as to its nature—is contained in the cytoplasmic ground substance which is physically differentiated into an inner "sol" (fluid) and cortical "gel" region. This differentiation is characteristic of very many cells, and reversible transitions from sol to gel states are common. Such transitions have been used as one explanation of amoeboid movement, but, as will be evident later, sols and gels are superficial descriptions of molecular states and a better, if not total, understanding of the process can be acquired from studies of specific molecular components.

Flagella and cilia are cellular projections distinctly specialized for effecting locomotion. Cilia are the shorter of the two and are much more numerous where present. Both structures are common in the animal kingdom. Flagella are particularly prominent in spermatazoa of animals and of lower plants. The motions of cilia and flagella are unique, the one propeller-like, the other whip-like. How the fibrillar elements which they clearly possess effect these motions is still a mystery. The arrangement of fibrils in these organelles is very different from that in muscle cells, which are undoubtedly the most specialized cells for performing mechanical work.

The common factor in all the above systems, whether or not they are structurally differentiated from other parts of the cell, is their ability to use chemical energy for the production of movement. This particular category of energy transformation was not included in earlier discussions of energy, mainly because contemporary interests are prejudiced in favor of molecular transformations, and partly because interpretations of mechanical work do not have the clarity and finality of those concerned with molecular syntheses. The phenomenon in nevertheless a fundamental attribute of living matter, and its importance was well recognized by the earliest students of cell biology. Historically two approaches were developed. One was aimed at the investigation of "simple" unicellular organisms, in the belief that their mechanism for effecting movement would be uncomplicated and hence easier to resolve. The other was aimed at the highly specialized muscle cells in the belief that the greater the specialization, the more evident the mechanism. Up to the present at least, those who followed the second approach proved to be the more prescient. Our basic understanding of how cellular molecules can use chemical energy to perform mechanical work is derived from studies of muscle tissues.

If muscle fibers (aggregates of muscle cells) are removed from an animal and stored in glycerin, the cells, of course, die, but as one of the outstanding biologists of this century, Albert Szent-Gyorgi, discovered, they retain the ability to contract. Contraction is effected by placing such fibers in a balanced salt solution and adding ATP. The energy for contraction is paid for by the ATP which is hydrolyzed to ADP. This now classical experiment is the most lucid demonstration that the most common carrier of chemical energy in the cell is an adequate source of energy for mechanical work. Viewed microscopically, the muscle cell is preponderantly an organization of fibrous structures. This fact is consistent with the degree of physiological specialization, and leads to the conclusion that the principal chemical component of muscle must be the substance of the contractile mechanism.

The major component of muscle may be extracted with concentrated salt solutions. It is a protein called "actomyosin." Unlike most proteins (see Chapter 4) it precipitates in dilute salt solutions, and if ATP is added to the precipitate a "contraction" ensues. A more appropriate model is obtained by squirting a solution of actomyosin through a narrow orifice into a dilute salt medium. In this way fibers are formed which can be induced to contract by addition of ATP. This *in vitro* experiment establishes the high probability that actomyosin is responsible for the contraction observed in glycerinated muscle fiber and also for the contractility of living muscle cells. The question then arises as to how actomyosin utilizes ATP to effect contraction. In none of the chemical reactions hitherto discussed has the hydrolysis of a pyrophosphate bond or of any other bond in a metabolite resulted in changing the shape of a molecular aggregate. The contrast could be intensified by including experiments in which a small weight is attached to an artificial fiber of actomyosin and ATP is used to lift the weight by contracting the fiber. The ability of protein molecules to perform mechanical work by utilizing ATP is a unique biological property.

Actomyosin is not a simple protein. It is readily separated into two components, actin and myosin. Neither of these components is capable of contracting separately even though myosin, but not actin, retains the capacity to hydrolyze ATP. Contractility is therefore a property which requires at least two molecular components. Even this, however, is a simplification. The actin molecule is globular, having a molecular weight of about 60,000 and measuring about 300 × 30 Å. In this respect, but in no other, it resembles an enzyme molecule and cannot form a contractile unit with myosin. If a solution of globular actin is treated with ATP in a medium of appropriate ion composition, the molecules aggregate into a fiber-like arrangement (or filament) which has a molecular weight of 40,000,000 and a length of 1 μ. The energy of the ATP molecule is used to form a filament, which in combination with myosin is capable of contraction; it is used for synthesis of a structure, not for mechanical work. Both the fact that energy can be directly used to form a structural element and the fact that this energy is in the common form of ATP have potential significance for cell behavior. For we may at least suppose that a nonspecialized cell has a relatively simple mechanism to form one of the components of a contractile system and, presumably, just as simple a mechanism to disperse it. This supposition is supported by the behavior of a group of protozoa called the "amoeboflagellates." For part of their life span, they are typically amoeboid in form and movement, but under certain conditions they develop flagella. This development undoubtedly involves much more than a single chemical transformation. Nevertheless, an understanding of the relatively simple conversion of G-actin to a fibrous structure, may ultimately lead to an understanding of the more complex transformation.

Some details of the process whereby G-actin is converted to F-actin are known. If G-actin is purified under appropriate conditions, each molecule is found to be associated with a molecule of ATP. The bonding between the two molecules is relatively weak and is dependent on the presence of Ca^{2+}. If Ca^{2+} is removed, the ATP is released and the G-actin cannot be polymerized. F-actin, on the other hand, contains ADP which is tightly bound. The inference drawn from these observations is that the energy of the terminal pyrophosphate bond is used in effecting a condensation of G-actin molecules, but neither the role of the remaining ADP nor the nature of the linkages binding the actin monomers is precisely understood. Superficially, actin behaves like an enzyme insofar as it hydrolyzes ATP, but the hydrolysis occurs only during the formation of F-actin, and stops when the formation stops. Whether the common requirement for ATP in this process and in actomyosin contraction is due to a direct relationship between the mechanism of contraction and the presence of ADP in the actin moiety of the system is a debated point. *In vitro*, however, the two roles of ATP are distinct (Fig. 16-2).

The myosin component of the contractile system is no less complex than the actin. If certain of the peptide bonds in the molecule are hydrolyzed by means of enzymes, two kinds of molecules are produced, heavy (H) and light (L) meromyosin. The H portion has the characteristic of an enzyme molecule; it catalyzes the hydrolysis of ATP. The L portion has the characteristics of a fibrous protein; it is rigid and regularly arranged. The myosin molecule is thus differentiated into two zones, one which resembles an enzyme, and the other which resembles a structural unit. It is soluble only in concentrated salt solutions, and under these conditions its molecular weight has been calculated as 450,000, and its dimensions as 25 $\times$ 1600 Å. If a solution of myosin is diluted, fibers or filaments are formed which may be visualized under the electron microscope and have a dimension of about 1.5-2.0 $\mu \times$ 100-150 Å. These filaments are similar in configuration to certain of the filaments identified in sectioned muscle cells under the electron microscope. Only the H components in the filament have an affinity for actin; combinations between actin filaments and H-meromyosin, but not L-meromyosin, have been demonstrated under the electron microscope.

The intensive studies of muscle by chemical and microscopic techniques, though they have yet to solve the problem of contractility, have provided biologists with a profound understanding of some aspects of the relationships between structure and function. One very important achievement has been the demonstration that the

(*a*) Formation of structural elements

(1) G-Actin + ATP → G-Actin − ATP
nG-Actin − ATP → [F-Actin-$(ADP)_n$] + nPi (inorganic phosphate)

(2) Myosin → [Myosin filaments]

(*b*) Contraction

(1) [F-Actin $(ADP)_n$] + [Myosin filament] → [Actomyosin structure]

(2) [Actomyosin] + ATP → [Contracted Actomyosin] + ADP + Pi

(3) [Contracted Actomyosin] + excess ATP → [Actin] + [Myosin]

Fig. 16-2. Molecular reactions associated with contractility.

two principal types of filaments in muscle as seen under the electron microscope, the "thick" (myosin) and "thin" (actin) can be formed in the test tube by virtue of properties inherent in the structures of the relatively simple and small molecular components. The total contractile structure of the muscle cell is not, of course, a random collection of these filaments. They are arranged and anchored in a regular pattern by factors still to be explored. Their spatial arrangement has, in fact, a direct bearing on how the contractile capacities demonstrated *in vitro* are utilized for cellular contraction. This last point has implications for cells in general. Various structural systems are known to effect motion. Are these systems all formed from the same type of molecules and, if so, are all types of motion produced by the same principle of molecular contractility?

Evidence for contractile proteins similar to actomyosin has been found in other types of cells, slime molds and fibroblasts. Analyses of such proteins are limited, however, and detailed comparisons are therefore impossible. Only the vague generalization can be made that various cell types do contain proteins which behave similarly to actomyosin when extracted. One major difficulty in pursuing the problem of mechanical work by cells is that for most cells the contractile mechanism constitutes a small fraction of their total substance, and that for all cells the contractile elements of cell division are transient structures. Despite these difficulties, which relate particularly to chemical investigations, some generalized features of contractile mechanisms have been made apparent through electron microscope studies. Striated muscle, which has been the principal object of study, is the poorest example for generalization. Its organization is unique (see Fig. 16-4) and so too, perhaps, is the specific contractile system. The most favored interpretation of muscle contraction is that the interdigitating filaments slide past each other and that no contraction of the filaments themselves occurs. A diagrammatic representation of some interpretations of the contractile system in striated muscle is given in Fig. 16-3. This should be compared with the diagram in Fig. 16-5 which represent other contractile or motility systems.

A comparison of the fine structure of muscle cells with that of cilia or flagella illustrates how differently the same basic unit, the fibrous and contractile protein molecule, may be organized. Even more impressive, however, is the fact that, despite the widespread occurrence of cilia or flagella in extremely different types of cells (e.g., the flagella of animal sperm and of the spermatazoa of lower plants, the cilia of protozoa), the pattern of organization of the basic filaments (or fibrils) which measure about 150-200 Å in diameter is universal. The filaments

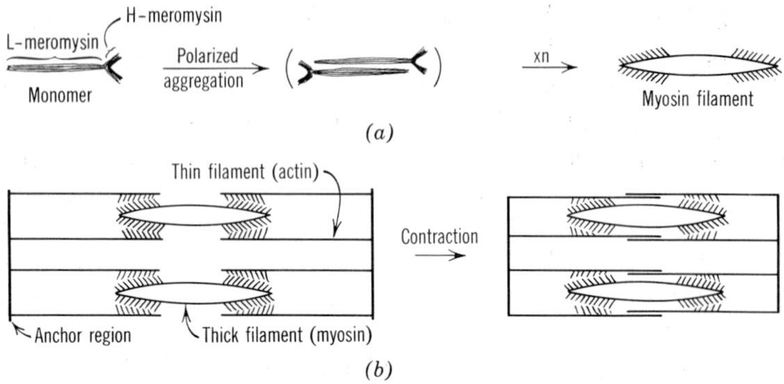

Fig. 16-3. A molecular interpretation of conduction in a striated muscle cell. (*a*) Polymerization of myosin monomer into myosin filament. (*b*) Contraction in a striated muscle fibril. The diagram is based on H. E. Huxley's theory of sliding filaments. Note that neither actin nor myosin undergoes contraction. Muscle fibers are presumed to contract because a shortening of the sarcomeres (the region between the anchor lines), which results from the actin filaments sliding over the myosin ones. Interaction between the filaments is pictured as occurring through the side chains. The diagram should be compared with the electron micrographs in Fig. 16-4.

themselves probably represent an association of fibrous protein molecules (see Chapter 4), but their chemical nature cannot be stated with certainty since electron micrographs do not reveal chemical composition. Such filaments, moreover, have been identified in the cytoplasm of various cell types where they are not components of ciliary or flagellar structures. Their possible role in intracellular movement, particularly in spindle function, is still a matter of speculation. However, when found in structures clearly specialized for motility, they are always arranged in a cylinder consisting of 11 filaments. Nine of the filaments lie close to the periphery, and two occupy the center. Minor morphological variations do exist, but in all cilia and flagella the cylinder is bounded by a membrane, the central filaments are enclosed in some type of sheath, and each of the peripheral filaments is compound (Fig. 16-7).

Another universal feature in the organization of these organelles is their relationship to "basal bodies." Under the light microscope these bodies appear as granules lying directly beneath the protruding flagella or cilia. Generally, the two are directly connected, but in some cells a direct connection has been questioned. Nevertheless, cilia or flagella do not occur in the absence of basal bodies, and the relationship between them is apparent in the similarity of their respective structures. Basal bodies also have nine peripheral filaments,

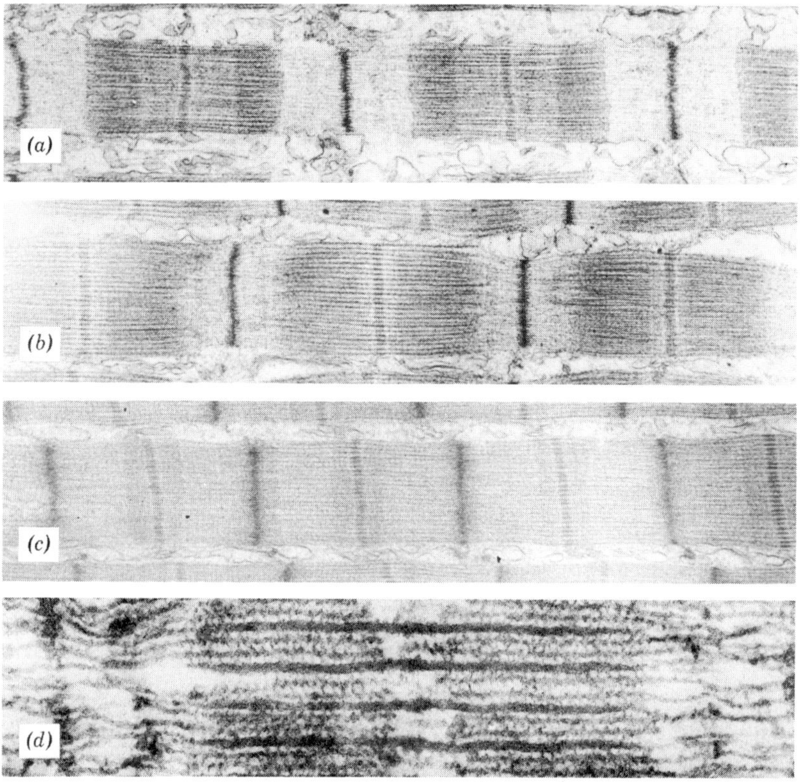

Fig. 16-4. Organization of striate muscle. (*a, b,* and *c*) Stages in contraction of a frog sartorius muscle. Contraction was induced by electrical stimulation and specimens were fixed at various intervals during the contraction process. The specimen shown in *c* was taken from a muscle which was shortened by about 50%. The thin filaments may be seen to overlap in the center of the sarcomere. All photographs are magnified about 18,000 times. (*d*) The central region of a sarcomere taken from a rabbit psoas muscle. The section photographed was much thinner than those used for *a, b,* and *c*. Thick and thin filaments are readily identifiable (100,000×). (Courtesy H. E. Huxley.)

although these are triplets rather than doublets, and the central region is occupied by one rather than two filaments. Since the basal body developmentally precedes the cilium or flagellum, the inference drawn is that the one gives rise to the other. Developmental relationships appear to be even more extensive than this. In animal cells pairs of bodies very similar in structure to the basal bodies are found just outside the nuclear membrane. These cylindrical bodies, which measure about 3000–5000 Å in length and 1200–1500 Å in diameter, are the "centrioles" referred to in the discussion of mitosis (Chapter 2).

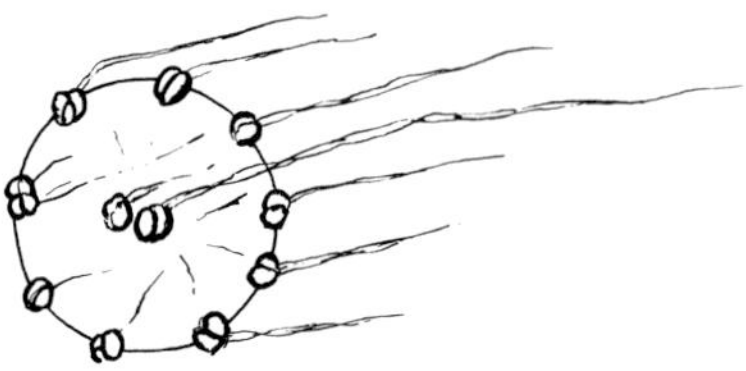

"9 + 2" pattern

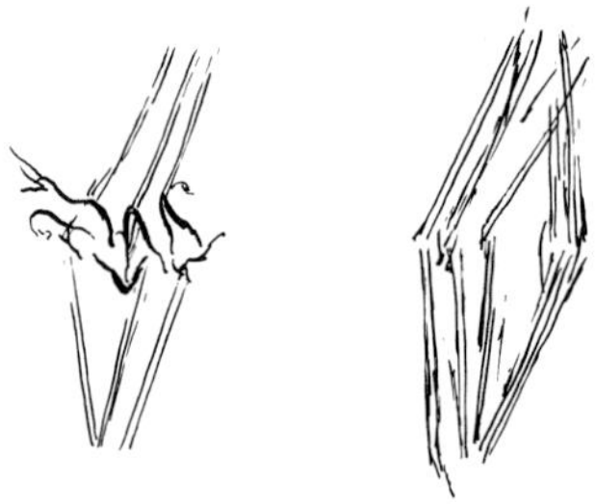

Mitotic fibers

Fig. 16-5. Patterns of fibrous structures associated with cell movements. The "9 + 2" pattern is observed in all flagella, cilia, and centrioles which are found in the polar region of spindles in animal cells. The fibers observed in mitosis are very different in arrangement from those found in cilia or in muscle. Two groups of similar fibers are present in the mitotic spindle: those that are entirely cytoplasmic extending from pole to pole, and those that connect chromosomes and polar regions. The fibers have the appearance of tubules and although no direct evidence is available to indicate that they actually contract, the individual tubules appear to have the same fine structure as those found in the 9 + 2 group arrangements.

In sperm the centrioles serve as the basal bodies for the flagellum. Of great interest developmentally is that following mitosis in animal cells a new centriole develops at right angles to the old centriole, and remains there until the initiation of division when one member of the pair migrates to the opposite side of the nucleus. Superficially at least, centrioles appear to be self-reproducing. And since no one has ever observed either centrioles or basal bodies forming in the absence of preexisting ones, the generalization has frequently been made that these bodies represent one of the self-replicating systems in the cytoplasm.

The supposition has frequently been made that just as there is a relationship between basal bodies and flagella and cilia, there must also be a relationship between centrioles and spindle fibers. But close examination of electron micrographs does not reveal any continuity between the spindle fibers and the centrioles. Moreover, centrioles have not been identified in plant cells. For these and other reasons chromosome movement during the anaphase cannot be attributed to a contractile system originating in these bodies. The only conclusion we may draw is that wherever cellular or intracellular movements occur, filaments or fibrils are found. The kinds of movements which occur must depend on the way in which the contractile elements are organized in a particular structure. The shortening of muscle fibers, the whip-like motion of flagella, the propellor-like movement of cilia, and the polar migration of chromosomes all attest to the fact that a basic unit of molecular organization may be modified and assembled in a variety of ways. Whether these different structures have a common derivation developmentally, or even phylogenetically, is a question that cannot yet be answered. See Figs. 16-6 and 16-7.

Partly from studies of muscle fibers, partly from studies of fibrous elements in general, and mainly from a predilection for natural uniformity, we are led to generalize that all cellular systems for contraction, movement or motility, have their origin in the fibrous, or partly

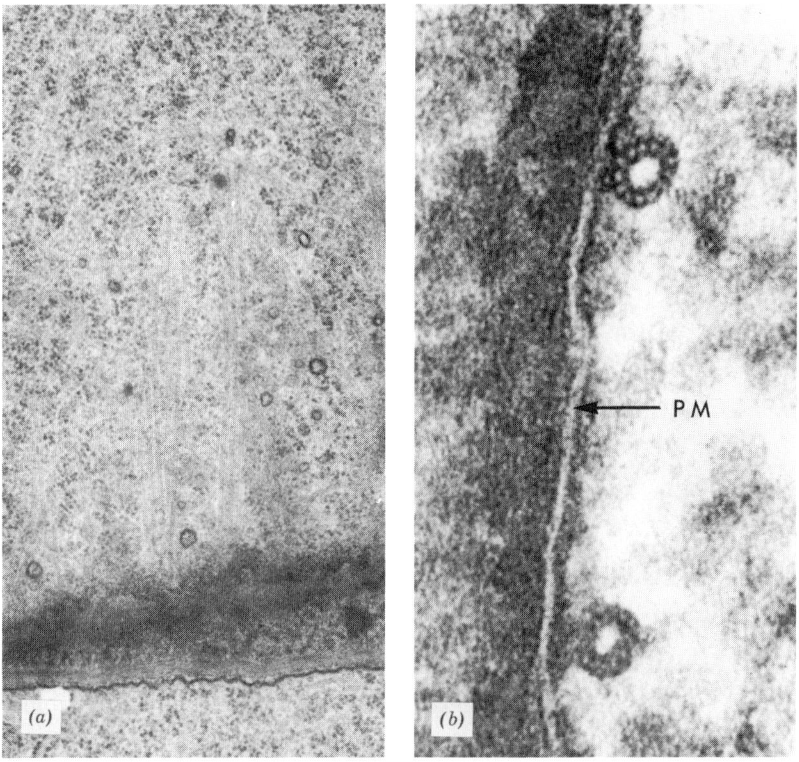

Fig. 16-6. "Microtubules" in plant cells. (*a*) Cell from a root tip of *Phleum pratense*. The slender tubules are approximately 230–270 μ in outside diameter and are especially prominent in the peripheral region of young cells. These tubules are similar in appearance to those found in cilia (38,000×). (*b*) Cross-section of microtubules in *Juniperus chinensis*. The tubule may be seen to consist of a thick wall composed of circular subunits. The number of such subunits has been found to be 13. They are undoubtedly filamentous in structure and probably composed of fibrous proteins. (From M. C. Ledbetter and K. R. Porter, *Science,* **144,** 872 (1964).)

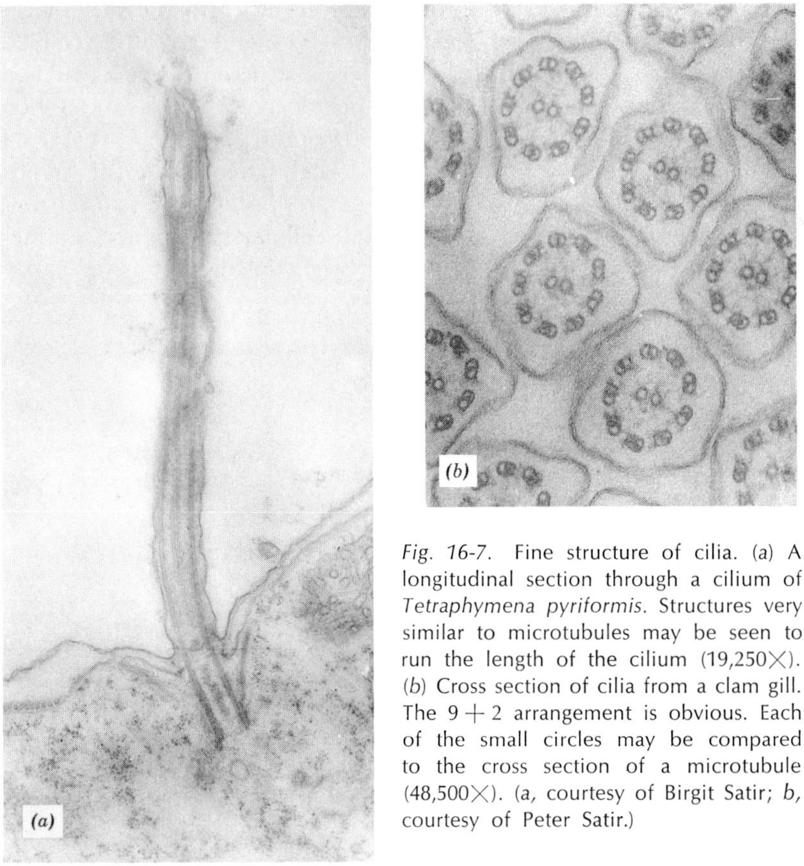

Fig. 16-7. Fine structure of cilia. (a) A longitudinal section through a cilium of *Tetraphymena pyriformis*. Structures very similar to microtubules may be seen to run the length of the cilium (19,250×). (*b*) Cross section of cilia from a clam gill. The 9 + 2 arrangement is obvious. Each of the small circles may be compared to the cross section of a microtubule (48,500×). (a, courtesy of Birgit Satir; *b*, courtesy of Peter Satir.)

fibrous, protein molecule. Even cells without a specialized apparatus for mechanical work probably have fibrous elements in the ground substance of the cytoplasm. For the common presence of variable gel regions in the cytoplasm can best be explained by the presence of fibrous proteins which form gels readily because of their susceptibility to cross-linking. Amoeboid movement is the best example of continuous transition in sol-gel states. That fibrous structures have not been morphologically demonstrated in all cells can be rationalized either on the basis of the relatively simple conditions required for their aggregation and disaggregation, or on technical difficulties of fixation.

The negative evidence is nevertheless less impressive than the limited findings recently made through electron microscopy. In certain plant cells microtubules have been found which measure approximately 150 Å in diameter and which are of variable length. They generally occur in the peripheral regions of the cytoplasm and appear to be oriented in the same direction as the cellulose microfibrils which are being laid down on the external side of the plasma membrane to form the cell wall. Whether the orientation of the tubule has a causal relationship to the orientation of the cellulose fibril is unknown. Both may reflect a more deep-seated mechanism of orientation within the cell. However, a few observations may be brought together which suggest that these tubules are fundamental structural aggregates associated with motion. The microtubules found in the plant cell are morphologically similar to the tubules found in cilia, flagella, and in the mitotic spindle. Moreover, the internal structures of the tubules of rat sperm flagella and of the microtubules (the two systems which have been studied in fine detail) are also similar. Both consist of filaments arranged in a ring. On the assumption that these filaments consist of contractile protein elements, the tubular arrangement of contractile filaments may be regarded as a basic unit of organization for effecting motility. The fact that the plant microtubules occur in the peripheral regions of the cytoplasm where streaming occurs is relevant to the generalization. Conceivably, with improvements in methods of fixation such microtubules may be found in other types of cells. Model mechanisms can be easily proposed to explain how tubules effect motion. If the filaments were of more than one kind and contracted differentially, the microtubule would be bent, and a sequence of such bendings and relaxation could produce motion. Models of this type are probably incorrect but they provide a basis for the design of meaningful experiments.

The specific configurations and relative stabilities of different fibrous structures may be regarded as variants of a basic molecular plan. That plan is the rod-like form of the molecule studded with lateral groups which are capable of forming cross links with similar groups in other molecules. The strength and dissociability of intermolecular cross links, the number of possible arrangements of such cross links, the capacity of the primary molecular unit to undergo changes in shape which affect the disposition of the side chains, as well as the number of arrangements of uncombined side chains remaining in the multimolecular aggregate which could react with similar or complementary aggregates, all interact to give the final structural unit its special mechanical characteristics.

One other property of mechanical systems which has its origin in molecular patterns is polarity. The poleward movement of chromosomes, the longitudinal contraction of muscle cells, and the axial asymmetry of unicellular organisms are selected examples of this characteristic. If the myosin molecule is used for illustration, though other examples could be chosen, we can see how the grouping of amino acid sequences in a polypeptide chain as specified by the gene could endow a molecule with the inherent capacity to form polarized structures not only in the morphological but also in the functional sense. Given a basic unit which is polarized, the elaborations which could flow from it to give polar properties to a cell are best estimated from a sweeping survey of the cells themselves, a survey left to the curiosity of the reader.

MODIFICATIONS OF THE NUCLEAR APPARATUS

The DNA content of bacterial cells is of the order of 2-3 $\times 10^{-14}$ g; that of diploid cells in higher organisms is of the order of 6-8 $\times 10^{-12}$ g. If the DNA in each of these cells were organized as a single duplex filament, the ratio of their lengths would also be 1:300. We could argue that the more elaborate organization of the nuclear apparatus in higher organisms is an adaptation to this simply expressed quantitative difference, especially with respect to the mechanical problems which arise in the partitioning of DNA at cell division. Although the argument might appear reasonable, it is not convincing. A mammalian cell of average size has about 8-10,000 times more volume in which to distribute its DNA. Other rationalizations could be provided, but none of them would be sufficiently founded in experimental evidence to enable one to explain the organization of the nuclear apparatus in higher organisms in terms of one general principle. The major difficulty is that the nature of the modifications and their functional implications are imperfectly understood.

Using the pure DNA filament as a standard for comparison, two categories of modification are evident in the nuclei of higher forms: a longitudinal differentiation of the chromosome with respect to microscopically identifiable regions and a structural association of DNA with certain proteins and RNA. The microscopically visible differentiation of the chromosome is distinct from, although related to, the universal differentiation of DNA filaments with respect to genetic coding. In general, chromosomes have three types of microscopically identifiable regions which are related to the behavior of the chromosome as a whole: centromeres, heterochromatic regions, and nucleolar regions.

Superimposed on these regional differences is the capacity of chromosomes to undergo a high degree of contraction, a process which is essential to the mechanics of their separation. Assuming, for example, that the largest chromosome in Trillium consists of one double strand of DNA, its extended length would be of the order of 10 cm; at metaphase, however, the length of that chromosome is about 30 μ. How this 3,000-fold reduction in length is accomplished is not entirely clear, but the final forms of packing may be seen with the light microscope. The chromatids within the undivided but already replicated chromosome are highly coiled, and the sister chromatids are loosely coiled about one another. In some cells this "major" chromatid coil can be seen to be formed from an even more tightly coiled filament, this smaller coil being designated as the "minor" one. How the DNA filament is packed in the minor coil is unknown. This question has thus far proved refractory to analysis by the electron microscope, either because techniques of fixation are inadequate, or because the DNA molecules are so tightly packed in metaphase chromosomes—and also in interphase nuclei—that their individual outlines are beyond the resolving power of the electron microscope. The extent to which chromosomes contract is variable, but for a given species under constant physiological conditions, it is a more or less fixed characteristic. Generally, meiotic chromosomes are more contracted than mitotic ones. The size of the contracted chromosomes is not necessarily related to their DNA content. In the genus Drosera, for example, the volume of the contracted chromosomes in one species may differ from that in another by a factor of 1000, but this is not matched by a corresponding difference in DNA. In corn a single gene mutation known as "elongate" markedly diminishes the degree of contraction of meiotic chromosomes. The known physicochemical properties DNA strands do not provide any clues as to how contraction might be effected. The process is more likely to be due to secondary associations between DNA and other macromolecules. Although such associations have not been identified, we may reasonably conclude that chromosomes not only provide informational sequences, but that they are also invested with the capacity to pack and unpack their DNA filaments under specific physiological conditions.

The centromere has already been discussed in relation to chromosome segregation (Chapter 2), and beyond the fact that polar movement of chromosomes does not occur without it, little is known about the nature of this organelle. Although it is commonly localized in a constricted region of the chromosome, it has been found to be diffusely spread over an appreciable length of the chromosome in some species. In such cases a profusion of fibers connecting the chromosomes to the poles of the spindle is seen under the microscope. Fragmented chromosomes of this type behave very differently during division from the usual type of chromosomal fragments. If a chromosome with a localized centromere is fragmented, only the fragment which contains the centromere moves poleward in regular fashion. For a chromosome with a diffuse centromere, most of the fragments are segregated normally, because they each form fibers extending to the pole of the spindle. Two important questions may be raised about centromere function, both of them still unanswerable. One bears on the general problem of polarity. The centromeres of the two chromatids in a metaphase chromosome behave as though each formed fibrillar connections with only one of the poles; some factor must therefore exclude the possibility of a single centromere forming connections with both poles of the cell. This factor may not be one of spindle polarity for if, as is generally supposed, the centromere of the original chromosome does not divide until after metaphase, the delayed and transverse division of the centromere makes possible polarized movement. The second question bears on the chemical nature of the centromere region. The fibrils originating from this region are proteins. Conceivably, the genes in this region may code these proteins, but if so, the translation process either occurs at the centromere site or the process occurs via the conventional ribosomal mechanism, and the proteins then diffuse back to the centromere site. These or any other mechanisms would require a structural differentiation of the chromosome, which is distinct from the differentiation considered at the genetic level.

Contraction and spindle fiber attachment are the two major modifications specifically related to meiosis and mitosis. The remaining modifications appear to be related to interphase functioning of chromosomes. The first of these is microscopically recognized as a difference in staining properties between different regions of a chromosome or between entire chromosomes. Chromosomes (usually, sex chromosomes) or regions of chromosomes which stain deeply during interphase and prophase are classified as "heterochromatic." The chemical basis for differential staining is not known, and may even be entirely due to condensation. Whatever the reason for the staining behavior, heterochromatic regions are known to have distinctive functional characteristics. Very few genes have been identified in heterochromatic regions by standard cytogenetic methods, and this may be attributed to either an absence of genes or to a suppression of gene expression. The alternatives are not mutually exclusive, but the second of these is more approachable experimentally. If heterochromatic segments of a chromosome are translocated in positions adjacent to nonheterochromatic ("euchromatic") regions either within the same or in a different chromosome, the expression of genes near the translocated heterochromatic segment is often affected. Generally, the closer the gene to the heterochromatic region, the stronger the effect. This type of experiment indicates that the heterochromatic state is not only localized and autonomous in origin, but also that its effect is transmitted by virtue of a structural association with neighboring parts of a chromosome. Other experiments further indicate that the heterochromatic state is not necessarily a permanent and self-perpetuating property of the chromosome filament. In various species of mammals, one of the XX chromosomes in a female is always heterochromatic. The property is acquired following zygote formation, an inference drawn from genetic data. Since the X chromosome derived from the male parent is euchromatic, that derived from the female would always have to be heterochromatic, if this condition were self-replicating. Such transmission would be detectable by genetic or cytogenetic analysis, but none has been detected. Futhermore, in cells which have acquired three X chromosomes, two are heterochromatic. A mechanism thus exists whereby one X chromosome induces the heterochromatic state in the others. One interesting parameter of this process has been observed. The euchromatic X chromosome always replicates ahead of the other X chromosomes. The few examples of heterochromatic effects do not cover all the known types of influences which heterochromatin exerts on chromosome behavior. They should nevertheless be sufficient to show that the arrangement of genes in a chromosome relates not only to the specific proteins formed for the cell as a whole, but also to some type of control on the activity of the genes themselves.

The second modification of the nuclear apparatus which is related to interphase function is the formation of nucleoli. Their morphology was discussed in Chapter 1. Although they appear to be distinct organelles, they are products of specific chromosomal regions. With few exceptions, they disappear during mitosis and are formed anew at the conclusion of division. Only certain chromosomes have nucleolar organizing regions, but the capacity to form them is not always expressed. Generally, the number of nucleoli in a nucleus is maximal in the young cell, but as the cell matures nucleoli tend to fuse. Fusion, however, is not the only factor which leads to a reduced number of nucleoli. A competition, or the equivalent of one, appears to exist between nucleolar organizing regions for the acquisition of nucleolar materials. This phenomenon is strikingly displayed by the plant Crepis, in which a number of species can be arranged according to their nucleolar organizing power. From such a listing we can predict which chromosomes in different hybrids will form nucleoli, because the nucleolus-forming chromosomes derived from one of the species in the hybrid always yield to the chromosomes derived from that species which is higher on the list.

Despite the variations in number, size, and chromosomal associations, the nucleolar complex has the constant characteristic of remaining attached to at least one of the specific nucleolar regions in the chromosome. The biochemical properties of the nucleolus are best explained in terms of this intimate structural association. The nucleolus is apparently the most intense site of RNA synthesis, and in mature cells nucleolar size is correlated with rate of protein synthesis. Since RNA is transcribed from DNA, the question arises why such transcription should occur in a highly localized region of the nucleus rather than along the dispersed chromosomes. The answer to this, although by no means complete, is that at least one type of RNA, ribosomal RNA, is formed in the nucleolus. Presumably, relatively few genes are required to transcribe this form of RNA, which is part of the general mechanism for protein synthesis, compared to the number of genes required to code sequences for each of the specific proteins found in a species. The simplest interpretation of nucleolar activity would be that all nucleolar organizing regions are associated with genes coding transfer and ribosomal RNA and that the nucleolar apparatus facilitates RNA production. In growing cells especially the number of RNA molecules coded by a ribosomal gene must greatly exceed the number coded by any other single gene. Whether this factor alone accounts for nucleolar organizations is debatable. The presence of ribosome-like particles and the occurrence of some protein synthesis in nucleoli suggest that these nuclear organelles may be specialized not only for RNA synthesis but also for combining RNA with protein prior to the transport of RNA into the cytoplasm. Alternatively, the nucleolus may be exclusively concerned with the packaging of RNA for transport.

Common to all chromosomes of higher organisms is the presence of a group of proteins called "histones." These proteins are characterized by a very high content of arginine and/or lysine, which gives them an unusually high positive charge at intracellular pH's. Since nucleic acids have a strong negative charge due to their phosphate groups, a clear physiocochemical basis exists for complex formation between these two types of macromolecules. Chemical analysis of cell fractions indicates that most of the histones are localized in the nuclei, and specific staining reactions further indicate that the histones are localized in the chromosomes. The nature of the chemical binding between histones and DNA within the cell is not entirely clear. DNA and histone combine readily in solution, and such combination is indifferent to the respective sources of the molecules. The isolation of DNA-histone complexes from disrupted cells is not compelling evidence for the existence of the same complexes within the cell. Nevertheless, even though the nature of the bonding is open to some question, the localized association is not.

Techniques have been developed by which the contents of a bacteriophage, a bacterial cell, or a nucleus may be directly dispersed on an electron microscope grid. In this way the expressed structures may be viewed with a minimum of structural alteration. DNA filaments from phage or bacterial cells provide a very different image from those of higher organisms. For phage and bacteria the filaments have a diameter of about 25 Å, which is the value expected for double-stranded DNA; for cells from higher organisms, the filament is much thicker, about 40 Å, thus indicating that some other component is associated with the DNA. This component is presumed to be histone. The 40 Å fibril may, in fact, be considered a subunit since it is always paired with another fibril to give a 100 Å fibril. One general and characteristic difference between the nuclear apparatus of bacteria and that of higher forms is therefore in the molecular association of the DNA filaments. So far no evidence has been adduced to indicate that the DNA filaments themselves differ in any significant way.

Since the amount of histone in a nucleus is approximately equal to the amount of DNA, various schemes have been proposed in which histones function as regulators of gene activity. Unambiguous demonstrations of such a role are still lacking. The concept is nevertheless attractive in some respects. The very fact that histones are universally present in higher organisms would suggest that they play some critical role in chromosome function, and this concept is at least made plausible by some evidence that the kinds of histone molecules found in different tissues vary to a limited degree according to the tissue of origin. An extreme example of such variation is provided by sperm in which the somatic types of histone are replaced by types with very much higher arginine contents. In some species the difference is so great that the positively charged proteins in the sperm have been classified separately as "protamines." Regulation, as already noted, may occur at various levels of organization, and the possible functions of histones in this respect cannot be properly explored without considering other defined examples of gene regulation. These will be considered in Chapter 18.

THE CONCEPT OF CELLULARITY

Most of what has been discussed in this chapter can be reduced to the proposition that given a number of basic molecular properties, elaborations may evolve which confer on the derived structures new behavioral characteristics. The same relationship applies to the behavior of individual cells and their behavior in multicellular organisms. We may ponder as to which aspect of the relationship is the more impressive—the continuity between the structures of the least and most specialized cells, or the emergence of distinctive structures and functions despite the continuity. Objectively, however, we accommodate to the fact that the possibilities for elaboration are virtually unlimited. The biologist classifies for convenience, but Nature elaborates for survival. The perennial discussions of what constitutes a cell have a somewhat artificial ring. A "typical cell" meticulously drawn with respect to all details of subcellular architecture does not exist, because the number of cell types which would correspond closely to the detailed drawing would be overwhelmed by the numbers of types which do not. Even if we omit all the specialized and accessory structures which cells develop, the organization of the major and basic components of cells is still subject to considerable variation.

The typical cell is uninucleate, but many types of cells are multinucleate. In some species of organisms (among algae, fungi, rotifers) the multinucleate condition is a constant feature of their vegetative life. In some organisms some tissues may have multinucleated cells whereas others do not. In almost all organisms some multinucleated cells are to be found. This seemingly major departure from typical organization is due to a relatively minor physiological factor—the failure of cytokinesis following nuclear division, a failure which is readily induced experimentally. The implications of the failure are probably more far-reaching than the cause. For although the absence of cell barriers facilitates the exchanges of substances throughout the multinucleated protoplasmic mass, it also makes difficult the establishment of localized environments which lead to differentiation.

Many unicellular organisms have forms which bear only a remote resemblance to the conventional image of a cell. The alga Acetabularia is large enough to be seen as a pinhead by the naked eye, and yet it contains only one nucleus. Many studies of nucleocytoplasmic relations have been conducted with this organism because of the ease with which the nuclear region can be removed and even replaced by grafting the nuclear piece from a different species. The ciliates (Paramecium, Tetrahymena) have an elaborately patterned cortical layer and are distinctive in their possession of a micro- and macronucleus. The macronucleus, which develops from a micronucleus in the course of sexual reproduction, consists of hundreds of chromosome sets and is the primary or sole source of genetic information for somatic activities. Another departure from typical cellularity is to be found in symbiotic cellular systems. *Paramecium bursaria,* for example, is green because of the algae present in its interior. The fact that two distinct genetic systems can coexist and interact within the confines of a single plasma membrane is of considerable evolutionary interest. At one extreme are the pathogenic viruses which, on penetrating a cell, multiply so rapidly that the host cell becomes disorganized and destroyed. At the other extreme are those viruses which ordinarily multiply in coordination with their host cell and are thus transmitted from one cell generation to another without pathological, and perhaps even with beneficial, consequences. The accumulating evidence for the presence of DNA in chloroplasts would suggest that the line of distinction between a subcellular organelle and an entrapped foreign organism can become thin to the point of vanishing.

The abstract concept of a typical cell is easier to defend and is operationally more meaningful than any generalized morphological model. For this concept merely states that the cell is the smallest biological unit which is capable of reproduction by extracting energy from its environment. To achieve this a cell must have a system for regulating diffusion, for converting energy, for coding molecular sequences, and for equipartitioning its genetic apparatus.

17

Functional Organization of Cells: Temporal

Since no cell is eternal, all cells must have a finite life history, and that history ends in one of three ways—division, fusion, or death. Just as the appearance of a cell bespeaks a high degree of spatial organization, its history bespeaks a high degree of temporal organization. Not only is there a beginning and end to the history of a cell, but its course is marked by characteristic patterns. Some features of the patterns are basic to all cell types; others are elaborations of a basic succession of intracellular events, and are prominent in the cells of higher forms. Whether simple or elaborate, such patterns are expressions of an inherent capacity of cells to regulate their activities along the axis of time. In a strict sense, no cell has the same structural and metabolic configuration for two intervals in its history, however close and however small these intervals might be. A fixed characterization of a mature cell is an approximation; all cells are continuously undergoing directional change.

The challenge of explaining how cells progressively alter with time is far from being met. This chapter can therefore contribute little in that direction. What can be done is to describe in the most general terms possible the principal characteristics of cellular life histories. This description may stake out the contours of the problem, but it will do no more than that. The inevitability of time and the inevitability of changes along the axis of time are aspects of reality which enrich the musings of the philosopher and deplete the confidence of the scientist. We should not expect, however, the temporal sequence of changes to be understood before the changes themselves, independent of their occurrence in time, are known. Such understanding is only now developing.

The simplest type of life history, and one which will serve as "reference pattern," is that of an asexually reproducing unicellular organism. Such an organism perpetuates itself by repeated cell divisions. Without any examination of the process we may predict that each division will be preceded by chromosome replication and will be associated with a development of daughter cells to a size and form identical with their parent. A brief inspection of such dividing cells will reveal that compensatory growth occurs in the daughter cells; a mature cell does not double its size prior to division. Asexually reproducing cell lines are thus clearly marked by a sequence beginning with the growth of a daughter cell to mature size followed by division. If this very simple sequence is studied in more detail, differences between cell types become apparent. As a group bacterial cells do not appear to organize the component events of the cell cycle into a sharply defined temporal sequence except for the mandatory requirement of DNA replication and segregation preceding cytoplasmic division. If we plot the changes in DNA, RNA, and protein against a time axis, the slopes turn out to be more or less parallel. Grossly interpreted, this means that nuclear and cytoplasmic elements are synthesized simultaneously. The only cytoplasmic event which is clearly timed with respect to DNA replication is the formation of a wall between daughter cells. Walls do not form until replication is complete. Wall formation, on the other hand, is not a signal for resumption of DNA replication and cytoplasmic growth. Various factors, some naturally occurring, others supplied experimentally, may inhibit wall formation without affecting any of the other processes. The dissociability of cytokinesis from all other phases of the division cycle is commonplace in the biological world and accounts for the variety of multinucleated forms discussed in Chapter 16.

In contrast to bacteria, dividing cells of higher forms show a distinct phasing of events. Following the telophase of division, daughter cells grow in size, both cytoplasm and nucleus enlarging in the process. During this initial growth period, however, chromosome replication does not occur. Once the cells reach their maximum size, a metabolic switch occurs and the synthesis of chromosomal components, notably DNA and histone, begins. This interval of chromosome replication may extend for about half the life span of the cell, and its completion signals a new metabolic phase which may be described as a "preparation for mitosis." The predominant metabolic events of this phase, if there are any, have not been characterized, but some evidence exists for the synthesis of certain nuclear proteins during this interval. Whatever its metabolic features, this interval is a prominent though variable characteristic of most cells, and because mitosis does not immediately follow chromosome replication, the interval has been interpreted as an essential prelude to the mitotic cycle which follows its completion. Cytologists use a set of symbols to designate the sequence of phases: G_1, first growth period; S, DNA synthesis period; G_2, second growth period; *M*, mitosis. The time spans of the respective periods are not the same for all cell types; neither are their relative durations within a cell cycle. Some actual time spans of cells which for this purpose may be considered as maintaining a somatic cell line are shown in Fig. 17-2.

That cells, protozoan, animal, or plant, temporally regulate subcellular processes such as cytoplasmic growth and nuclear division is patent from these observations. In multicellular organisms, however, this basic pattern may become considerably modified. The modifications reveal many additional capacities of cells to regulate their behavior in time (Fig. 17-1). The first cycle discussed applied to cells which give rise to identical replicates, but this is the exception rather than the rule in multicellular organisms. In all sexually reproducing organisms, cells must be present which, on division, do not give rise to similar progeny; gametes are distinct from their somatic counterparts. And even in nonsexually reproducing multicellular organisms, such cells must be present if the organism is composed of different cell types. The mechanisms underlying cell differentiation is so challenging a subject that it is reserved for a special chapter. What will be considered here are two primary cellular characteristics which accompany this process: Changes in time patterns and disruption of the "typical" cell cycle.

THE SEXUAL CYCLE

A logical starting point is fertilization because it is also the starting point of an organism's life. No attempt will be made here to cover the details of this event.* Patterns of fertilization are varied and complex, but in all cases a new cell cycle begins as a result of nuclear fusion. The sequel to fertilization is not a typical cell cycle, even though a series of cell divisions follows immediately. In very many egg cells nuclear

* See William Telfer and Donald Kennedy, *Biology of Organisms*, John Wiley & Sons, 1965.

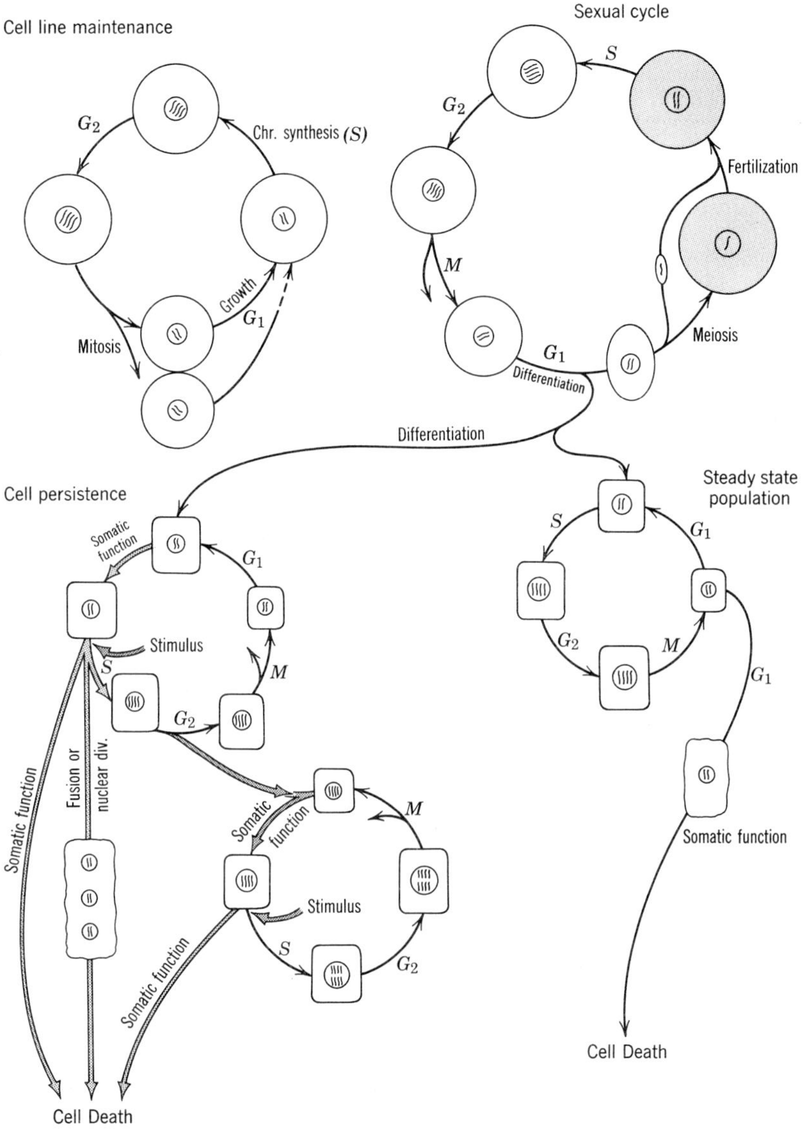

Fig. 17-1. Types of cell cycles.

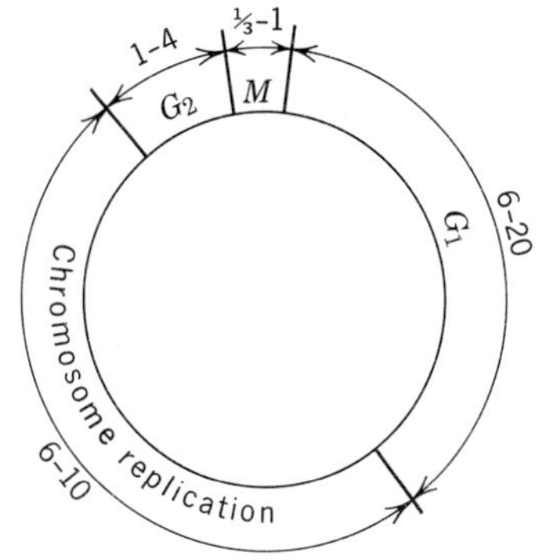

Fig. 17-2. Typical cycle of dividing somatic cells. The numbers assigned represent the duration of each phase in hours. Most somatic cells of higher plants and animals fall into this pattern. One feature of the cycle which may not be evident in the figure is the relatively high variability of the G_1 phase. Such variability has been observed both within groups of similar cells and between cells of different species. By contrast, the interval of chromosome replication is more or less constant for a broad range of organisms. For a particular strain of cells, the periods of chromosome replication, G_2, and mitosis (M) appear to be constant within the limits of experimental error. The most accurate determination of the mitotic interval is by direct observation of living cells under the phase microscope. The other intervals cannot be so determined because no morphological criteria are available to distinguish between them. The most commonly used technique to determine these intervals is autoradiography. Its success depends on the fact that if a population of dividing cells is exposed to tritiated thymidine for a restricted period of time, only those that are in the replication phase become labeled. By suitable variations of exposure intervals and fixation times following exposure, the data obtained on the numbers of labeled and unlabeled cells, and on the numbers in mitosis, may be used to calculate the intervals.

divisions occur in rapid succession without any cytoplasmic growth. The Drosophila egg is remarkable in this respect (see Table 17-1); "cell divisions" occur every nine minutes, and two-thirds of this interval is reserved for mitosis. In the eggs of insects generally, the term cell division is a misnomer. What actually happens is a series of nuclear divisions without any accompanying cytoplasmic divisions. In the first stages following fertilization, nuclei behave analogously to parasites multiplying freely in a host cytoplasm. The occurrence of cytoplasmic divisions, which is common in other animal eggs, does not represent a deep-seated distinction between the two modes of egg development, for no cytoplasmic growth takes place in either. Thus not only is the typical sequence (G_1SG_2M) unnecessary for dividing cells of higher organisms, but a very different sequence can occur and may be written as $G_1(SG_2M)^n$, in which G_1 represents the growth of the egg prior to fertilization. Nuclear replication and cytoplasmic replication are dissociable processes. The division of a nucleus does not necessarily induce the cytoplasm to grow, nor does growth of the cytoplasm necessarily induce the nucleus to divide. Cells must therefore have distinct regulatory mechanisms which can associate or dissociate the two events.

In the fertilization cycle (Fig. 17-1) a stage is reached in the division of the egg cell at which mitosis is followed by a metabolic shift in the daughter cells. The result is a differentiation of cell types. The G_1 interval is thus no longer a phase of cytoplasmic replication, but is one of cytoplasmic modification. The modification provides a variety of consequences for the future histories of the daughter cells. One line of development, the production of gametes, leads to a completion of the formal cycle. Despite the impression that gametes are the most unspecialized of cells because they are the potential source of all cells in an organism, they are, in fact, highly specialized. Not only are they differentiated into distinctive cell forms, but they are products of a unique type of chromosomal division. The basic design of meiosis has already been described in Chapter 14 and its regulatory features will be discussed later. In the context of this discussion, the significant point is that one consequence of differentiation can be the production of a temporary arrest in daughter-cell development. This arrest, like all other events in a sequence, shows the characteristic of dissociability. Normally, the arrest is broken by fertilization, but the experiments in which unfertilized egg cells have been induced to develop are now classics in the history of biology. Not only has such "parthenogenetic" development been induced experimentally, but it is a natural occurrence in some organisms.

TURNOVER OF SOMATIC CELLS IN MULTICELLULAR ORGANISMS

Alongside the development of gametes as a consequence of differentiation, is the development of a variety of cell types which may be grouped under two headings; those which continue to divide and those which do not. Members of the first group are found in both plants and animals, but their behavior in each of these classes of organisms differs in some respects. Plant growth is usually characterized as "indeterminate" because of the fact that individual plants do not grow to a predetermined size but continue to elaborate roots, stems, and leaves as long as they are in a healthy state. Indeterminate growth is made possible by the retention of populations of cells in characteristic regions of the plant (e.g., root and apical meristems) which undergo cell divisions continually. Unlike the pattern found in the maintenance of a cell line, daughter cells in meristematic regions have divergent fates. One group of daughter cells retains its meristematic character and divides again; the other group undergoes differentiation into one of several cell types. Cells in meristematic regions may therefore be regarded as maintaining a steady state population by releasing half their progeny for differentiation and retaining the other half for self-perpetuation. For any individual cell which undergoes division, the cycle of changes is very similar to that described for cell line maintenance. The similarity does not apply to each set of daughter cells. For reasons which are embedded in the phenomenon of differentiation, the G_1 period of one of the daughter cells leads to a replicate of the parent, while that of the other daughter cell leads to a new cell type.

Although higher animals also maintain steady-state populations of cells, the relationship of daughter cells to parent is distinctive. Whereas in plants meristematic cells can give rise to all the cell types found in the adult organism, their counterpart in animals do not do so. Adult animals are perpetually losing certain specialized types of cells—those found in the skin, in the lining of the digestive tract, and in the blood. A skin cell, for example, has an average life of 18 days; a red blood cell survives for 85. Such cells are not replaced from a common pool. Each tissue has its own source of already differentiated cells which are dividing at a rate equal to cell loss. Such divisions, like those in plants, do not lead to identical daughter cells. The divergence between them, however, is more restricted—one cell matures to a single cell type or a group of related types, performs its newly acquired somatic functions, and dies; the other matures, without acquiring these functions, enters an S phase and ultimately divides again. Thus despite the fact that adult animals may not grow, the lack of growth is true only in an arithmetic sense. The survival of individual animals is dependent on populations of continuously dividing cells. "Cell turnover" is an attribute of every higher multicellular organism.

PERSISTENT CELLS

Although the near immortality of some trees would suggest that plant cells have a much better capacity for prolonged survival than do animal cells, the opposite is true. The comparatively short life of an animal is due to the long life of some of its cells; the long life of a tree is due to the comparatively short life of all of its cells. Most cell types in a mammal do not ordinarily divide once they have differentiated, and some cell types (e.g., nerve, muscle) do not divide at all. Both groups have one characteristic in common. Following the first growth period they enter a prolonged mature phase. This phase is frequently referred to as the "steady-state" phase. It may be considered analogous to the steady-state cell population with the subcellular components representing the cells as units of turnover. This characterization is no more than an approximation, for cells cannot escape the irresistible drag of time. The approximation is nevertheless useful, because it distinguishes between that phase in the life of a cell during which its performance is more or less constant, and the one which is clearly marked by a deterioration of function or "senescence."

Even as an approximation, the steady state is alien to a population of cells which is undergoing repeated divisions. In such cells each phase is characterized by progressive changes which ultimately yield to a new phase similarly characterized. Accretion of components, not their maintenance, is the chief characteristic. To a bacteriologist, for example, a healthy culture is one in logarithmic growth, one in which every cell undergoes division. In higher organisms persistent divisions of cells in one of many tissues reflect a pathological rather than a healthy condition. The physiological disposition of cells in a proliferating system is very different from that in a fixed one. And the problem which arises with respect to persistent cells concerns the mechanisms of cell maintenance. At one time it was believed that molecules which are functioning actively in cells are highly susceptible to breakdown, and that cells must therefore be constantly renewing their macromolecules in order to maintain themselves. This extreme view is incorrect. DNA filaments are being continually transcribed, but on the whole they do not undergo breakdown and resynthesis. Proteins and ribonucleic acids are not inherently unstable molecules, and their behavior in a nongrowing cell can only be correctly evaluated by defining cell maintenance in specific terms.

Very few cells are known which function purely to maintain themselves. The only types of cells to which this description fully applies are those which have been induced to undergo a distinctive set of transformations resulting in dormancy. The durability of the dormancy depends on the nature of the transformations. Dehydration of cells such as occurs in seeds and spores generally leads to the most durable form of dormancy. Cell maintenance is effected by a virtual suspension of all activities. Theoretically, if the suspension is complete, survival should be indefinite. Whether a population of seed or spores can thus survive is obviously unknown, but many are known to have survived long intervals of storage, and at least as many to have undergone deterioration in the course of storage. The mechanisms by which cells become dehydrated without damage to their structure remain a subject of inquiry. Bacterial strains are artificially dehydrated and stored in sealed tubes; cells of higher organisms have not yet been preserved in this way. Some types of cells, sperm, for example, have been successfully stored at low temperatures which, like the removal of water, causes a virtual suspension of all activities.

In multicellular organisms persistent cells perform specialized functions which are incidental to their maintenance but are critical to the survival of the organism. The nature of these functions varies with cell type. Some cells (liver, pancreas) synthesize proteins for secretion; others synthesize hormones; and still others channel their energy resources to effect mechanical movements or to control the transport of solutes across membranes. A difficult question to answer precisely is the extent to which the molecular framework of subcellular systems which perform these functions undergoes destruction and resynthesis. In bacteria the average life span of a message RNA molecule may be about 2-3 minutes, and the average number of protein molecules transcribed about 15. The number of copies of each particular message made during a single cell cycle probably varies considerably. The suggestion has been made that some messages may be copied only once, whereas others may be copied as many as 200 times during a single cycle. Presumably, this turnover of RNA is a device which enables the genome to control the composition of the molecular population during growth. A similar turnover probably exists in cells of higher organisms. In the nucleolus, particularly, the very rapid synthesis of RNA appears to be associated with a rapid destruction. On the other hand, there is some evidence, though insufficient to provide for generalization, that some RNA messages are relatively stable so that they persist as sites of translation for extended periods of time.

Ribosomal RNA is on the whole far more stable than message RNA. This characteristic is highly prominent in bacterial cells where even invading viruses use the unaltered host ribosomes as structures for their synthesis. Ribosome persistence may be supposed for other cell types, but whether this persistence is absolute is highly doubtful. An induced enzyme such as B-galactosidase has been found to persist in bacterial cells after its synthesis has stopped, the enzyme being lost only after progressive dilution in daughter cells. We may infer that if protein molecules persist in bacterial cells, they also persist in other cell types. Such persistence, however, cannot be universal, for as will be seen in the discussion of regulatory mechanisms (Chapter 18), removal of a particular protein may be as necessary to a cell as its synthesis. One example of this is the enzyme changes which occur in liver cells when an animal is starved. Various enzymes which catalyze the breakdown of organic molecules (phosphates, arginase, esterase) increase markedly in concentration. The changes are adaptations to the novel nutritional situation created by starvation.

Thus one aspect of maintenance in nongrowing cells must be related to the continuous adaptation of cells to environmental fluctuations. Ultimately, the net constancy of the internal environment within a mammal, for example, must be related to the sensitivity of adaptive mechanisms within the component cells. How much the novel synthesis of macromolecules is part of these adaptive mechanisms, and hence the source of molecular turnover in cells, is an unsettled question which will be discussed in Chapter 18. Yet in addition to such adaptive responses, the evidence, though limited, points to a wear and tear of macromolecules. Some data, for example, indicate a high rate of mitochondrial renewal in nondividing cells. Since the magnitude of renewal is open to question, generalizations about molecular renewal are of little value. We may nevertheless conclude that some renewal does occur and that if cells lose their capacity to synthesize essential macromolecules, they also lose their capacity for survival.

Superimposed upon all questions concerning the magnitude of molecular renewal which accompanies the maintenance of a steady state in mature nondividing cells, is the fact that all such cells age. All non-dividing cells eventually die, and the approach to death is manifested in the process called senescence. Whether senescence should be regarded as a more obvious manifestation of a cumulative process beginning with cell maturation is a matter of definition. Young tissues differ in appearance from older tissues; abrupt transformations have not been observed. Various biochemical studies have been performed to determine the nature of the aging process, but on the whole they confirm the morphological picture rather than supply insights into mechanisms. One of the very striking manifestations of aging in humans is the progressive accumulation of pigmented granules in nondividing cells. In heart muscle cells the concentration of pigment has been measured and found to be approximately a linear function of age. The composition of the granules is complex; lipids and proteins (including some enzymes) are found to be present. Although we can draw very few, if any, conclusions, about the effects of these apparently nonfunctional granules on cell performance, their steady accumulation in a variety of cells is sufficient proof that cells do not maintain a true steady state. Progressive transformation with time is an inescapable characteristic of cells.

Regardless of the mechanisms responsible for the modifications of cells and tissues which accumulate with age, the fact that certain cell types cannot divide assures the mortality of the individual. In spite of their considerable capacity for self-repair and the protective effect of the internal environment of the organism, all cells are vulnerable to accident such that irreplacable cells will be lost at some finite rate. Histological studies demonstrate that the number of brain cells in man progressively declines with age. At some point the loss of certain kinds of cells becomes crucial and the organism succumbs. Mortality may well be the price which organisms pay for efficient cellular differentiation; evolution has not yet provided a mechanism for the renewal of all tissues.

CELL PERSISTENCE AND CAPACITY FOR DIVISION

The phenomenon of cell persistence would be easier to understand if daughter cells either lost their capacity to divide or continued in division. This is true for many but not all specialized cell types. In both groups the characteristic accompaniment of differentiation is a cessation of DNA synthesis, and an induction of novel syntheses which result in a set of specialized properties. Some specialized cells, however, retain the capacity to undergo further divisions and manifest this capacity under appropriate conditions of stimulation. As much as two-thirds of a rat liver, for example, may be removed and yet the liver will be quickly restored to its original size, because the cutting has stimulated the remaining cells to divide. In many lower forms (amphibians, for example) entire limbs are regenerated following removal. Chromosome replication and mitosis are phases in cell development which are subject to reversible repression. What mechanisms an organism uses to effect or remove such repression is one more phenomenon still marked by ignorance. The general supposition is that these controls are exercised by hormonal factors; in higher plants this type of control is demonstrable in many ways. Frequently the capacity of an otherwise nondividing cell population to divide is marked by the occurrence of polyploidy which results from chromosome replication unaccompanied by mitosis. In some tissues, depending on species, polyploidy may be extensive; in others it may be relatively rare.

Of great interest both medically and biologically is the fact that the division of cancerous cells usually cannot be suppressed by the host organism. Originally it was supposed that cancerous tissues have an extremely different metabolic pattern from that of their normal counterparts, but comparative biochemical studies have failed to reveal any outstanding differences which might account for the disease. What has been clearly lost by cancerous cells is the capacity to suppress and induce differentiation. The loss has the characteristics of a mutation, for once acquired by a cell it is transmitted to succeeding generations.

THE RECOGNITION OF TIME

Cellular recognition of time is manifest in the fact that different species of cells have characteristic life spans. This is true both for cells which undergo division and those which do not. Table 17-1 has been drawn up to show that even where senescence is not a factor, life spans of cells may show extreme differences. Not only are differences manifest with respect to total life spans, but also with respect to the duration of comparable phases in a life span. Despite the uniformity of mitosis, the relative duration of phases may vary widely between species (Fig. 17-3). Both aspects of temporal regulation cannot be explained in terms of environmental parameters such as temperature and nutrients, even though a factor like temperature may markedly affect developmental rates.

Table 17-1 Approximate Generation and Mitotic Times for Different Cells

	Total Generation Time (Min.)	Mitotic Times (Min.)
Drosophila egg	9	6
Bacterium	20	—
Grasshopper Neuroblast	210	180
Rat Corneal Epithelium	14,000	70

An extremely interesting expression of time recognition is found in some cells which manifest prominent rhythms. These rhythms have been called "circadian rhythms" because they are patterns which recur approximately every twenty four hours, and are unaffected by temperature. The rhythms have been found in a variety of cell types and generally encompass several distinct physiological activities. Temperature may affect the activities by making them more or less intense (amplitude modification) but it does not markedly affect their periodicity.

The unicellular marine organism *Gonyaulax polyhedra*, for example, is capable of luminescence. It has a metabolic system which converts a fraction of its stored energy to the production of light, a property for which no biological function is yet clearly recognized. Normally

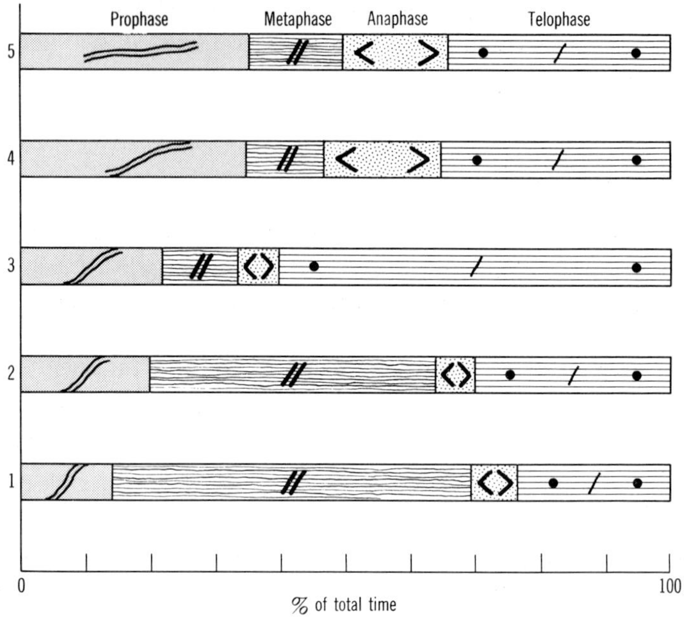

Fig. 17-3. Temporal patterns in mitosis. The metaphase and anaphase stages are commonly the shortest in both meiosis and mitosis. The main factor which determines the duration of metaphase is the time taken by the chromosomes to align at the metaphase plate. Once such alignment is complete, anaphase movement begins. Extended metaphases, such as shown in 1 and 2, are not common but they are characteristics of some species of cells. The patterns drawn are based on observations made on living cells by means of a phase microscope. The technique has been employed by a number of investigators. Cell types 1 and 2 are from cancerous animal tissue; 3 and 5 are from plant endosperm; 4 is from rat spleen.

such luminescence occurs with maximal intensity late at night. If cells are transferred from their natural light-dark-light environment and maintained continuously under dim light, they continue to show that same rhythm. A similar pattern of behavior may also be shown with respect to photosynthesis. Under natural conditions cells display a greater capacity to utilize light for photosynthesis during the day. Cells exposed during the night to a given intensity of light photosynthesize much less than cells exposed during the day to identical conditions. Cells transferred to continuous dim light retain the characteristic periodicity. In Gonyaulax and many other cells there appears to be some built-in clock mechanism; the cells somehow manage to recognize elapsed time even under conditions where the external environment is kept constant. The clock is not permanently set. By repeatedly exposing cells to artificial dark-light cycles which are reciprocals of the natural ones, the cells acquire a pattern of periodicity which corresponds to the artificial cycle. Once the periodicity is acquired it perists under constant environmental conditions.

An attempt to explain how cells effect temporal regulation would be premature, since we do not yet understand the various mechanisms which control cell development. It is not premature, however, to consider briefly one mechanism, which does have the built-in characteristic of recognizing time. The discovery of this characteristic is due to studies of bacterial viruses which have relatively simple life histories. When viral DNA enters a bacterial host, it induces the synthesis of a number of enzymes essential to the production of virus progeny. By carefully timed application of protein inhibitors, it can be shown that the enzymes and other proteins essential to virus reproduction are synthesized sequentially and not simultaneously. The coat protein which covers the virus is synthesized last, whereas the enzymes necessary to DNA synthesis are synthesized first. Moreover, the sequential appearance of proteins can be shown to be due to a sequential formation of RNA messages. The orderly development of the virus can be said to be taped in the DNA filament.

The last conclusion, however, is open to question unless it can be shown that the orderliness of reading the genetic tape is inherent in the structure of the DNA itself and not to accessory factors operating in the bacterial cell. This question has been resolved by a set of brilliant *in vitro* experiments. The transcription of DNA into RNA by means of the enzyme, RNA-polymerase, can be performed in a test tube if the mixture contains double stranded DNA, a supply of ribonucleotide triphosphates, and the enzyme. One important question emerged as a result of such experiments. Analysis of the ribonucleic acid thus synthesized indicated that both strands of the DNA were being transcribed, whereas the evidence from *in vivo* analyses indicated that only one of the strands was being transcribed. The most convincing evidence for this was obtained from an unusual bacterial virus which consisted of only one strand of DNA in the vegetative state but became two-stranded on infection. By determining the base composition of message RNA following infection, it was found that the RNA was not complementary to the strand of DNA which infected the bacterium, but to the complementary strand of DNA formed after infection. When the double-stranded replicating form of the DNA was isolated by standard procedures and used as a template for RNA synthesis *in vitro*, the RNA formed was complementary to both strands of DNA. Further experimentation, however, yielded the explanation. First, the

DNA was in the form of a closed circle, a form in which it also occurs in bacteria. If the circles are broken on isolating the DNA, both members of the DNA double helix are transcribed. If, on the other hand, the circles are maintained intact, only one of the strands is transcribed, the same type of strand as is transcribed *in vivo*. Clearly, the structure of the DNA molecule inherently determines how the DNA should be read.

To suppose that this simple picture explains the developmental characteristics of all cells would be most presumptuous. We must recall the remarks made in connection with spatial regulation: given a basic functional element, the elaborations which may flow from it are virtually unlimited. In this respect S. Spiegelman's brilliant discovery of temporal regulation by intact DNA circles represents the first molecular model of how living systems may recognize time.

18

Mechanisms of Regulation: Environmental

Cell behavior is consistent but not constant, and the principal problem of regulation is the mechanism which determines *when* a cell will perform a particular set of activities. In earlier chapters we have considered how a cell synthesizes its components, how it performs mechanical work, and how it transmits signals along membrane layers. These capacities are expressed neither uniformly nor simultaneously. Along the axis of time their expression changes progressively and largely irreversibly in conformity with inherent developmental patterns. At right angles to the axis of time their expression changes discretely and largely reversibly in adaptive response to environmental fluctuations. In both cases some form of signal must exist to induce the changes, and some regulated response system must exist to yield optimal behavior. This chapter is concerned with the mechanisms of response to environmental signals.

No natural environment is constant in all its parameters. Some cell types have evolved which can only survive if one or even two parameters of their environment remain constant (e.g., temperature and blood pH for the internal cells of a mammalian organism), but no cell type has evolved, or could evolve, which required a completely constant environment for its survival. Natural environments undergo seasonal and diurnal cycles of change. Superimposed on these regular cycles are irregular fluctuations which may persist for minutes, hours, days, or weeks. Some environmental parameters such as temperature, humidity, and, to a lesser extent, light intensity are subject to extreme and irregular fluctuations. Other parameters such as light period,

CO_2/O_2 concentration in air, osmotic concentration of sea water are virtually constant either for any given point in the seasonal cycle or throughout the year. Environmental differences, however, exist not only with respect to time but also with respect to space. Latitudinal differences, such as between polar and equatorial regions, are obvious, but the responses of cells to these geographic extremes is primarily a function of genotype selection. Polar and tropical organisms cannot survive an interchange of environments. Far more subtle spatial differences do exist to which cells within any particular geographic region are able to respond. At the peak of photosynthetic activity the layers of air adjacent to a leaf are different in temperature, composition, and humidity from those prevalent in the atmosphere at large. The temperature of shallow water fluctuates much more than that of deep water. In both terrestrial and aquatic environments nutrient sources are unevenly distributed and subject to continual change because of utilization. Many other examples could be given, and they would all lead to the same conclusion. For any individual cell the external environment is inconstant, and to the extent that environmental variations affect cell function, cell survival is dependent on mechanisms which enable a cell to accommodate to these variations.

The capacity to accommodate to environmental variations is a property even of mammalian cells which, on the whole function in a fairly constant organic environment. The constancy of that environment, however, is due to the regulatory responses of the cells themselves. The constancy of body temperature depends on heat production by metabolic activity, and such activity must be regulated according to external temperature. Much more energy must be expended by cells for heat production when the external temperature is low than when it is high. Although the thermostatic mechanism in mammals is not understood, cells in the hypothalamus sense temperature changes and signal (presumably by hormones) other cells to alter their metabolism in the direction of increasing or decreasing heat production. Temperature, pH, and osmotic concentration of the circulatory system are kept fairly constant by mechanisms classed as "homeostatic." Other parameters of the internal environment, however, are subject to a great deal of fluctuation. Food is ingested periodically, and to accommodate to ingestion, hydrolytic enzymes must be periodically secreted into the alimentary tract, the degradation products absorbed and transported to various tissues for metabolic transformation. Energy must be stored for periodic muscular activity, and hormones must be synthesized and secreted according to instructions from the nervous system. Thus not even in the most constant of natural cellular environments do we find cells whose behavior is constant from one time interval to another. All cells have the capacity to alter their activities to a greater or lesser degree in response to environmental fluctuations.

Regulatory systems may be formally described by the simple diagram shown in Fig. 18-1. The initiation of any change must ultimately be related to a cause, and this cause is defined as the stimulus. A stimulus must be recognized by at least one component of a cell in order to be acted on, and such a component is designated as the receptor. The activities which follow sensing of the stimulus may be restricted to the activities of the receptor system itself, or they may be restricted to an entirely separate system (the "effector") which acts on receiving a message from the receptor. This formal description provides some useful terminology, but little else. To understand the nature of cell regulation, we must identify the actual regulatory mechanisms.

NONSPECIFIC RESPONSES

Any factor which can affect the activity of one or more components of a cell is a potential stimulus. It becomes an actual stimulus at the time that it changes magnitude. Cells react to *changing* environmental conditions (temperature, light, nutrient, etc.), but once the change is complete, the particular environmental factor is no longer a stimulus to the cell, but a characteristic of the environment to which it has adapted. The responses of a cell to environmental change may be either nonspecific or specific. Nonspecific responses are elicited by environmental factors for which the cell has no organized response system. Specific responses are those for which the cell has an organized reception-effector system. Of all environmental parameters, water availability and temperature are the principal agents of nonspecific responses. Both these factors are ubiquitous in their effects. Water is so essential a component of cell organization that any concentration changes within the living matrix indiscriminately affect all activities. Enzyme catalysis is so intimately tied to the activities of all subcellular systems that any change in temperature directly affects the rates of all cellular processes. Even if cells have specific receptors to sense

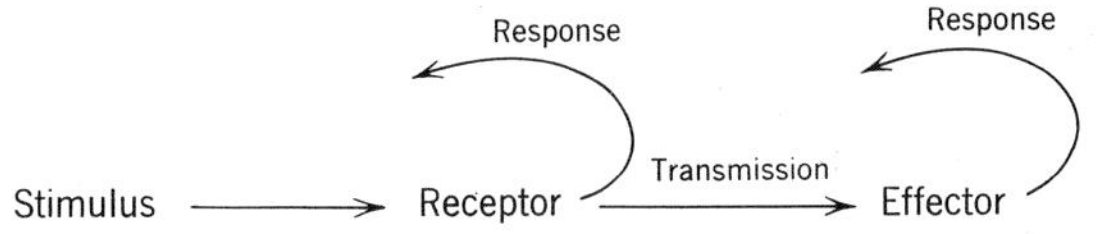

Fig. 18-1. Formal description of a regulatory system.

humidity or temperature change, as do some cells of higher organisms, they cannot escape the generalized effects which these two environmental factors produce.

In aquatic organisms water content is generally determined by osmotic factors. Cells with rigid or nearly rigid walls can maintain their water balance even with high intracellular concentrations of solutes because the walls limit the volume of water which can be absorbed. Such cells, on the other hand, cannot withstand hypertonic environments because there is no mechanism to prevent water loss. Among aquatic organisms protozoa are unusual in their possession of a contractile vacuole which actively regulates the water content of the cells. Protozoa do not burst in hypotonic environments because the contractile vacuole can pump out water which has diffused in by osmotic forces (see Chapter 3). Generally, water diffuses passively through cell membranes in the direction of the osmotic gradient. In multicellular organisms, particularly terrestrial ones, water balance is achieved by various specialized structures including those which function in water absorption and those which act to reduce losses by evaporation.

Although cells are not equipped to respond rapidly to fluctuations in water content, various devices have evolved to obviate the perils of dehydration. Many species of bacteria and blue-green algae can form spores which are resistant to virtually all degrees of environmental desiccation. Protozoan cells may also become encapsulated as spores. Such spores (which have been given different names for different organisms) are distinct from reproductive spores. They represent cells which have been transformed into a dormant state and can thus survive extreme environmental conditions. Many reproductive cycles have dormant stages, the most complex dormant form being the seeds of higher plants. The achievement of dormancy cannot, however, be classified as a direct response to dehydrating conditions. It is a developmental response which has evolved to anticipate such conditions. Individual cells cannot adjust their functions to low and high intracellular water concentrations. Progressive dehydration is accompanied by a progressive drop in metabolic activity. When such dehydration is an organized developmental process, the cell survives in a dormant condition; when dehydration is caused by a sudden environmental change, the cell dies.

Temperature variation, like water variation, is an environmental factor to which individual cells can accommodate within a very limited range. Mammalian cells, as already discussed, are adapted to function within an extremely limited temperature range. Cells of poikilothermic animals and of plants in general can survive over a fairly broad range of temperature. The metabolic rates within any particular type of such cells vary approximately in the same way as chemical reactions do in response to temperature change. Metabolic rates generally increase two- or three-fold for every 10°C rise in temperature. The range of temperatures within which such changes can occur have both upper and lower limits. Very few cell types can survive temperatures above 40°C. Just what factors enter into disruption of cell function by elevated temperatures is not clear. An absolute upper limit is set by the fact that proteins denature at elevated temperatures (see Chapter 4), but denaturation is not necessarily the immediate cause of lethality. Subcellular processes do not respond uniformly to temperature changes. This fact, even though it is not well understood, has been taken advantage of in artificial synchronization of cell populations. Certain strains of Tetrahymena, for example, do not undergo division at a temperature of 34°C, but other cellular growth processes continue at this temperature for a limited period of time. If a population of Tetrahymena cells which is heterogeneous with respect to developmental stage is exposed to a temperature of 34°C, those cells which are about to enter division are prevented from doing so, whereas those cells which are at an earlier developmental stage continue to grow until the divisional stage is reached. If the temperature is now lowered to 28°C all cells simultaneously enter division. In practice several cycles of temperature changes are necessary to achieve a high degree of synchrony.

Temperature may thus disrupt cell function by differentially altering the rates of subcellular processes or by selectively inhibiting some of them. Individual cells are not equipped to reorganize their metabolism in response to temperature fluctuations; metabolic changes are imposed and cells are adapted to survive greater or lesser degrees of such change. Yet, as in dehydration, cells have evolved developmental response to temperature variation, and these are particularly apparent in plants. Perennial plants in temperate zones anticipate the cold season by a process known as "frost-hardening," whereby cells become resistant to damage from temperatures well below freezing. In certain types of plants meiotic divisions occur only at low temperatures. Most bulb plants require a cold period for their bulbs to develop, and many types of seed require a cold period to break their dormancy. Acclimation to cold is not, however, restricted to the plant world; animals too can become cold-adapted. Regardless of the nature of the adaptive process, most cells can perform their normal functions only within a limited temperature range.

SPECIFIC RESPONSES: SPATIAL ACCOMMODATION

All viable cells adapt to their environment, but some cells have response mechanisms which appear to direct them toward an optimal environment. The property of cells moving toward a favorable environment may seem trivial. Cells in multicellular organisms do not commonly move, even though most organisms do. Yet the capacities of organisms to move in response to stimuli are elaborations of the much simpler capacities found in single cells. And for many unicellular organisms such capacities are not secondary physiological features, but primary adaptive mechanisms.

If the movements of an amoeba in relation to a food particle are observed under the microscope, they do not appear to be random. The amoeba moves toward the food, eventually surrounding it by a pseudopod and then engulfing it. Amoebae do not entirely depend on chance collision with food particles for their nutrition; they are equipped to sense the presence of food at a distance, and such sensing elicits a movement in the direction of the stimulus (see Fig. 18-2). If a beam of bright light is focussed on a suspension of Euglena, the cells swim away from the lighted region; the cells sense the intense light and respond negatively to its presence. Motile algal spores will also move away from very intense or very weak illumination. An amoeba which is touched with a glass rod, recoils from the rod, moving away from the site of mechanical disturbance. Numerous examples could be chosen to illustrate the fact that motile unicellular organisms respond either positively or negatively to chemical, photic, and mechanical stimuli. These responses may be related to nutrition, to the mating of motile gametes in reproduction, to the selection of optimal light conditions for photosynthetic organisms, to the avoidance of injurious environments, or (as will be discussed later) to particular phases of development.

More so than any other aspect of regulation, the pattern of mechanisms governing the direction of cellular movements follows the formal scheme described in Fig. 18-1. A specific environmental stimulus may be identified for which there is a correspondingly specific receptor in the cell. The receptor responds to the stimulus by transmitting a message to some other cellular component. In this particular category of regulation the cellular component is ultimately of one kind—a fibrous system responsible for locomotion. One of the most studied types of locomotory responses is that of phototaxis. Many species of micro-

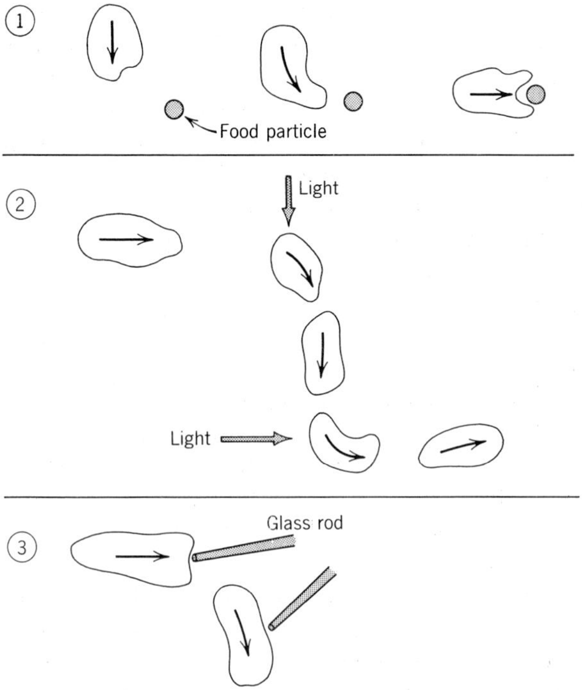

Fig. 18-2. Spatial accommodation in Amoeba in response to various environmental factors. The directions of movement are approximate. Many unicellular organisms have specialized structures to sense light (for example, the "eye spot" in *Euglena*), to ingest food (the "mouth" of *Paramecium*), and to effect movement (cilia or flagella).

organisms, most of them photosynthetic, respond to changes in light intensity. The magnitude of the intensity change required to elicit a response varies with the particular type of response and with the organism. In general, however, the magnitude is small and far below that required for a process such as photosynthesis. The stimuli to which cells respond fall into two classes: a change in general intensity of illumination such as might be produced by a passing cloud; and a gradient in light intensity such as might be obtained by illuminating an aquarium tank from one side. Purple photosynthetic bacteria have been observed to respond only to the first type of stimulus. If the light intensity in their habitat is suddenly altered, they reverse their direction of motion. Other types of phototactic organisms not only can respond in this way but, more prominently, move in the direction of a light gradient. The movement may be either toward or away from the light source depending on the intensity. This movement is not direct. If, for example, a particular algal cell is observed, its line of locomotion does not adhere closely to the axis of the light gradient. The movement is somewhat erratic and appears to be generated by the tendency of the cell to vary its degree of motion according to light intensity. The net result of a differential rate of locomotion would be directional movement. The trapping of phototactic cells in a particular illuminated area occurs because the random movements of the cells decrease as they move from optimal to suboptimal intensities and increase as they move in the opposite direction.

Although the locomotory response of phototactic microorganisms is different in nature from the conscious responses of mammals, the basic ingredients of receptor and effector must be present. The effector is, of course, the locomotory mechanism which generally consists of cilia or flagella. Some phototactic organisms—blue-green algae, diatoms—do not possess specialized structure for motility; the extent to which movements in these organisms may be explained by postulating an action of "microtubules" or analogous fibrous systems has been discussed in Chapter 16.

Even though most phototactic cells contain chlorophyll, the receptor appears to be some other type of pigment. If photosynthetic activity and phototactic response in purple bacteria are compared with respect to the effectiveness of different wavelengths of light, the resulting "action spectra" do not have the same profiles (see Fig. 9-3). We may infer from such experiments that the pigment responsible for photosynthesis is not the same as the one responsible for the phototactic response. In the green alga, Euglena, blue light is the most effective phototactic stimulus, whereas red light is the strongest promoter of photosynthesis. In general, the amount of pigment necessary to produce a phototactic response is very small, and the pigment is not necessarily localized in a special region of the cell. No evidence has so far been obtained which might point to a localization of phototactic pigment in purple bacteria. However, a demonstration of localization is not easily achieved unless some obvious structure is present. The colored eye spot present in flagellates appears to be the site of photic reception. However, in Euglena this turns out not to be so. The photoreceptor lies in a thickening at the base of the flagellum. Mutants which have lost the thickening but not the eye spot are incapable of phototactic response. Pigment is not evident in the basal thickening, but the amount of light essential to phototaxis is very small, and, theoretically at least, relatively few molecules would be required to perceive the stimulus.

The responses of cells to light stimuli are not restricted to photosynthesis and phototaxis. Leaving aside the phenomenon of vision in which a high degree of cell specialization is involved, the growth patterns of plants in particular indicate that apparently unspecialized cells may respond in specific ways to light stimuli. Some fern spores, for example, germinate into filaments of cells, the direction of filamentous growth being controllable by unilateral illumination (Fig. 18-3). No specialized light absorbing structure has been observed in these cells. Neither has any such structure been observed in higher plants generally even though their growth patterns are much affected by light periodicity. These observations are consistent with what is gradually being learned about the nature of light responses other than those concerned with photosynthesis. Phototaxis, photoperiodism, vision all require very low levels of energy to trip the response mechanism.

The specificity of chemical receptors, like that of photic receptors, may be rationalized on theoretical grounds, but here too experimental evidence has been obtained to support the conclusion. The olfactory sense of animals is, of course, the most impressive demonstration of chemoreceptors. That a male European gypsy moth can respond to about

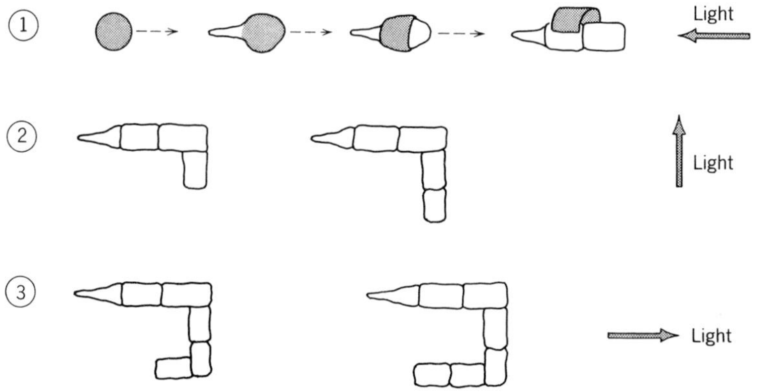

Fig. 18-3. Photoresponses in a germinating fern spore. Diagrammatic representation of a germinating spore of a species of *Dryopteris*. The plane of cell division and direction of growth is determined by the source of light if the spores are germinated in an otherwise dark environment. Different wavelengths of light have different developmental effects. Red light, for example, inhibits division and a spore grown in the presence of only red light produces a long, unicellular filament. In the phenomenon illustrated above, white light was used. The shading represents the extracellular coat ("exine") of the spore; it breaks open on germination and is finally shed. (Courtesy of Y. Hotta.)

10^{-18} moles of the alcoholic substance secreted by the female is a striking example not only of receptor specificity (since other moths do not perceive this compound) but also of receptor sensitivity. Chemoreception has also been studied in unicellular organisms, and although a number of observations have been made on food attractants, some of the most interesting observations have been made on developmental behavior. *Dictyostelium discoideum* is unicellular and amoeboid for one part of its life cycle. The amoebae feed on bacteria, but once the food supply becomes exhausted they begin to aggregate and fuse into a multinucleated mass. Once the mass reaches a certain size, it begins to differentiate a vertical stalk which eventually develops a fruiting body at is apex. A compound named "acrasin" has been isolated from the aggregating amoeba, and it appears to be the chemical agent which directs amoeboid movement. Just how the initial center of aggregation is established is not entirely clear, but once established the gradient of acrasin concentration orients amoeboid movement. The orientation of the stalk is sensitive to CO_2 concentration; although ordinarily it grows perpendicularly, it may be induced to develop at various angles by controlling the gas composition in its immediate environment.

The subject of cell movement has evoked considerable interest in recent years both with respect to chemical attractants as such and also with respect to the behavior of cell populations. That a variety of natural attractants and repellents exists—some species specific, others with a broad range of effectiveness—is clearly evident. Biological responses to chemical factors not only encompass those which diffuse through the environment and effect directional movement, but also those which are localized at cell surfaces and effect selective associations. The aggregation of animal cells is of extreme interest with respect to embryogenesis, although the mechanism is probably different from that described for amoebae. Animal tissues may be dissociated into their component calls by means of the proteolytic enzyme trypsin, which ruptures the connections between them. If the dissociated products of several tissues are mixed and grown together in culture, the cells gradually reassociate in a definite pattern. Cells not only aggregate into homogeneous clusters, but each newly aggregated "tissue" occupies a specific position relative to the others (see Fig. 18-4). Cell movements also occur normally in growing embryos. Thus cell locomotion is not only an adaptation to fluctuations of the inanimate environment, but is also a mechanism utilized in higher organisms for achieving orderly development.

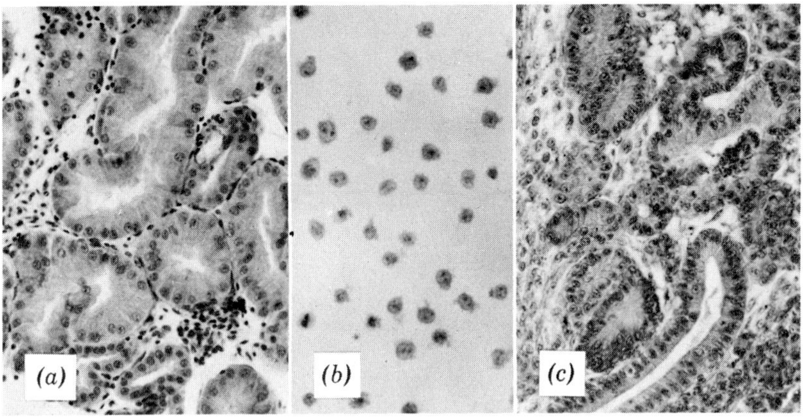

Fig. 18-4. The aggregation of a population of free cells into an organized pattern of association. (*a*) A section through the kidney tissue from a chick embryo. Two types of cells may be seen in the photomicrograph, those forming the tubules and those forming the connective tissue. (*b*) On treatment with trypsin, the cells disaggregate and may be cultured as such *in vitro*. The two types of cells are virtually impossible to distinguish in culture. (*c*) Under appropriate conditions, the cells reaggregate into masses which, as may be seen in the photograph, become organized into typical kidney structures. Thus, the cells reassert themselves by mechanisms that involve certain properties of their surfaces. Just what these mechanisms are remains a mystery to be resolved. (Courtesy of A. A. Moscona.)

The principal gap in knowledge concerning regulation of cell movement lies in the mechanisms which translate the original stimulus into the locomotory response. Membranous systems have been considered to be the agents of stimulus transmission partly because membrane responses have been noted in some cases, and partly because the membrane appears to be the natural vehicle for carrying such signals. Mechanical disturbance of cell membranes, for example, are invariably accompanied by a propagation of changes in electrical potential (Chapter 16). All structurally organized photoreceptor systems are membranous in nature. There is no direct evidence for the localization of chemoreceptors in the cell membrane, although this is generally supposed to be the case. Assuming, however, that membranes are the seats of receptors and are the route by which signals are transmitted, the major problem of explaining how the signal is translated by the effector system remains unsolved. This gap, as pointed to in somewhat different connections in earlier discussions, is a fundamental one in cell biology. The transformation of chemical energy into mechanical work, and the regulatory factors which govern the initiation of such transformation, are aspects of cell behavior which have not yet lent themselves to the kind of clear-cut experiments performed in studies of molecular transformations.

ACTIVE REGULATION: METABOLIC ACCOMMODATION

Regulation of Enzyme Activities

Any molecular transformation, whether associated with energy production, energy utilization, biosynthesis, autolysis, or solute transport, falls under the heading of metabolic activity. All cells, except dormant ones, are continuously active metabolically, but very few, if any, cells have a uniform metabolic pattern over extended intervals of time, even if such intervals are within a single developmental phase of the life cycle. Nutrient intake is periodic in both unicellular and multicellular organisms. Heterotrophic organisms have to find food; autotrophic organisms which utilize light as a source of energy can synthesize their primary foodstuffs only during the light period. Whereas cells of autotrophs have a food supply of more or less constant composition, those of many heterotrophs must not only regulate with respect to periodicity of supply but also with respect to variable composition. The substrate molecules on which the metabolic apparatus of a cell acts are therefore variable both in concentration and composition, the range of variability depending on the nature of the organism and of the environment in which it functions.

Metabolic patterns have already been adequately discussed in preceding sections, and it should therefore be sufficient to point out that such patterns may be related to two fundamental mechanisms—enzyme catalysis and genetic coding. To the extent that cells adjust their metabolic activities in response to environmental fluctuations without recourse to gene-mediated syntheses, the adjustment may be called a regulation of enzyme activities. Such regulatory behavior does not require the synthesis of new protein; if it did, code translation would be necessary and gene action would be directly involved. This sharp distinction between regulation of the products of gene transcription and regulation of the transcription process itself will be retained through the remainder of the chapter. Nevertheless, the distinction may not be as sharp as described. That RNA messages must be transcribed from the genome is taken for granted, but in many cell types such messages may persist for much longer periods of time than is common in bacteria. When messages are stable, protein may be synthesized without direct participation of the genome. Indeed, in the development of the enucleate red blood cell hemoglobin is synthesized by virtue of the RNA messages which persist after nuclear disappearance. The role of stable RNA messages in the behavior of cells requires, however, much more clarification. In this discussion protein synthesis, whether occurring via stable or unstable messages, will be considered as a facet of gene regulation.

All physical and chemical parameters of the environment affect cell metabolism to a greater or lesser degree. Of these, the passive response of cells to changes in temperature and water availability have already been discussed; the remainder may be grouped under one heading—fluctuations in the intracellular pool of small molecules. Simple cases of fluctuation are those which arise from the periodic availability of nutrients or the periodic consumption of metabolites in irregular cellular activities such as locomotion, contraction, restoration of membrane potentials, and secretion. More complex cases are those which arise in multicellular organisms due to intercellular reactions. Specialized metabolic products from cells of one tissue type are absorbed by cells of other types; hormones, in particular those which are produced by specialized cell groups, find their way to a variety of tissues and affect their metabolism in varying degrees; in animals nerve signals effect rapid and often profound metabolic changes in receptor tissues. Regardless of the causal agent, the immediate effect is an alteration in the metabolic pool of the cell. To specify the kind of alteration produced by each of the causal agents is at the present time impossible. This is especially true for the translation of nerve messages. However, even if we omit the subtle mechanisms which provoke metabolic change, a broad and general source of metabolic change is the fluctuation of molecular components which are neither part of the structural apparatus nor of the enzymatic complex of the cell.

Even "simple" fluctuations, such as the availability of carbohydrate, are by no means simple in their consequences. For inasmuch as the products of carbohydrate metabolism are partitioned among various pathways, each of these pathways is affected by carbohydrate availability. If enzymes catalyzed substrate transformations solely in response to substrate availability, metabolic coordination would be impossible. If, for example, the cell metabolized all the carbohydrate available to it at a particular instant into CO_2 and water, it would theoretically have only ATP available during the period when a source of nutrient was unavailable. If such a cell were synthesizing any type of carbon compounds, all activity would come to a halt until a new source of nutrient presented itself. This, of course, does not occur. Cells generally store surplus energy as polysaccharides, fats, or somtimes protein. Cellular enzymes do not therefore indiscriminantly metabolize available substrates. In some way the enzymatic activities are regulated with respect to the fluctuations in concentration of a particular substrate and with respect to the partitioning of metabolic products along required metabolic pathways. Such partitioning becomes especially prominent in multicellular organisms, where cellular responses to nervous or chemical messages result in the induction of specific biosynthetic activities.

The impulse to metabolic adjustment is thus either a molecule which is itself a substrate for enzyme activity or a molecule, such as a hormone, which influences the activity of the metabolic complex. Several facts are known about the nature of cellular responses to substrates but relatively few are know about responses to hormones. One of the major questions which must be answered in order to clarify mechanisms of response is the nature of the receptor. If a compound affects cellular metabolism, either because it has undergone a change in concentration or because it is a novel component of the cellular pool, some constituent of the cell must recognize it, and this constituent in turn must be able to effect a response either directly or by transmitting a message to some other cell constituent. For hormones no generalization can yet be made about the nature of receptor molecules. But for metabolic substrates the general belief is that the enzyme itself is the receptor. This belief is based to some extent on negative evidence—that no other group of cell components is known which displays a highly specific affinity for substrate molecules. Nucleic acids are not known to interact directly with substrate molecules, and although certain antibiotics are known to combine in the cell with RNA or DNA, such combinations have a limited specificity relating to the configuration of the nucleic acid molecules as a whole rather than to specific cistrons. On the other hand, the evidence for specific complexes between enzymes and substrates is so overwhelming that, for the present at least, they furnish the only category of cell component which can be presumed to sense the fluctuations within the intracellular pool of metabolites.

Responses to Substrate Concentration. The most direct type of metabolic response to fluctuations of the intracellular pool of metabolites is that between the enzyme and the substrate it transforms. The general relationship between the two is illustrated in Fig. 18-5.1. For any given concentration of enzyme, there is a maximum velocity of substrate transformation beyond which increases in substrate con-

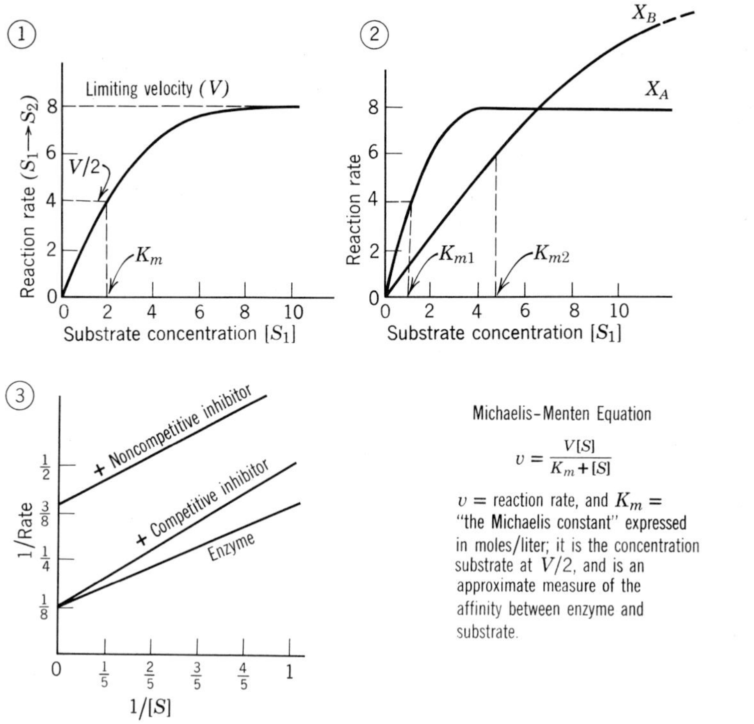

Fig. 18-5. Regulatory capacities of enzymes with respect to changes in substrate concentration. Graphs 1 and 2 are typical plots of the rates of enzyme activity versus substrate concentration. The plots can be derived theoretically by assuming that the enzyme forms a complex with its substrates, and that the equation governing the formation follows the law of mass action. Michaelis and Menten formulated the relationship in 1913. Graph 3 is a plot of the reciprocals of reaction rates and substrate concentration, and is frequently used to distinguish between competitive and noncompetitive inhibition. The bearing of these relationships on regulation is discussed in the text.

centration have no effect. Cells may therefore limit the amount of nutrient metabolized by the capacity of their enzyme system. Furthermore, if one enzyme in a sequence of reactions (e.g., hexokinase in glycolysis) has a maximum velocity below that of the other enzymes, it effectively controls the entire sequence of reactions by virtue of being the rate-limiting step. This relationship is modified in those cases where the reaction catalyzed is readily reversible. The hexokinase reaction, which is not readily reversible, follows the curve shown in Fig. 18-5.1; two substrates are involved, glucose and ATP, and their concentrations are the main determinants of enzymatic rate. The phosphorylase reaction (see Chapter 6), on the other hand, is readily reversible and in this case both the rate and direction of reaction mirror the fluctuations in the metabolic pool. A high concentration of inorganic phosphate within the cell promotes the conversion of starch or glycogen to glucose phosphate; high concentrations of glucose and ATP oppose polysaccharide conversion. As pointed out earlier (Chapter 6), cells do not depend on this equilibrium for polysaccharide synthesis. They use a different enzymatic system for such synthesis, and presumably the phosphorylase equilibrium largely controls the breakdown of the stored form of energy. Reversibility of reactions, quite apart from its significance in thermodynamic efficiency, ties the rate and direction of enzyme activity to other metabolic activities within the cell. As long as other enzymes remove the product, the reversible reaction cannot reach equilibrium. A good example of this is the phosphoglyceraldehyde dehydrogenase step in glycolysis. Commonly, the reaction proceeds in the direction of aldehyde oxidation, but in chloroplasts, where phosphoglyceric acid, ATP, and NADPH accumulate as a consequence of photosynthesis, the reaction proceeds in the direction of phosphoglyceric acid reduction.

The limiting velocity of an enzyme reaction is basically a function of the affinity between enzyme and substrate. The lower the K_m value of an enzyme, the greater the affinity of the enzyme for its substrate (Fig. 18-5.2). Thus if two enzymes have an affinity for the same substrate molecule, the extent to which that substrate is transformed along each of the alternate pathways is a function both of the K_m values and substrate concentration. If one of the enzymes has a much lower K_m value than the other, it will pull most of the reaction in its direction at very low substrate concentrations. As substrate concentration increases, a point is reached where the two reactions proceed at equal rates, and beyond that concentration, the enzyme with the lower affinity (or higher K_m) will pull an increasing proportion of the reaction in its direction. This model has been frequently used to explain the partitioning of ADP in energetic reactions. The compound is essential as a phosphate acceptor for the oxidative step in glycolysis and for the sequence of oxidations in mitochondria. Although cells may synthesize ADP *de novo*, this synthesis is a minor factor in determining the intracellular ADP level except during the growth period of the cell. The main determinant is the extent to which ATP is utilized in other reactions. Sudden performance of mechanical work, muscle contraction for example, causes a correspondingly rapid shift in the ADP/ATP ratio. *In vitro* studies of the kinetics of glycolysis and mitochondrial oxidation indicate that the mitochondrial system responds much more readily to low concentrations of ADP than does the glycolytic system. In the complete absence of ADP neither phosphoglyceraldehyde oxidation or mitochondrial oxidation occurs. Mitochondrial oxidation rises very rapidly when very low amounts of ADP are supplied whereas phosphoglyceraldehyde oxidation does not. The respective responses of the two systems follow the diagram in Fig. 18-5.2 where X_A represents the mitochondrial system and where S_1 is the ADP concentration. Thus in a cell where the two systems coexist, as is commonly the case, a marked depletion of the ADP pool preferentially induces a high rate of mitochondrial oxidation. Once the mitochondrial system reaches a limiting velocity, glycolysis continues to increase its utilization of ADP if the concentration of that compound also increases. Because many metabolic pathways utilize ATP, the details of the dynamics of ATP formation and utilization are, for the present at least, beyond our powers of description.

Responses to Inactivating and Activating Factors. During the period immediately following the demonstration that enzymes are protein molecules, most studies were directed at elucidating the nature of enzyme specificity. The concept of the active site was eventually developed to account for the fact that the protein molecule is much too big to interact in its entirety with the substrate. The relatively small dimensions of the substrate molecule require only a small region of the protein chain to effect complete interaction, and this region was termed the "active site." An important question lingering in the minds of many biologists was the functional significance of the molecular size of enzymes in view of the limited size of the active catalytic area. That size itself is an insufficient explanation may be surmised because amino acid sequences throughout the polypeptide chain are distinctive for different enzymes. Moreover, parts of the polypeptide chains of some enzymes have been removed experimentally without affecting catalytic activity. Thus the region of the protein molecule which recognizes its specific substrate and responds to it in the ways discussed accounts only partially for the specific configuration of each enzyme molecule. The problem, then, is to account for the high degree of specificity in regions other than the active site of enzymes. This problem is related to the observations made in recent years that enzymes recognize not only the substrate molecules whose transformation they catalyze but also other molecules which are not transformed by the activities of the enzyme but affect them.

We may make the broad generalization that enzymes carry information not only about their catalytic substrates but also about other molecules in the cell. The evidence for this is still fragmentary, but it is unambiguous. Although we cannot yet properly evaluate the role of enzymes in metabolic regulation, there is no doubt that enzymes can respond to factors in the metabolic pool which are not evident in the equations describing their chemical reactions. At present we may point to at least three responses of enzymes other than those already discussed which indicate an extensive capacity for metabolic accommodation: natural inhibitors; end product inhibition; and hormonal activation.

NATURAL INHIBITORS. Specific inhibitors were once extensively sought in the hope that their presence would explain how cells turned enzymatic activity on and off. The search was largely unsuccessful in that most enzymes do not appear to have specialized inhibitors. Some inhibitors of this type have been presumed to exist because extracts from certain tissues have occasionally been found to inhibit the activities of certain enzymes. Since such inhibitors have not been fully characterized, they may fall under one of the other two categories. Various antibiotics, however, have specific affinities for certain enzyme systems, but these cannot be considered as components of cellular regulatory mechanisms. General interest in inhibitory mechanisms has nevertheless provided basic information about the characteristics of enzyme molecules, and this information has made possible a clearer understanding of the regulatory mechanisms which will be discussed below.

Kinetic studies of enzyme inhibition revealed two types of inhibitory action, competitive and noncompetitive. Plots typical for each of these types are shown in Fig. 18-5.3. In competitive inhibition the activity of an enzyme in the presence of a fixed concentration of inhibitor approaches that of the enzyme in the absence of inhibitor as substrate concentration is increased. With a noncompetitive inhibitor convergence does not occur. This difference is interpreted as being due to the fact that competitive inhibitors compete with the normal substrate for the active site of the enzyme, whereas noncompetitive inhibitors combine with some other portion of the enzyme molecule. This interpretation is strongly supported by the fact that competitive inhibitors are generally similar in their molecular configuration to the normal substrates and would therefore be expected to have a chemical affinity for the active site. The term "competitive" has been chosen to emphasize this point; inhibitor and substrate compete for the active site, and as the relative concentration of substrate is increased, so is the proportion of enzyme molecules which complexes with it. Just how combination at a site other than the active one inhibits enzyme activity is not yet generally understood. One well established example of noncompetitive inhibition is that of sulfhydryl group oxidation. Many enzymes contain one or more molecules of cysteine in their polypeptide chain, which must be in the reduced state for the enzyme to function. If the —SH groups of such cysteine residues are oxidized, the enzyme becomes inactive. In classical studies of this reaction the naturally occurring tripeptide, glutathione (which itself contains an —SH group), was used in the reduced form as an activator of the enzyme and in the oxidized form as an inhibitor. Since many enzymes in the pathway of carbohydrate metabolism have —SH groups, it was believed that oxidative processes in the cell could be regulated by glutathione which was itself subject to fluctuations in oxidation-reduction conditions. In some instances this may indeed be the case, but the more recent observations on enzyme inhibition have revealed much more subtle regulatory mechanisms with far more specific attributes.

END PRODUCT INHIBITION. Virtually all compounds synthesized in a cell are products of a sequence of reactions. In recent years the discovery has been made that in a number of such sequences, the end product has a specific inhibitory effect on the first enzyme in the reaction sequence. Several sequences of this type are illustrated in Fig. 18-6. As more phenomena of this type have been sought, more have been found. Although relatively few reaction sequences have been fully examined, this form of metabolic control is widespread. Two features of end product inhibition are of fundamental interest. The first is that the inhibitory substance is different in its chemical configuration from the normal substrate. The combination between end product and first enzyme in the sequence is of the noncompetitive type, although the kinetics of some reactions follow a course which is neither competitive nor noncompetitive. Some enzymes must therefore possess at least two recognition sites—one for the substrate molecule, the other for the inhibitory metabolite. The second feature of interest is the physiological relationship between end product and enzyme. The control is of a feedback type so that a cell may regulate a metabolic pathway not only by the availability of substrate, as was the case in direct substrate-enzyme responses, but also by the level of the product which functions as such in the cell (a coenzyme, for example) or leads into other synthetic channels. This form of regulation is effective both with respect to extracellular and intracellular fluctuations in metabolities. Moreover, end product inhibition is closely tied to gene regulation, a topic which will be considered later.

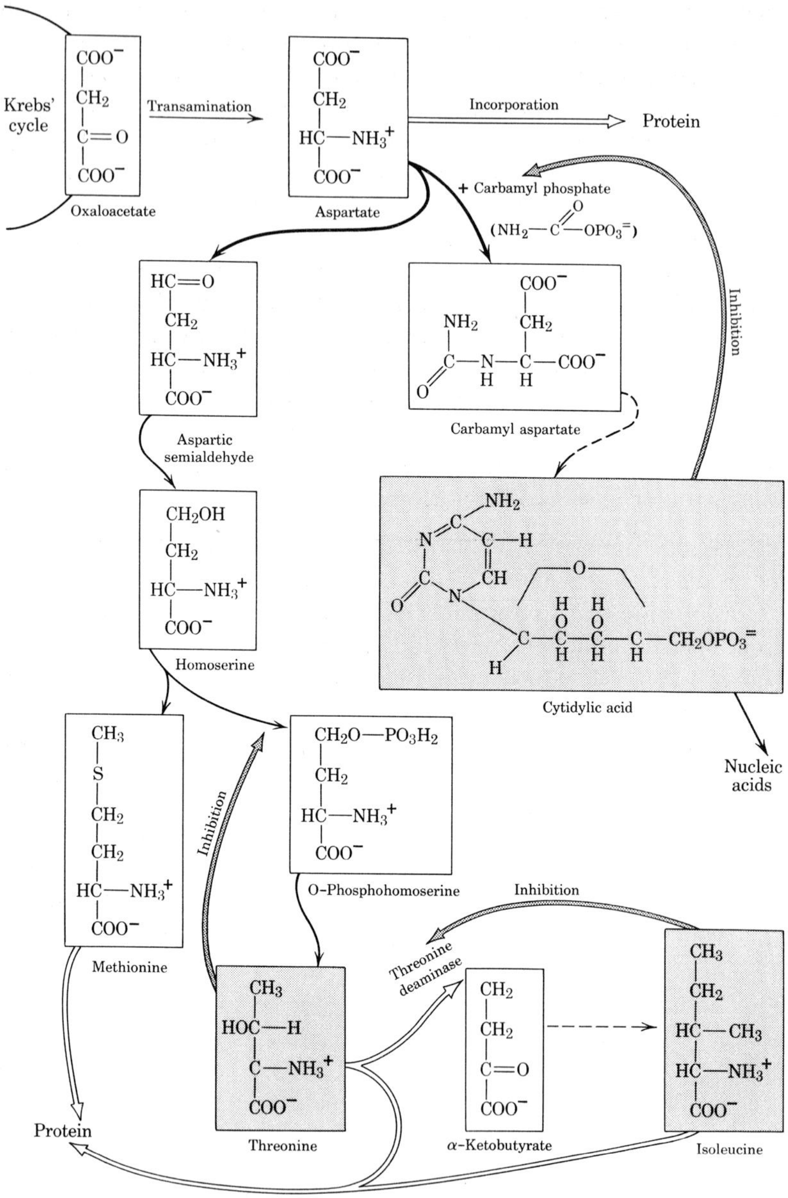

Fig. 18-6. End-product inhibition of enzyme activity. Regulation of amino acid and pyrimidine synthesis. Several interlocking reactions are shown here in order to emphasize the interrelationships of end-product inhibitions. In this set of reactions the primary source of carbon chains is aspartic acid, which is derived from the Krebs' cycle. Note that aspartate itself is required for protein formation

HORMONAL ACTIVATION. Given the capacity of enzymes to recognize molecules other than their own substrates, broad possibilities exist for phenotypic regulation of metabolism. If enzymes evolve which can respond to metabolites formed at the end of a reaction sequence, we may reasonably suppose that some evolve which can respond to molecules outside the reaction sequence. The essential requirement for either type of response is some region in the enzyme molecule which specifically interacts with a substance chemically distinct from the enzymatic substrate and which, on interaction, alters the catalytic activity of the enzyme. Studies of the effects of hormones on enzyme behavior are still in their infancy but, even so, there is little doubt that hormones may have profound effects on enzymes. Some of these effects fall into the category of gene regulation because, directly or indirectly, hormones induce the synthesis of new enzyme. Other effects are at the enzyme level, and one example of such effects will be discussed here.

The characteristics of the enzyme phosphorylase, which converts glycogen into glucose phosphate, have already been described. This particular enzyme has evoked considerable interest because it controls the flow of stored energy into the metabolic channels of the cell. Certain diseases of the human liver are in fact traceable to malfunctioning of this enzyme, for liver cells accumulate relatively large deposits of glycogen following feeding and use the reserves as a source of energy for their various organismic functions. That liver cells do not continuously degrade glycogen was known from physiological studies as was also the fact that degradation was in some way subject to hormonal control. In the 1940's biochemical studies of the enzyme led to the discovery that it could exist in an active and an inactive form; the two forms were respectively designated A and B. Since then the nature of the difference between the two forms has been clarified. The active form of the enzyme consists of two protein molecules of equal molecular weight. When the protein subunits are dissociated, the enzyme is no longer active. The linking of the subunits is under hormonal influence.

A detailed picture of the structure of the active enzyme is still unavailable, but one important factor in subunit linkage appears to be the amino acid, serine ($CH_2OHCHNH_2COOH$). The R group of this amino acid serves to form a chemical bridge between the two

---

and that the inhibitory end products act at points leading to alternate pathways of metabolism. Many other examples could be provided and among these are cases in which it has been established that the inhibitory end product also represses enzyme synthesis.

protein molecules. There are several serine residues in each of the protein chains, but how many of these are involved in the bridge formation is not yet clear. However, the bonding between the chains appears to be due to the fact that the hydroxyl groups of the R portions are phosphorylated, the phosphate groups forming the chemical bridge between the hydroxyls of the serine residues. An enzyme capable of phosphorylating the serine residues of phosphorylase in the presence of ATP has been identified in liver cells. The same enzyme can also hydrolyze the phosphate linkages in the absence of ATP. By phosphorylating the serine residues, this enzyme activates phosphorylase; conversely, by dephosphorylating the serine residues, phosphorylase is inactivated simultaneously with its disaggregation into two subunits. The enzyme which phosphorylates serine residues is subject to hormonal activation. Hormonal control of glycogen breakdown is thus effected via an enzyme which can activate phosphorylase by catalyzing the union of inactive subunits.

This example not only illustrates a mechanism by which a hormone may control the activity of an enzyme, but also reveals a facet of enzyme behavior not hitherto considered. Several enzymes have now been found which consist of two or more protein subunits. Whether all such subunits are catalytically inactive in the disaggregated form is not certain; the subunits of some enzymes have been found to retain catalytic activity. Many of the compound enzymes, however, do become activated on aggregation. Some consist of identical subunits; others consist of different ones. In some of the cases studied different forms of the same type of enzyme may arise by various combinations of the different subunits. Such "isoenzymes" catalyze the same substrate transformation, but each may have certain distinctive kinetic properties. As more is learned about the nature of enzyme behavior, the early picture of the enzyme as a protein molecule with one specific site is replaced by a more complex one in which the main feature is one of functional flexibility. The flexibility is essentially a property of adaptiveness to variable intracellular metabolic conditions. At any given developmental stage of the cell, such variability may generally be traced to environmental fluctuations. But whether or not these are the ultimate generators of enzyme response, the ability of enzymes to sense different aspects of their chemical environment provides cells with an extensive mechanism for regulating metabolic activities.

Summary. The principal theme of this discussion has been the communication of a cell with its environment. Such communication is a system in which the cell perceives environmental changes and responds to these by a set of reactions which lead to novel physical or metabolic activities. The biological value of the response lies in the effectiveness with which it enables a cell to function optimally in an inconstant environment. Cells may or may not sense the different parameters of their environment by specific receptors. Temperature and water are so ubiquitous in their influence on intracellular functions that even if specific sensing sites did exist, other portions of the cell would still be directly affected by fluctuations in these factors. By contrast, light and chemicals cannot be recognized except by specific receptors. The nature of the response mechanism is in part determined by whether stimuli are sensed by virtually all components of a cell or by a specific receptor body. If the stimulus is sensed generally, as with water availability or temperature change, the character of the response mechanism is difficult to trace. There is no doubt that certain types of cells are adapted to function under conditions of varying temperatures, but just how the different components of a cell respond to such variations is unresolved. At present, the only reasonable generalization is that different metabolic processes do not respond identically to temperature shifts and that the continued function of a cell at elevated or depressed temperatures is not necessarily determined by the continued function of specific enzymes. A likely cause of functional failure at lethal temperatures is a loss of coordination in metabolic activities. Extracted enzymes from mammalian cells which can function only within a very narrow temperature range and those from plant cells which can function over a broad temperature range, do not show corresponding differences in sensitivity to temperature changes.

With respect to stimuli which are sensed by specific receptors, we may identify two principal types of intracellular communication. Impulses transmitted through membranes carry yes-no information, and the specificity of regulation depends on a specific receptor or effector or both. The chief characteristic of such communication is its rapidity and low energy of stimulus. Communication through molecules is much less rapid, but the specificity is very high. The distinctive values of each of these communication systems is best revealed in the intricate intercellular relationships of higher organisms.

The Regulation of Gene Action

Regulation of enzyme activities has one outstanding limitation with respect to accommodating cells to environmental changes. Such regulation can modulate the activities of existing enzymes, but it cannot provide novel ones. Conceivably, a cell could synthesize all the enzymes encoded in its genetic apparatus, and regulate their activities by the mechanisms already discussed. The facts, however, are otherwise. Cells do not individually express their full genetic potential. This is evident in comparing the enzymatic activities of different cell types in a multicellular organism; it is evident in comparing the enzymatic activities of the same cell at different stages of its development cycle; and it is evident in the responses of unicellular organisms to novel metabolic conditions. The plasticity of cell behavior is due not only to a capacity for controlling the activities of existing enzymes, but also to a capacity for determining if and when a particular enzyme is to be synthesized. Why cells should thus partition their sources of plasticity is not entirely clear. One factor may be energy expenditure. A cell which synthesized all possible proteins would be investing much of its energy in unproductive channels. We cannot presume, however, that such an energy drain would necessarily be greater than the investment a cell must make in providing regulatory mechanisms for gene action. As discussed in Chapter 11, too little is known about the quantitative aspects of intracellular ordering to draw any meaningful conclusions about energy investments in ordering processes. Perhaps the only reasonable statement one may make about the two categories of regulation is that cells not only control the synthesis of specific proteins in an all-or-nothing sense, but also in a quantitative sense. Specific cell types have characteristic complements of proteins both qualitatively and quantitatively. The mechanisms by which cells govern the amount of proteins made have much in common with those which govern whether or not a particular protein is to be made.

With this preamble out of the way we may now proceed to an analysis of some control systems. Ever since techniques were developed for assaying the amounts of particular enzymes in microorganisms, numerous studies have demonstrated that the amounts vary from cell to cell and that the variations are correlated with the conditions of growth. Some enzymes show relatively large variations, even at times becoming undetectable, but others show only slight and insignificant variations. These differences were striking enough to lead to a classification of enzymes into two major groups—constitutive enzymes, those produced under all known conditions, and inducible enzymes, those produced only under certain prescribed conditions. More recent studies show, however, that the classification is superficial; the behavior is not a special characteristic of the enzyme itself, but of the control systems; precisely the same enzyme may be inducible in one strain and constitutive in another. In any case the demonstration of quantitative variations in enzymes under controlled conditions provided the opportunity for studying the regulation of protein synthesis.

Many of the inducible enzymes studied were those involved in sugar utilization. Generally the enzymes were produced when, and only when their substrates were present. This observation established clearly the fact of regulation and also demonstrated its adaptive character. To probe the properties of the inducer, cells were often presented with an array of compounds similar to the substrate, some of which were never to be found in the natural environment and which could not be attacked by the enzyme. Certain analogs of the normal substrate were just as efficient in inducing enzyme formation as the substrate. The analog studies demonstrated that the signals to the enzyme-forming system did not require the entire substrate molecule, small though it was, but were effectively transmitted by specific atomic groupings of even smaller size. These studies did not, however, provide much information about the system receiving the signals and new approaches had to be developed.

Among the more significent of these later studies were those of J. Monod, F. Jacob and their collaborators on the bacterium *Escherichia coli.* They undertook a comprehensive analysis of the genetics and biochemistry of the enzyme systems responsible for the utilization of lactose (the beta-galactosidase system). In this analysis they assembled a large variety of mutant strains which showed significant differences in their behavior. One group of mutants differed from the standard type in that beta-galactosidase was constantly produced at a high rate, regardless of the presence or absence of the substrate; the strains containing the mutant gene were no longer inducible, but constitutive. Genetic analysis provided two significant facts concerning these mutants. First, they were found in crossing experiments to be located in the chromosome at a locus at some distance from the gene specifying the beta-galactosidase enzyme. Second, although *E. coli* is haploid through most of its life cycle, transitory heterozygotes may be established and the dominance relations of genes can be explored. Cells which are heterozygous for mutants of this class are found to behave like wild-type cells, that is, they are inducible. This observation demonstrates that the normal allele at the locus is dominant and suggests that it is active in preventing beta-galactosidase synthesis when the inducer is absent. The mutant alleles are interpreted as defective genes which are no longer capable of repressing enzyme synthesis. The locus of these mutants is termed the *R* (or regulator) locus and the agent it produces is called a *repressor.*

Not all constitutive mutants behave in the manner described above. Another class is found in breeding experiments to be localized very close to the gene for beta-galactosidase production; indeed, the mutants are clustered abruptly against one end of the structural gene. These mutants also differ from the other constitutive mutants in that they are completely dominant; normal genetic material for this region elsewhere in the cell does not convert the cell into an inducible enzyme former. The mutants have lost the capacity to be repressed; the repressor produced by the *R* locus has no effect on enzyme synthesis. The chromosome region defined by these studies is designated as the *O* (or operator) region and is interpreted as the region of the chromosome which ordinarily serves as the target for the repressor. (Other mutants with somewhat different properties have been described at both the *R* and the *O* regions, but we may neglect these in the generalized account.)

The system of enzyme regulation suggested by these studies may be summarized as follows (Fig. 18-7). Two main genetic regions are involved, the region of the regulator gene and the region of the structural gene (including the *O* segment). Furthermore, two kinds of regulator substances are indicated, the low molecular weight inducer and the product (probably a macromolecule) of the *R* locus designated as the repressor. The chemical nature of the repressor has not yet

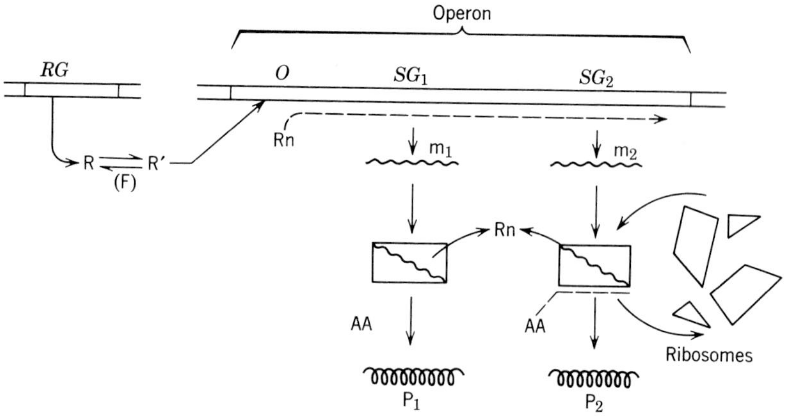

Fig. 18-7. The Jacob-Monod model for the regulation of enzyme synthesis. *RG* = regulator gene; *R* = repressor converted to R^1 in presence of effector *F* (inducing a repressing metabolite); *O* = operator gene; SG_1 and SG_2 = structural genes; rn = ribonucleotides, m_1 and m_2, messengers made by SG_1 and SG_2; aa = amino acids; P_1P_2 = proteins made under control of SG_1 and SG_2 via specific messenger RNA. (From F. Jacob and J. Monod, *Cold Spring Harbor Symp. Quant. Biol.*, **26** (1961).)

been established with certainty. Moreover, the interactions of the various elements have not been completely elucidated, but plausible schemes are available. One of these considers that the repressor attaches to the *O* region, and that this union prevents the transcription of the structural gene. The inducer's action is rationalized as a union with the repressor which modifies its properties sufficiently to prevent interference at the *O* region.

To return for a moment to the question of economics, the savings accomplished by this system of regulation are not entirely obvious. If the repressor substance is a protein, as some studies suggest, the regulated system consists of two genetic regions and two protein products, one of which is produced continuously and the other only in the presence of the inducer. In contrast, a hypothetical unregulated system could consist of one genetic region (the structural gene) and one protein product which is produced continuously. The latter case would appear economically preferable and much simpler. An advantage of the regulated system could be imagined, however, if the amount of protein produced by the regulator gene is very low.

This system must, however, be viewed in a larger context. Regulator genes have now been discovered for a number of inducible systems, and many of these are compound systems. As an example we may consider the system for arginine biosynthesis in *E. coli*. Seven biochemical steps catalyzed by separate enzymes have been identified. Ordinarily all the enzymes are produced and arginine is produced from glutamic acid. If, however, the cells are supplied with an exogenous source of arginine, the enzymes are no longer produced. Note that the addition of the end product of the sequence acts in much the same manner as the substrate in the previous example, except that the direction is different; the end product blocks enzyme synthesis while the substrate releases synthesis. In both cases, however, a low molecular weight compound signals a change in the synthetic machinery of the cell. Genetic analysis of the arginine system shows that the structural genes for the various enzymes are scattered throughout the genetic map, even though they behave more or less as a unit in repression and de-repression. Their coordinated behavior is related again to a regulator gene located at still another position. Apparently this entire complex of structural genes is under the control of a single regulator gene and a single repressor substance, and presumably each of the structural genes has an operator region capable of combining with the repressor. The regulation of arginine synthesis is thus very similar to that for lactose utilization except that the system is more compound, and presumably more economical.

Not all compound regulatory systems are as dispersed as that for arginine synthesis. The genes for certain biosynthetic pathways (those for tryptophane and histidine synthesis in Salmonella, for example) are tightly clustered on the genetic map and, interestingly enough, are arranged in a sequence paralleling the biochemical steps. Although information is not yet available on many such sequences, genes arrayed in a group and related by a common biological function may behave as a unit in repression and have a single operator region at one end. Such a unit of repression (or a unit of genetic transcription) would be called an "operon."

Other patterns of organization of regulatory and structural elements have been established or suggested, and other principles of regulation may be expected to emerge. But our purpose is not to provide a complete survey of this field but only an introduction. We should return now to an attribute of the beta-galactosidase system which we ignored previously. The structural gene for beta-galactosidase has associated with it another element whose function is clear, but whose mode of action is still problematical. This is a gene associated with the mechanism (permease system) responsible for the accumulation of lactose inside the cell. Cells with the normal allele concentrate lactose to a level many times higher than that in the surrounding medium, but mutant alleles have been identified and the cells bearing them have no higher concentrations inside the cell than outside. The permease gene is located beside the structural gene for beta-galactosidase, and is controlled by the same operator. The beta-galactosidase gene and the permease gene constitute, with their operator, an operon.

The functioning of the permease system provides an added dimension to the regulatory system. If an inducible cell is exposed to a critical concentration of its inducer, both the beta-galactosidase and the permease genes begin to act. With permease activity the intracellular concentration of the inducer rises far above the critical level. Even if the external concentration of inducer is allowed to fall below the critical level, the internal level remains high and the cell remains induced. Only if the external level falls so low that even an active permease system is insufficient to maintain high internal levels, does repression set in.

These facts permit the following experiment. A population of cells can be divided into two groups, one of which is exposed to a critical level of inducer long enough to derepress the lactose operon. These cells are then washed and placed with the unexposed cells in a subcritical concentration of the inducer. Under these conditions the preinduced cells continue to concentrate inducer and to produce beta-galactosidase; in contrast, the unexposed cells are incapable of de-repression and produce none of the enzyme. These differences between the two groups of cells persist indefinitely under the proper conditions. The differences are fully hereditary by the criterion of dilution; the differences persist not just for 40 to 50 cell divisions but for hundreds. Yet the differences do not reflect fundamental differences in the cells' capacities, as may be readily demonstrated by either raising or lowering the concentration of inducer. The significant point here is that regulatory systems have, in some cases at least, a capacity to maintain their settings with a considerable degree of stability. The demonstration of persistent differences may be sufficient to establish the hereditary nature of the differences (if that is the basis of the definition), but it may not be sufficient to establish a difference in the primary genetic material.

The example of hereditary variations persisting in the face of a constant environment and a common genome is not the only one available. One other may be cited. If a culture of the protozoan *Paramecium aurelia* is injected into a rabbit, the rabbit responds by forming antibodies against the protozoan. The antibodies have the ability to immobilize cells of the kind against which they have been developed, primarily by causing the cilia to stick together. If paramecia are immobilized by sublethal concentrations of their specific antibodies, they will recover after a period of time and begin swimming actively again. These cells are not only insensitive to the antiserum which immobilized them, but all their progeny over a period of many cell divisions are also insensitive to the antiserum. An hereditary alteration has been induced by a single exposure to a specific agent; since the alteration may occur in all the cells of a population without killing any of them, selection is not involved.

Although the molecular basis for the antigenic transformations has not been solved, sufficient information is available to demonstrate that regulatory systems similar in principle to those discussed above are probably involved. If a single "transformed" cell from a treated culture is isolated and permitted to give rise to a population of cells, all these cells will be resistant to the original antiserum; if a portion of these cells is injected into another rabbit, a new antiserum is developed which will immobilize the derived strain but not the original strain. Thus the transformation does not involve a general development of resistance to antibodies, but rather the formation of alternative antigenic determinants. We may designate the critical antigens in the original culture as type *A* determinants, those in the derived culture as type *B*. Actually, the treatment of an *A* culture with anti-*A* serum may yield variants of several sorts; some will change to type *B*, some to type *C*, others to type *D*, *E*, or *F*, each of the types being distinguished from the others by specific antisera. The array of antigenic types is not, however, unlimited. A particular strain treated in this way will yield a characteristic and limited array of cellular types; other strains will produce other arrays which may overlap to some extent. Of considerable significance in evaluating these changes is that they are all reversible. That is, a type *B* derivative of a type *A* culture is again capable of producing type *A* cells, as well as the rest of the array under suitable conditions. In a sense, the array of antigenic types is an hereditary property of a strain; a particular cell produces only one of the antigens, but it maintains the capacity to produce the other types. We must again distinguish between the inheritance of a capacity and the inheritance of an expression.

Further studies provide insight into both types of hereditary mechanisms, but it is sufficient for our purposes to indicate that each type in the antigenic array of a strain is represented in the genetic material by a specific gene; the genes behave in perfectly regular ways in breeding analyses. Each antigenic type also produces a different protein component, so that changes in antigenic type may be interpreted as changes in the activities of a set of genes leading to changes in protein synthesis. The genes governing the antigenic properties are locked in a system of mutual repression; when one of the genes is expressed, the others are repressed. Quite possibly some product of each gene is capable of repressing all the others. No cell may simultaneously express all its potentialities; only a portion of the genetic material is employed at any one time.

In the face of observations such as these, discussions about the economics of regulatory systems are superfluous. Cells clearly do have the capacity to "turn on" and "turn off" specific syntheses, and hence we must conclude that regulatory systems are economically sound. Since various cellular states may persist in the same environment, we must conclude that many regulatory systems possess a stability which might be useful in maintaining a phenotypically heterogeneous population of microorganisms, and which would be essential for achieving cellular differentiation in a multicellular form.

We should be careful, however, in assuming that all changes in the properties of cells reflect changes in the activity of the genetic material. Certain hereditary differences between cells might be maintained not only against a common genetic background, but in the face of an identical pattern of activity of the genes. Consider, for example, a study of hereditary variation in Difflugia. This is a sarcodinid protozoan that characteristically builds a shell out of sand grains which it cements together with a cellular secretion. The shell is roughly globular with an opening in the bottom through which the pseudopodia are protruded. Into the opening project parts of the shell in the form of "teeth." The number of teeth varies in different individuals, ranging from about 6 to about 20. However, when an individual divides, the two cells produced have the same number of teeth. And a clone of individuals produced by successive divisions from a common progenitor are very similar in this respect. The number of teeth in the shell appears to be a reasonably good hereditary trait.

The physical basis of the hereditary differences is not conventional, however. When a cell divides, the naked daughter cell emerges from the original shell, and while still in contact with its sister begins to secrete a shell of its own. In the region of the original opening the new shell is formed in the interstices of the old; the gaps in the original shell provide a model for the teeth of the new. Since the original structure is radially symmetrical, the same number of teeth is found in each successive shell which is formed. The mode of shell formation suggests that some details of shell architecture are determined by the parental shell and not by the parental genes. This suggestion is strongly reinforced by the observation that deliberate modifications of the parental shell, by removing teeth, result in a modification in the progeny and new hereditary characteristics.

In this particular case a template very different from a nucleic template is responsible for transferring hereditary information. In one case the template is essentially one-dimensional; in the other case it is three-dimensional. Yet in both cases specificity is transferred, and the traits involved are hereditary. We do not know to what extent two-dimensional and three-dimensional templates may be involved in the perpetuation of the architectural details of membranes and of various subcellular organelles. But we should be prepared for new mechanisms of regulation and should not assume that our present knowledge is complete.

19

Mechanisms of Regulation: Developmental

Development is an expression of the irreversible flow of biological events along the axis of time. Whether the synthesis of a protein molecule, the transformation of a proplastid into an chloroplast, the division of a cell into two daughter cells, or the divergence of a cell from its parental type—each event has a history and a succession. In one very important sense, all cells undergo "differentiation." A newly formed daughter cell is different from its parent, and can only come to resemble its parent by undergoing a sequence of changes. Whether or not a cell becomes differentiated in the usual sense of the term depends on whether the various syntheses induced during its postmitotic growth are the same or different from those which were induced in the parent cell at comparable stages of the life cycle. The structural and functional organization of an individual cell at any given moment is different from that at any other moment in its past. The specific kinds of changes and their stability provide the contrast between a population of cells which is proliferating true to type, and one which is differentiating concomitantly with proliferation.

The capacity of a single cell to yield progeny of various types is an impressive phenomenon. That the term "development" has been commonly applied to differentiating organisms and not to intracellular cycles, despite the fact that both are developmental processes, reflects the strong emphasis placed on their distinguishing features. Their common features should nevertheless not be overlooked. Both processes begin after completion of mitosis in the parent cell. How soon divergence from parental type begins after mitosis is an aspect of the phenomenon that is still being explored. In certain types of cells, however, the pattern is clear: divergence begins with the reconstitution of interphase nuclei. The most striking example of this is in microspore mitosis where the two daughter nuclei already differ from one another with respect to size and density at the time of reconstitution. (Fig. 19-1) The dense nucleus divides once again, and its daughters migrate into the egg cell. The other nucleus does not divide, but serves to control the physiological events in the germinating pollen cell.

DEVELOPMENTAL DIFFERENTIATION

Two of the most productive biological sciences in the twentieth century have been those concerned with genetics and development. These disciplines have many common concerns and mutual problems, but a syn-

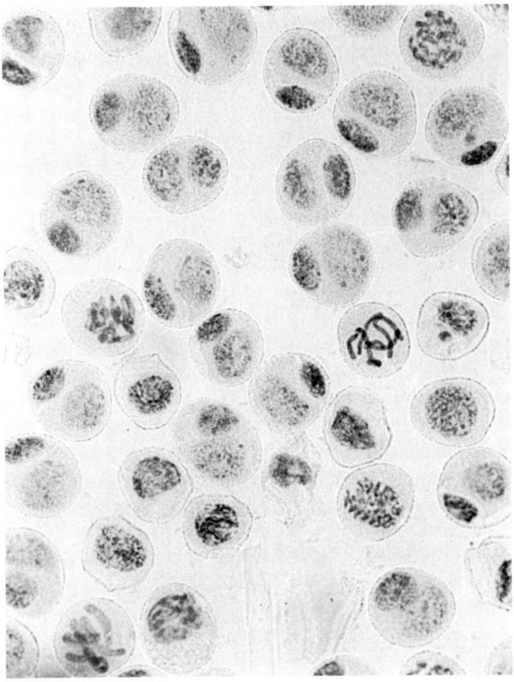

Fig. 19-1. Microspores of *Trillium erectum* at various stages of mitosis. Note the differences in size and stainability between the two nuclei. In those cells that have just completed mitosis the nuclei are similar. As development proceeds, one nucleus becomes highly condensed and later participates in the fertilization process. The larger nucleus serves as the "tube nucleus" and functions in regulating the growth of the pollen tube.

thesis is still awaited. The primary objective of geneticists has been an understanding of biological constancy, of the conservation of specificity generation after generation. They have been concerned with change, but as an exception to the rule, and the changes with which they dealt have been the haphazard errors called mutation or the random combinatorial changes associated with sexual processes. In contrast, developmentalists have been concerned primarily with change, and changes of a very different nature than those produced by mutation and recombination. A major factor in the development of a multicellular organism is the establishment of a heterogeneous population of cells from a single cell. And the changes encountered in this process are not rare, haphazard or random, but directed and patterned changes which occur with great reliability. Developmentalists have also been concerned with constancy, of course, but in a different way than the geneticists. They have assumed, with good reason, that the genetic constitution remained constant during development; they have been understandably impressed with the constancy of the patterns of change. And the end products of cellular differentiation were often found to possess great stability.

Much of the apparent discord between the observations of developmentalists and of geneticists has been resolved in recent years, primarily through the recognition that genetic activity, quite unlike genetic structure, is readily susceptible to regulation. The larger problems of development transcend the emphasis of this book, but for historical perspective we should discuss briefly some of the ideas and observations which link developmental studies with those on cellular activities.

Developmental biologists, observing the transformations occurring in developing embryos, early recognized that distinctive cellular types emerge during the ontogeny of the individual. Particular cell types appeared in specific locations at precisely the same time in each embryo. This diversification of cell types could be easily rationalized, because the phenotype of a cell, as of an organism, depends not only on its constitution but also on its environment; and the cells of an embryo do not occupy identical environments. Even in the beginning, the contents of eggs are not usually distributed uniformly; food stuffs and cytoplasmic organelles of various kinds are often visibly stratified in eggs so that as soon as the first or second cell division has occurred the cells have different kinds or amounts of inclusions. In part because of the differences in cytoplasmic composition, the rates of cell division in the different parts of the multicellular mass diverge. Since little growth takes place during the early cell divisions, the parts of the embryo with more rapid cell division come to have much smaller cells. With differences in size come also differences in surface-volume relations, and differences in the relative amounts of nuclear and cytoplasmic material. At still later stages, some of the cells are located on the periphery of the embryo and others are embedded in solid cellular masses, or face into the internal cavities of the developing embryo. The more complex the embryo becomes, the greater the variety of its internal environments and the greater the possibilities for environmentally directed modifications. It appeared reasonable to assume, therefore, that cellular differentiation was only in small part a cellular problem; the cells were "indicators" of the evolving internal environment of the embryo. Of course, the cells had to play a major role in the elaboration of the embryonic environment and were required to respond to it in a suitable fashion, but the patterning of change was an "organismic" function in which the role of the individual cells was essentially passive.

This interpretation presented no conflict with genetic ideas, but existing genetic concepts also contributed little to an understanding of development. Another set of observations did create some strain between the disciplines. Embryologists discovered that they could remove bits of tissue in developing embryos and transplant them to abnormal sites. If these operations were carried out early in development (how early depending on the tissue and the organism) the tissues departed from their normal developmental route and conformed to the patterns set by their new location. This observation supported strongly the idea that the cellular environment was the decisive factor in differentiation and that the role of the cells was passive. However, after a critical time in development, transplanted tissues lost their capacity to conform; they became "determined" and proceeded through the final stages of development along the previously prescribed pathways or remained fixed with a specific assortment of properties. The cells appeared to have undertaken a new and more active role and were capable of maintaining their differentiated state without environmental reinforcement. The problem posed by stable differentiations was this: heredity and environment were recognized as the two factors which determine a cell's characteristics. If all the cells have the same genetic constitution, they should behave alike when placed in the same environment—but they did not. Hence it appeared that hereditary differences might have been imposed on the cells during the course of development.

If genetically identical cells in the same environment could manifest different phenotypes and persist in maintaining their differences, the conceptual framework for explaining cellular variation was inadequate.

Alternatively, the basic assumptions were incorrect. Several attempts were made to surmount these difficulties before a satisfactory solution appeared. One possible explanation was that the inference of nuclear equivalence was wrong. August Weismann, perplexed by these problems, suggested, even before the rediscovery of Mendel and the modern formulation of the chromosome theory, that differentiation was due to the sorting out of chromosomes and chromosome segments into different tissues; he supposed that a complete chromosome set was retained only in the "germ plasm," those cells set aside to form the sex cells for subsequent generations. Weismann's theory seemed reasonable, but it soon fell before the accumulated data on chromosomal constancy. With rare exceptions, the cells of an organism have the same numbers and kinds of chromosomes. Moreover, later chemical studies indicate no gross changes in the kinds or amounts of DNA in variously differentiated cells. These tests are crude, and the possibility still existed that specific mutations were induced during development. This interpretation, however, was almost universally rejected because of the uniform failure to achieve directed changes by a variety of agents in the genetic materials of the germ line. Although direct analyses of the genetic constitutions of somatic cells were not possible, most students were convinced that the basic genetic equipment of the cells was identical.

A second possibility for escaping the dilemma was that the differences in somatic cells, though stable, were not "hereditary." The cells could perhaps sequester materials during their developmental history which rendered them passively distinctive. Because of the limitations imposed on the growth of an organism, a sufficient number of cell divisions does not always occur between differentiations and the cessation of growth to permit an adequate dilution test. Hence the hereditary nature of the cellular differences was not firmly established. With the development of tissue explantation and cell culture, hopes arose that this question might be settled decisively. Differentiated cells could be grown in tissue culture through unlimited numbers of cell generations. Unfortunately, this approach has not produced unequivocal results. The techniques so far perfected are suitable for the indefinite propagation of only a limited number of kinds of cells. Moreover, most cells when grown in artificial media lose many of their distinguishing characteristics (which are often based on cellular relationships in multicellular masses) and conform to a common morphology that seems to be required for their existence in thin sheets on glass. Even so, distinctive cellular types have been cultured from the same organism and have maintained distinctions long enough for them to be characterized as hereditary. Although few investigators are convinced that differentiations are categorically irreversible, most believe that the stability of the cell types is sufficiently great to require special explanation.

Having accepted nuclear equivalence and the stability of cell types, at least as a working interpretation, investigators were then faced with the requirement for rationalizing a conflict. If the cells were different and the nuclei were alike, obviously the differences must reflect cytoplasmic mechanisms. Hence most theoretical discussions of cellular differentiation focused attention on cytoplasmic mechanisms. The concept of the plasmagene, a gene-like cytoplasmic element capable of differential assortment and replication during development, provided a possible means of rationalization. These were considered in some cases to be copies of nuclear genes, fabricated perhaps under special conditions and released into the cytoplasm where they became semiautonomous. By attributing particular properties to the plasmagenes, an enormously flexible interpretation could be developed which would theoretically account for most of the facts observed. Moreover, several studies on microorganisms carried out in the 1940's and 1950's appeared to provide substantial evidence for the existence of such elements. Nevertheless, this interpretation was never generally accepted with enthusiasm, mainly perhaps because direct evidence for plasmagenes was almost totally lacking from embryonic studies. And later reevaluations of the studies on microorganisms demonstrate that many are less decisive than once imagined. In some cases the plasmatic elements have been shown to be parasitic or symbiotic organisms closely integrated in their activities with their hosts; in other cases alternative interpretations of the observations not based on gene-like elements have either been established or rendered equally probable.

A rejection of the plasmagene hypothesis as a sufficient explanation of cellular differentiation does not constitute a categorical rejection of gene-like elements in the cytoplasm. We have already discussed the episomes of bacteria which on occasion behave like plasmagenes. We may also note that certain cytoplasmic organelles contain DNA. Chloroplasts, for example, have long been known to manifest some degree of autonomy, and the DNA which they contain may carry genetic information essential to their function. Also, in certain trypanosomes a large DNA-containing body (the blepharoplast) is associated with the formation of mitochondria and may contain genetic determinants for mitochondrial components. The fact remains, however, that most DNA is nuclear, and variations in cytoplasmic DNA have not been associated with cellular differentiation. We might postulate that gene-like elements composed of substances other than DNA exist in the cytoplasm, but with the exception of viral RNA, none of these other compounds has yet been found to be capable of autosynthesis.

Although the plasmagene hypothesis was never generally accepted, it persisted as a possible explanation until a more satisfactory alternative was proposed. We have seen from microbial studies that differences in nuclear function may persist in cells of the same genotype. Hence the existence of two or more stable cell types in the same environment is not incompatible with an equivalence of the primary genetic information of the cell types. This explanation may now seem simple and obvious, but it ran contrary to the older view that neither gene function nor genetic constitution was capable of regulation, and it was accepted only after irrefutable proof was produced. The evidence for nuclear regulation in microorganisms has now removed all theoretical grounds for rejecting it as a major factor in development, but it leaves the practical problem of obtaining direct evidence for nuclear regulation in the differentiation of multicellular organisms.

The demonstration that cells in different tissues have their individual arrays of enzymes and other proteins is *prima-facie* evidence for genic regulation since, as we have seen, proteins are characteristic gene products. However, proteins are not the primary gene products, but the results of translating genetic specificity through an RNA intermediate. If regulation occurred at some step beyond the first, differences in proteins might be produced without interfering with primary gene action. If differences in the message RNA's of differentiated cells could be established, the assumption of regulation of primary gene action would be more firmly based. The techniques for such analyses have only recently been developed, but characteristic differences in message RNA's have been reported. Hence chemical analyses do support the interpretation and should provide even firmer evidence in the near future.

Other kinds of evidence also support the hypothesis of differential gene action and nuclear differentiation. Giant cells of a special kind are known in the tissues of many insects and particularly in the Diptera. In these cells the chromosomes continue to replicate after division has stopped and produce in some cases a thousand or more "copies" of each chromosomal strand. The chromosomes are not scattered randomly in the nucleus, but homologs are aligned by point in bundles many times longer than ordinary chromosomes. The effect of this extension and multiplication of fibers is to magnify slight variations which might exist along the lengths of the strands. A slight thickening of a single strand may be difficult to detect, but the same thickening in a thousand bands may appear as a definitive and characteristic marking on the chromosome. Actually, such multiple (polytene) chromosomes appear as a series of well marked bands of various sizes and spacings specific for a particular chromosome type. (Fig. 19-2)

When polytene chromosomes are examined in different tissues or at different times in the development of a single tissue, the chromosome types are readily equated, but they show characteristic changes from time to time and from tissue to tissue. At some times (or in some tissues) a particular band or group of bands grows larger and more diffuse, forming structures designated as "bulbs," "puffs," or "rings." (Fig. 19-3) Cytochemical studies demonstrate that the enlargement of a band is often correlated with an increased rate of RNA synthesis in the area, and thus suggest that the genic materials in these areas are unusually active. The correlation between puffing patterns and developmental status strongly support the idea that different genes actually function at different rates in different tissues and at different times in multicellular organisms. The general inference is supported by several specific observations. The treatment of Chironomus larvae with the hormone ecdysone, for example, leads to an immediate sequence of changes in puffing pattern in the chromosomes of the salivary glands and after a period of time to an early pupation. The exact mechanisms regulating gene activity in multicellular forms have not yet been determined, but the mechanisms suggested by work on unicellular organisms provide a useful working basis for their exploration.

We may note in passing that in some puff regions (but not in most), not only large amounts of RNA but also some additional DNA appears to be synthesized. The meaning of this observation is not clear, but perhaps additional "copies" of some genes may be produced to augment their functions.

INTRACELLULAR DEVELOPMENT

Only one interval in the life cycle of a cell is marked by a virtual absence of protein synthesis: the prometaphase, metaphase, and anaphase stages of meiosis or mitosis. The precise borderline between no synthesis and some synthesis is difficult to define, but the general correlation between the absence of appreciable protein synthesis and the highly condensed state of chromosomes has been verified by both chemical and autoradiographic techniques. A simple, but by no means demonstrated, explanation of this relationship is that the extremely close packing of DNA filaments, which is fortified by the presence of chromosomal proteins, precludes the movement of extra-chromosomal

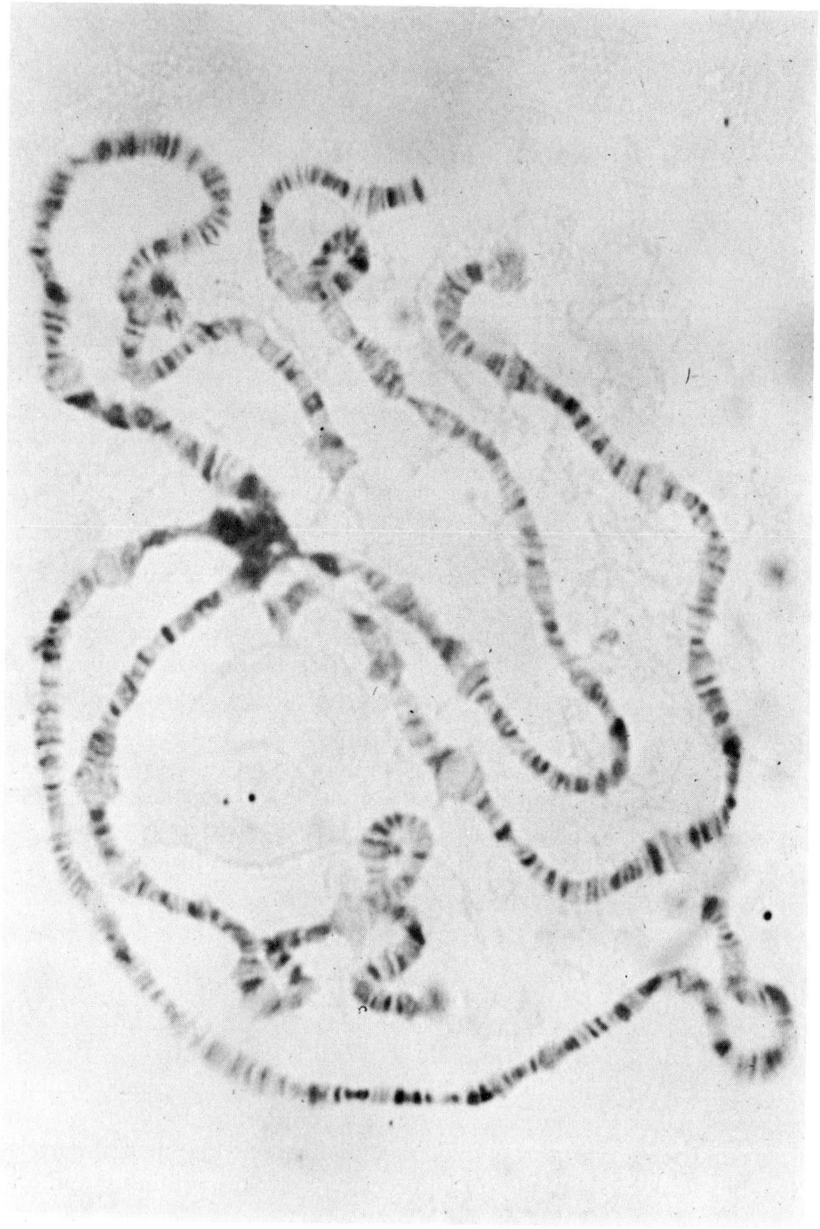

Fig. 19-2. Polytene chromosomes of *Drosophila virilis*. The chromosomes have been stained specifically for histones. The bands represent regions of high histone concentration. A similar band pattern would be found if the chromosomes were stained for DNA. (Courtesy of Dr. H. Swift (prepared by M. Gorovsky).)

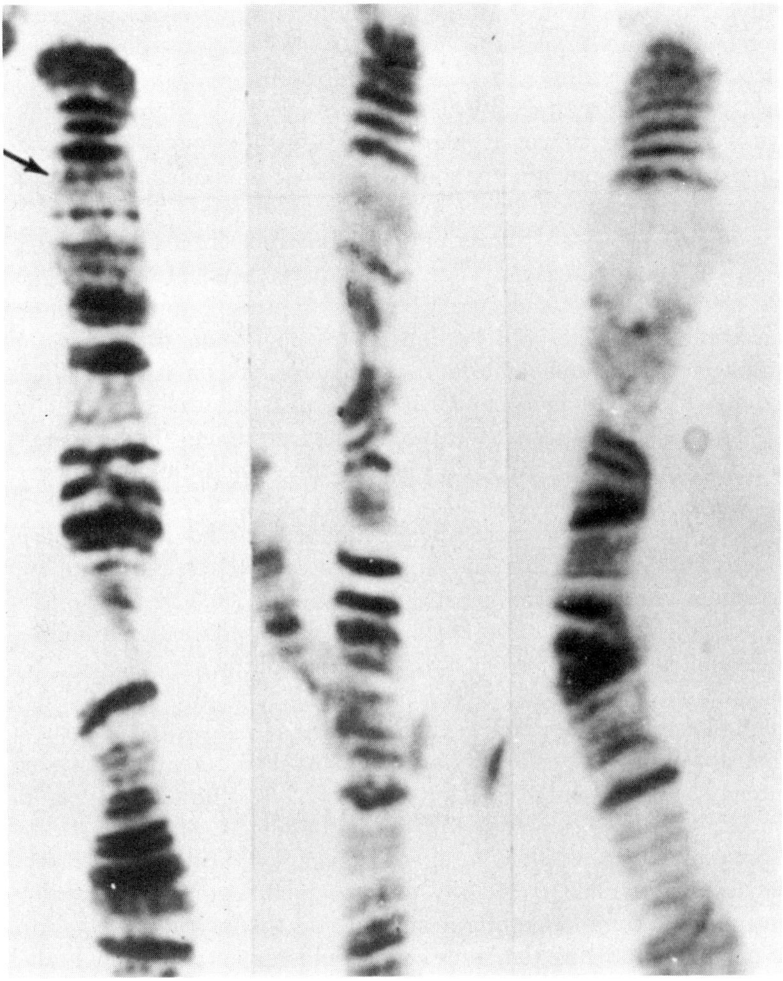

Fig. 19-3. "Puffing" in a salivary gland chromosome of *Drosophila virilis.* The dark bands are regions of DNA and RNA. The region where puffing begins consists mainly of RNA. In this particular puff, the associated DNA undergoes no change. In puffs of other genera, however, DNA has been observed to undergo synthesis and degradation. (From H. Swift, *Molecular Control of Cell Activity,* ed. by J. M. Allen, McGraw-Hill, 1962.)

substances (enzymes or substrates) into the vincinity of the genes. This same explanation has been given for the apparently sluggish activity of heterochromatic regions. With respect to the developmental cycles of cells, the nature of the inactivity of condensed chromosomes is perhaps less significant a problem than the nature of chromosomal activity at other intervals. Marked quantitative differences in levels of protein and RNA synthesis exist between different developmental intervals, the mitotic or meiotic interval representing the lowest level, the post-telophase interval usually representing the highest. Quantitative regulation is, however, only one aspect of the mechanisms underlying cell development; qualitative differences in metabolic patterns also exist between different developmental stages.

Evidence for gene transcription at different intervals of the cell cycle is being increasingly accumulated as chemical techniques for the identification of message RNA are improved. Such evidence is consistent with older observations that enucleation of a unicellular organism has an immediate and profound effect on its physiological activity. Although, as pointed out in the discussion of differentiation, the direct evidence for different gene messages being produced in the course of differentiation is still meager, the circumstantial evidence is so strong that the exclusion of differential gene transcription as part of the regulatory process seems most unlikely. A far more perplexing issue is how such differential transcription is effected. Intricate though the linear reading of the genetic tape may be, such reading must be supported, induced, or repressed by agents of the extrachromosomal environment. Of such agents we still know little.

Commonly, a daughter cell in developing to maturity synthesizes a weight of substance equal to that which it has inherited. Departures from a doubling of cell mass are frequent in differentiating systems, but in all cells a synthesis of enzymes, which in turn promotes other syntheses, follows completion of the mitotic cycle. Since genetic messages must be transcribed to effect enzyme syntheses, genes must be read during the cell cycle. Any theory which required that genes be read prior to completion of mitosis and that the messages thus formed be partitioned among daughter cells would conflict with genetic data. Most probably, if not certainly, the period of growth immediately following mitosis is accompanied by gene activity. That the activity is not randomly distributed within the genome may be inferred from a variety of experimental observations. Few, if any, cells express their full genetic potential during their life cycle. This is obvious for cells of differentiated systems, but it is also evident in bacteria where nutrient conditions may induce the formation of novel proteins. Thus even if all gene messages were produced just following mitosis, such production would have to be selective. Moreover, in cells of higher organisms at least, the proteins which will be formed by a cell during its life cycle do not appear simultaneously, and the evidence which is accumulating favors the hypothesis that most novel proteins, regardless of the time of their appearance, are produced as the result of the activation of specific gene transcriptions. There are thus two major issues with respect to the relationship between gene reading and cell development: the factors regulating *which* genes are to be read and the controls determining *when* these factors are to be effective.

Little can be added to what has already been said in discussing environmental regulation and developmental differentiation which would throw more light on these two aspects of intracellular regulation. The genome codes and reads codes; the extragenic components sense the intracellular environment, respond adaptively within defined limits to environmental changes, and transmit signals to the genome. Yet even this complex statement is an oversimplification. On *a priori* grounds we cannot exclude the possibility that genes can send messages to nonadjacent genes without the intervention of extragenic components. This possibility needs to be considered because intracellular development is an orderly process. The environment may modify, disturb, or accelerate development, but it does not randomize the characteristic sequence. The model of the circular chromosome introduced in the chapter on temporal regulation is by itself inadequate to explain the orderly sequence of events in cell development. For cells of higher organisms have many chromosomes, and even if each chromosome were a giant circle of DNA, some orderly communication would have to exist between the chromosomes. But the fact that pieces of chromosomes may be translocated and inverted without ordinarily disrupting developmental sequences suggests that directional reading of DNA filaments may be only one mechanism for coding intracellular sequences along the axis of time. The ordering of cell development must somehow be encoded, but how the coding is achieved remains unknown. One important fact is that from the instant of daughter cell formation the intracellular environment undergoes progressive change. Whether the change is entirely endogenous in origin or is caused by extracellular factors originating from other cells in a multicellular system is incidental to the mechanism by which a changing intracellular environment selectively activates genes.

In studies of cell cycles major attention is commonly focused on the regulation of chromosome reproduction and nuclear division. The critical role which the equipartitioning of chromosomes among cell progeny plays in the maintenance of life is sufficient justification for such emphasis. Nevertheless, other aspects of cell behavior should be taken into account even though our lack of knowledge about them permits little more than a brief mention. One aspect is the coordination which must exist between the replication of the chromosomal complex and that of the population of cytoplasmic inclusions. The fact that chloroplasts cannot arise *de novo* has been known for some time. A cell which does not inherit chloroplasts (or their progenitors, the proplastids) cannot develop chloroplasts even though such a cell may have the full complement of chromosomal genes essential to chloroplast development. Chloroplasts thus have a basic characteristic of self-replicating systems in that replicates do not arise in their absence. In this respect chloroplasts (and possibly mitochondria) are similar to viruses. There must be a coordination between the rate of chloroplast replication and that of chromosomal replication. In a line of cells which is reproducing true to type, the net replication rate of the two subcellular systems must remain equal. Since the number of chloroplasts in a cell is usually large, a random segregation of chloroplasts during cell division, as well as the presence of some mechanism in daughter cells which in the course of growth would restore the parental ratio of chloroplasts to nucleus is sufficient to account for the constancy of cell type. In such cells we could regard the mechanisms which regulate chloroplast numbers as similar to those which regulate the numbers of other cell components. In some algal cells, however, there may be no more than one chloroplast, and in these cells the chloroplast divides prior to cell division and its progeny migrate to opposite poles of the cell. Chloroplasts are therefore capable of precise segregation. Whether an orderly segregation of other cytoplasmic components occurs is not so clear, except for centrioles which are associated with the spindle body.

The maintenance of cellular characteristics thus requires a regulation of the relative proportions of various subcellular organelles. Another essential factor, which is only now beginning to be appreciated, is the pattern of association of the organelles. This pattern is often difficult to discern or to describe, but in some forms results in geometrically obvious organizations in which specific variations can be detected. Ciliated protozoa, for example, have a highly structured cortex in which are associated a variety of specialized elements. By appropriate experimental procedures new cortical patterns can be constructed; cells can be formed which are duplex (double cells with two oral apparatuses), with altered numbers of ciliary rows, changed polarities of ciliary rows, and various other modifications. Significantly, T. M. Sonneborn and his associates have shown that many of these new patterns, once established, are reproduced at cell division for an indefinite period of time. The cellular components, specified by nuclear instructions, are capable of being associated in a number of self-reproducing patterns. The mechanisms of pattern perpetuation

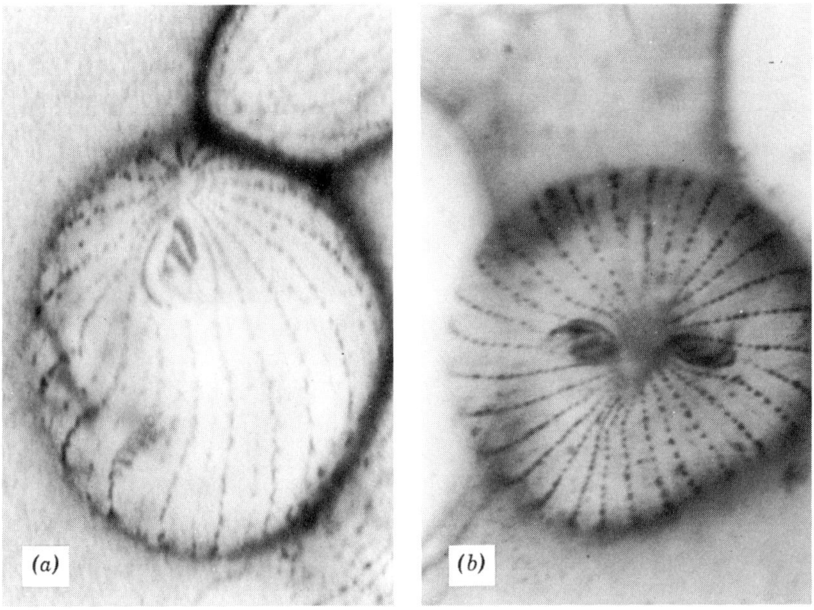

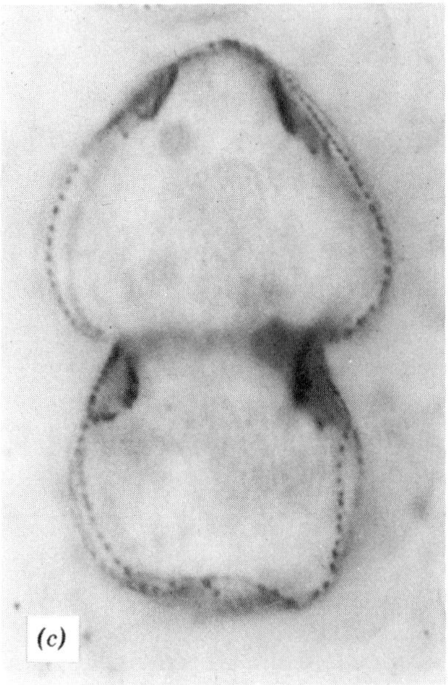

Fig. 19-4. Silver impregnation preparations of the ciliate *Tetrahymena pyriformis.* (*a*) "Standard" form with a single oral apparatus and 18 ciliary rows. (*b*) Polar view of a "duplex" form with two sets of oral structures and 29 ciliary rows. (c) Longitudinal optical section of a duplex form in division showing two sets of oral structures on both the anterior and posterior daughters.

are still obscure, and the extent to which preexisting structure is required in the maintenance of cellular order is an open question. However, just as the order of nucleotide bases is an ancient evolutionary heritage, so the patterns of association of subcellular components —transmitted through unbroken cellular bridges to a distant past— may provide essential information for cellular structure. The information encoded in nucleic acids, though quantitatively the most important for cellular function, may not in fact be sufficient to specify a cell.

However important the transmission of cytoplasmic components and organellar patterns may be, there can be no doubt that the reproduction and segregation of chromosomes are the major events in cell division. Yet our understanding of these events is largely limited to the molecular nature of DNA replication and to the geometry of chromosome segregation. How these events are regulated is still mainly a matter of speculation. At best, therefore, the problems surrounding such regulation can only be listed and briefly discussed.

In cells of most organisms chromosome reproduction requires not only the replication of DNA but also that of protein. DNA and protein, which is largely histone, are synthesized almost simultaneously; at no time are chromosomes found to be histone-free. Yet the evidence from biochemical studies clearly indicates that unlike DNA, histones are not conserved. Apart from the fact that they are located in the chromosomes, they are presumably similar to other proteins with respect to mode of synthesis. There must be a special relationship between histone synthesis and DNA replication, for histone synthesis is confined almost entirely to the interval of DNA replication. Since DNA is probably not transcribed during replication, the simultaneity must be apparent rather than real. Presumably, at the molecular level localized regions of the DNA replicate and then combine with histone. Failure so far to identify a sequence in the two events may be attributed to the fact that present techniques of temporal analysis are too insensitive. The significance of histones to chromosome function remains unresolved. The evidence that histones are not conserved implies that during the replication of DNA the histones become dissociated from the DNA filaments. Although they appear to be associated with most if not all of the DNA in the chromosome, it is most unlikely that each gene codes its own histone. Such a relationship would require that very many genes code for two very different kinds of protein, and our present interpretations of coding systems precludes this type of mechanism. Moreover, in many spermatazoa the histones found in somatic tissues are absent, and instead are replaced by a much more basic type of protein, protamine. Differences of a much lesser magnitude are also found between the kinds of histones found in different tissues. If our present interpretation of protein synthesis is universal, histones must be coded by specific genes. Whether such genes are transcribed during the replication process or whether they anticipate histone requirements by forming stable messages prior to DNA replication is unknown. Even more obscure are the factors which activate the transcription of histone-coding genes.

In both meiosis and mitosis chromosomes undergo contraction, and there must be some regulatory mechanism which induces the process at the end of G_2 period. The nature of this mechanism is also unknown. Like other processes, it must be subject to the influences of the intracellular environment and also to the activities of specific genes. Chromosomes contract more in meiotic than in mitotic cells (Fig. 19-5). Mutants are also known in which the contraction of meiotic chromosomes is incomplete. These few remarks, however, more or less cover the extent of our knowledge about this aspect of chromosome behavior.

Perhaps the most intriguing questions which may be raised concerning chromosome reproduction and nuclear division relate to the differences between meiosis and mitosis. In meiotic cells sister chromosomes pair, and although somatic pairing is known, it is the exception rather than the rule. Just how this pairing comes about remains a mystery. Meiotic cells are derived from somatic ones in

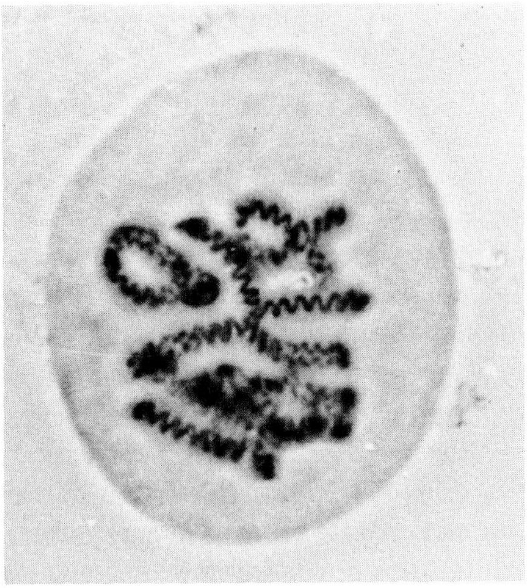

Fig. 19-5. Contracted meiotic chromosomes of *Trillium erectum*. A coiling of the chromosomal filament (the "major" coil) is evident. In some regions the chromatids have separated and their respective coils may be seen.

which pairing is presumably absent. The chromosomes must therefore assort themselves during early meiosis. Since pairing is gene specific and occurs over microscopically visible distances, a molecular explanation of the phenomenon is not obvious. The hydrogen bonding between complementary strands of DNA could not possibly operate over the distance maintained between sister chromosomes. Nevertheless, at some interval in meiosis sister chromosomes must align themselves with respect to one another according to the base sequences in the DNA components. Whether such alignment occurs by direct gene-gene interaction at a stage prior to the microscopically evident pairing, or whether it occurs via some intermediary process is still a matter of speculation. Indeed, almost any question which might be raised concerning the mechanisms underlying chromosome behavior during meiosis and mitosis would go unanswered. The timing of centromere division, the chemical nature of the centromere and its associated fibrils which extend to the poles of the spindle, the nature and function of the other spindle body fibrils which extend from pole to pole without contacting the chromosomes—these and many other related questions cannot be answered because we do not understand the mechanisms underlying the physical behavior of chromosomes during nuclear division.

In attempting to understand nuclear division as a phase in the development of a cell, we must explain not only the nature of the individual events which comprise the division process, but also their coordination. In a typical meiotic or mitotic cycle chromosome replication is followed by the appearance of two distinct chromatids. The paired chromatids (and in meiosis the paired chromosomes also), contract, and when maximally contracted they align at the metaphase plate. In mitotic cells centromere division occurs after metaphase alignment, and the chromosomes then segregate equationally by poleward movement. In meiotic cells centromere division is delayed until the second division. In both types of cells interphase nuclei are reconstituted at telophase, and the daughter nuclei become separated either by cell cleavage or by formation of a cell plate. Although such a coordinated sequence might suggest that each event in the division cycle is the direct cause of the succeeding event, the evidence does not support so simple an interpretation. In polytene cells chromosomes replicate, but no separate chromatids are formed; polyploid cells may arise because the steps beyond chromatid separation fail to occur, or they may arise due to a disturbance of the spindle mechanism which is followed by a reconstitution of a single interphase nucleus from all the chromosomes at the metaphase plate. Multinucleate cells arise by a failure of the final step in the division sequence—cytokinesis. Clearly, the component events of the nuclear cycle are not obligatorily connected. Each event must be tied to its succeeding event by a mechanism which is subject to independent regulation.

The partial autonomy of the component events of nuclear replication is evident not only at the microscopic level but also at the molecular one. All the chromosomes in a cell do not replicate simultaneously. Some chromosomes (the X chromosome in certain cells) typically lag behind others, and within a single chromosome some regions are characteristically replicated ahead of others. Thus the presence of the catalytic mechanism (the polymerase enzyme, etc.) for chromosome replication does not automatically assure such replication. Moreover, cells do not necessarily confine the regulation of chromosome replication to the chromosome itself. In some cells the interval of chromosome replication is marked by the appearance of various enzymes specifically concerned with the process. Not only do enzymes appear at a given time, but they are also removed from the cell after their function is completed. In this respect the line of distinction between the mechanisms underlying differentiation and those underlying cyclical replication is thin indeed. In both developmental phenomena specific proteins appear to be formed at defined intervals.

Mitosis and meiosis share a common requirement with all other developmental phenomena: the need to synthesize specific substances at certain intervals. To the extent that these substances must be coded by genes, the requirement is gene regulated. But to the extent that the activation of genes is effected by the extrachromosomal portion of the cell, the requirement is cytoplasmically regulated. Given the various mechanisms by which cells can sense even the most subtle changes in their environment, and the various kinds of response systems which cells possess, little imagination is required to perceive the many possibilities for interactions within communities of cells. Just as each of the highly specialized processes which are displayed by higher organisms can be traced to less elaborate but homologous processes in primitive cell types, so too should we be able to trace the complex regulatory processes in higher organisms to the basic and universal regulatory mechanisms in individual cells. The barrier to understanding, however, does not lie solely in resolving these basic mechanisms, for we must also resolve the manifold ways in which cell communities may exploit these mechanisms. The simplest unicellular organism is subject to the variables in its inanimate environment; it continuously senses variations and responds to them. But even the most variable of inanimate environments cannot match the environment of a multicellular association with respect to the numbers and kinds of stimuli which pass between the cells. Molecules may be described as aggregates of atoms, cells as aggregates of molecules, and organisms as aggregates of cells. Yet, although the behavior of an aggregate expresses the potential of its components, that potential remains unknown until the properties of the aggregate are recognized. Even as our knowledge of cells enhances our knowledge of organisms, so must our understanding of organisms enhance our understanding of cells.

INDEX